Advances in Intelligent Systems and Computing

Volume 563

Series editor

Janusz Kacprzyk, Polish Academy of Sciences, Warsaw, Poland
e-mail: kacprzyk@ibspan.waw.pl

The series "Advances in Intelligent Systems and Computing" contains publications on theory, applications, and design methods of Intelligent Systems and Intelligent Computing. Virtually all disciplines such as engineering, natural sciences, computer and information science, ICT, economics, business, e-commerce, environment, healthcare, life science are covered. The list of topics spans all the areas of modern intelligent systems and computing.

The publications within "Advances in Intelligent Systems and Computing" are primarily textbooks and proceedings of important conferences, symposia and congresses. They cover significant recent developments in the field, both of a foundational and applicable character. An important characteristic feature of the series is the short publication time and world-wide distribution. This permits a rapid and broad dissemination of research results.

More information about this series at http://www.springer.com/series/11156

Khalid Saeed · Nabendu Chaki
Bibudhendu Pati · Sambit Bakshi
Durga Prasad Mohapatra
Editors

Progress in Advanced Computing and Intelligent Engineering

Proceedings of ICACIE 2016, Volume 1

 Springer

Editors
Khalid Saeed
Faculty of Computer Science
Białystok University of Technology
Białystok
Poland

Nabendu Chaki
Department of Computer Science
and Engineering
University of Calcutta
Kolkata
India

Bibudhendu Pati
C.V. Raman College of Engineering
Bhubaneswar, Odisha
India

Sambit Bakshi
Department of Computer Science
and Engineering
National Institute of Technology, Rourkela
Rourkela, Odisha
India

Durga Prasad Mohapatra
Department of Computer Science
and Engineering
National Institute of Technology, Rourkela
Rourkela, Odisha
India

ISSN 2194-5357   ISSN 2194-5365   (electronic)
Advances in Intelligent Systems and Computing
ISBN 978-981-10-6871-3   ISBN 978-981-10-6872-0   (eBook)
https://doi.org/10.1007/978-981-10-6872-0

Library of Congress Control Number: 2017955277

Printed on acid-free paper

This Springer imprint is published by Springer Nature
The registered company is Springer Nature Singapore Pte Ltd.
The registered company address is: 152 Beach Road, #21-01/04 Gateway East, Singapore 189721, Singapore

Preface

This volume contains the papers presented at the International Conference on Advanced Computing and Intelligent Engineering (ICACIE 2016) that was held during December 23–25, 2016, at the C.V. Raman College of Engineering, Bhubaneswar, India (www.icacie.com). There were 638 submissions, and each qualified submission was reviewed by a minimum of two Technical Program Committee members based on the criteria of relevance, originality, technical quality, and presentation. The committee accepted and published 136 full papers in the proceedings for oral presentation at the conference, and the overall acceptance rate was 21.32%.

ICACIE is an initiative focusing on research and applications on several topics of advanced computing and intelligent engineering. The focus was also to present state-of-the-art scientific results, to disseminate modern technologies, and to promote collaborative research in advanced computing and intelligent engineering.

The accepted papers were chosen based on their research excellence, presentation quality, novelty, and state-of-the-art representation. Researchers who presented their work had an excellent opportunity to interact with eminent professors and scholars in their area of research. All participants benefitted from discussions that facilitated the emergence of innovative ideas and approaches. Many distinguished professors, well-known scholars, industry leaders, and young researchers participated in making ICACIE 2016 an immense success.

We organized a special session named as Women in Engineering (WiE) on the topic "Empowerment of Women in the Field of Engineering and Management" to encourage young women in the field of engineering and management to participate in the discussion. We also had industry and academia panel discussions, and we invited people from software industries such as TCS and Infosys.

We thank the Technical Program Committee members and all the reviewers/sub-reviewers for their timely and thorough participation in the reviewing process.

We express our sincere gratitude to Shri Sanjib Kumar Rout, Chairman, C.V. Raman Group of Institutions, for allowing us to organize ICACIE 2016 on campus. We also thank Prof. B. Bhattacharya, Principal, C.V. Raman College of

Engineering, for his moral support. We thank Dr. Manmath Narayan Sahoo, NIT Rourkela, Program Chair, for his valuable and timely support. We especially thank Dr. Chhabi Rani Panigrahi, C.V. Raman College of Engineering, for her support in local arrangements that helped in making ICACIE 2016 a grand success. We appreciate the time and efforts put in by the members of the local organizing team at C.V. Raman College of Engineering, Bhubaneswar, especially the student volunteers, administrative staff, account section staff, and hostel management staff, who dedicated their time and efforts to ICACIE 2016. We thank Mr. Swagat Ranjan Sahoo for designing and maintaining ICACIE 2016 Web site.

We are very grateful to all our sponsors, especially DRDO and other local supporters, for their generous support toward ICACIE 2016.

Finally, we acknowledge the help of EasyChair in the submission, review, and proceeding creation processes. We are very pleased to express our sincere thanks to Springer, especially Mr. Anil Chandy, Mr. Harmen van Paradijs, Mr. Aninda Bose, and the editorial staff, for their support in publishing the proceedings of ICACIE 2016.

Białystok, Poland	Khalid Saeed
Kolkata, India	Nabendu Chaki
Bhubaneswar, India	Bibudhendu Pati
Rourkela, India	Sambit Bakshi
Rourkela, India	Durga Prasad Mohapatra

Organization

Advisory Board:

Laxmi Narayan Bhuyan, FIEEE, FACM, FAAAS, University of California, Riverside, USA
Shyam Sundar Pattnaik, SMIEEE, FIETE, Biju Patnaik University of Technology, Odisha, India
Israel Koren, FIEEE, University of Massachusetts, USA
Katina Michael, SMIEEE, University of Wollongong, Australia
L. M. Patnaik, FIEEE, FINSA, FIETE, FIE, Indian Institute of Science, India
Rajib Mall, SMIEEE, Indian Institute of Technology Kharagpur, India
Prasant Mohapatra, FIEEE, University of California, USA
Abhay Bansal, SMIEEE, FIETE, FIET, Amity School of Engineering and Technology, India
Arun Somani, FIEEE, Iowa State University, USA
Atulya Nagar, Liverpool Hope University, UK
Brijesh Verma, SMIEEE, Central Queensland University, Australia
Debajyoti Mukhopadhyay, SMIEEE, FIE, Maharashtra Institute of Technology, India
George A. Tsihrintzis, University of Piraeus, Greece
Hugo Proenca, SMIEEE, University of Beira Interior, Portugal
Janusz Kacprzyk, FIEEE, Polish Academy of Sciences, Poland
Kenji Suzuki, SMIEEE, The University of Chicago, USA
Khalid Saeed, SMIEEE, Białystok University of Technology, Poland
Klaus David, University of Kassel, Germany
Gautam Das, FIEEE, University of Texas at Arlington, USA
Ganapati Panda, SMIEEE, IIT Bhubaneswar, India
Nabanita Das, SMIEEE, Indian Statistical Institute, Kolkata, India
Rama Krishna Challa, SMIEEE, NITTTR, Chandigarh, India
Biswanath Mukherjee, FIEEE, University of California, Davis, USA
Subhankar Dhar, FIEEE, San Jose State University, USA

Ashutosh Dutta, SMIEEE, AT&T Labs, USA
Kuan-Ching Li, FIET, SMIEEE, Providence University, Taiwan
Maode Ma, FIET, SMIEEE, Nanyang Technological University, Singapore
Massimo Tistarelli, FIAPR, SMIEEE, University of Sassari, Italy
Mohammad S. Obaidat, FIEEE, Monmouth University, USA
Sudip Misra, SMIEEE, Indian Institute of Technology Kharagpur, India
Michele Nappi, University of Salerno, Italy
Nishchal K. Verma, SMIEEE, Indian Institute of Technology Kanpur, India
Ouri E. Wolfson, FIEEE, FACM, University of Illinois at Chicago, USA
Pascal Lorenz, SMIEEE, FIARIA, University of Haute-Alsace, France
Pierre Borne, FIEEE, Central School of Lille, France
Raj Jain, FIEEE, FACM, FAAAS, Washington University in St. Louis, USA
Rajkumar Buyya, SMIEEE, LMACM, University of Melbourne, Australia
Raouf Boutaba, FIEEE, University of Waterloo, Canada
Saman Halgamuge, SMIEEE, University of Melbourne, Australia
Sansanee Auephanwiriyakul, SMIEEE, Chiang Mai University, Thailand
Subhash Saini, The National Aeronautics and Space Administration (NASA), USA
Arun Pujari, SMIEEE, Central University of Rajasthan, India
Sudhir Dixit, FIEEE, HP Lab, USA
Sanjay Mohapatra, Vice President, CSI, India

Chief Patron

Shri. Sanjib Kumar Rout, Chairman, C. V. Raman Group of Institutions, India
Patron
Smt. Shailja Rout, Managing Director, SSEPL Skills Pvt Ltd, Odisha, India

Honorary General Chairs

Prasant Mohapatra, University of California, Davis, USA
Rajib Mall, Indian Institute of Technology Kharagpur, India
Sudip Misra, Indian Institute of Technology Kharagpur, India

Steering Committee

Kartik Chandra Patra, C.V. Raman College of Engineering, Odisha, India
Bhabes Bhattacharya, C.V. Raman College of Engineering, Odisha, India
Debdas Mishra, C.V. Raman College of Engineering, Odisha, India

General Chairs

Bibudhendu Pati, C.V. Raman College of Engineering, Odisha, India
Pankaj K. Sa, NIT Rourkela, Odisha, India

Organizing Chairs

Chhabi Rani Panigrahi, C.V. Raman College of Engineering, Odisha, India
Sambit Bakshi, NIT Rourkela, Odisha, India

Special Session Chairs

Rachita Mishra, C.V. Raman College of Engineering, Odisha, India
Brojo Kishore Mishra, C.V. Raman College of Engineering, Odisha, India

Program Chairs

Manmath Narayan Sahoo, NIT Rourkela, Odisha, India
Subhas Chandra Misra, Indian Institute of Technology Kanpur, India

Publication Chairs

Sukant Kishoro Bisoy, C.V. Raman College of Engineering, Odisha, India
Soubhagya S. Barpanda, C.V. Raman College of Engineering, Odisha, India

Finance Chair

Mohit Ranjan Panda, C.V. Raman College of Engineering, Odisha, India

Website Chair

Swagat Ranjan Sahoo, C.V. Raman College of Engineering, Odisha, India

Registration Chair

Priyadarshini Nayak, C.V. Raman College of Engineering, Odisha, India

Publicity Chair

Tanmay Kumar Das, C.V. Raman College of Engineering, Odisha, India

Organizing Committee

Amardeep Das
Abhaya Kumar Sahoo
Amrut Ranjan Jena
Amulya Kumar Satpathy
Babitarani Garanayak
Banee Bandana Das
Bijaylaxmi Panda
Biswajit Upadhyay
Chhabirani Mohapatra
Chandra kanta Mohanty
Debasis Mohanty

Debapriya Panda
Harapriya Rout
Himansu Das
Jyotiranjan Swain
Kartik chandra Jena
Khitish Kumar Gadnayak
Lalat Kishore Choudhury
M. Priyattama Sahoo
Madhusmita Mishra
Mamata Rani Das
Mamata Rath
Manas Ranjan Mishra
Monalisa Mishra
Nilamadhaba Dash
Prakash Chandra Sahu
Prashanta Kumar Dash
Rashmiprava Sahoo
Rojalin Priyadarshini
Sharmistha Puhan
Sasmita Parida
Satyashree Samal
Soumya Sahoo
Shreela Dash
Sujit Mohapatra
Sunil Kumar Mohapatra
Sushruta Mishra
Suvendu Chandan Nayak

Technical Program Committee

Chui Kwok Tai, City University of Hong Kong, Hong Kong
Bernd E. Wolfinger, University of Hamburg, Hamburg
Amin Al-Habaibeh, Nottingham Trent University, UK
Carlo Vallati, University of Pisa, Italy
Rajendra Prasath, University College Cork, Ireland
Chi-Wai Chow, National Chiao Tung University, Taiwan
Mohammed Ghazal, Abu Dhabi University, UAE
Felix Albu, Valahia University of Targoviste, Romania
Vasanth Iyer, Florida International University, USA
Victor Govindaswaormy, Concordia University Chicago, USA
Priyadarshi Kanungo, C.V. Raman College of Engineering, Odisha, India
Sangram Mohapatra, C.V. Raman College of Engineering, Odisha, India

Saikat Charjee, C.V. Raman College of Engineering, Odisha, India
Chakchai So-In, Khon Kaen University, Thailand
Cristina Alcaraz., University of Malaga, Spain
Barun Kumar Saha, Indian Institute of Technology Kharagpur, India
Pushpendu Kar, Nanyang Technological University, Singapore
Samaresh Bera, Indian Institute of Technology Kharagpur, India
Ayan Mandal, Indian Institute of Technology Kharagpur, India
Tamoghna Ojha, Indian Institute of Technology Kharagpur, India
Subhadeep Sarkar, Indian Institute of Technology Kharagpur, India
Somanath Tripathy, Indian Institute of Technology Patna, India
George Caridakis, University of the Aegean, Greece
Carlos Alberto Malcher Bastos, Universidade Federal Fluminense, Brazil
Laizhong Cui, Shenzhen University, China
Srinivas Prasad, GMRIT, Rajam, India
Prasant Kumar Sahu, Indian Institute of Technology Bhubaneswar, India
Mohand Lagha, University of Blida, Algeria
Vincenzo Eramo, University of Rome, La Sapienza, Italy
Ruggero Donida Labati, Università degli Studi di Milano, Italy
Satyananda Rai, SIT, Bhubaneswar, India
Dinesh Bhatia, North Eastern Hill University, Meghalaya, India
Vasilis Friderikos, King's College London, UK
C. Lakshmi Devasena, IFHE University, India
Arijit Roy, Indian Institute of Technology Kharagpur, India
Roberto Caldelli, Universita' degli Studi Firenze, Italy
Christos Bouras, University of Patras, Greece
Iti Saha Misra, Jadavpur University, India
Salil Kumar Sanyal, Jadavpur University, India
J. Joshua Thomas, School of Engineering, KDU Penang University College, Penang
Shibendu Debbarma, Tripura University, India
Angelo Genovese, Università degli Studi di Milano, Italy
Marco Mussetta, Politecnico Di Milano, Italy
Radu-Emil Precup, Politehnica University of Timisoara, Romania
Debi Acharjya, VIT University, Vellore, India
Samaresh Mishra, KIIT University, Bhubaneswar, India
Rio D'Souza, St Joseph Engineering College, Mangalore, India
Yogesh Dandawate, Vishwakarma Institute of Information Technology, Pune, India
Sanjay Singh, Manipal Institute of Technology, Manipal, India
Rajesh R., Central University of Kerala, India
Abhishek Ray, KIIT University, Bhubaneswar, India
Lalat Indu Giri, NIT Goa, India
Debdas Mishra, C.V. Raman College of Engineering, Odisha, India
Ameresh Panda, C.V. Raman College of Engineering, Odisha, India
Tripti Swarnakar, SOA University, Bhubaneswar, India

Judhistir Mohapatro, NIT Delhi, India
Manas Khatua, SUTD, Singapore
Sujata Pal, University of Waterloo, Canada
Sumit Goswami, DRDO, New Delhi, India
Rabi Narayana Sathpathy, HIT, Bhubaneswar, India
Harihar Kalia, SEC, India
Hari Saran Dash, Infosys, Bhubaneswar, India
Siba Kumar Udgata, University of Hyderabad, India
Mu-Song Chen, Da-Yeh University, Taiwan
Félix J. García, University of Murcia, Spain
Prasant Kumar Pattnaik, KIIT University, India
Poornalatha G, MIT, Manipal, India
Nishant Doshi, MEFGI, Rajkot, India
V. N. Manjunath Aradhya, JCE, Mysore, India
Prabhakar, C. J., Kuvempu University, Karnataka, India
Enrico Cambiaso, National Research Council, CNR-IEIIT, Italy
Gianluigi Ferrari, University of Parma, Italy
Elena Benderskaya, Saint-Petersburg State Politechnical University, Russia
Josep Domènech, Universitat Politècnica de València, Spain
Himansu Das, KIIT University, India
Vivek Kumar Sehagl, Jaypee University of Information Technology, Waknaghat, India
Monish Chatterjee, Asansol Engineering College, Asansol, India
Teresa Gomes, Universidade de Coimbra—Polo II, Portugal
Chandralekha, DRIEMS, India
Haoxiang Wang, Cornell University, USA

Contents

About the Editors

Khalid Saeed received B.Sc. in Electrical and Electronics Engineering in 1976 from University of Baghdad and M.Sc. and Ph.D. from Wroclaw University of Technology, Poland, in 1978 and 1981, respectively. He received his D.Sc. (Habilitation) in Computer Science from Polish Academy of Sciences, Warsaw, in 2007. He is Professor in Computer Science Department, Białystok University of Technology, Poland. He has authored more than 190 publications including 23 edited books, journals, and conference proceedings and 8 text and reference books. He has supervised more than 110 M.Sc. and 12 Ph.D. theses. His areas of interest are Biometrics, Image Analysis and Processing, and Computer Information Systems. He has given 39 invited lectures and keynote in different universities in Europe, China, India, South Korea, and Japan. The talks were on Biometric Image Processing and Analysis. He has received about 16 academic awards. He is a member of the editorial boards of over 15 international journals and conferences. He is an IEEE Senior Member and has been selected as IEEE Distinguished Speaker for the periods 2011–2013 and 2014–2016. He is the Editor-in-Chief of International Journal of Biometrics with Inderscience Publishers.

Nabendu Chaki is Professor in the Department of Computer Science and Engineering, University of Calcutta, Kolkata, India. He completed his undergraduation in Physics from the legendary Presidency College, Kolkata, and then his postgraduation in Computer Science and Engineering from the University of Calcutta, Kolkata. He completed his Ph.D. in 2000 from Jadavpur University, India. He is sharing two US patents and one patent in Japan with his students. He is quite active in developing international standards for Software Engineering. He represents the country in the Global Directory (GD) for ISO/IEC. Besides editing more than 20 books in different Springer series including LNCS, he has authored 5 text and research books and about 130 peer-reviewed research papers in journals and international conferences. His areas of research interest include Distributed Computing, Image Processing, and Software Engineering. He has served as Research Assistant Professor in the Ph.D. program in Software Engineering in Naval Postgraduate School, Monterey, CA, USA. He has strong and active collaborations with the USA, Europe, Australia, and few institutes and industries in India. He is a visiting faculty member of many universities in India and abroad. He has been the Knowledge Area Editor in Mathematical Foundation for SWEBOK project of IEEE Computer Society. Besides being in the editorial board of several international journals, he has also served in the committees of more than 50 international conferences. He is the Founder Chair of ACM Professional Chapter, Kolkata.

Bibudhendu Pati is Associate Professor in the Department of Computer Science and Engineering at C.V. Raman College of Engineering, Bhubaneswar, Odisha, India. He has a total of 19 years of experience in teaching, research, and industry. His areas of interest include Wireless Sensor Networks, Cloud Computing, Big Data, Internet of Things, and Network Virtualization. He completed his M.E. from NITTTR, Chandigarh, in 2008, M.B.A. from PunjabTechnological University in 2010, and Ph.D. from IIT Kharagpur in 2014. He is a Life Member of the Indian Society of Technical Education (ISTE), Member of IEEE, ACM, CSI, and Computer Science and Engineering Research Group, IIT Kharagpur. He has got several papers published in journals, conference proceedings, and books.

Sambit Bakshi is Assistant Professor in the Department of Computer Science and Engineering, NIT Rourkela, Odisha, India. He completed his M.Tech. and Ph.D. from NIT Rourkela in 2011 and 2014, respectively. His areas of research interest are Biometric Security and Visual Surveillance. He has several journal publications, book chapters, two authored books, and six edited volumes to his credit. He has been teaching subjects like Biometric Security, Statistical Analysis, Linear Algebra and Statistical Analysis Laboratory, Digital Image Processing. He has also been involved in many professional and editorial activities.

Durga Prasad Mohapatra received his Ph.D. from Indian Institute of Technology Kharagpur, India. He joined the Department of Computer Science and Engineering at the NIT Rourkela, India, in 1996, where he is presently serving as Associate Professor. His research interests include Software Engineering, Real-Time Systems, Discrete Mathematics, and Distributed Computing. He has published over 30 research papers in these fields in various international journals and conferences. He has received several project grants from DST and UGC, Government of India. He has received the Young Scientist Award for the year 2006 by Orissa Bigyan Academy. He has also received the Prof. K. Arumugam National Award and the Maharashtra State National Award for outstanding research work in Software Engineering for 2009 and 2010, respectively, from the Indian Society for Technical Education (ISTE), New Delhi. He is about to receive the Bharat Sikshya Ratan Award for significant contribution in academics awarded by the Global Society for Health and Educational Growth, Delhi.

Part I
Advanced Image Processing

A Framework for Pixel Intensity Modulation Based Image Steganography

Srijan Das, Saurav Sharma, Sambit Bakshi and Imon Mukherjee

Abstract Secured data transmission is one of the real issues faced in the world of Web. As the measure of data on Web is expanding daily, the importance of data security is also increasing. Several techniques like cryptography, watermarking, steganography are used to enhance the data to be transmitted. This paper uses a novel steganographic algorithm in the spatial domain using the concept of pixel modulation which diminishes the changes that occur in the stego image generated from the cover image. Experimental results and analysis of the observations show the effectiveness of the proposed algorithm. Different metrics like mean square error (MSE), peak to signal ratio (PSNR), bit-plane analysis, and histogram analysis have been used to show the better results of the proposed algorithm over the existing ones.

Keywords Steganography · Data privacy · Pixel intensity modulation · Spatial domain · Data hiding

1 Introduction

Data privacy has become an important issue in the world of Web. Transmission of data over the Web is involved in our everyday life. Therefore, different techniques are implemented on the data to be transmitted, so that the data transmission becomes a

S. Das (✉) · S. Sharma · S. Bakshi
Department of Computer Science & Engineering, National Institute of Technology,
Rourkela 769008, India
e-mail: srijandas07@gmail.com

S. Sharma
e-mail: srv902@gmail.com

S. Bakshi
e-mail: sambitbaksi@gmail.com

I. Mukherjee
Department of Computer Science & Engineering, Indian Institute
of Information Technology, Nadia 741235, India
e-mail: mukherjee.imon@gmail.com

© Springer Nature Singapore Pte Ltd. 2018
K. Saeed et al. (eds.), *Progress in Advanced Computing and Intelligent Engineering*,
Advances in Intelligent Systems and Computing 563,
https://doi.org/10.1007/978-981-10-6872-0_1

3

secured process. The different techniques include cryptography, watermarking, and steganography. In cryptography, the concept of public and private keys are used. Moreover, it modifies the original data into some other form. In watermarking, the data is embedded in an object which is retrieved later in order to conclude certain facts about the object. Steganography is the art of concealing secret messages in a cover file in a way that no one other than the intended receipts can understand the existence of the hidden message. There are different types of steganography, which includes image steganography, audio steganography, video steganography. In image steganography, the secret message is embedded in a cover image which results into a generation of a stego image. This stego image is transmitted to the receiver side where the extraction algorithm is applied on the stego image to retrieve the secret message. This type of data hiding mechanism has been used widely in recent days.

Least significant bit (LSB) substitution is the most primitive steganographic technique used for securing data. Here the LSB of the intensity values of the image is modified according to the data. So, the overall changes in the resultant stego image is very low since only a small change is incurred due to the changes in the LSB of the intensity values. But these changes can be easily captured on performing bit-plane analysis. Bhattacharyya et al. [1, 2] have performed the modulation technique in steganography in the frequency domain using discrete cosine transformation(DCT) and discrete wavelet transformation(DWT). Mukherjee et al. [3] have used the concept of DWT and applied an adjacent pixel modulation technique in all the matrix of the cover image other than the approximation matrix which results in a better MSE and PSNR values but low embedding capacity. Their algorithm involves generation of seed matrix, identifying the positions where the data is to be embedded and the adjacent pixels in the seed matrices are modified in such a way that the difference between them represents the data to be embedded. All these operations are performed in the frequency domain after applying DWT. On applying inverse DWT, the stego image is generated. In the similar way, the secret embedded data is extracted from the stego image by checking the modulation between the neighboring pixels in the seed matrix window in the frequency domain. But their methods though ensures high security in terms of retrieving the data but generates high distorted stego images.

Sravanthi et al. [4] have proposed a steganographic algorithm in the spatial domain using the concept of encryption and decryption after using LSB operator. An adaptive steganographic technique is proposed by Mandal and Das [5] where the pixel values in stego image are adjusted to be within gray scale range, but at the same time the difference must be same as it was in the stego image without the adjustment. Wang et al. [6], proposed a genetic-algorithm-based steganography method to prevent the RS attack which is a type of steganalysis by modifying the pixel values in stego image in such a way that the pixel values conform to a statistical pattern. But genetic algorithm provides suboptimal solution and hence the embedding may not always results to an optimal solution. Ker [7] introduced the LSB matching scheme which modifies the LSBs of the cover image as LSB substitution does by comparing it with a secret bit. On a mismatch, another bit is added or subtracted on a random basis to the pixel value in the cover image. But this method decreases the embedding capacity as compared to primitive LSB substitution method.

Da-chun and Wen-Hsiang [8] have developed a steganographic algorithm by the method of pixel modulation. The cover image is divided into blocks of two adjacent pixels which are not overlapping. The difference between the pixels in each blocks is computed. The possible differences are categorized into a number of ranges which are selected based on the perceptibility of the human eye to different variations of gray values. A sub-stream of the data stream is embedded by using the new value as the difference value. The width of the range in which the difference value belongs to decides the number of bits to be embedded in a pixel pair. This method is advantageous in terms of producing a more imperceptible results compared to the other LSB replacement methods. But the storage and complexity of the algorithm is huge which restricts the usage of this algorithm. As the spatial domain algorithms became popular, Joo et al. [9] proposed an Adaptive Steganographic Method Using the Floor Function and Modulus Function with Practical Message which provides better resistance to spoof attacks and thus enhancing the security of the hidden message.

In this paper, we propose a modulation-based embedding technique in the neighboring pixels of the cover image so as to generate the stego image. The modulation technique is generally used in frequency domain, but using this technique in the spatial domain minimizes the changes expected in the stego image. It is believed that the spatial domain algorithms are less secured as compared to frequency domain algorithms. But most of the frequency domain algorithms use modulation techniques to embed and extract data. So, in this method also, the concept of modulation has been used which not only maintains the security but also diminishes the distortion in the stego images.

2   Proposed Method

While modulation method has been widely used in frequent, proposed approach utilizes modulation in spatial domain to achieve the same. The method involves adjusting the adjacent pixels based on the data to be embedded. This adjustment ensures minimum deviation from the cover image. The embedding technique is discussed below.

2.1   Embedding Method

The embedding method involves generation of sub-matrices called seed matrix and embedding the equivalent binary data pair-by-pair based on their values by modulating the adjacent pixel values such that the difference corresponds to the pair value.

Fig. 1 Embedding the data

A1	A2	A3
(Embed d1)	(Embed d2)	(Embed d3)
A4	**A5**	**A6**
(Embed d4)		(Embed d5)
A7	**A8**	**A9**
(Embed d6)	(Embed d7)	(Embed d8)

Table 1 Adjacent pixels difference for modulation and embedding

Bit Pattern	Difference Value
00	−1
11	+1
10	−2
01	+2

Hence, at first the secret message is converted into a binary string d based on the ASCII value of each character where each character is represented in eight-bit notation. Then each character or eight bits are embedded in each seed matrices.

(i) Seed Matrix Generation:
From the cover image 3×3 sub-matrices called seed matrices are taken. Each window of 3×3 size starting from top left corner of the image matrix, pixel values are stored in an array in row major order.

(ii) Embedding Location:
Before starting the embedding process, the positions of embedding the data are to be identified. The position of embedding the data can be taken in any order but the same convention should be maintained throughout the process and in the extraction method too. The identification of the position where the data is to be embedded can be identified from Fig. 1.

(iii) Data Embedding:
For embedding a single character, adjacent pixel values are modulated to embed a pair of bits from binary string d in a window. After embedding, the window slides in a row major order with no overlapping. Table 1 shows the encoding scheme to specify bit value pair from 00, 01, 10, 11 based on the difference of the adjacent pixel values, and this encoding is repeated for each pair of bits from binary string d. The modified pixel values forms the stego image.

Table 2 Bit pair values for decoding

Sign of Intensity Difference	Magnitude of Difference	Extracted Message bits
Negative	−1	00
Positive	+1	11
Positive	−2	10
Negative	+2	01

2.2 Extraction Method

The extraction algorithm takes the stego image as the input and retrieves the data embedded by checking the modulation of the neighboring pixels in the seed matrices.

(i) Seed Matrix Generation and Extraction Location:
The stego image is scanned from left to right from top left corner, and 3×3 seed matrices are generated such that no overlapping occurs. Once the seed matrices are selected, the extraction location is to be identified based on the positions identified for embedding in the embedding strategy.

(ii) Data Extraction:
The difference between the neighboring pixels in the seed matrices is computed and checked from Table 2. Based on a mapping from the difference computed to a pair of binary bits, the data is extracted.

3 Results and Analysis

The proposed method for embedding secret data is tested with several images which includes calculating the MSE, PSNR and analyzing the histograms and bit planes of the cover and stego images. All these parameters used to calculate the efficiency of a steganographic algorithm are giving satisfactory results for the proposed algorithm. All these parameters are computed for three types of stego images of the same cover image, one by embedding 1000 characters, second by embedding 5000 characters, and the other by embedding 10000 characters. Figure 2 shows some pair of stego and cover images. The stego images are so less deviated that it cannot be detected easily by naked eyes which is the ultimate goal of any steganographic algorithm.

3.1 Histogram Analysis

Histogram of an image is a capacity that refers to the quantity of pixels corresponding to a gray level. The histogram representation of the cover and stego images with

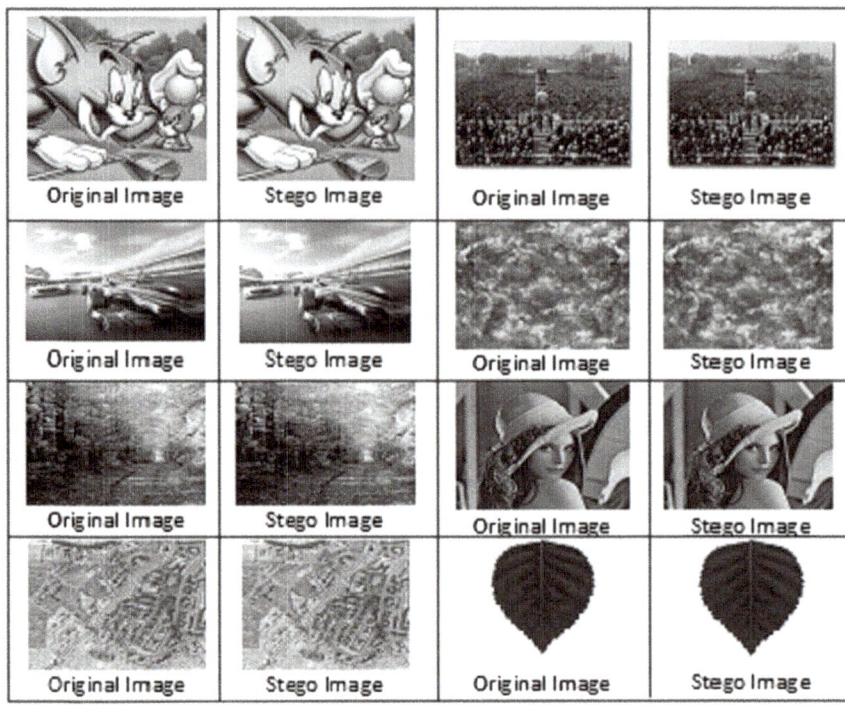

Fig. 2 Cover image (1st and 3rd column) and stego image (2nd and 4th column)

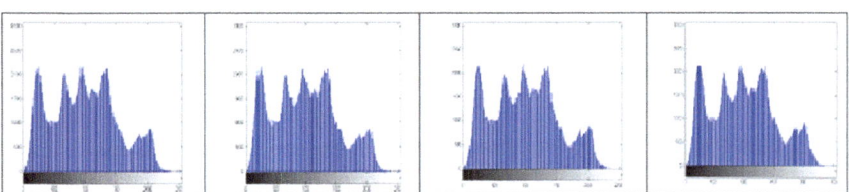

Fig. 3 Histogram of cover image (at left), stego image (1000 characters), stego image (5000 characters) and stego image (10000 characters)

1000, 5000, and 10000 characters embedded separately is shown in Fig. 3. The deviation in the histograms of the cover image and the stego images is very low. Slight deviation at the peaks of the histogram can be visualized. Such changes are more visible when the number of characters embedded increases.

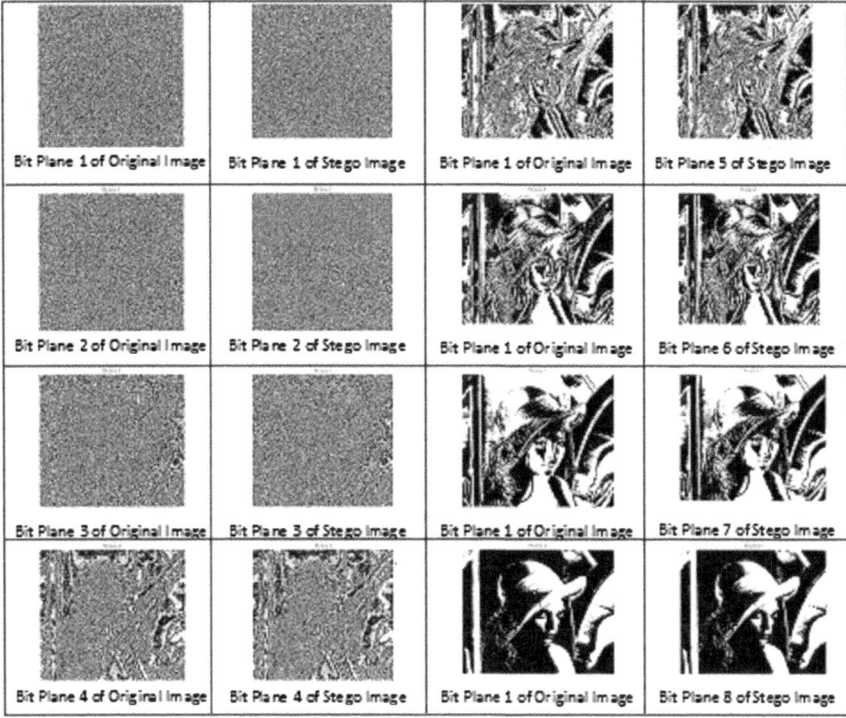

Fig. 4 Bit Planes of cover image (1st and 3rd column) and stego image (2nd and 4th column)

3.2 Bit-Plane Analysis

Bit-plane slicing is the phenomenon of representing an image into its eight equivalent binary images by slicing each bit of the image pixels of 256 gray levels. It can detect how much stego images are inconsistent from their cover images in LSB substitution. This slicing technique can represent the changes of each bit planes of a pair of stego and cover images. Figure 4 depicts the bit planes of the cover and stego images after applying the proposed method, and the distinction between these images cannot be easily distinguished.

3.3 Capacity Analysis

For an $R \times C$ Image, the existing algorithms in frequency domain take 8×8 window within which four 3×3 seed matrices are taken for embedding four characters, since one 3×3 window can accommodate only one character. Hence the embedding capacity (in bits per pixel) in such algorithms can be evaluated as

Table 3 Comparative study of embedding capacities of existing works and our proposed algorithm

Sl. No.	Existing Work	Avg. embedding capacity (bpp)
1	HUGO [14]	0.4
2	Ternary ±1 embedding [15]	<1
3	Feng [16]	<1
4	LSB based steganography [10–13]	1
5	Proposed algorithm	0.88

$$capacity = \frac{R \times C}{16} \qquad (1)$$

But for our proposed algorithm, the embedding capacity is enhanced, since each 3×3 window accommodates one character, so the capacity is evaluated as:

$$capacity = \frac{R \times C}{9} \qquad (2)$$

Thus, the embedding capacity is increased by 4.8 times which is a very important feature of image steganography.

Table 3 shows the comparative study between our proposed algorithm and the existing works with respect to the capacity of storing information in a single 3×3 window. Though our proposed algorithm has lower average embedding capacity (bpp) when compared to some of the LSB-based steganographic algorithms, but the PSNR values of the stego images produced from our proposed algorithm are better than those generated from these LSB-based steganographic algorithms [10–13]. Efficiency of image steganographic algorithms not only depends on the embedding capacity, but also on the PSNR values.

3.4 Mean Square Error (MSE) and Peak to Signal Noise Ratio (PSNR)

MSE and PSNR are some of the parameters in image processing to evaluate the efficiency and performance of the proposed algorithms. Both the parameters indicate the deviation of the stego image from the cover image. MSE is the sum of difference between the intensities of the pixels between the cover and the stego image. More the MSE value, more deviated will be the stego image from the cover image. It is always desired to have less MSE values for a given pair of cover and stego images.

Table 4 MSE and PSNR values of some standard images

Images	1000 characters		5000 characters		10000 characters	
	MSE	PSNR	MSE	PSNR	MSE	PSNR
Cartoon	0.630	50.133	4.266	41.824	8.895	38.639
Crowd	0.023	64.365	0.521	50.961	6.574	39.952
Fast	0.122	57.237	0.670	49.870	1.764	45.663
Fire	2.279	44.553	11.754	37.428	23.116	34.491
Landscape	2.653	43.893	13.028	36.982	26.239	33.941
Lena	0.417	51.922	2.592	43.944	6.891	39.747
Map	2.181	44.743	11.747	37.431	24.169	34.298
Object	0.261	53.959	2.393	44.341	5.492	40.733

$$MSE = \frac{1}{RC} \sum_{x=1}^{R} \sum_{y=1}^{C} (g(x,y) - g'(x,y))^2 \qquad (3)$$

In the above equation, R is the number of rows, C is the number of columns, g is the cover image, and g' is the stego image. PSNR can be computed from the obtained MSE values. This is another parameter used to trace the deviation of the modified stego image obtained from the embedding algorithm from the cover image which is initially taken as the input. Higher PSNR values indicate less deviation.

$$PSNR = 10 \log \frac{255^2}{MSE} \qquad (4)$$

Table 4 shows the MSE and PSNR values of some standard images. These values are computed for small (1000 characters), medium (5000 characters), and large text (10000 characters). These values are quite better as compared to the values obtained in the frequency domain algorithm proposed by Mukherjee et al [3].

Tables 5 and 6 show a comparison between the MSE and PSNR values of some stego images obtained on applying the existing frequency domain algorithm proposed by Bhattacharyya and Sanyal [2]. and the stego images obtained by the proposed method. The Tables show the MSE and PSNR values for stego images obtained on embedding 100, 500, 1000, 3000, 5000, and 10000 characters. As the number of embedded characters increases, the MSE and PSNR values deteriorate as the distortion in the stego images increases. The values of MSE and PSNR for all number of embedded characters clearly indicate the less deviation of the stego images obtained from the proposed method as compared to the existing one.

Table 5 Comparison of MSE values of different images between the proposed algorithm and the existing frequency domain algorithm

Images	Algorithm	Embedding Capacity (in characters)					
		100	500	1000	3000	5000	10000
Lena	Existing	0.063	0.625	1.813	9.723	20.957	38.917
512×512	Proposed	0.029	0.178	0.369	1.177	2.372	6.092
Lena	Existing	0.993	12.482	32.799	108.660	169.860	N/A
256×256	Proposed	0.073	0.352	0.683	2.460	5.169	13.403
Lena	Existing	13.392	27.418	23.669	N/A	N/A	N/A
128×128	Proposed	0.362	1.884	5.142	22.579	43.157	46.751
Peppers	Existing	1.078	37.992	40.200	51.316	65.569	97.246
512×512	Proposed	0.039	0.208	0.332	0.797	1.268	2.352
Peppers	Existing	23.354	58.178	87.737	203.210	240.730	N/A
256×256	Proposed	0.209	0.918	1.717	4.507	7.150	16.088
Peppers	Existing	81.847	242.678	434.526	N/A	N/A	N/A
128×128	Proposed	0.936	4.232	7.815	26.842	42.656	44.962

Table 6 Comparison of PSNR values of different images between the proposed algorithm and the existing frequency domain algorithm

Images	Algorithm	Embedding Capacity (in characters)					
		100	500	1000	3000	5000	10000
Lena	Existing	60.107	50.167	45.545	38.252	34.917	32.229
512×512	Proposed	69.549	61.641	58.496	53.460	50.415	46.320
Lena	Existing	48.157	37.168	32.972	27.769	25.829	N/A
256×256	Proposed	59.493	52.659	49.783	44.221	40.996	36.858
Lena	Existing	36.862	27.418	23.669	N/A	N/A	N/A
128×128	Proposed	46.480	39.325	34.964	28.538	25.725	25.378
Peppers	Existing	47.800	32.333	32.088	31.028	29.963	28.252
512×512	Proposed	68.242	60.988	58.950	55.152	53.134	50.453
Peppers	Existing	34.447	30.483	28.699	25.051	24.315	N/A
256×256	Proposed	54.921	48.499	45.783	41.591	39.587	36.065
Peppers	Existing	29.000	24.280	21.750	N/A	N/A	N/A
128×128	Proposed	42.359	35.810	33.146	27.787	25.776	25.547

4 Conclusion

The proposed spatial domain algorithm based on pixel modulation technique produces better MSE, PSNR values over the existing frequency domain algorithms. The distortion in the stego image produced by the proposed algorithm is less which is evident from the histogram and bit-plane analysis. One of the major drawbacks of steganography is the embedding capacity which is somewhat improved by this proposed algorithm. The proposed algorithm ensures an increase in the embedding capacity by 4.8 times compared to the embedding capacity of the existing algorithms [1–3]. We have also found out the comparative performance of our proposed algorithm with the existing works. The embedding capacity of our proposed algorithm is close to 1 bits per pixel, and at the same time, the stego image generated has good PSNR values.

References

1. Bhattacharyya, S., Khan, A., Sanyal, G.: DCT difference modulation (DCTDM) image steganography. Int. J. Comput. Inf. Net. Secur. **3**, 40–63 (2014)
2. Bhattacharyya, S., Sanyal, G.: A robust image steganography using DWT difference modulation (DWTDM). Int. J. Comput. Netw. Inf. Secur. **4**, 27–40 (2012)
3. Mukherjee, I., Datta, B., Banerjee, R., Das, S.: DWT Difference Modulation Based Novel Steganographic Algorithm. In: Jajodia, S., Majumdar, C. (eds.) ICISS 2015, vol. 9478, pp. 573–582. LNCS, Springer, Kolkata, Dec 2015
4. Sravanthi, G.S., Devi, B.S., Riyazoddin, S.M.: A spatial domain image steganography technique based on plane bit substitution method. Glob. J. Comput. Sci. Technol. Graph. Vis. USA **12** (2012)
5. Mandal, J.K., Das, D.: Steganography using adaptive pixel value differencing(APVD) of gray images through exclusion of overflow/underflow. In: The second International Conference on Computer Science, Engineering and Applications (CCSEA-2012) (2012)
6. Wang, C., Yang, B., Niu, X.: A secure steganography method based on genetic algorithm. J. Inf. Hiding Multimed. Signal Proc. **1**, 28–35 (2010)
7. Ker, A.D.: Steganalysis of LSB matching in grayscale images. IEEE Signal Process. Lett. **12**(6), 441–444 (2005)
8. Da-Chun, W., Wen-Hsiang, T.: A steganographic method for images by pixel-value differencing. Pattern Recognit. Lett. **24**, 613–626 (2003)
9. Joo, J.C., Oh, T.W., Lee, H.Y., Lee, H.K.: Adaptive steganographic method using the floor function with practical message formats, international. J. Innov. Comput. Inf. Control **7**(1), 161–175 (2011). ISSN 1349-4198
10. Dagar, E., Dagar, S.: LSB based image steganography using X-Box mapping. In: IEEE International Conference on Advances in Computing, Communications and Informatics (ICACCI), New Delhi, India, 24–27 Sept 2014
11. Das, S.K., Dhara, B.C.: A new secret image sharing with arithmetic coding. In: Proceedings of 2015 IEEE International Conference on Research in Computational Intelligence and Networks, Kolkata, India, pp. 395–399, Nov 2015
12. Deshmukh, P.U., Pattewar, T.M.: A novel approach for edge adaptive steganography on LSB insertion technique. In: IEEE International Conference on Information Communication and Embedded Systems (ICICES), Chennai, India, 27–28 Feb 2014

13. Paul, G., Davidson, I., Mukherjee, I., Ravi, S.S.: Keyless steganography in spatial domain using energetic pixels. In: Venkatakrishnan, V., et al. (eds.) Proceedings of the 8-th International Conference on Information Systems Security (ICISS), Guwahati, India, vol. 7671, pp. 134–148. LNCS, Springer, 15–19 Dec 2012. ISBN: 978-3-642-35129-7
14. Bas, P., Filler, T., Pevny, T.: Break our steganographic system, the ins and outs of organizing BOSS. Information Hiding, Czech Republic, vol. 6958/2011, pp. 59–70. LNCS, May 2011
15. Fridrich, J., Lisonek, P., Soukal, D.: On Steganographic Embedding Efficiency, Information Hiding. 8th International Workshop, Alexandria, VA, vol. 4437, pp. 282–296. LNCS (2008)
16. Feng, B., Lu, W., Sun, W.: Secure binary image steganography based on minimizing the distortion on the texture. IEEE Trans. Inf. Forensics Secur. **10**(2), 243–255 (2015). February

Higher-Order Nonlinear Analysis with Core Tensor and Frame-Wise Approach for Dynamic Texture Synthesis

Premanand P. Ghadekar and Nilkanth B. Chopade

Abstract Dynamic texture synthesis is the process of generating artificial frames from the original ones. Three different innovative approaches are proposed to analyse the chaotic behaviour of dynamic texture. In the proposed method, an operation on the core tensor by considering a fewer number of principal components requires less number of model coefficients. The proposed approach of separating the colour tensor and applying kernel principal component analysis (KPCA) on all the dimensions of different colour cuboids helps to reduce the time complexity with the best quality. The method of creating a tensor with a frame-wise approach is more dynamic. Each frame of the dynamic texture is being taken in 3-D tensors, and then, KPCA is applied on each dimension of the tensor. A frame-wise approach is more flexible in terms of time complexity, PSNR, model coefficients and compression ratio as compared to existing algorithms.

Keywords HOSVD · KPCA · HOKPCA · Nonlinear · Tensor

1 Introduction

Dynamic texture [1] is having a redundant pattern within inter frames of images. It is necessary first to analyse the behaviour of the dynamic texture for processing and to check the linearity and nonlinearity in the input data. In the synthesis process [2], an artificial dynamic texture is created from the original for the applications like video gaming, patching and pattern recognition.

P. P. Ghadekar (✉)
Vishwakarma Institute of Technology, Pune, India
e-mail: ppghadekar@gmail.com

N. B. Chopade
Department of E & TC Engineering, Pimpri Chinchwad College of Engineering,
Pune, India
e-mail: nbchopade@gmail.com

© Springer Nature Singapore Pte Ltd. 2018   15
K. Saeed et al. (eds.), *Progress in Advanced Computing and Intelligent Engineering*,
Advances in Intelligent Systems and Computing 563,
https://doi.org/10.1007/978-981-10-6872-0_2

There are two approaches for the synthesis [3, 4]: one is physics based, and the other is an image based. In the physics-based approach, the model is derived from physical laws within the particles of frames, while in the image-based approach, there are two methods: parametric and nonparametric. Different clips are extracted from the video in the nonparametric or patch-based approach, and they are patched together to generate a number of synthesized frames. The parametric method considers different parameters for synthesis. It produces more compact videos with better visual quality. In the proposed techniques, a parametric approach is used.

A tensor is a generalized term used for arrays. Higher-order tensors have an order of more than two. The colour dynamic texture is a fourth-order tensor, where the four dimensions are: height, width, number of frames and colours as shown in Fig. 1.

There exist two basic approaches for tensor decompositions [5]; those are Candecomp (canonical decomposition)/Parafac (parallel factors) [6] and Tucker decomposition. Tucker decomposition is formed by multiplying the super diagonal core tensor with the number of orthonormal matrices of every dimension. Higher-order SVD uses Tucker decomposition, so they are well known as higher-order singular value decomposition (HOSVD). HOSVD [7] deals with an extended version of the singular value decomposition (SVD) [8] by using a higher-order tensor. Decomposing a tensor by Tucker decomposition and then finding the best rank is its fundamental idea. HOSVD takes its input in four-dimensional arrays, and that means there is no need to convert the video frames into column vectors. SVD provides dimension reduction in the temporal domain only, but HOSVD allows dimension reduction in the spatial as well as the temporal domain. Thus, it better exploits the relation between every frame. However, the HOSVD does not capture nonlinear motion; it captures only linear motion. It requires high time complexity, and its analysis part is more expensive [7].

SPIHT-based threshold filtering approach with GPU [9] is used for dynamic texture synthesis. It represents more valuable information by using less number of model coefficients and reduces the time complexity. In linear motion, the motion of any object can be easily predicted, but in nonlinear motion, it is very difficult to predict each and every coordinate of the trajectory. There are various methods for predicting the nonlinear motion of a dynamic texture like nonlinear PCA and the

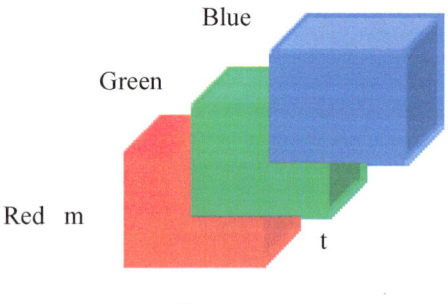

Fig. 1 Colour video tensor (m*n*3*t)

Gaussian process latent variable model (GPLVM), but these methods require high computational complexity.

Principal component analysis (PCA) [10] is one of the linear techniques for dimensionality reduction. This method does not capture nonlinear motion. For capturing nonlinear motion, an additional kernel function has been used that maps principal components, which are the eigenvectors related to the eigenvalues in feature space. There are various kernel functions like a polynomial, and Gaussian [11, 12]. Unlike linear PC, the KPCA used to extract a number of principal components, which can exceed the input dimensionality. KPCA reduces the mean approximation error, and it also minimizes the entropy and gives maximum mutual information of the provided input. Different types of KPCA are given in [13].

The article is structured as follows. Section 2 represents the synthesis model based on higher-order kernel principal component analysis (HOKPCA) with a less number of coefficients for reconstructing the core tensor and hence the original tensor. Section 3 describes the separate colour tensor approach. The frame-wise approach is given for reconstruction in Sect. 4. In Sect. 5, the performances of the various proposed algorithms are evaluated, and the comparisons between the existing and proposed methods based on visual quality and the numbers of model coefficients are calculated. Finally, the last section, Sect. 6, concludes the article.

2 Higher-Order KPCA

The proposed HOKPCA with the core tensor creation approach is used to decrease the number of model coefficients. As mentioned in the previous algorithms every time, a core tensor is being generated from row feature vectors obtained by applying the KPCA. In [14], the algorithm for tensor decomposition with Tucker decomposition and candelinc decomposition is described, which first applies to Tucker tensor decomposition and later for core tensors it applies to candelinc decomposition. The HOKPCA with an operation of the core tensor algorithm primarily focuses on the core tensor. It generates a core tensor with the least number of principal components. Figure 2 shows that the core tensor creates an original tensor to achieve more compression with good quality.

Algorithm

1. Take the input as dynamic texture frames.
2. Create a 4-D tensor from input dynamic textures as shown in Fig. 1.
3. Unfold the tensor in the 2-D matrix of all dimensions of the tensor.
4. Apply KPCA on each dimension.
5. Select principal components from the respective matrices of each dimension.
6. Create a core tensor and again apply Tucker tensor decomposition with KPCA on the core tensor.

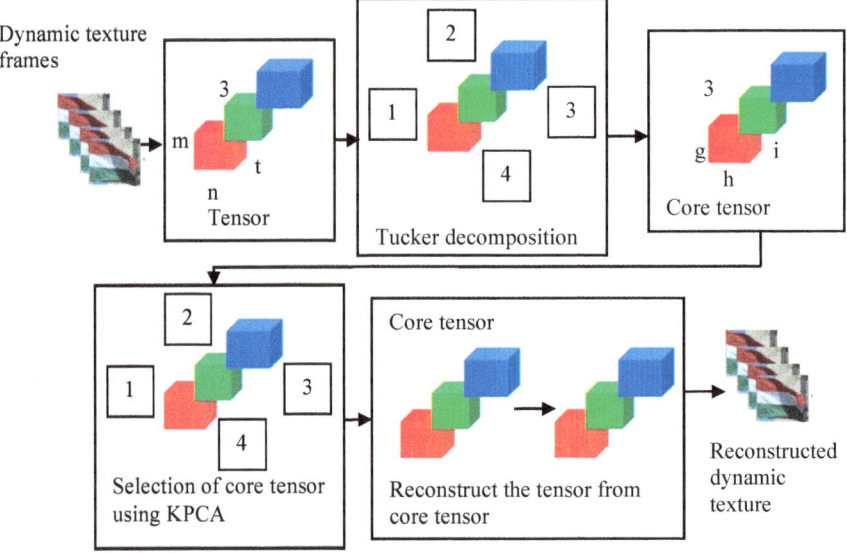

Fig. 2 Block diagram of HOKPCA (core tensor)

$$A_2 = A_1 * U_1^T * U_2^T * U_3^T * U_4^T \tag{1}$$

7. Unfold the matrix of each dimension for every core tensor.
8. Apply KPCA on each dimension.
9. Select principal components from the respective matrices of the core tensor of each dimension.
10. Reconstruct the core tensor by selecting minimum principal components.

$$A_1 = A_2 * V_1 * V_2 * V_3 * V_4 \tag{2}$$

From the reconstructed core tensor, reconstruct the original tensor.

$$S_1 = A_1 * U_1 * U_2 * U_3 * U_4 \tag{3}$$

11. Generate dynamic texture frames.

3 Separate Colour Tensor Processing

The proposed HOKPCA with separate colour tensor approach decreases the time complexity. The 4-D tensor array consists of three colour cuboids of the same size. So, instead of applying KPCA on a 4-D tensor here, KPCA is used on every 3-D array (that is colour cuboids). Individually, every 3-D array is reconstructed, and then, a 4-D array is being reconstructed as shown in Fig. 3.

Algorithm

1. Take the input as dynamic texture frames.
2. Create three colour cuboids that are red, green, and blue.
3. Unfold the tensor in the 2-D matrix for all dimensions of three colours tensors.
4. Apply KPCA on each dimension.
5. Select principal components from the respective matrices of each dimension.
6. Create a core tensor by using Eq. 1 of every colour cuboids.
7. Reconstruct the original three tensors using Eq. 3.
8. Generate dynamic texture frames.

The proposed approach helps to select principal components from a different colour tensor, so if any image has a particular colour in a significant amount, then we select that colour component more to get better results. Also, compared to HOKPCA, it requires less time to execute.

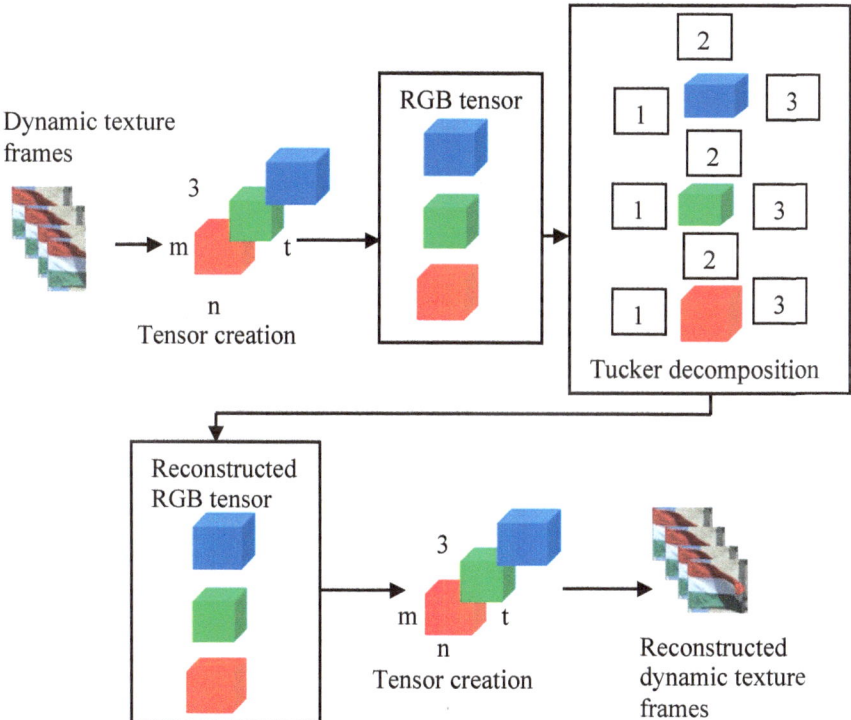

Fig. 3 Block diagram of HOKPCA (separate colour tensor processing)

4 Frame-Wise Approach

By taking the frame into 3-D arrays, i.e. third-order tensor, [15] and choosing the minimum number of components from a single frame, it gives a better quality of the image. Also, it is analysed that if there is more difference between the height and width of the image, HOSVD and HOKPCA do not provide a good-quality result by choosing the equal number of height and width components. After varying selection of the number of principal components from each frame by keeping the average number of components same, the quality of the video goes on increasing. However, this type of logic works only for videos having a small number of frames. So, as we go on increasing the number of frames, the task of entering the number of components from each frame becomes difficult. So, to make it more general before selecting the components, the mean squared error is calculated. If the mean squared error is greater than the average mean squared error, then more number of components are selected. Otherwise, less number of components are selected. Then, KPCA is applied on each frame. This approach provides excellent visual quality as well as less time complexity. Figure 4 depicts the overall project flow of the proposed frame-wise approach.

Algorithm

1. Take the input as nonlinear dynamic texture frames.
2. Extract the number of frames from the video.
3. Calculate the mean squared error between two consecutive frames.

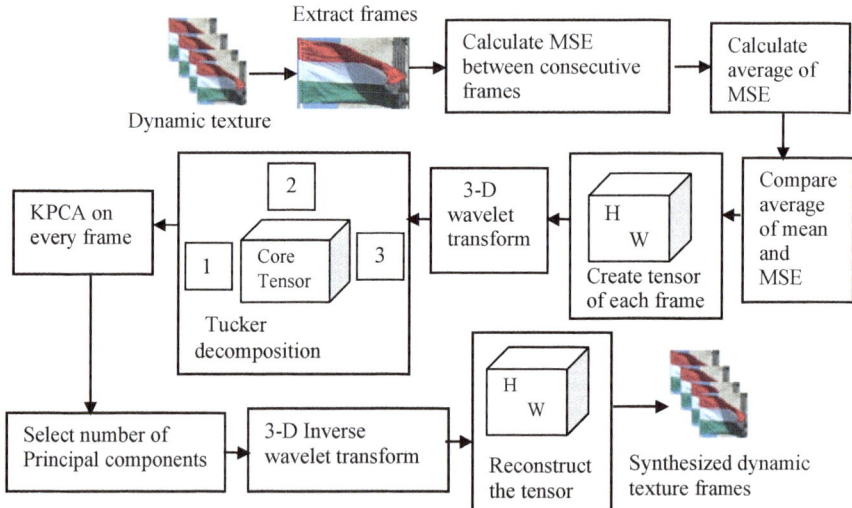

Fig. 4 Block diagram of HOKPCA with a frame-wise approach

$$\text{MSE}(i) = \frac{1}{n} \sum_{i=1}^{n} (Oi - Oi + 1)^2 \qquad (4)$$

4. Calculate the average value of mean squared error.

$$\text{Average of mean} = \frac{1}{n} \text{MSE}(i) \qquad (5)$$

5. If (average of mean >MSE)
   Select a more number of components.
   else
   Select less number of components.
6. Create a 3-D tensor of each frame with the first dimension as height, second as width, and the third as colour components.
7. Apply 3-D DWT on created tensor so that every dimension will get halved.
8. Unfold the tensor in the 2-D matrix for all dimensions of three colours tensors.
9. Apply Tucker decomposition on the tensor.
10. Use the KPCA on each dimension of a tensor.
11. Select the principal components of each dimension according to MSE between two frames.
12. Apply inverse 3-D wavelet transform and reconstruct the tensor.
13. Reconstruct the synthesized frames of dynamic textures.

If the disk size is very less, then for a less number of model coefficients, the 3-D wavelet [16, 17] can be used. The discrete 3-D wavelet transform directly applied to videos in tensor form to exploit spatial, temporal correlation. 3-D DWT is efficient for videos having more number of frames, and the time required for the execution is less as there is no need to apply the wavelet transform for every frame.

5   Experimental Results

Dynamic texture sequences, which are used for testing purposes, were taken from the MIT temporal texture database and dynamic texture database (Renaud et al.) [18].

The peak signal-to-noise ratio (PSNR) [19] is used to check the quality of reconstructed dynamic texture frames. Compression ratio (CR) calculates the reduced data size in percentage. If the CR is more, more compression is achieved; if it is less, the compressed image has less CR. Model coefficients are the number of parameters required to reconstruct the dynamic texture. By keeping the constant PSNR and comparing all the algorithms, it is noticed that the frame-wise approach is overall best because it requires a less number of model coefficients with high PSNR. Also, it is observed that after changing the resolution of every frame by keeping the small difference between the height and width, the quality of dynamic texture increases with a less number of model coefficients.

Table 1 represents the comparison of the proposed models with the existing SVD [8] and HOSVD models [7] by using dynamic texture (flame) with constant PSNR. The proposed approach HOKPCA core and frame-wise approach provide better results as compared to SVD and HOSVD approaches.

The comparison of the proposed models with the existing SVD model [8] by using nonlinear dynamic texture (Brownian motion) with constant PSNR is given in Table 1. In the case of Brownian motion dynamic texture [20], HOSVD requires a long time to execute the code. Therefore, the results of HOSVD [7] are not mentioned in Table 2. For constant PSNR, SVD needs more number of components, and hence, its compression ratio is in the negative form.

Table 1 Comparison of the proposed models with the existing SVD and HOSVD models (flame)

Parameters	SVD	HOSVD	HOKPCA[a]	HOKPCA core[a]	HOKPCA colour tensor[a]	Frame-wise approach[a]
PSNR (dB)	35.53	35.11	35.14	35.14	35.10	36.06
CR %	4.94	77.48	89.25	99.29	89.50	85.28
Time complexity	3.70 s	522 s	12.14 s	9.94 s	10.40 s	3.31 s
Model coefficients	14059742	603913	598662	60152	606674	972134
Original size MB	19.56	19.56	19.56	19.56	19.56	19.56
Compressed size MB	18.59	4.40	2.10	0.138	2.11	2.87

[a]Proposed approach

Table 2 Comparison of the proposed models with the existing SVD model (Brownian motion)

Parameters	SVD	HOKPCA[a]	HOKPCA core[a]	HOKPCA colour tensor[a]	Frame-wise Approach[a]
PSNR(dB)	21.45	21.72	21.72	21.78	21.02
CR %	−155	90.78	99.35	90.78	73.04
Time complexity (Sec)	7.05	131.96	89.97	19.29	5.97
Model coefficients	26834002	958759	14188	1051250	2325600
Original size (MB)	37.08	37.08	37.08	37.08	37.08
Compressed size (MB)	94.68	3.41	0.24	3.78	9.99

[a]Proposed approach

In Fig. 5, by increasing the number of components, PSNR, the number of model coefficients of the dynamic texture (flame) increases. The HOKPCA frame-wise approach requires a significant number of model coefficients, but its PSNR is also more. It provides improved results as compared to HOSVD approach [7].

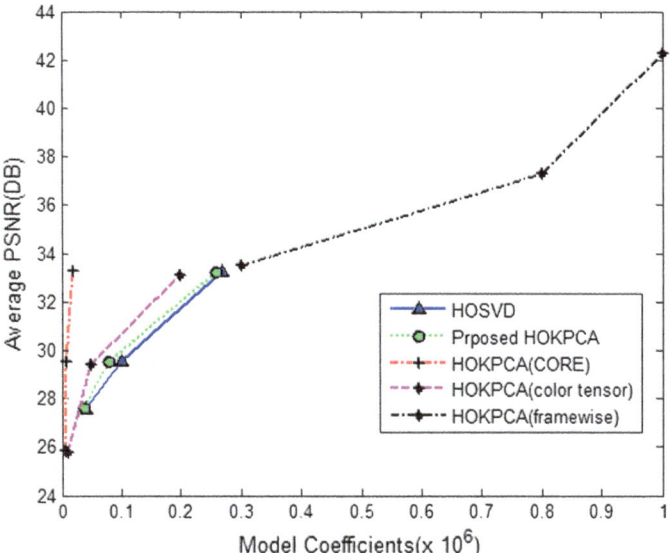

Fig. 5 Comparison graph of average PSNR versus model coefficients for dynamic texture (flame)

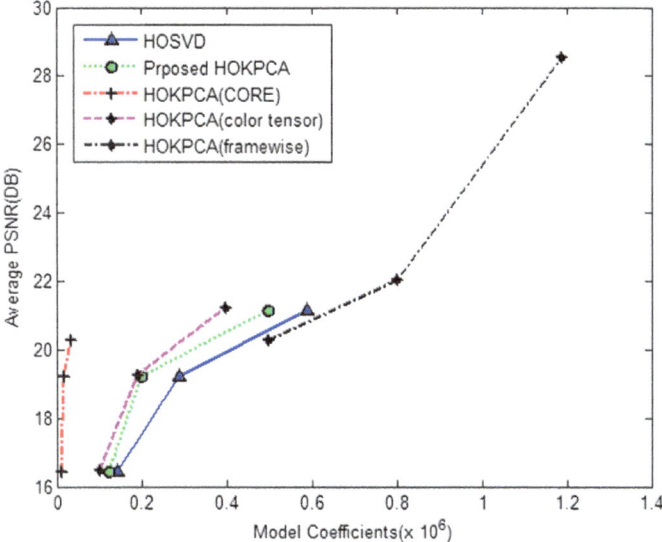

Fig. 6 Comparison graph of average PSNR versus model coefficients for nonlinear dynamic texture (Brownian motion)

In Fig. 6, by increasing the components, PSNR, the number of model coefficients of the dynamic texture (Brownian motion) increases. The HOKPCA frame-wise approach requires a large number of model coefficients, but its PSNR is also more. It provides better results as compared to HOSVD approach [7].

6 Conclusion

The proposed algorithms are based on the tensor algebra and Tucker decomposition with KPCA. It captures linear as well as nonlinear motion along with a less number of parameters. Tensor decomposition with an operation on the core tensor gives a more compact model as it requires a less number of parameters for synthesizing the core tensor and hence a reconstructed tensor. Tucker decomposition with the separate colour tensor approach, by creating three colour cuboids and applying Tucker decomposition on every cuboid, helps to improve the quality of dynamic texture. Representing the frame in a 3-D tensor and using KPCA on every dimension of the 3-D tensor reduce the time complexity and give better results in terms of quality and compression. The time complexity of the frame-wise algorithm decreases because of a reduction in the number of multiplications. Tensor algebra along with 3-D DWT is helpful to get a more compact form and better quality of the dynamic texture model. Thus, here, a better exploitation in the temporal, as well as the spatial domain, is achieved along with less execution time.

The proposed algorithms are useful for linear and nonlinear dynamic texture analysis and synthesis. It requires very less model coefficients. It has less time complexity, better compression ratios and less computational complexity with better visual quality.

References

1. Tuceryan, M., Jain, A.: Texture analysis Chapter 2.1. The Handbook of Pattern Recognition and Computer Vision (2nd Edition), pp. 207–248. World Scientific Publishing Co. (1998)
2. Chen, H., Ruimin, H., Mao, D., RuiZhong, Zhongyuan W.: Video coding using dynamic texture synthesis. IEEE Int. Conf. Multimed. Expo 203–208 (2010)
3. Costantini, R.: Compact Representation for Static and Dynamic Texture Synthesis, pp. 1–160 (2007)
4. Baker, K.: Singular Value Decomposition Tutorial, pp. 1–24 (2005) (Revised January 14, 2013)
5. Kolda, G., Bader, B.: Tensor decompositions and applications. SIAM Rev. 51(3), 455–500 (2009)
6. HuyPhan, A., Tichavský, P., Cichocki, A.: CANDECOMP/PARAFAC decomposition of high-order tensors through tensor reshaping. IEEE Trans. Signal Proc. 61 (2013)
7. Costantini, R., Sbaiz, L., Usstrunk, S.: Higher Order SVD analysis for dynamic texture synthesis. IEEE Trans. Image Process. 17(1), 4–52 (2008)

8. Doretto, G., Chiuso, A., Wu, Y., Soatto, S.: Dynamic textures. Int. J. Comput. Vis. **51**(2), 91–109 (2003)
9. Ghadekar, P., Chopade, N.: Content-based dynamic texture analysis and synthesis based on SPIHT with GPU. J. Inf. Proc. Syst. (JIPS) Korea **12**(1), 46–56 (2016)
10. Smith, L.: A Tutorial on Principal Components Analysis, pp. 1–27 (2002)
11. Müller, K., Mika, S., Ratsch, G., Tsuda, K., Scholkopf, B.: An introduction to kernel-based learning algorithms. IEEE Trans. Neural Netw. **12**(2), 181–202 (2001)
12. Ghadekar, P., Chopade, N.: Modelling nonlinear dynamic textures using hybrid DWT-DCT and Kernel PCA with GPU. J. Inst. Eng. Ser.-B Springer **97**(4), 549–555 (2016)
13. Ghadekar, P., Parmar, N., Chopade, N.: Nonlinear dynamic texture analysis and synthesis using different types of KPCA. IEEE Int. Conf. Inf. Proc. 739–744 (2015)
14. Kiers, H., Harshman, R.: Relating two proposed methods for speedup of algorithms for fitting two and three-way principal component and related multilinear models. Chemometr. Intell. Lab. Syst. **36**, 1–10 (1997)
15. George, J., Indus, P., Rajeev, K.: Three-Dimensional ultrasound image enhancement using local structure tensor analysis. IEEE Region 10 Conference, pp. 19–21 (2008)
16. Bhoopathi, G.: An image compression approach using wavelet transform and modified self-organizing map. IJCSI Int. J. Comput. Sci Issues **8**(5), 323–330 (2011)
17. Malik, S., Verma, V.: Comparative analysis of DCT, haar and daubechies wavelet for image compression. Int. J. Appl. Eng. Res. **7**(11), 1–6 (2012). ISSN 0973-4562
18. Renaud, P., Fazekas, S., Huiskes, M.: DynTex: a comprehensive database of dynamic textures. Pattern Recogn. Lett. **31**(12), 1627–1632 (2010)
19. Peak Signal-to-Noise Ratio as an Image Quality Metric pdf Publish, White Paper, pp. 1–2 (2013)
20. https://www.physics-animations.com/

Biometric Identification (Analysis Based on Fingerprints and Faces)

Annapurna Mishra, Niharika Modi and Monika Panda

Abstract Biometric system used as a pattern recognition system helps in personal identification by considering specific physiological or behavioral characteristics of an individual. It is divided on the basis of the authentication medium used such as fingerprint, face, and iris. In our paper, we have considered fingerprint and face recognition process. Fingerprint is used as a medium of identification on account of its uniqueness, permanence, universal acceptability, and ease of acquisition. We have proposed a hybrid technique which combines minutia-based technique with the global features to obtain a better recognition rate especially for poor-quality fingerprints. Face being one of the unique, non-intrusive, and non-contact physical traits which can be captured without user cooperation finds its application in the recognition system. Every face is illustrated as a linear combination of singular vectors of set of faces. Hence, we have used principal component analysis (PCA) for the recognition. PCA using Eigen face approach has been used as it reduces the dimensionality of data set, thereby enhancing the computational efficiency.

Keywords Principal component analysis (PCA) · Eigen faces
Eigen vector · Local features · Global structures · Euclidean distance

A. Mishra (✉) · N. Modi · M. Panda
Department of Electronics and Communication Engineering,
Silicon Institute of Technology, Bhubaneswar, India
e-mail: annapurnamishra12@gmail.com

N. Modi
e-mail: niharika.modi25@gmail.com

M. Panda
e-mail: monikapanda07@gmail.com

© Springer Nature Singapore Pte Ltd. 2018
K. Saeed et al. (eds.), *Progress in Advanced Computing and Intelligent Engineering*,
Advances in Intelligent Systems and Computing 563,
https://doi.org/10.1007/978-981-10-6872-0_3

27

1 Introduction

Identification using biometrics is the process of identifying an individual on the basis of his/her distinguishing features [1]. It utilizes several methods for uniquely recognizing the humans on the basis of one or more behavioral or physical traits. The biometric recognition system basically comprises two phases namely:-

 (i) Enrollment phase.
(ii) Authentication phase.

These two phases can be seen in the block diagram of the recognition system using biometrics shown in Fig. 1.

During enrollment phase, the training images are obtained using the sensor like optical sensor, capacitive sensor. From these images, features are extracted and the templates created are retained in the database. Within authentication phase, sensors retrieve test image. It then undergoes preprocessing and finally the features are extracted. The features so obtained are compared with the training set features to obtain the final result. Thus, the process of recognition is accomplished using the behavioral or physical traits under consideration.

There are basically four attributes which are considered while choosing the biometric data which are as follows [2]:-

1. Universality:- It means every individual must have these characteristics.
2. Uniqueness:- It means that the characteristic under consideration must not be the same for any two persons.
3. Permanence:- It means that the characteristics must be time invariant.

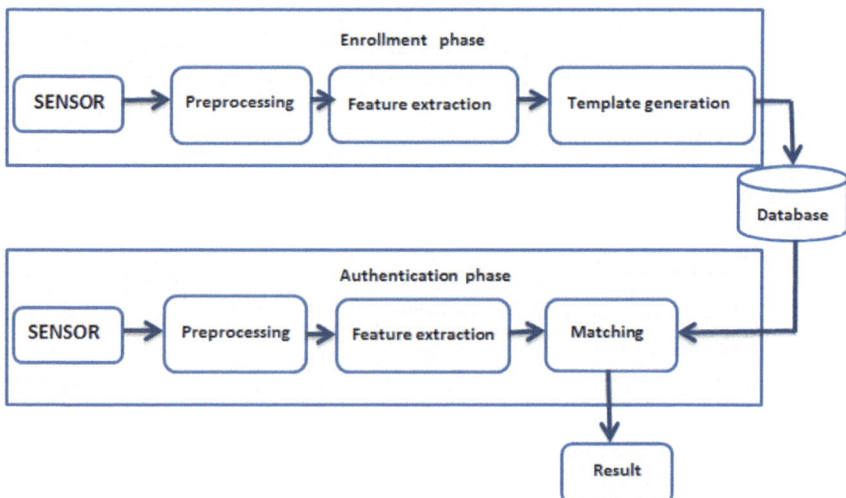

Fig. 1 Basic block diagram of biometric recognition system

4. Collectability:- It means that the quantitative measurement of the characteristic must be possible.

There are various biometric indicators which are extensively used in the current scenario. Some of them are face, fingerprint, iris, retina, voice, and so on.

2 Fingerprint Recognition

Fingerprint recognition is a verification process using which a match between two fingerprints is obtained. Fingerprints find its use in vast range of application in the field of biometrics. Fingerprint refers to distinct pattern of ridges with valleys on the surface of a finger of an individual [3]. Automatic Fingerprint Recognition system uses the two elementary types of minutiae features namely ridge terminations which refers to the immediate ending of the ridge and ridge bifurcations are the points on ridge from which it is divided into two branches. These features are illustrated in Fig. 2a [4].

In our proposed technique, we have used global orientation information along with minutiae points for the recognition purpose. In this hybrid technique, we use additional feature besides minutiae to implement matching so that better results can be obtained especially in the case of poor-quality images [5]. The orientation field estimate is included in this hybrid technique to obtain more accurate matching rates which can be analyzed using singular points and directional field as in Fig. 3.

Figure 2b shows the directional field (triangle), core point (big circle–indicating center pixel), singular field, and the minutiae points (small circles).

The block diagram shown in Fig. 3 shows the different steps involved in the proposed technique. The fingerprint was acquired using the optical scanner which

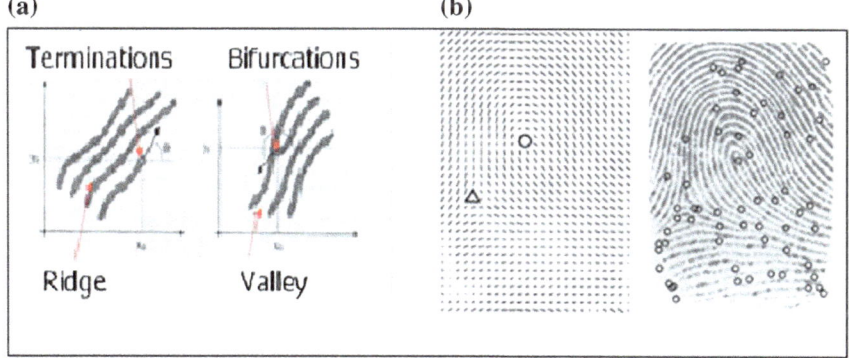

Fig. 2 **a** Terminations and Bifurcations **b** Fingerprint with singular point, directional field (triangle), and minutiae (small circles)

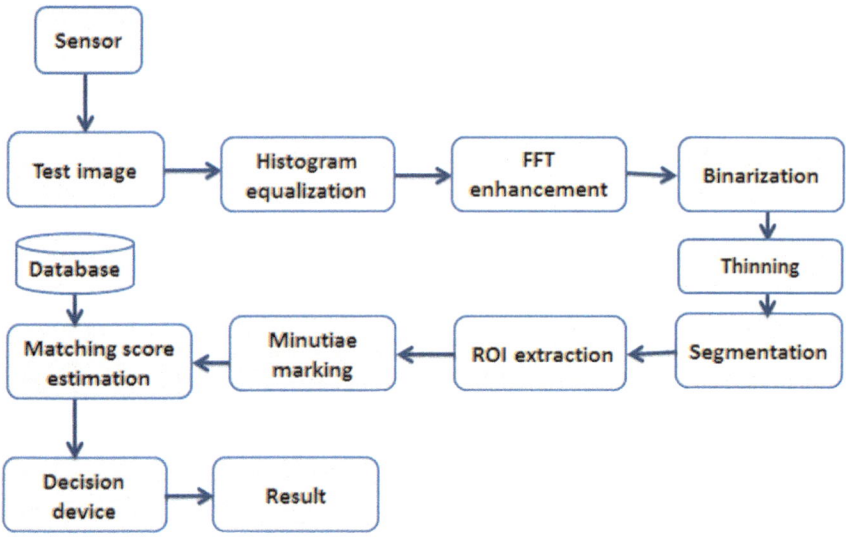

Fig. 3 Block diagram of fingerprint recognition system

was taken as the test image. The database used here is the standard database FVC 2000 and FVC 2002.

Then, the histogram equalization of the test image is obtained followed by the fast Fourier transform (FFT) to obtain the enhanced test image [3]. Now the gray scale image so obtained is converted to the binarised image [6]. The orientation field estimate is then obtained by the segmentation process which also removes the noise. Since on a ridge or valley, the intensity of image is expected to be closer to each other, thus the corresponding direction is assigned to that image point. Thus, the directional field [5, 6] and singular points are obtained which is shown in Fig. 2b.

Now the region of interest is extracted from which the minutiae points can be marked. These can then be compared with the training vectors retained in the database to obtain matching score and then the final decision is made to obtain the final result.

The basic algorithm for fingerprint recognition can be understood from following steps:

Step 1: The database is taken from FVC 2000 and FVC 2002.
Step 2: The test image is taken using a scanner.
Step 3: Histogram equalization is performed on the gray scale image so that the gray level or pixel values are distributed over a wider range. This process helps in enhancing the perceived information [7].
Step 4: After histogram equalization is done, image is then divided into smaller blocks (32 by 32 pixels) for further processing and fast Fourier transform is performed using Eq. (1).

$$F(u,v) = \sum_{x=0}^{M-1}\sum_{y=0}^{N-1} f(x,y) \times \exp\left\{ -j2\pi \times (\frac{ux}{M} + \frac{vy}{N}) \right\} \tag{1}$$

Where $u, v = 0, 1, 2 \ldots 31$

Then enhanced block is obtained using Eq. (2).

$$g(x,y) = F^{-1}\left\{ F(u,v) \times |F(u,v)|^{K} \right\} \tag{2}$$

where $F^{-1}\{F(u,v)\}$ is done using following equation.

$$f(x,y) = \frac{1}{MN} \sum_{x=0}^{M-1}\sum_{y=0}^{N-1} F(u,v) \times \exp\left\{ j2\pi \times (\frac{ux}{M} + \frac{vy}{N}) \right\} \tag{3}$$

Where $x, y = 0, 1, 2 \ldots 31$

where k in Eq. (2) is a constant obtained from experiment having value k = 0.45.

Step 5: Binarization transforms gray scale image into single bit binarized image where one value holds for valleys and zero for ridges [5, 6].

Step 6: Then, the thinned image is obtained by reducing the binarized image to the strokes of one-pixel width.

Step 7: After thinning, segmentation using orientation field is performed. In the process of direction mapping of each pixel, the thinned image is divided into the blocks of 16 by 16 pixels. Block direction is estimated for each block using Eq. (4).

$$\tan 2\beta = 2 \sum\sum (g_x^* g_y) / \sum\sum (g_x^2 - g_y^2) \tag{4}$$

where g_x and g_y is the gradient value of each pixel along x and y direction, respectively.

After obtaining direction of each block, the ones without significant information (ridges) get discarded using Eq. (5).

$$E = \left\{ 2 \sum\sum (g_x^* g_y) + \sum\sum (g_x^2 - g_y^2) \right\} / W*W* \sum\sum (g_x^2 + g_y^2) \tag{5}$$

Step 8: Region of interest (ROI) depicts region containing effective ridges and furrows. The estimation is based on morphological operations namely 'OPEN' and 'CLOSE.' 'OPEN' operation expands images while removing the peaks introduced by background noise. 'CLOSE' operation shrinks images along with elimination of small cavities.

Step 9: Minutiae points are then marked by using window technique which can then be used for recognition purpose.

Step 10: The matching score is then evaluated by comparing the test and the training vectors.

3 Face Recognition

Face recognition, an integral aspect of biometrics, uses basic traits of human face for matching with existing data. The result so obtained aids in identification of a human being. Face represents a complex multidimensional structure which requires good computing technique for its recognition [8]. Each face can be represented as linear combination singular vectors of set of faces. Normally, 80 nodal points are used for the recognition process. Some of them are distance between eyes, jaw lines, position of cheek bones, wrinkles, shadows, etc. These features can be seen in Fig. 4.

Here, we have used principal component analysis (PCA) [9] using Eigen faces for face recognition which is a simple and efficient way of recognition. PCA uses an orthogonal transformation to convert a set of linearly correlated variables into a set of linearly uncorrelated variables known as principal components. The first principal component has largest variance (consisting of most variability in data) and each successive component has highest possible variance provided that it is orthogonal to the preceding component and non-overlapping in nature.

The objective of PCA is to reduce the dimensionality of data [10]. This technique transforms the faces into a small set of characteristics called Eigen faces

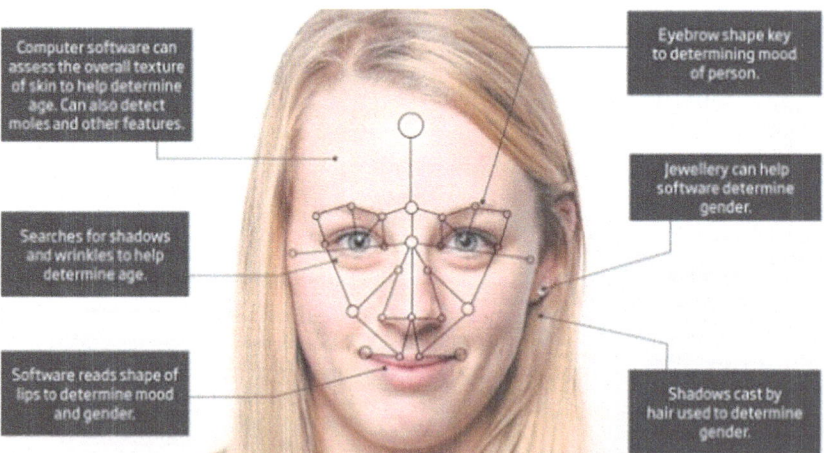

Fig. 4 Points of recognition

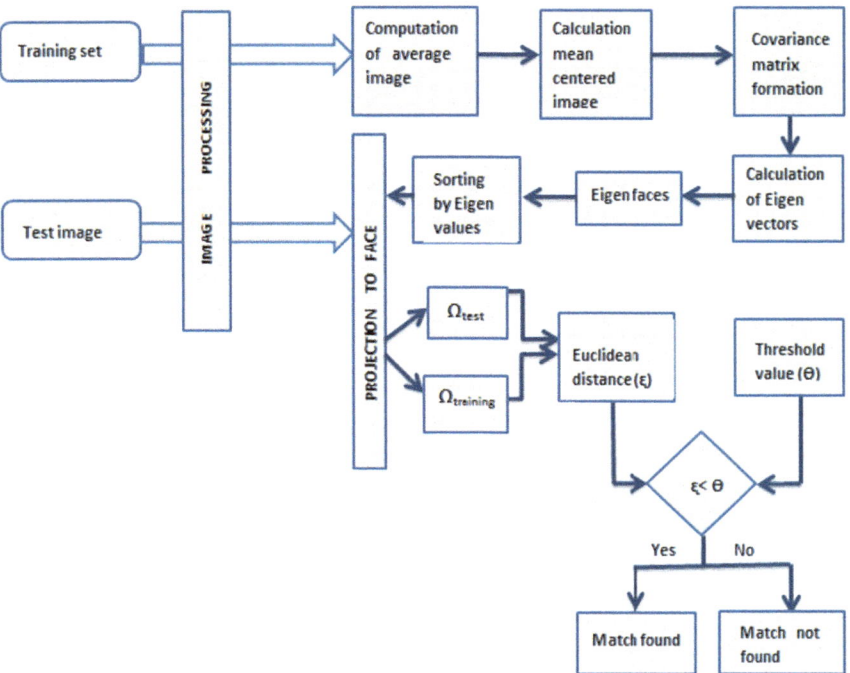

Fig. 5 Block diagram of face recognition system

which forms the training set. Then, the recognition process is achieved by projecting test image onto Eigen subspace and comparing it with template stored in the database [11].

The different steps involved in the process of face recognition can be seen in Fig. 5. The training set of images is given as input to find the Eigen space. These images are used to compute the average face image. Then, the mean centered image is calculated using which covariance matrix is formed. From the covariance matrix, Eigen vector and Eigen value are calculated, which is used to form the Eigen faces. Then, the Eigen values are sorted according to their variance values. The ones with the highest variance contain more information and thus are considered. These form Eigen space which has lower dimension than the original image space.

Now a test image is taken and projected onto the face space which gives the weight vector also known as face key. Then, Euclidean distance [12] is calculated between the weight vectors of the test and training set image. If the Euclidean distance is less than some threshold value, then the test image is said to be matched else no match is found.

The basic algorithm for face recognition can be observed in steps given below:

Step 1: The training set is prepared by obtaining the face images.

$$\text{Image set} = (I_1, I_2, \ldots I_M)$$

Step 2: The data set is prepared by transforming each face image I_i in the database into a vector which is placed in the training set.

$$\text{Data set } S = (T_1, T_2, \ldots T_m)$$

Step 3: The average face vector (Ψ) is computed using the formula given in Eq. (6).

$$\Psi = \frac{1}{M} \sum_{n=1}^{M} T_n \tag{6}$$

Step 4: The average face vector (Ψ) is subtracted from the data set faces (T_i) to obtain the mean centered image (Φ_i) [13] which can be seen in Eq. (7).

$$\Phi_i = T_i - \Psi \tag{7}$$

Step 5: The covariance matrix (C) is obtained from Eq. (8).

$$C = 1/M \sum_{n=1}^{M} \Phi_n \Phi_n^T$$
$$= AA^T \quad (N^2 \times N^2 \text{matrix}) \tag{8}$$
$$\text{Where } A = \Phi_1, \Phi_2 \ldots \Phi_M$$

Step 6: Eigen vector is computed for the covariance matrix. This is done using Eq. (9) which represents Eigen vector (v_i) for a matrix (L) of reduced dimension having the same Eigen vector (u_i) as that of the covariance matrix(C). This equation showing Eigen value is given below

$$u_i = \sum_{i=1}^{N} V_i \Phi_i \tag{9}$$
$$\text{Where } V_i = \text{Eigen vector of } L = A^T A$$

Step 7: Eigen face space is obtained by considering only K Eigen vectors corresponding to K largest Eigen values.

Step 8: The training sample is projected onto the Eigen face space to obtain the feature weight (W_i) for training set as shown in Eq. (10) and the final weight vector (Ω_i) obtained is given in Eq. (11).

$$W_i = u_i^T(T_i - \Psi) \qquad (10)$$

where W_i is the weight for the ith Eigen face u_i, i = 1, 2, 3......... K.

$$\Omega_i^T = [W_1, W_2, W_3 . . . . W_K] \qquad (11)$$

Step 9: Now the test image is obtained from the sensor and feature weight is calculated using Eq. (12).

$$W_{test} = u_i^T(T_{test} - \Psi) \qquad (12)$$

The weight vector for Eq. (12) is represented in Eq. (13).

$$\Omega_{test}^T = [W_1, W_2, W_3 . . . . W_K] \qquad (13)$$

Step 10: The average distance (Euclidean distance) between the test weight vector and training set weight vector is computed using Eq. (14).

$$\varepsilon_u{}^2 = \|\Omega_{test} - \Omega_i\|^2 \qquad (14)$$

Step 11: Threshold value is calculated using Eq. (15).

$$\Theta = \frac{1}{2}\max \, [\Omega i - \Omega i] \qquad (15)$$
$$\text{Where } i = j = 1, 2, 3 . . . . . . .K$$

Step 12: To obtain the final result, the Euclidean distance is compared with the threshold value .

$$\varepsilon_u < \Theta, \text{ match is found.}$$
$$\varepsilon_u >= \Theta, \text{ match is not found.}$$

4   Result Analysis

4.1   Fingerprint Recognition

The result obtained after the successful execution of the above-mentioned algorithm for fingerprint recognition is shown below:

The test image as shown in Fig. 6a is obtained using a sensor. It is then equalized using a histogram as seen in Fig. 6b. This image is then enhanced using fast Fourier transformation as viewed in Fig. 6c. Now the enhanced gray scale image is converted to a binarized black and white image as in Fig. 6d.

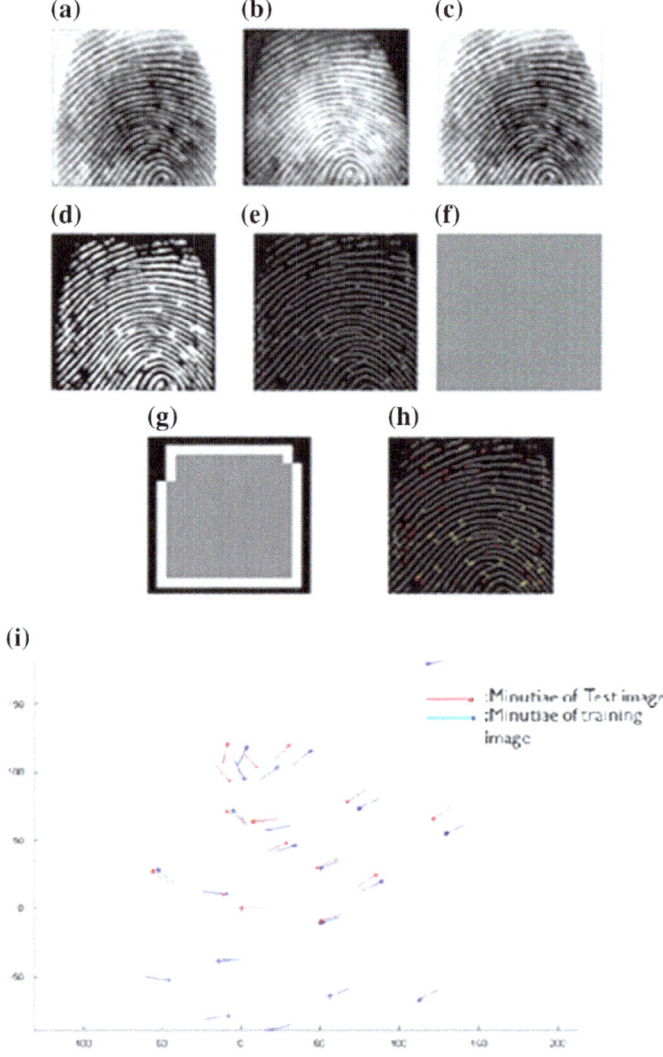

Fig. 6 Results obtained in fingerprint recognition process

The binarized image so obtained is converted into a thinned image of one-pixel width which is given in Fig. 6e. Then, the direction of each pixel is obtained as in Fig. 6f. The region of interest (ROI) is obtained as per Fig. 6g. The minutiae points are marked as shown in Fig. 6h. The matching score obtained = 0.77067 by comparing the test and training sets which is visible in Fig. 6i. Thus, the match of the test image is present in the database.

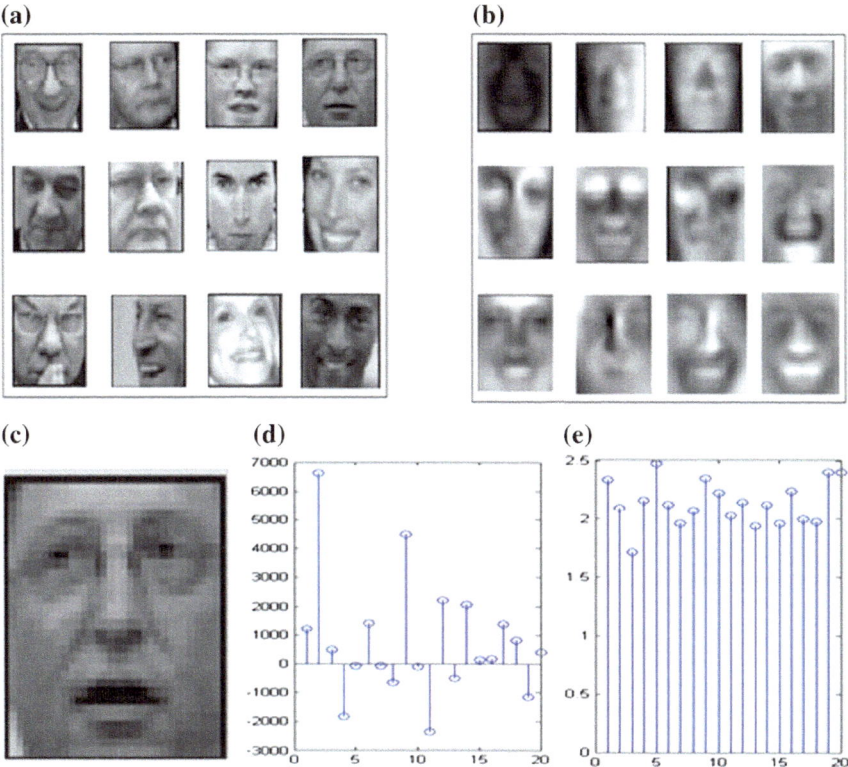

Fig. 7 Results obtained for face recognition

4.2 Face Recognition

The training set which is considered in our analysis is shown in Fig. 7a. Eigen faces extracted from the training set is given in Fig. 7b. The test image taken for recognition purpose is given in Fig. 7c. The weight assignment can be seen in Fig. 7d. The Euclidean distance which is calculated between training and test vectors is shown in Fig. 7e. The final result shows that the input image matches with one of the images of the training set.

5 Conclusion

Biometrics refers to the specific characteristics which are considered for the identification of an individual. Here, we have successfully used fingerprint and face as biometric traits for obtaining a match. For fingerprint recognition, initially minutia-based technique was used to obtain a match between the training and test image. This technique takes into consideration the local features only, hence does not work well in the presence of various factors such as noise, large displacement. Hence, we have also included orientation field estimate as an additional feature which aids in the achievement of better results especially for the poor-quality images. Thus, by using the hybrid technique, the similarity scores of 0.77067 were obtained. But the hybrid technique used for fingerprint recognition is not rotational invariant in nature and is time consuming; hence, more work needs to be done by taking other features into consideration. The Face Recognition using PCA technique was successfully used to obtain a match from the database image. Several challenges were encountered such as imaging condition, occlusions, and acquisition geometry during the recognition process. More effective result can be obtained by using 3D face recognition.

References

1. Majekodunmi, T.O., Idachaba, F.E.: A review of the fingerprint, speaker recognition, face recognition and iris recognition based biometric identification technology. In: Proceedings of the World Congress on Engineering, vol. II, London, UK, 6–8 July 2011
2. Ko, T., Krishnan, R.: Fingerprint and face identification for large user population. Syst. Cybern. Inf. **1** (2003)
3. Modi, N., Mishra, A., Singh, N.: Fingerprint Recognition using local and global structures. Int. J. Innov. Res. Electr. Electron. Instrum. Control Eng. **4**(3), Mar 2016
4. Garg, M., Bansal, H.: Fingerprint Recognition using minutiae estimation. Int. J. Appl. Innov. Eng. Manag. (IJAIEM) **2** (5), May 2013
5. Gu, J., Zhou, J., Yang, C.H.: Fingerprint Recognition by combining global structures and local cues. IEEE Trans. Image Process. **15** (7), July 2006
6. Khindre, A.A., More, V.A.: An approach to touchless Fingerprint Recognition using matlab. Int. J. Emerg. Trends Technol. Comput. Sci. **3**(4), July-Aug 2014
7. Chouthmal, P.P., Bhosale, S.A., Kale, K.V.: A novel approach for Fingerprint Recognition. Int. J. Adv. Res. Comput. Sci. Softw. Eng. **4**(8), Aug 2014
8. Mishra, A., Swain, M., Dash, B.: An approach to Face Recognition of 2-D images using Eigen Faces and PCA. Signal Image Process. Int. J. (SIPIJ) **3**(2), Apr 2012
9. Mane, A.V., Manza, R.R., Kale, K.V.: Human face recognition using superior principal component analysis (SPCA). Int. J. Comput. Theory Eng. **2**(5), Oct 2010
10. Oommel, A.A., Singh, C.S., Manikandan, M.: Design of Face Recognition system using principal component analysis. Int. J. Res. Eng. Technol. **3**(1), Mar 2014
11. Paul, L.C., Sumam, A.A., Face Recognition using principal component analysis method. Int. J. Adv. Res. Comput. Eng. Technol. **1**(9), Nov 2012
12. Bhensle, A.C., et al.: An efficient Face Recognition using PCA and euclidean distance. Int. J. Comput. Sci. Mob. Comput. **3**(6), 407–413, June 2014
13. Shivastava, A., Sad, S.: Face Recognition for different Facial expressions using principal component analysis. Int. J. Innov. Res. Adv. Eng. (IJIRAE) **1**(6), July 2014

Low-Resolution Image Recognition Using Cloud Hopfield Neural Network

Neha Soni, Narotam Singh, Amita Kapoor
and Enakshi Khular Sharma

Abstract Our brain can with ease recognize images despite size variance, a capability that has been attributed to the normalization of images and feature extraction by biological computational models. In this paper, we study the recognition and retrieval of ultra-low-resolution facial images using Hopfield neural network, a biologically inspired recurrent artificial neural network. We employ two variations of Hopfield, one, proposed by Hopfield and Tank, the asynchronous Hopfield neural network. Second, cloud Hopfield neural network proposed by Singh and Kapoor for its better performance both in terms of retrieval rate and convergence time. Our results show that even when the distortion in the presented facial images is up to 35%, cloud Hopfield neural network is able to give 75.4% successful retrievals, while the asynchronous Hopfield network gives 63.4% successful retrievals. The results for these very low-resolution face images show promise and can help to achieve size invariance for computer vision.

Keywords Biologically inspired · Hopfield neural network · Low-resolution image recognition

N. Soni · E. K. Sharma (✉)
Department of Electronic Science, University of Delhi South Campus, Delhi, India
e-mail: enakshi54@yahoo.co.in

N. Soni
e-mail: soni.neha2191@gmail.com

N. Singh
India Meteorological Department, Information Communication and Instrumentation Training Centre, Ministry of Earth Sciences, Delhi, India
e-mail: narotam.singh@gmail.com

A. Kapoor (✉)
Shaheed Rajguru College of Applied Sciences for Women, University of Delhi, Delhi, India
e-mail: dr.amita.kapoor@ieee.org

© Springer Nature Singapore Pte Ltd. 2018
K. Saeed et al. (eds.), *Progress in Advanced Computing and Intelligent Engineering*,
Advances in Intelligent Systems and Computing 563,
https://doi.org/10.1007/978-981-10-6872-0_4

39

1 Introduction

Natural computing is a contemporary branch of computer science that takes inspiration from nature to design complex computational models and algorithms. One of the most popular approaches in natural computing is using artificial neural networks (ANNs), inspired from biological neural networks in the human brain. ANNs have found application in varied intelligent tasks like prediction, modelling and control [1–3]. In this paper, we study the recognition and retrieval of ultra-low-resolution images (ULRI) using Hopfield neural network (HNN), a biologically inspired recurrent artificial neural network.

Our brain can with ease recognize images despite size variance, a capability that has been attributed to the normalization of images and feature extraction by biological computational models [4]. An important question when trying normalization of images is what is the minimum size to which images can be normalized? This is important since a low-resolution image will reduce both computational complexity and size of the network. In this paper, we study the retrieval of two different-sized images, viz. an image with 60×60 pixels and an ultra-low-resolution image with 10×10 pixels, in the presence of occlusions and/or distortions. We employ two variations of Hopfield, one, proposed by Hopfield and Tank, the asynchronous Hopfield neural network (AHNN) [5, 6]. Second, cloud Hopfield neural network (CHNN) proposed by Singh and Kapoor [7] for its better performance both in terms of retrieval rate and convergence time. Our results show that even when the distortion in the presented facial images is up to 35%, cloud Hopfield neural network is able to give 75.4% successful retrievals, while the asynchronous Hopfield network gives 63.4% successful retrievals. The results for these very low-resolution face images show promise and can help to achieve size invariance for computer vision.

There are four main features that describe an ANN [8, 9]: the network architecture (arrangement of neurons and their connectivity), the activation function (determines the output of a single neuron as a function of its input), a learning algorithm for the calculation or updating of connections between neurons also called weights (using training data) and the retrieval algorithm (the network evolution to a stable state in the presence of unknown or test inputs). Asynchronous Hopfield neural network (AHNN), given by Hopfield and Tank [5, 6] in the year 1982, has a non-layered architecture, uses 'sign' activation function, calculates weight using Hebb rule (Fig. 1) and updates single neuron at a time. The cloud Hopfield neural network (CHNN), proposed by Singh and Kapoor [7], is a modification of AHNN; here the retrieval algorithm is changed so as to simultaneously update a cloud of r neurons (distributed either randomly or uniformly across the entire network) at a time. Figure 2 shows the retrieval algorithm of CHNN. CHNN offers both better convergence rates and retrieval rates as compared to AHNN [7].

Face recognition is an important component of computer vision. In the last three decades, extensive research has taken place on face recognition. Approaches like independent component analysis (ICA) [10], principal component analysis (PCA) [11, 12], linear discriminant analysis (LDA) [13, 14] have been explored for

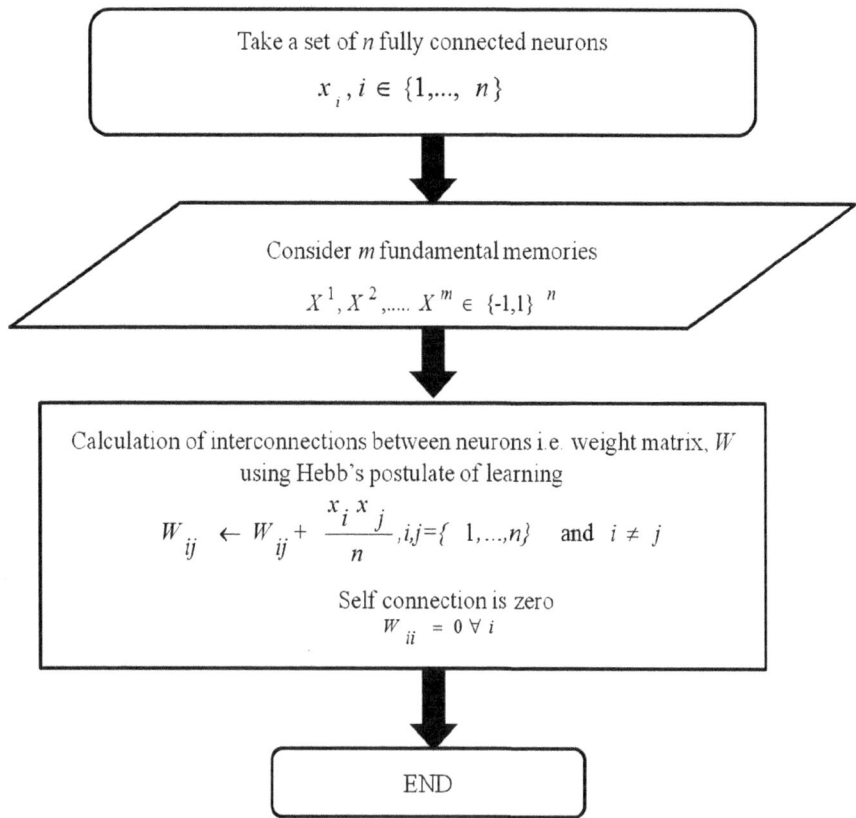

Fig. 1 Weight calculation using Hebb rule in Hopfield neural network

face recognition. Several factors that degrade the performance of most of the face recognition systems [10, 11, 13] are varied face expressions [10, 12, 14, 15], age [10, 14, 15], glasses [12, 15], rotation [15], size variations [11, 15], changing lighting conditions [11, 12], etc.

In this paper, we study the retrieval of two different-sized images, one of normal size that human eye can perceive (60×60 pixels) and other an ultra-low-resolution image (ULRI) that human eye cannot identify (10×10 pixels) in the presence of both occlusions and distortions. The motivation to study ULRI arises from the biological computational models of human vision [4]; these models propose that human eyes have size invariance because it either reduces the perceived image to a normalized size or extracts features from the image (or perhaps both). The reduction in image size offers advantage both in terms of computational complexity and network size (for 60×60 pixels images the network consists of 3600 neurons as compared to only 100 neurons for a 10×10 pixel image). For 45% randomly distorted 60×60 facial images, our results show at least 3172 (90%) and 2713 (77.5%) correct retrievals (out of 3500) by using CHNN

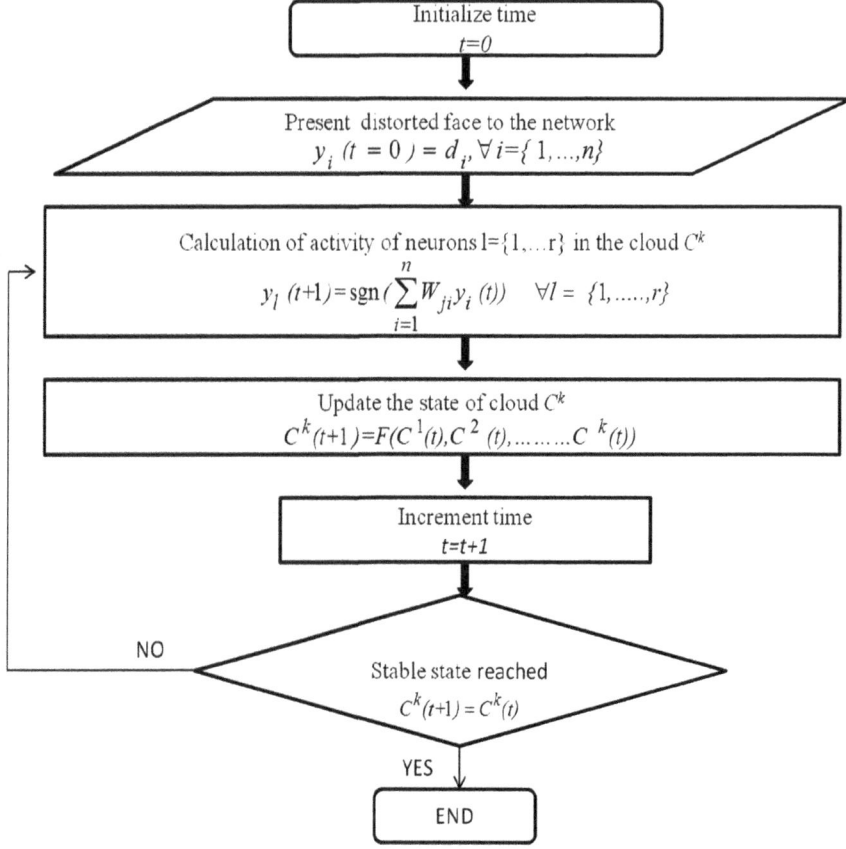

Fig. 2 Retrieval algorithm of cloud Hopfield neural network

and AHNN, respectively. For 35% randomly distorted ULRI's, our results show 1884 (75.4%) and 1585 (63.4%) correct retrievals (out of 2500) by using CHNN and AHNN, respectively. We have also performed experiments for occluded images. The results for these very low-resolution face images show promise and can help to achieve size invariance for computer vision.

2 Data set Used and Experiment

We have used seven grey scale facial images of varied pixel size (202×249–60×60). These images are reduced to 60×60 facial images using row and column elimination. These reduced images are then converted to binary images using Otsu's method [16, 17] by choosing an optimum threshold value.

Fig. 3 Seven grey scale facial images and their binary images (60 × 60 pixels)

Figure 3 represents seven reduced facial images (60 × 60) and their binary images. Out of seven reduced facial images (60 × 60), five facial images are further reduced to ULRI (10 × 10). Figure 4 represents five reduced facial images (10 × 10) and their binary images. These binary images are used by the network for the calculation of weights using the flow chart presented in Fig. 1.

Next, we investigated the retrieval of distorted and occluded images, using both AHNN and CHNN (Fig. 2). For CHNN retrieval algorithm $r = 4$ in the simulations. Firstly, original grey scale facial images from varied (202 × 249–60 × 60) pixel size are reduced to constant 60 × 60 pixel size using row and column elimination. Then, these facial images are further reduced to ULRI (10 × 10 pixels). Secondly, by using Otsu's method [16, 17], these ULRI are converted to bipolar facial images by choosing an optimum threshold value. These ULRI form the training data for the storage phase of Hopfield algorithm, which involves the calculation of weight matrix W via Hebb rule.

To test the network, it is presented with distorted and occluded facial images. Figures 5 and 6 represent corrupted facial images and the facial images retrieved by

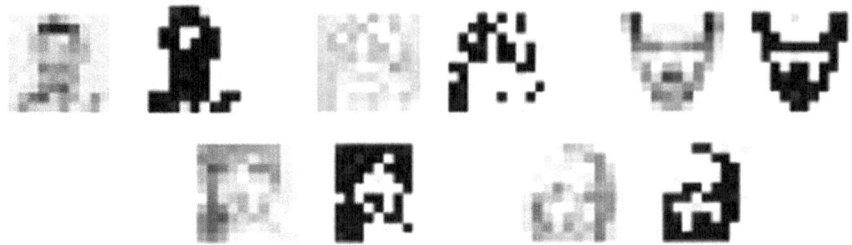

Fig. 4 Five grey scale facial images and their binary images (10 × 10 pixels)

(a)   **(b)**

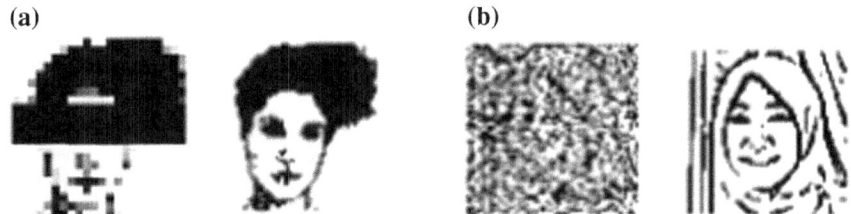

Fig. 5 **a** Occluded and retrieved face 7 (60 × 60 pixels), **b** 35% distorted and retrieved face 2 (60 × 60 pixels)

the CHNN. For the computer simulations, we consider $r = 4$ for CHNN, thus we have 25 clouds for ULRI faces, each with four unique neurons; these clouds are updated randomly and asynchronously till a stable image is obtained. We then calculate the Hamming distance between the original image (undistorted) and the stable image pattern, and an Hamming distance of zero indicates successful retrieval [18, 19]. To validate our results, we repeat the procedure 500 times for each distortion, every time a new randomly generated distorted facial is presented to the network.

3   Results and Conclusions

Table 1 summarizes the response of the network for seven facial images (60 × 60 pixels). The results show that for distortion of 45% when a total of $7 \times 500 = 3500$ distorted facial images are presented to the network, CHNN is able to successfully retrieve 3172 facial images and AHNN is able to successfully retrieve only 2716 facial images.

Table 2 shows the results for the five ULRIs. For a distortion of 35%, CHNN is able to successfully retrieve 1884 facial images out of the 2500 distorted facial images presented to the network. AHNN, on the other hand, could successfully retrieve only 1585 facial images. Changes introduced due to ageing, light

(a)   (b)

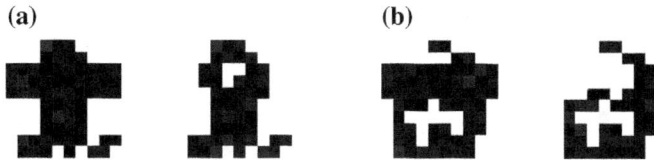

Fig. 6 a Occluded and retrieved face 1 (10 × 10 pixels), **b** Occluded and retrieved face 5 (10 × 10 pixels)

Table 1 Number of successful retrievals for AHNN (H) and CHNN (C) when presented with seven distorted facial images (60 × 60 pixels) with distortion in range 30–45%

Number of successful retrievals								
Faces	Distortion percentage							
	30%		35%		40%		45%	
	H	C	H	C	H	C	H	C
1	500	500	499	500	472	492	361	414
2	500	500	500	500	499	499	444	479
3	500	500	500	500	497	499	431	483
4	500	500	500	500	500	500	426	469
5	500	500	500	500	480	495	345	452
6	500	500	488	498	455	487	315	418
7	500	500	497	499	480	497	391	457

Table 2 Number of successful retrievals for AHNN (H) and CHNN (C) when presented with five distorted ULRI's (10 × 10 pixels) with distortion in range 10–35%

Number of successful retrievals								
Faces	Distortion percentage							
	10%		20%		30%		35%	
	H	C	H	C	H	C	H	C
1	500	500	497	499	437	447	324	390
2	500	500	499	497	414	444	346	377
3	500	500	485	491	414	433	297	373
4	500	500	492	491	405	429	307	367
5	500	500	494	497	425	441	311	377

variations, expressions and accessories lie within the range of 5–15%, with the help of CHNN we have been able to achieve successful retrieval of 99% for up to 20% distortions for ULRI facial images, this is a significant improvement over previous reported results [15].

Thus, for ULRI, while image capacity is reduced from seven to five, the network size is significantly reduced (1/36 of the first case), this small 100 neuron network can be easily embedded on a chip and used for various security and recognition applications for personal electronic devices. The results for these ultra-low-resolution facial images are favourable to achieve size invariance for computer vision.

References

1. Gardner, M.W., Dorling, S.R.: Neural network modelling and prediction of hourly NO_x and NO_2 concentrations in urban air in London. Atmos. Environ. 709–719 (1999)
2. Kapoor, A., Sharma E.K.: Neural network modelling of EDFA. In: 2013 IEEE Workshop on Recent Advances in Photonics (WRAP), pp. 1–3 (2013)
3. Liu, Y.J., Chen, C.P., Wen, G.X., Tong, S.: Adaptive neural output feedback tracking control for a class of uncertain discrete-time nonlinear systems. IEEE Trans. Neural Netw. 1162–1167 (2011)
4. Wiskott, L.: How does our Visual System Achieve Shift and Size Invariance. 23 Problems in Systems Neuroscience, Chapter 16 (2005)
5. Hopfield, J.J., Tank, D.W.: Computing with neural circuits—a model. Science 233(4764), 625–633 (1986)
6. Tank, D., Hopfield, J.J.: Collective computation in neuron like circuits. Sci. Am. 257(6), 104–114 (1987)
7. Singh N., Kapoor, A.: Cloud Hopfield neural network: analysis and simulation. In: 2015 IEEE International Conference on Advances in Computing, Communications and Informatics (ICACCI), pp. 203–209 (2015)
8. Wasserman, P.D.: Neural Computing. Van Nostrand Reinhold, New York (1989)
9. Haykin, S.: Network N. A Comprehensive Foundation. Neural Networks (2004)
10. Bartlett, M.S., Movellan, J.R., Sejnowski, T.J.: Face recognition by independent component analysis. IEEE Trans. Neural Netw. 13(6), 1450–1464 (2002)
11. Mohammed, A.A., Minhas, R., Wu, Q.J., Sid-Ahmed, M.A.: Human face recognition based on multidimensional PCA and extreme learning machine. Pattern Recogn. 44(10), 2588–2597 (2011)
12. Turk, M.A., Pentland, A.P.: Face recognition using eigenfaces. In: IEEE Computer Society Conference on Computer Vision and Pattern Recognition, 1991. Proceedings CVPR'91, pp. 586–591 (1991)
13. Martínez, A.M., Kak, A.C.: Pca versus lda. IEEE Trans Pattern Anal Mach Intell 23(2), 228–233 (2001)
14. Zhao, H., Yuen, P.C.: Incremental linear discriminant analysis for face recognition. IEEE Trans. Syst. Man Cybern. Part B: Cybern. 38(1), 210–221 (2008)
15. Dai, Y., Nakano, Y.: Recognition of facial images with low resolution using a Hopfield memory model. Pattern Recogn. 31(2), 159–167 (1998)
16. Sezgin, M.: Survey over image thresholding techniques and quantitative performance evaluation. J. Electron. Imaging 13(1), 146–168 (2004)
17. Otsu, N.: A threshold selection method from gray-level histograms. Automatica. (285-296), 23–27 (1975)
18. Hamming, R.W.: Error detecting and error correcting codes. Bell Syst. Tech. J. 29(2), 147–160 (1950)
19. Robinson, D.J.: An Introduction to Abstract Algebra. Walter de Gruyter, pp. 264–276 (2003)

Adaptive Multi-bit Image Steganography Using Pixel-Pair Differential Approach

Uttiya Ghosh, Debanjan Burman, Smritikana Maity
and Imon Mukherjee

Abstract With the increase of communication over Internet, the issue of security
has become an important factor. Steganography is a consequence of the increasing
degradation of reliability. The roots of steganography lie in ancient Greek civiliza-
tion. With time, steganography has moved a long way in the path of advancement. It
started off with least significant bit (LSB) embedding which mainly focussed on data
security. With time, many algorithms have been designed using multi-bit steganography which takes into account both security and capacity of data embedded. In this
paper, a new technique has been introduced where various parameters determine the
number of bits embedded. This helps to improve the robustness of this method. We
show empirically that our method withstands the statistical attacks and benchmark.

Keywords Multi-bit steganography · Information hiding · Spatial domain
Data security

U. Ghosh (✉) · D. Burman · S. Maity
Department of Computer Science and Engineering, St. Thomas' College
of Engineering and Technology, Kolkata 700023, India
e-mail: uttiyaghosh@gmail.com

D. Burman
e-mail: burmandebanjan@gmail.com

S. Maity
e-mail: smritikanamaity94@gmail.com

I. Mukherjee
Department of Computer Science and Engineering, Indian Institute
of Information Technology, Kalyani 741235, West Bengal, India
e-mail: mukherjee.imon@gmail.com

© Springer Nature Singapore Pte Ltd. 2018
K. Saeed et al. (eds.), *Progress in Advanced Computing and Intelligent Engineering*,
Advances in Intelligent Systems and Computing 563,
https://doi.org/10.1007/978-981-10-6872-0_5

47

1 Introduction

The technique of hiding a secret data into another file in a secured manner, such that it can be transmitted from sender to receiver without being suspected by any other party who are not the intended recipients, can be defined as *Steganography*. Mainly digital images are used as cover versions, and the image carrying the secret data is known as stego version.

Mandal et al. [3] used pixel value differencing (PVD) method to embed data in every component of a colour image. Moreover, the number of bits embedded differs with the different pixel components to improve the security of the embedded data. In the process introduced by Mukherjee et al. [4], frequency domain of the image is used to select the potential pixels, whereas the spatial domain is used for embedding data. Veron et al. [10] have shown how algebraic coding theory can be used in various ways to formulate secure steganographic techniques. Yang et al. [11] introduced a technique to improve reversible hiding of data based on histogram of greyscale images. Zhang et al. [12] have formulated a new method where data is hidden in last three bits of cover sample using a triple-layered construction. Paul et al. [5] have introduced keyless steganographic algorithms based on the lattice spin glass model of Ising [1] of physics. This method has been further improved in [6] where the maximum number of bits embedded in one pixel has been increased to five bits. Song et al. [9] aim at reducing histogram fluctuation by using affine transformation which are special functions that are used to maintain collinearity and ratio of distances.

In this paper, a new technique is being presented for embedding data in cover image where bits embedded in each pixel vary based on different parameters. This helps to increase the robustness of this method and security of the embedded data. Our algorithm helps to make it difficult for attackers to retrieve the hidden data and can also withstand various attacks.

2 Proposed Method

In this method, we have embedded different number of bits in different pixels considering various parameters. This whole process is described below.

2.1 Embedding of Data

The novel method proposed here is used for embedding different number of bits in greyscale images as shown in Table 1 using a dual parameter approach.

We first select non-overlapping 4×4 block from Table 1 as shown in Table 2. We introduce a pixel-pair differential multi-set $d = \{y | y = (\omega_{i,j} \sim \omega_{i,j+1}) \wedge y \geq \alpha\}$ from Table 2 where $1 \leq i \leq 4$, $j = \{1, 3\}$ and $0 \leq \alpha \leq 255$. The number of data bits(n)

Table 1 Greyscale image of size $h \times w$

$I_{1,1}$	$I_{1,2}$	...	$I_{1,w}$
$I_{2,1}$	$I_{2,2}$	...	$I_{2,w}$
...	...	...	...
$I_{h,1}$	$I_{h,2}$	...	$I_{h,w}$

Table 2 4×4 block of a digital image

$\pi_{1,1}$	$\pi_{1,2}$	$\pi_{1,3}$	$\pi_{1,4}$
$\pi_{2,1}$	$\pi_{2,2}$	$\pi_{2,3}$	$\pi_{2,4}$
$\pi_{3,1}$	$\pi_{3,2}$	$\pi_{3,3}$	$\pi_{3,4}$
$\pi_{4,1}$	$\pi_{4,2}$	$\pi_{4,3}$	$\pi_{4,4}$

Table 3 Text message arrangement

T_1	T_2	...	T_i	...	T_l

Table 4 Binary representation of the message

$b_{1,7}$	$b_{1,6}$	...	$b_{1,0}$	$b_{2,7}$	$b_{2,6}$	...	$b_{i,j}$	...	$b_{l,0}$

which will be hidden in Table 2 is obtained from Eq. (1) where $|d|$ denotes the number of elements in d and $2 \le \beta \le 6$.

$$n = \begin{cases} \beta \,, \text{if } |d| \ge 4; \\ \beta - 1 \,, \text{otherwise} \end{cases} \tag{1}$$

Consider a text message with l characters as shown in Table 3.

We then convert the values of Table 3 to the binary representation of their corresponding ASCII code as shown in Table 4 to obtain a 8-bit code for each character where $b_{i,j}$ denotes the jth bit of the ith character. Sequential n bits are taken from Table 4 to form a number m that is to be embedded in the pixels of Table 2. Let the pixel in which embedding is to take place is denoted as $\pi_{i,j}$. We first adjust the higher order $8 - n$ bits ($\Omega_{i,j}$) to minimize distortion. This is done with the help of Eq. (3) by an adjustment factor (δ) obtained from Eq. (2) where MSB denotes the most significant bit and abs denotes absolute value.

$$\delta = \lfloor \frac{mod(\pi_{i,j}, 2^n)}{2^{n-1}} \rfloor - MSB(m) \tag{2}$$

$$\Omega^*_{i,j} = \begin{cases} abs(\Omega_{i,j} + \delta) \text{, where } \lfloor \log_2(\Omega_{i,j} + \delta) \rfloor \leq \lfloor \log_2(\Omega_{i,j}) \rfloor; \\ \Omega_{i,j} \text{, otherwise.} \end{cases} \quad (3)$$

The value of stego pixel $(\pi^*_{i,j})$ is obtained with the help of Eq. (4).

$$\pi^*_{i,j} = \Omega^* \times 2^n + m \quad (4)$$

Finally, we store a status bit (S) in the $(\beta + 2)$th bit of $\pi_{1,1}$ so that it can be used during the retrieval process for determining the value of n. We calculate the value of S from Eq. (5).

$$S = \begin{cases} 0 \text{, if } |d| \geq 4; \\ 1 \text{, otherwise.} \end{cases} \quad (5)$$

The steps for embedding the message are described in Algorithm 1.

Input: An image of size $h \times w$, a text message of length l, α and β.
Output: The stego image containing the secret data.

1 Convert the text message into a binary stream as shown in Table 4;
2 $c \leftarrow 0$;
3 $f \leftarrow 0$;
4 **while** $(c \leq 8 \times l \vee f \leq \frac{h \times w}{16})$ **do**
5   Obtain non overlapping 4×4 block from Table 1 as shown in Table 2;
6      $f \leftarrow f + 1$;
7   Create multiset d;
8   Calculate n from Equation (1);
9   Sequential n bits are taken from Table 4 to form a number m;
10     $c \leftarrow c + n$;
11   Calculate δ, $\Omega^*_{i,j}$ and $\pi^*_{i,j}$ from Equation (2), (3) and (4) respectively;
12   Calculate S from Equation (5) and store in the $(\beta + 2)^{th}$ bit of $\pi_{1,1}$;
    end
13 Output the embedded stego image;

Algorithm 1: Algorithm for embedding data bits into the stego image.

2.2 Retrieval of Data

The process of retrieving the embedded message from the stego image starts with partitioning the stego image into non-overlapping 4×4 blocks as shown in Table 5. We retrieve status bit (S) from the $(\beta + 2)$th bit of $\pi^*_{1,1}$. It is to be noted that the value of β must be same during the process of embedding and retrieval. The number of bits (n) that is embedded in Table 5 is obtained from Eq. (6).

$$n = \begin{cases} \beta \text{, if } S = 0; \\ \beta - 1 \text{, otherwise} \end{cases} \quad (6)$$

We then obtain the value of m from the pixel $\pi^*_{i,j}$ with the help of Eq. (7). Then, m is converted into its binary equivalent, concatenated with its corresponding value

Table 5 4×4 block of stego image

$\pi^*_{1,1}$	$\pi^*_{1,2}$	$\pi^*_{1,3}$	$\pi^*_{1,4}$
$\pi^*_{2,1}$	$\pi^*_{2,2}$	$\pi^*_{2,3}$	$\pi^*_{2,4}$
$\pi^*_{3,1}$	$\pi^*_{3,2}$	$\pi^*_{3,3}$	$\pi^*_{3,4}$
$\pi^*_{4,1}$	$\pi^*_{4,2}$	$\pi^*_{4,3}$	$\pi^*_{4,4}$

Table 6   Binary representation of the message

b_1	b_2	$\ldots$	b_i	$\ldots$	$b_{8 \times l}$

obtained from $\pi^*_{i,j+1} \cdots \pi^*_{4,4}$ to form Table 6.

$$m = \quad \mod (\pi^*_{i,j}, 2^n) \tag{7}$$

Sequential 8 bits are taken from Table 6, and they are converted into their decimal equivalent ASCII code which is finally used to generate the corresponding character of the secret data. The process is repeated throughout the stego image to generate the secret data. The steps for retrieving the message are described in Algorithm 2.

Input: A stego image of size $h \times w$ and β.
Output: The hidden data.

1 $f \leftarrow 0$;
2 **while** $(f \leq \frac{h \times w}{16})$ **do**
3   Obtain non overlapping 4×4 block as shown in Table 5;
4      $f \leftarrow f + 1$;
5   Retrieve status bit (S) from the $(\beta + 2)^{th}$ bit of $\pi^*_{1,1}$;
6   Calculate n from Equation (6);
7   Calculate m from Equation (7);
8   Convert m into its binary equivalent and add it to the binary stream as shown in Table 6;
   end
9 Obtain 8 sequential bits from Table 6 and converted into their decimal equivalent ASCII code;
10 Repeat Step (9) until the entire message is extracted;
11 Output the secret data;

Algorithm 2: Algorithm for extracting hidden data bits from the stego image.

3   Experimental Results

The first and foremost target of any steganographic algorithm is to have the least visible distortion in stego images. The proposed method is tested on various standard images, and the results are produced below.

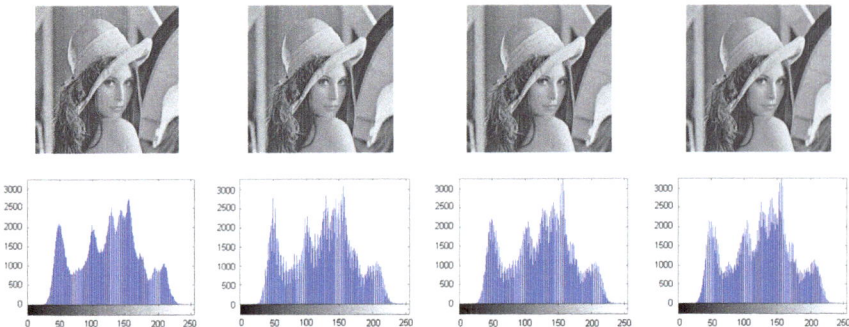

Fig. 1 Cover and stego versions (top row) and their histograms (bottom row) having ($\alpha = 4, \beta = 2$), ($\alpha = 8, \beta = 2$) and ($\alpha = 16, \beta = 2$) of Lena, respectively

3.1 Visual Perceptibility Analysis

Visual analysis of the tested images is shown in Fig. 1 which depicts that there is almost no visual distortion.

3.2 MSE, BER and PSNR

Mean squared error (MSE) is a statistical quantity that is measured by computing the squares of the errors between the original and stego images and then finding their average. It is calculated with the help of Eq. (8) where w and h denote the breadth and height of cover image (say, D) and the stego image (say, D').

$$MSE = \frac{1}{w \times h} \sum_{i=1}^{h} \sum_{j=1}^{w} (D_{i,j} - D'_{i,j})^2 \tag{8}$$

Peak signal-to-noise ratio (PSNR) is a statistical quantity that finds the ratio between the highest possible power of an original signal and the power of debasing noise that affects the reliability of its representation. It is calculated with the help of Eq. (9).

$$PSNR = 20 \log_{10}(\frac{MAX}{\sqrt[2]{MSE}}) \tag{9}$$

Here, MAX is equal to 255 (in case of images).

Bit error rate (BER) is a statistical quantity that measures the number of bit having errors divided by the total number of bits sent or received during a time interval under consideration. It is calculated with the help of Eq. (10).

$$BER = \frac{Total\ number\ of\ bits\ changed}{Total\ number\ of\ bits\ present} \qquad (10)$$

The results of these tests are shown in Table 7.

3.3 Normalized Correlation

Normalized correlation (*NC*) denotes the extent of similarity between a cover image and its stego image. As the difference between the two versions tends to 0, its value tends to 1. The value of *NC* is obtained from Eq. (11).

$$NC = \frac{\sum_{u=1}^{p} \sum_{v=1}^{q} [D(u, v) \cdot D^*(u, v)]}{\sum_{u=1}^{p} \sum_{v=1}^{q} D(u, v)^2} \qquad (11)$$

where $D(u, v)$ denotes the pixel intensities of the cover image and $D^*(u, v)$ denotes the pixel intensities of stego image. The values of *NC* for various images are shown in Table 7.

3.4 Histogram Analysis

Histogram of an image can be defined as the graphical representation of the frequency of every pixel value versus the pixel value. Histogram is one of those basic tools of quality control which gives an overview of the pixel value distribution of the image at one glance. It gives the frequency distribution of each distinct value in a set of data, which is the number of pixels for each intensity value. The histograms of the cover and the stego versions of different standard images are shown in Fig. 1.

3.5 StirMark Analysis

Along with the efficiency, the robustness of all steganographic algorithms requires a proper analysis with respect to some standardized methods. These analyses were performed using StirMark 4.0 as described in [7, 8].

Our proposed method was successful in providing some excellent results. The very minute deviation between the values of the original and the stego image shown in Table 8 is an indication to the robustness of the proposed algorithm.

Table 7 NC, MSE, PSNR, BER and capacity values for sample Lena, Baboon and Cameraman images

Name		NC			BER			MSE			PSNR			Capacity (in bpp)		
		$\beta=2$	$\beta=3$	$\beta=4$	$\beta=2$	$\beta=3$	$\beta=4$	$\beta=2$	$\beta=3$	$\beta=4$	$\beta=2$	$\beta=3$	$\beta=4$	$\beta=2$	$\beta=3$	$\beta=4$
Lena	$\alpha=4$	1.008	1.0039	1.0026	1.87	1.97	2.09	5.38	6.31	7.98	40.82	40.13	39.11	1.5539	2.5539	3.5539
	$\alpha=8$	1.0015	1.0028	1.0097	1.92	2.11	2.21	6.18	6.58	8.93	40.22	39.95	38.62	1.2458	2.2458	3.2458
	$\alpha=16$	1.0008	1.0006	1.0012	2.05	2.27	2.35	6.74	8.71	12.13	39.84	38.73	37.29	1.0965	2.0965	3.0965
Baboon	$\alpha=4$	0.9982	1.0001	1.0141	1.93	2.41	2.90	10.28	11.17	13.24	38.01	37.65	36.91	1.8346	2.8346	3.8346
	$\alpha=8$	1.0003	1.0008	1.0115	1.78	2.16	2.75	8.65	9.79	10.71	38.76	38.22	37.83	1.5516	2.5516	3.5516
	$\alpha=16$	1.0029	1.0016	1.0086	1.62	2.20	2.60	7.98	8.18	9.20	39.11	39.00	38.49	1.2195	2.2195	3.2195
Cameraman	$\alpha=4$	1.0025	1.0014	1.0069	1.17	2.53	2.76	7.83	8.55	9.50	38.01	37.65	36.91	1.2856	2.2856	3.2856
	$\alpha=8$	1.0034	1.0017	1.0058	1.85	2.24	2.69	6.54	9.79	8.31	38.76	38.22	37.83	1.1616	2.1616	3.1616
	$\alpha=16$	1.0040	1.0019	1.0052	1.73	2.33	2.57	6.05	8.18	7.74	39.11	39.00	38.49	1.0808	2.0808	3.0808

Table 8 StirMark analysis of proposed technique on cover and stego version of Baboon (512 × 512)

Test name		Cover	Stego		
			$\beta = 2$	$\beta = 3$	$\beta = 4$
Add noise (factor = 20)	$\alpha = 4$	7.59261	7.58297	7.60819	7.51232
	$\alpha = 8$		7.5767	9.20584	7.53868
	$\alpha = 16$		7.58218	9.14098	7.54559
PSNR (factor = 10)	$\alpha = 4$	37.6544	37.7299	37.6922	38.2764
	$\alpha = 8$		37.7299	37.485	38.5543
	$\alpha = 16$		37.7674	37.3956	38.5544
Median cut (factor = 5)	$\alpha = 4$	23.3523	23.316	23.0922	22.4585
	$\alpha = 8$		23.3221	20.4089	22.4511
	$\alpha = 16$		23.2916	20.4253	22.5601
Conv filter (factor = 1)	$\alpha = 4$	9.37115	9.37031	9.36836	9.36589
	$\alpha = 8$		9.37152	9.07607	9.36597
	$\alpha = 16$		9.37175	9.07542	9.3684

4 Comparison with Existing Algorithms

Several attributes of the original image and stego image obtained by the proposed technique are calculated for visual quality analysis, viz. bit error rate (BER), mean square error (MSE), embedding efficiency and peak signal-to-noise ratio (PSNR). This method is compared with various other existing methods. The fact that our algorithm is better can be seen from the comparative study shown in Table 9.

Table 9 Comparison of capacity and PSNR with existing techniques applied on images of size 512 × 512

Existing technique	Avg. capacity (bits)	Avg. PSNR (dB)
A [2]	2097152	41.026
B [3]	1161915	41.663
C [13] (threshold value 12)	370000	36.19
C [13] (threshold value 16)	120000	33.20
C [13] (threshold value 20)	260000	32.27
D [11] (column method)	64882	48.72
D [11] (chessboard method)	71462	48.80
Proposed method	2359296	38.196

5   Conclusion

This paper introduces a novel steganographic algorithm where the number of bits embedded in each pixel varies based on the value of α and β. This helps to increase the robustness of the algorithm along with increasing the security of secret data. Due to the change of value of α and β, various other parameters also change. It allows the user to choose which parameter he or she wants to focus on while transmitting the secret data. This method can successfully withstand several attacks as shown earlier.

References

1. Cipra, B.: An introduction to the Ising model. Am. Math. Mon. **94**(10), 937–959 (1987)
2. Ghosh, U., Maity, S., Mukherjee, I.: Statistical attack resistant multi-bit steganography using mobile keypad character encoding. In: International Conference on Telecommunication Technology and Management (ICTTM-2015), IIT Delhi, India, pp. 19, 11–12 Apr 2015. ISBN: 987-0-9926800-5-3
3. Mandal, J.K., Das, D.: Colour image steganography based on pixel value differencing in spatial domain. Int. J. Inf. Sci. Tech. (IJIST) **2**(4), 83–93 (2012)
4. Mukherjee, I., Podder, A.: DCT based robust multi-bit steganographic algorithm. In: 2nd International Conference on Advanced Computing, Networking and Informatics, vol. 28. Springer (2014)
5. Paul, G., Davidson, I., Mukherjee, I., Ravi, S.S.: Keyless steganography in spatial domain using energetic pixels. Int. Conf. Inf. Syst. Secur. (ICISS) **7671**, 134–148 (2012)
6. Paul, G., Davidson, I., Mukherjee, I., Ravi, S.S.: Keyless dynamic optimal multi-bit image steganography using energetic pixels. In: Multimedia Tools and Applications, vol. 75, pp. 1–27. Springer (2016). https://doi.org/10.1007/s11042-016-3319-0
7. Petitcolas, F.A.P., Anderson, R.J., Kuhn, M.G.: Attacks on copyright marking systems. In: Second International Workshop on Information Hiding, IH98, Portland, Oregon, U.S.A., Proceedings. LNCS 1525, pp. 219–239. Springer (1998). ISBN 3-540-65386-4
8. Petitcolas, F.A.P.: Watermarking schemes evaluation. IEEE Signal Process. **17**(5), 58–64 (2000)
9. Song, X., Wang, S., Niu, X.: An integer DCT and affine transformation based image steganography method. In: IEEE 2012 Eighth International Conference on Intelligent Information Hiding and Multimedia Signal Processing, pp. 102–105, 18–20 July 2012
10. Veron, P.: Code based cryptography and steganography. In: 5th International Conference on Algebraic Informatics. LNCS, vol. 8080, pp. 9–46. Springer (2013). ISBN 978-3-642-40662-1. https://doi.org/10.1007/978-3-642-40663-8_5
11. Yang, C.H., Tsai, M.H.: Improving histogram-based reversible data hiding by interleaving predictions. IET Image Process. **4**(4), 223–234 (2010). https://doi.org/10.1049/iet-ipr.2009.0316
12. Zhang, X.: Efficient data hiding with plus-minus one or two. IEEE Signal Process. Lett. **17**(7), 635–638 (2010)
13. Zhang, X., Wang, S., Zhou, Z.: Multi-bit assignment steganography in palette images. IEEE Signal Process Lett. **15**, 553–556 (2008). https://doi.org/10.1109/LSP.2008.2001117

OCR-Assessment of Proposed Methodology Implications and Invention Outcomes with Graphical Representation Algorithmic Flow

Santosh Kumar Henge and B. Rama

Abstract The OCR innovative techniques are used to create digital formed text from the basic handwritten and printed text papers. Once it is converted then it can be reused for data processing and reprocessing purpose. OCR system provides the approach of reproducing the editable text-result which near to the approximated unique required digitalized page with same orientations and alignments. The OCR algorithm derives the huge set of learned letters, characters, symbols and their desirable properties. It is best suited for pattern-cum-symbolic image-based recognition and digitalizing the passive mode characters into active mode characters. These inventions are widely executing in private and public sector for various data processing purposes. In OCR, the normal page imprecise text can be processed through various recognition level stages, such as Text-Classification, Level of Pre-cum-Postprocessing, Segmented Processing and Characteristic Mining. Researchers invented the new ideas and approaches for solving critical problems in innovative OCR mechanism. This paper contains the detail descriptive assessment of proposed methods, methodologies, steps handled and invention outcomes in the discovery of optical character recognition approaches and also described graphical representation of OCR algorithmic variations with their handled steps for processing various levels of text and the flow of methodology. This descriptive graphical representation will be helpful to all upcoming researchers in the innovative OCR field. Graphical representation flow is easy to understand and simple to gain the basic important knowledge to focus in their OCR research for further innovations.

Keywords Digitized-retrievable form (DRF) · Optical character recognition (OCR) · Feed-forward back propagation neural network (FFwBPNN) Neural network-based algorithm (NNbA) · Algorithm based on support vector

S. K. Henge (✉) · B. Rama
Computer Science Department, University College, Kakatiya University,
Hanamkonda, Telangana, India
e-mail: hingesanthosh@gmail.com

B. Rama
e-mail: rama.abbidi@gmail.com

© Springer Nature Singapore Pte Ltd. 2018   57
K. Saeed et al. (eds.), *Progress in Advanced Computing and Intelligent Engineering*,
Advances in Intelligent Systems and Computing 563,
https://doi.org/10.1007/978-981-10-6872-0_6

networks (ASVN) · Template matching algorithm (TMA) ·
Neural-network-feed-forward-method (NNFwFM) · Top-down-hierarchical his-
togram (THH) · Neural-network-feed-backward-method (NNFBwM)
Freeman chain code (FCC) · Mirror Image Learning (MIL) · Statistical algo-
rithm (STA) · Syntactic algorithm (SA) · Back-propagation training algorithm
(BPTA) · Decision tree classifier algorithm (DTCA)

1 Introduction

The OCR innovative techniques are used to create digital formed text from the basic
handwritten and printed text papers. Once it is converted then it can be reused for data
processing and reprocessing purpose. OCR system is used to build the approach with
the features of conversion of static image based text to editable form of text with the
advanced services like as text alignments and text orientations. The OCR algorithm
derives the huge set of learned letters, characters, symbols and their desirable properties. It is best suited for pattern-cum-symbolic image-based recognition and digitalizing the passive mode characters into active mode characters. These inventions are
widely executing in private and public sector for various data processing purposes.
In OCR, the normal page imprecise text can be processed through various recognition
level stages, such as Text-Classification, Level of Pre-cum-Postprocessing, Segmented Processing and Characteristic Mining. Researchers invented the new ideas
and approaches for solving critical problems in innovative OCR mechanism.

2 Article Objectives

This paper contains the detail descriptive assessment of proposed methods,
methodologies, steps handled and invention outcomes in the discovery of optical
character recognition approaches and also described graphical representation of
OCR algorithmic variations with their handled steps for processing various levels of
text and the flow of methodology. This descriptive graphical representation will be
helpful to all upcoming researchers in the innovative OCR field. Graphical representation flow is easy to understand and simple to gain the basic important
knowledge to focus in their OCR research for further innovations.

3 Graphical Representation of Innovations of OCR Algorithmic Flows

Researchers invented the new ideas and approaches for solving critical problems in
innovative OCR mechanism. Different OCR algorithms are designed and successfully implemented for required human processing data. Each and every

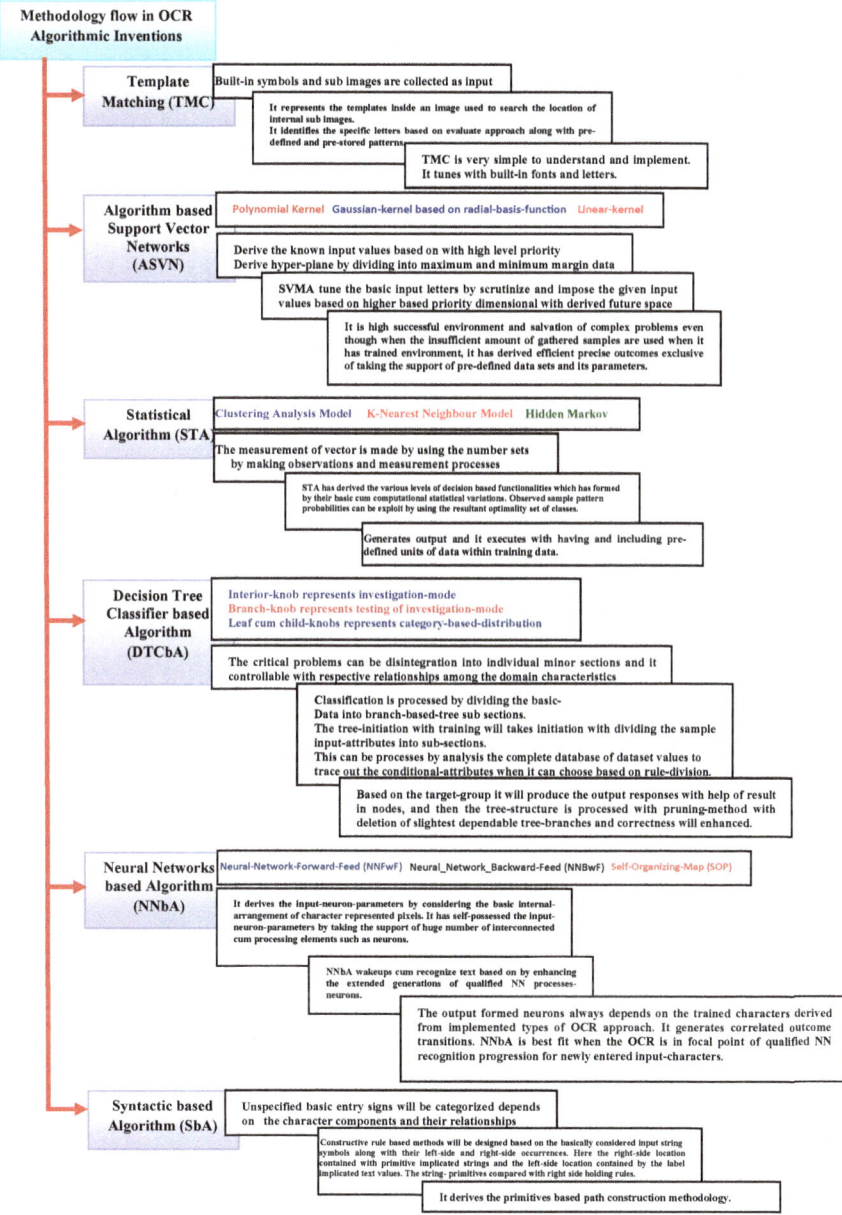

Fig. 1 Graphical representation of OCR algorithmic flows and their variations

OCR-based algorithm consists compensations and drawbacks. The graphical representation of OCR algorithm-based variations with their handled steps for processing various levels of text and the flow of methodology is shown in Fig. 1.

Table 1 Assessment of proposed approaches, intension, invention result with output responses

Reference/ author	Derived technology tools	Type of approach	Intentions	Implication of flow of approach	Result with output responses (RORs)
Pal et al. [1]	ANN	Fourier descriptors with back propagation network	Handwritten English character recognition	The frontier-based environmental approach has been implicated for obtaining the characteristic extraction	The RORs are up to 94%
Gupta et al. [2]	ANN	Four Fourier (FF) method	Offline handwritten character recognition	It has derived FF implications which are employed for detecting the offline letters and symbols	Here, the RORs are 98.75%
Perwej et al. [3]	NNbA	Neural network	Handwritten alphabets in English language	Here, the input sequences are formed based on the binary digits for processing the general sequence of English letters to execute the mannerism extraction structure	Here, the RORs detection rate is 82.5%
Asthana et al. [4]	NNbA	Multiple hidden layers markov model	Handy written multi-variation-scripted-numeral detection	Here, three concealed coating layers are employed and two self-characterized methods are implicated by employing the various levels of five different scripts	The RORs recognition rate is in between 94–95%
Graves et al. [5]	NNbA	A novel connectionist system	For unconstrained handwriting recognition	Used the recurrent NN and the result has been compared with HMM-based system	HMM-based performance output
Ping et al. [6]	NNbA	Appliance thirteen-point feature skeleton-based approach	NN-environmental text detection method	Here, the expected data has been composed based on the thirteen-point zone division from derived-based input characters. The back-propagation training algorithmic implications are used to detect the letters	Author articulated that derived effective RORs recognition rate
Ganapathy et al. [7]	NNbA	Multiscale neural network training technique	Handwritten character recognition	In this NNbA, the choice-based threshold minimum distance techniques have been implicated to achieve desirable outcomes	The researcher has mentioned that he computed successful RORs detection rate
Kaur et al. [8]	NNbA	Levenberg-Marquardt BPTA used in NN to train to train	Natural and simple CR	This implication has contained seven advanced flow of features: covered space,	Here, the RORs is up to 93% of accuracy rate

(continued)

Table 1 (continued)

Reference/ author	Derived technology tools	Type of approach	Intentions	Implication of flow of approach	Result with output responses (RORs)
				boundaries in sequence, minor-cum-major axis span values and solidity prolog and their orientation occurrences	
Shaaban [9]	NNbA	Text-based approach	Machine-printed Arabic texts based	It has derived new technical multiple-based parallel approach through NNbA mechanism implicated for detection of Arabic language-based text sequences employed by multi-formation method	The RORs rate is 98%
Kumar et al. [10]	NNbA	Multilayer-FFwBPNN without feature extraction	Handwritten OCR machine	It has designed with the 70 × 50 pixels sizes of input characters, implicated as input for preparation-based system. Here, the NNbA contains four implicated layers, first and last layers are used for input, output and middle layers called as hidden layers which are used for computational purpose	The researcher has obtained up to 81% of recognition rate of RORs
Arora et al. [11]	ASVN and NNbA	Performance comparison of ASVN to ANuNet	Devanagari-handy-character-detection methods	This approach derives the permutation mechanism of ANN-ASVN primary features are implicated such as shadow, CC-histogram, longest-run and view-based environmental features	The researcher has mentioned that the computed successful RORs detection rate
Sheth et al. [12]	ANN and ASVN	Chain code-based approach	Chain code environmental handy written text-based detection system	CC-based approach feature set can be implemented along with FFwBPNN and SVM	The researcher has mentioned that the exactness of recital of diverse feature-based set techniques
Gohil et al. [13]	NN and ASVN	CC with holistic-based feature-based OCR	Devanagari language-based text	It derived the recital scrutiny environmental comparisons by holding the features of CC, holistic and binary-based approaches	Comparison-based analysis presentations are described

(continued)

Table 1 (continued)

Reference/author	Derived technology tools	Type of approach	Intentions	Implication of flow of approach	Result with output responses (RORs)
Kavallieratou et al. [14]	Structural analysis algorithm	Structural characteristics and lexical support	Handwritten word recognition based on	Here, two hundred and eighty dimension-based feature vector has been implicated	achieved effective accuracy range of 72.8–98.8%
Jonathanet al. [15]	Structural algorithm	Structural analysis multiple algorithms	Numerous algorithmic implications for Handy written character recognition	This approach has completely involved with the full structural analysis based algorithmic sequences for implementing the planned test cases	The RORs rate is 79% has been obtained.
Singh et al. [16]	Statistical method	K-Nearest neighbour algorithm	Handwritten Gurumukhi script	In this approach, the K-adjacent node or volume has detected through the dimension of distance-based Euclidean in-between the examine node–node testing and to other referenced-node orders	The RORs detection rate is 72.54%
Arora et al. [17]	Statistical model	Statistical approach	Application of statistical features in handy written Devanagari language-based text detection	Here, the two various self-implicated features such as moment-based and CC-histograms are generated for further progress	The researcher has obtained the RORs recognition rate in between 65–99%
Nadeem et al. [18]	TMC	Comparing-cum-matching methodology	Handy and type-written text detection methods	The mathematical-based geometrical possessions are implicated to enumerate basic input letters into mathematical-based vector formed entities	The RORs recognition rate is in between 75–95%
Mohammad et al. [19]	PMC	Comparing-cum-matching methodology	Type-written and handy written text detection methods	Here, the track-based-sector implicated-matrix forms are represented for inspection the identical approaches along with the preceding prototypes	The researcher has obtained the RORs recognition rate up to 80%
Som et al. [20]	Fuzzy Logic Systems	Fuzzy membership function", international	Handwritten character recognition	The fuzzy environmental membership computational parameters, functions are functional on various unidentified lettering	Few mis-match occurrences will happen when there hug changes in between the sample test cases of a variety of variations of letters considered

(continued)

Table 1 (continued)

Reference/author	Derived technology tools	Type of approach	Intentions	Implication of flow of approach	Result with output responses (RORs)
Blumenstein et al. [21]	A novel feature extraction technique	Novel-based feature extraction method	Segmented handwritten text recognition	Transition evaluation implicated and track-based characteristic origin techniques were implicated for detecting the desirable characters	The researcher has mentioned that he computed successful RORs detection rate
Pal et al. [22]	Comparative study	Comparative learning of different feature-based and classifiers-based scripted recognition	Relative learning of Devanagari handy written lettering detection by using the various stages-based feature and classifiers	Here, the output responses are obtained by implicating MIL and other twelve various classifiers along with four features based sets. The gradient-based and variations were implicated for lettering detection	The RORs average detection rate is 94.94%
Patel et al. [23]	Euclidean distance metric	Multiresolution technique and Euclidean distance metric	Handy written lettering detection method	Here, the wavelet-renovate Multienvironmental decision techniques implicated for feature-based extraction	The RORs average detection rate is 90%
Gajjar et al. [24]	Top and down based hierarchical histogram (THH) variations	Top and down based hierarchical histogram methods	Printed formed Devanagari language-based scripted lettering Isolation	Here binary based feature implicated set has employed	The researcher has mentioned that he computed successful RORs detection rate.
Deshpande et al. [25]	Standard expressions and minimum-editable-distance-based method	Minimum-editable-distance-based process implications	Fine-classification and detection of handy written Devanagari Lettering	The prototype identical methods are used for further processing. The evaluation process executed based on the minimal-editable-distance environmental filters	An achieved overall accuracy rate up to 82%
Sheth et al. [26]	Correlation coefficient	Correlation coefficient	Handy written lettering detection system by using correlation coefficient	Here, the correlation approache coefficient implicated HCR system has used for further processing	The researcher has mentioned that he computed successful RORs detection rate
Phokharatkul et al. [27]	Ant-Miner algorithm(AMA)	Ant-Miner algorithm	Offline-Handy written Thai text detection	Here, the AMA-based algorithmic sequences are implicated for text detection purpose	The RORs are 97%

(continued)

Table 1 (continued)

Reference/ author	Derived technology tools	Type of approach	Intentions	Implication of flow of approach	Result with output responses (RORs)
Jannoud [28]			Automatic based Arabic language handy written letter detection machine	Used discrete wavelet transform method	The RORs recognition rate in between 91–99%
Sharma et al. [29]	Zone environmental hybrid implicated feature extraction design	Euler number method	Handwritten alphabets recognition	Here, the letters classified into a range of clusters for further detection through the zone environmental hybrid implicated feature origin representation	Here the RORs are 98% of detection rates has gained
Das et al. [30]	Hidden Markov model (HMM)	Universal and confined characteristic extraction	Detection of offline mode based handwritten-self-governing-English letters	Here, the researchers experimented 12,500 handy written samples which are formed from various test cases of different writers	HMM sequences has employed to get qualified character-based symbols and the RORs are the 98.26%
Sastr et al. [31]	Decision tree (DT) classifier based algorithm (DTCbA)	DTCbA	Telugu language handwritten letters haul out from palm plant-leaves	Here the DT implication has composed and employed by taking major support of SEES-based algorithm	Here, the RORs are 94%

Descriptive graphical representation will be helpful to all upcoming researchers in the innovative OCR field. This approach based graphical representation described the various concerns such as how the inputs are formed and processed; sequence cum flow of the methodology; usage-cum-type of the research techniques and tools are implemented; past designed individual methodology best outcome consequences.

4 Assessment of Proposed Methodologies, Invention Outcomes

The detail descriptive assessment of proposed methods, methodologies, steps handled and invention output responses in the discovery of optical character recognition approaches is shown in Table 1.

5 Conclusion

This paper contained the detail descriptive assessment of proposed methods, methodologies, steps handled, and invention outcomes in the discovery of OCR approaches and also described the graphical representation of OCR algorithmic variations with their concern involved steps for further processing in various levels of text executions. Through the graphical representation described: how the inputs formed and processed, flow of the methodology, usage-cum-type of the research techniques and tools implemented, range of deriving the output responses from input data and output-reflections. This descriptive graphical representation will be helpful to all upcoming researchers in the innovative OCR field. Graphical representation flow is easy to understand and simple to gain the basic important knowledge to focus in their OCR research for further innovations.

References

1. Pal, A., Singh, D.: Handwritten English character recognition using neural network. Int. J. Comput. Sci. Commun. 1(2), 141–144 (2010)
2. Gupta, A., Srivastava, M., Mahanta, C.: Offline handwritten character recognition using neural network. In: The 2011 IEEE International Conference on Computer Application and Industrial Electronics, pp. 102–107, Penang, Malaysia, Dec 2011
3. Perwej, Y., Chaturvedi, A.: Neural networks for handwritten English alphabet recognition. Int. J. Comput. Appl. 2(7) (2011)
4. Asthana, S., Haneef, F., Bhujade, R.K.: Handwritten multiscript numeral recognition using artificial neural networks. Int. J. Soft Comput. Eng. 1(1), 1–5 (2011)

5. Graves, A., Liwicki, M., Fernandez, S., Bertolami, R., Bunke, H., Schmidhuber, J.: A novel connectionist system for unconstrained handwriting recognition. IEEE Trans. Pattern Anal. Mach. Intell. **31**(5), 855–868 (2009)
6. Ping, N.S., Yusoff, M.A.: Application of 13-point feature of skeleton to neural networks-based character recognition. In: IEEE International Conference on Computer & Information Science, vol. 1, pp. 447–452, Kuala Lumpur, Malaysia, June 2012
7. Ganapathy, V., Liew, K.L.: Handwritten character recognition using multiscale neural network training technique. In: Proceedings of World Academy of Science: Engineering & Technology, vol. 41, pp. 32–37 (2008)
8. Kaur, G., Aggarwal, M.: Artificial intelligent system for character recognition using Levenberg-Marquardt algorithm. Int. J. Adv. Res. Comput. Sci. Softw. Eng. **2**(5), 220–223 (2012)
9. Shaaban, Z.: A new recognition scheme for machine-printed Arabic texts based on neural networks. World Acad. Sci. Eng. Technol. **41**, 706–709 (2008)
10. Kumar, P., Sharma, N., Rana, A.: Handwritten character recognition using different Kernel based SVM classifier and MLP neural network. Int. J. Comput. Sci. **53**(11) (2012)
11. Arora, S., Bhattacharjee, D., Nasipuri, M., Malik, L., Kundu, M., Basu, D.K.: Performance comparison of SVM and ANN for handwritten Devnagari character recognition. IJCS Issues **7** (3), no. 6, 18–26 (2010)
12. Sheth, R., Chauhan, N., Goyani, M.: Chain code based hand written character recognition system using neural network and support vector machine. In: International Conference on Recent Trends in Information Technology and Computer Science, pp. 62–67, Mumbai, India, 09–10 Dec 2011
13. Gohil, G., Teraiya, R., Goyani, M.: Chain code and holistic features based OCR system for printed devanagari script using ANN and SVM. Int. J. Artif. Intell. Appl. (IJAIA) **3**(1) (2012)
14. Kavallieratou, E., Sgarbas, K., Fakotakis, N., Kokkinakis, G.: Handwritten word recognition based on structural characteristics and lexical support. In: IEEE 7th International Conference on Document Analysis and Recognition, vol. 1, pp. 562–566 (2003)
15. Hull, J.J., Commike, A., HO, T.-K.: Multiple algorithms for handwritten character recognition
16. Singh, P., Budhiraja, S.: Feature extraction and classification techniques in O.C.R. systems for handwritten Gurmukhi script. Int. J. Eng. Res. Appl. (IJERA) **1**(4), 1736–1739
17. Arora, S., Bhattacharjee, D., Nasipuri, M., Basu, D.K., Kundu, M.: Application of statistical features in handwritten Devanagari character recognition. Int. J. Recent Trends Eng. **2**(2), 40–42 (2009)
18. Nadeem, D., Rizvi, S.: Character Recognition Using Template Matching. Department of Computer Science, JMI
19. Mohammad, F., Anarase, J., Shingote, M., Ghanwat, P.: Optical character recognition implementation using pattern matching. Int. J. Comput. Sc. Inf. Technol. **5**(2) (2014)
20. Som, T., Saha, S.: Handwritten character recognition using fuzzy membership function. IJETSE **5**(2), 11–15 (2011)
21. Blumenstein, M., Verma, B., Basli, H.: A novel feature extraction technique for the recognition of segmented handwritten characters. In: IEEE 7th International Conference on Document Analysis and Recognition, vol. 1, pp. 137–141, August, 2003
22. Pal, U., Wakabayashi, T., Kimura, F.: Comparative study of Devnagari handwritten character recognition using different feature and classifiers. In: IEEE 10th International Conference on Document Analysis and Recognition, pp. 1111–1115, Barcelona, July 2009
23. Patel, D.K., Som, T., Yadav, S.K., Singh, M.K.: Handwritten character recognition using multiresolution technique and Euclidean distance metric. JSIP **3**, 208–214 (2012)
24. Gajjar, T., Teraiya, R., Gohil, G., Goyani, M.: Top down hierarchical histogram based approach for printed Devnagari script character isolation. In: International Conference on Digital Image Processing and Pattern Recognition (DPPPR), vol. 205, pp. 55–64. Springer, Tirunelveli, Tamilnadu, India, 23–25 Sept 2011

25. Deshpande, P.S., Malik, L., Arora, S.: Fine classification and recognition of handwritten Devanagari characters with regular expressions and minimum edit distance method. J. Comput. **3**(5), 11–17 (2008)
26. Sheth, R., Chauhan, N., Goyani, M.: A handwritten character recognition system using correlation coefficient. In: International Conference on Innovative Science and Engineering Technology, pp. 395–398, 08–09 Apr 2011. ISBN: 987-81-906377-56
27. Phokharatkul, P., Sankhuangaw, K., Somkuarnpanit, S., Phaiboon, S., Kimpan, C.: Offline handwritten Thai character recognition using ant-miner algorithm. World Acad. Sci. Eng. Technol. **8**, 276–281 (2005)
28. Jannoud, I.A.: Automatic Arabic hand written text recognition system. Am. J. Appl. Sci. **4** (11), 857–864 (2007)
29. Sharma, O.P., Ghose, M.K., Shah, K.B.: An improved zone based hybrid feature extraction model for handwritten alphabets recognition using Euler number. IJSCE **2**(2), 504–508 (2012)
30. Das, R.L., Prasad, B.K., Sanyal, G.: HMM based offline handwritten writer independent English character recognition using global and local feature extraction. Int. J. Comput. Appl. (0975 8887), **46**(10), 45–50 (2012)
31. Sastry, P.N., Krishnan, R., Ram, B.V.S.: Classification and identification of Telugu handwritten characters extraction from palm leaves using decision tree approach. ARPN J. Eng. Appl. Sci. **5**(3) (2010)

Local Diagonal Laplacian Pattern
A New MR and CT Image Feature Descriptor

Praveen Kumar Reddy Yelampalli and Jagadish Nayak

Abstract Feature extraction of medical images is a challenging task due to variation in modalities and imaged objects (organs, tissues and specific pathologies). This article presents a simple feature descriptor, local diagonal Laplacian pattern (LDLP), devised on the idea of local diagonal extrema pattern (LDEP). LDEP is implemented using first-order diagonal derivatives, whereas LDLP uses Laplacian operation. In order to reduce computational complexity, the intensities at each diagonal element are correlated to the rest of the three diagonal elements together with the centre pixel. Furthermore, the corner elements are compared with the centre pixel to improve the quality of feature description. In addition, the dimension of the pattern is lowered by a half. LDLP is applied to various MR, CT images of the TCIA database and CT images of the NEMA database that results in better feature description at less computational cost.

Keywords Medical image analysis · Feature description · Local diagonal patterns · Laplacian operator

1 Introduction

Medical image analysis is a major area of image processing applications and is being widely used in clinical diagnosis and early detection of tumours. A physician not likely to judge the patient's condition without thorough observation and understanding of the data acquired. Analysis methods enhance the diagnostic information and greatly improve the interpretation of medical images and thus are potentially complex and the computation burden for accurate evaluations due to different

P. K. R. Yelampalli (✉) · J. Nayak
Department of Electrical and Electronics Engineering, BITS Pilani-Dubai Campus, 345055 Dubai, UAE
e-mail: praveenyelampalli@gmail.com

J. Nayak
e-mail: jagadishnayak@dubai.bits-pilani.ac.in

© Springer Nature Singapore Pte Ltd. 2018   69
K. Saeed et al. (eds.), *Progress in Advanced Computing and Intelligent Engineering*,
Advances in Intelligent Systems and Computing 563,
https://doi.org/10.1007/978-981-10-6872-0_7

modalities, sensing mechanisms and the presence of noise in acquisition methods. MR, CT, PET and SPECT are a few in the list of medical imaging [1]. Features are primary elements in image that portray variation in shape, size and objects present in the image. The aim of a feature descriptor is extracting the best possible features for superior analysis [2]. These features further classified as general- and domain-specific features. In general, both of them are used in medical image analysis to focus on all the shapes and avoid the difficulty in perception.

An enormous amount of methods has been developed to characterise medical image features due to their potential impact in clinical studies. A detailed survey on feature description and point detection methods is published in [3]. The early method for binary patterns is local binary pattern (LBP) proposed by Ojala et al. [4], that defines the texture by encoding the contrast and patterns. This scheme navigated researchers towards binary patterns. Some variants of LBP that predominantly used in medical image analysis include local ternary pattern (LTP) [5], rotation invariant LBP (RILBP) [6], volume LBP (VLBP) [7] for 3D dataset, centre symmetric LBP (CSLBP) [8] and data driven LBP (DDLBP) [9]. Further, local neighbouring intensity relationship pattern (LNIRP) [10], local mesh patters (LMeP) [11], and local ternary co-occurrence pattern (LTCoP) [12] are being formulated for biomedical texture classification and retrieval. Xianbiao et al. [13] proposed multi-scale co-occurrence of LBPs (MCLBP) for obtaining an effective correlation between LBPs in different scales. However, all the methods resulted in higher dimensional patterns. In the works Murala et al. [14] and Kaya et a. [15], the pattern is obtained by encoding spatial relationships between neighbours with a distance parameter or a reference pixel. In addition, the local wavelet pattern [16] and local bit-plane decoded pattern [17] are recently investigated for biomedical image indexing and retrieval.

Recently, Dubey et al. [18] have presented local diagonal extrema pattern (LDEP), a new feature descriptor that relates diagonal neighbours with centre pixel, for CT image retrieval. Diagonal neighbours consist of majority of the local information and reduce the dimensionality of the descriptor [19]. LDEP encoded the relationships among diagonal neighbours and between centre pixel and each diagonal neighbour using first-order diagonal derivative, to a 24-bit length binary pattern. Also, efficiently retrieved CT images compared to other binary descriptors. Nonetheless, the vector length can further be reduced for faster image retrieval schemes. In this article, we formulated a novel descriptor, local diagonal Laplacian pattern (LDLP), using the Laplacian of diagonal elements and compared them with the centre pixel. As a result, the number of computations is scaled down and the dimension of the feature vector became half. Also, the indices of extremas are calculated in a single iteration in contrary to three iterations in LDEP and to increase the discrimination capacity, these are further compared with the centre pixel.

The scope of this paper is to develop LDLP, describing features of MR and CT images and comparing results with the LDEP. This paper is organised into three components: the computation of LDLP in Sect. 2, results in Sect. 3 and conclusions and future work in Sect. 4.

2 Local Diagonal Laplacian Transform

Laplacian operator is the simplest linear and isotropic derivative operator. The approach consists of defining a discrete formulation of the second-order derivative. The discrete Laplacian of a two variable function $\xi(x, y)$ can be defined as below,

$$\Delta^2 \xi(x, y) = \xi(x + 1, y) + \xi(x - 1, y) + \xi(x, y + 1) + \xi(x, y - 1) - 4\xi(x, y). \quad (1)$$

The Laplacian highlights rapidly changing intensity regions of an image and is therefore often used for edge detection [20]. First, the relationship of each diagonal element (diagonal elements along with centre pixel as given in Fig. 1a) is obtained using the Laplacian Kernel shown Fig. 1b. This generates the maxima and minima (extremas), and these are further related with the centre pixel to encode to a 12-bit length feature vector (the LDLP descriptor).

2.1 Laplacian Operator/Second-Order Diagonal Derivatives

Let Γ be an image of rows υ and columns ϑ with centre pixel position as $P^{m,n}$ and $P^{m,n}_t$ is tth diagonal neighbour of $P^{m,n}$ at a distance ρ where $t \in [1, 4]$. The intensities at $P^{m,n}$, $P^{m,n}_t$ are $I^{m,n}$ and $I^{m,n}_t$, respectively. Here, $I^{m,n}_t$ can be defined as,

$$I^{m,n}_t = I^{m+\delta, n+\eta}. \quad (2)$$

where δ and η are constants with values either $(-\rho)$ or $(+\rho)$ depending upon the t value.

$$\delta, \eta = \begin{cases} -\rho, +\rho \text{ for } t = 1 \\ -\rho, -\rho \text{ for } t = 2 \\ +\rho, -\rho \text{ for } t = 3 \\ +\rho, +\rho \text{ for } t = 4. \end{cases} \quad (3)$$

An image of size 5-by-5, shown in Fig. 2a, has been considered for our example. A block of 3-by-3 dimension (Fig. 2b) is read and the diagonal neighbours are extracted

Fig. 1 a Matrix contains diagonal elements and a centre pixel b Laplacian kernel

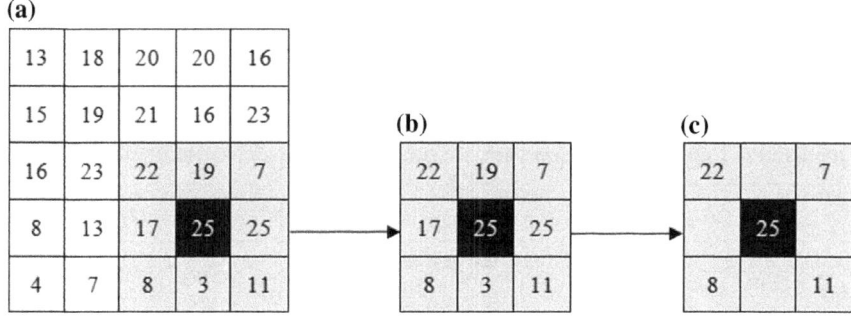

Fig. 2   a An image of size 5×5 **b** A block of (a) of order 3×3 **c** diagonal neighbours of the centre pixel

as shown in Fig. 2c. Second-order derivative $\tilde{I}_t^{m,n}$ is applied to each individual diagonal element, the resulting value replaces the original intensity value and this repeats for the rest of the elements. The process flow is given in Fig. 3 that depicts all the diagonal elements are now filled with newly evaluated values.

$$\tilde{I}_t^{m,n} = I_{mod(t,4)+1}^{m,n} + I_{mod(t,4)+2}^{m,n} + I_{mod(t,4)+3}^{m,n} + I^{m,n} - 4I_t^{m,n}. \tag{4}$$

where $t \in [1, 4]$ and $mod(\Omega_1, \Omega_2)$ is the remainder of the division of Ω_1 by Ω_2. The local diagonal maxima (LD-M) and local diagonal minima (LD-m) are calculated as shown in Fig. 3.

Now, the relationship between the centre pixel intensity $I^{m,n}$ with the maximum intensity $I_{\tau max}^{m,n}$ and minimum intensity $I_{\tau min}^{m,n}$ can be obtained using (5) and (6).

$$\Delta_{max}^{m,n} = I_{\tau max}^{m,n} - I^{m,n}. \tag{5}$$

$$\Delta_{min}^{m,n} = I_{\tau min}^{m,n} - I^{m,n}. \tag{6}$$

where $\Delta_{max}^{m,n}$ and $\Delta_{min}^{m,n}$ are the local diagonal extrema-centre difference factors for the intensities $I_{\tau max}^{m,n}$ and $I_{\tau min}^{m,n}$, respectively. In addition to that, τ_{max} and τ_{min} are the index values of LD-M and LD-m, which are computed as follows,

$$\tau_{max} = \underset{t}{argmax} (\tilde{I}_t^{m,n}). \tag{7}$$

$$\tau_{min} = \underset{t}{argmin} (\tilde{I}_t^{m,n}). \tag{8}$$

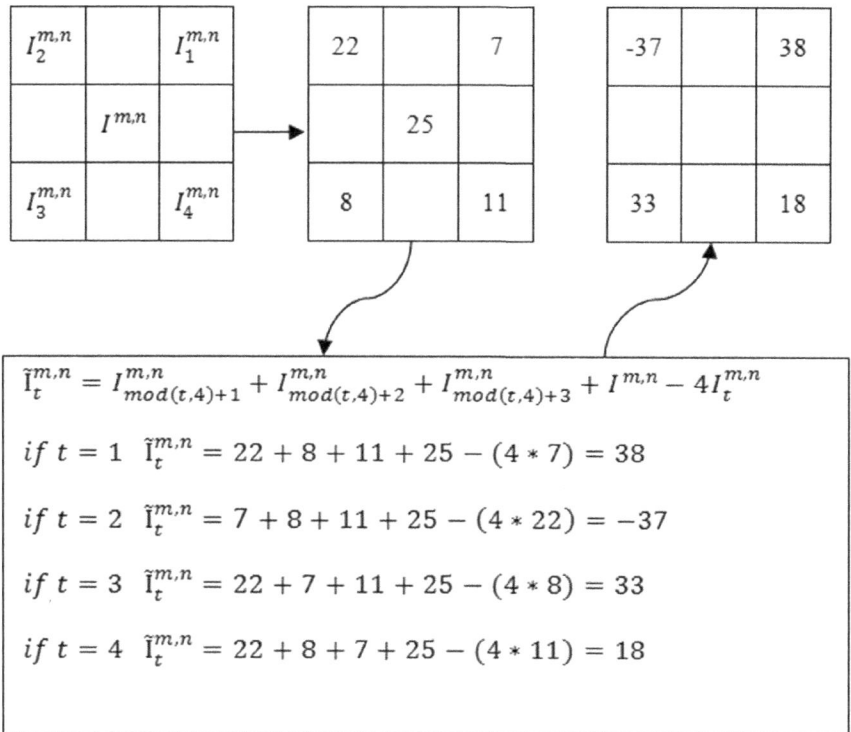

Fig. 3 Evaluation of second-order diagonal derivatives

The calculation of second-order diagonal derivative values, indices and diagonal extremas is completed. Next, by using this data we have to compute the local diagonal Laplacian pattern.

2.2 Computation of LDLP

In the above section, the indices τ_{max} and τ_{min} are calculated for the LD-maxima $I^{m,n}_{\tau_{max}}$ and LD-minima $I^{m,n}_{\tau_{min}}$ of the centre pixel $P^{m,n}$ with intensity value $I^{m,n}$. This section discusses the LDLP evaluation as illustrated in Fig. 4. The LDLP of the centre pixel $P^{m,n}$ of a 3-by-3 image is represented as $LDLP^{m,n}$. Here, we assumed the local diagonal neighbourhood is located at a distance ρ.

$$LDLP^{m,n} = [LDLP^{m,n}_1, LDLP^{m,n}_2, \ldots, LDEP^{m,n}_{dim}]. \qquad (9)$$

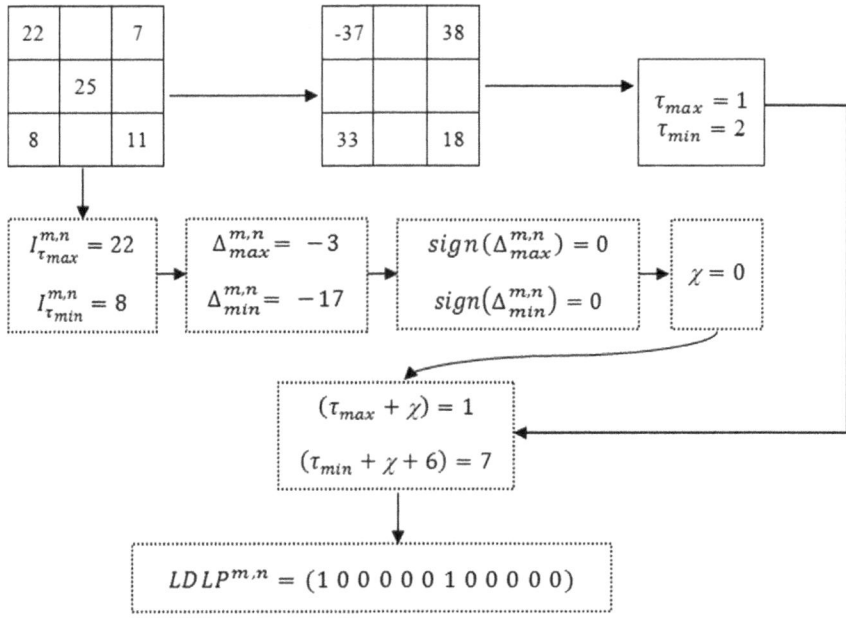

Fig. 4 Evaluation of local diagonal laplacian pattern

where *dim* is the dimension/length of the LDLP binary pattern, which is 12 and $LDLP_k^{m,n}$ is the *k*th element of the $LDLP^{m,n}$ and is presented as below,

$$LDLP_k^{m,n} = \begin{cases} 1 & \text{if } k = (\tau_{max} + \chi) \text{ or } k = (\tau_{min} + \chi + 6) \\ 0 & \text{else} \end{cases}. \qquad (10)$$

where, χ is called the extrema-centre pixel relationship factor and $LDLP^{m,n}$ is '1' if $k = (\tau_{max} + \chi)$ or $k = (\tau_{min} + \chi + 6)$ and '0' at rest of the positions. LDLP contains 1s, only at two positions, i.e. maxima is at $\tau_{max} + \chi$ and minima is at $\tau_{min} + \chi + 6$. The extrema-centre relationship factor χ is evaluated as,

$$\chi = \begin{cases} 0 & \text{if } (sign(\Delta_{max}^{m,n}) = 0 \text{ and } (sign(\Delta_{min}^{m,n}) = 0) \\ 1 & \text{if } (sign(\Delta_{max}^{m,n}) = 1 \text{ and } (sign(\Delta_{min}^{m,n}) = 1) \\ 2 & \text{else} \end{cases}. \qquad (11)$$

The function *sign* is used for the sign estimation of a given number. Where

$$sign(\psi) = \begin{cases} 1, & \psi \geq 0 \\ 0, & \psi < 0 \end{cases}. \qquad (12)$$

The dimension of the pattern $LDLP^{m,n}$ is 12-bit which is half of the LDEP, and it is maximum possible value of the k which successively depends upon the value of τ_{min} (which is 6) and χ (which is 2). The proposed method has lower dimension pattern yet entrusts better feature description. Figure 4 shows how to calculate the parameters τ_{max}, τ_{min}, $\Delta_{max}^{m,n}$, $\Delta_{min}^{m,n}$ and χ by which, the LDLP pattern has been formed.

In the above example, the local diagonal Laplacian pattern $LDLP^{m,n}$ is computed only for one pixel $P^{m,n}$. The LDLP over the entire image Γ is given below:

$$LDLP = (LDLP_1, LDLP_2, \ldots, LDLP_l, \ldots, LDLP_{dim}). \tag{13}$$

where $LDLP_l$ is the LDLP of the lth centre pixel element of the image. This can be defined as follows:

$$LDLP_l = \frac{1}{(\upsilon - 2\rho)(\vartheta - 2\rho)} \sum_{m=\rho+1}^{\upsilon-\rho} \sum_{n=\rho+1}^{\vartheta-\rho} LDLP_l^{m,n}. \tag{14}$$

3   Results

We have verified our method, local diagonal Laplacian pattern, with knee MRI of DICOM library [21], prostate MRI of TGCA-PRAD [22], lung CT image of NEMA [23], breast MRI of TCIA [24] and brain MRI of RIDER NEURO [25] in that order, shown in Fig. 5a. All the DICOM images of unsigned-integer-16 (uint16) are first converted to greyscale image before applying LDEP (b) and LDLP (c).

The DICOM images in Fig. 5a are of size 256×256. Figure 5b and c show descriptor output for LDLP and LDEP for knee MRI, prostate MRI, lung CT, breast MRI and brain MRI. It can be observed that the LDLP highlights the regions with smoother edges, and the introduction of the Laplacian operation elevates the intensity variations significantly. The LDEP results of prostate, brain MR and lung CT images demonstrated that certain image features are not precisely visible due to conflict in the index values (τ_{max} and τ_{min}) of LD-M and LD-m. On the other hand, LDLP evaluates the indices of extrema based on the maximum and minimum values of the second-order diagonal derivative. Hence, conflicts occur rarely or almost nil. Also, the boundaries of knee MR image are better noticeable in LDLP. However, in case of breast MR image both methods offer almost similar features. In computational efficiency, the proposed scheme converged in 1024 arithmetic operations for obtaining diagonal derivatives, while the LDEP needs 3072 (three times that of LDLP, one for each iteration of first-order diagonal derivative). In addition, the resulting LDLP (12-bit) is half the size of the LDEP (24-bit), as stated earlier.

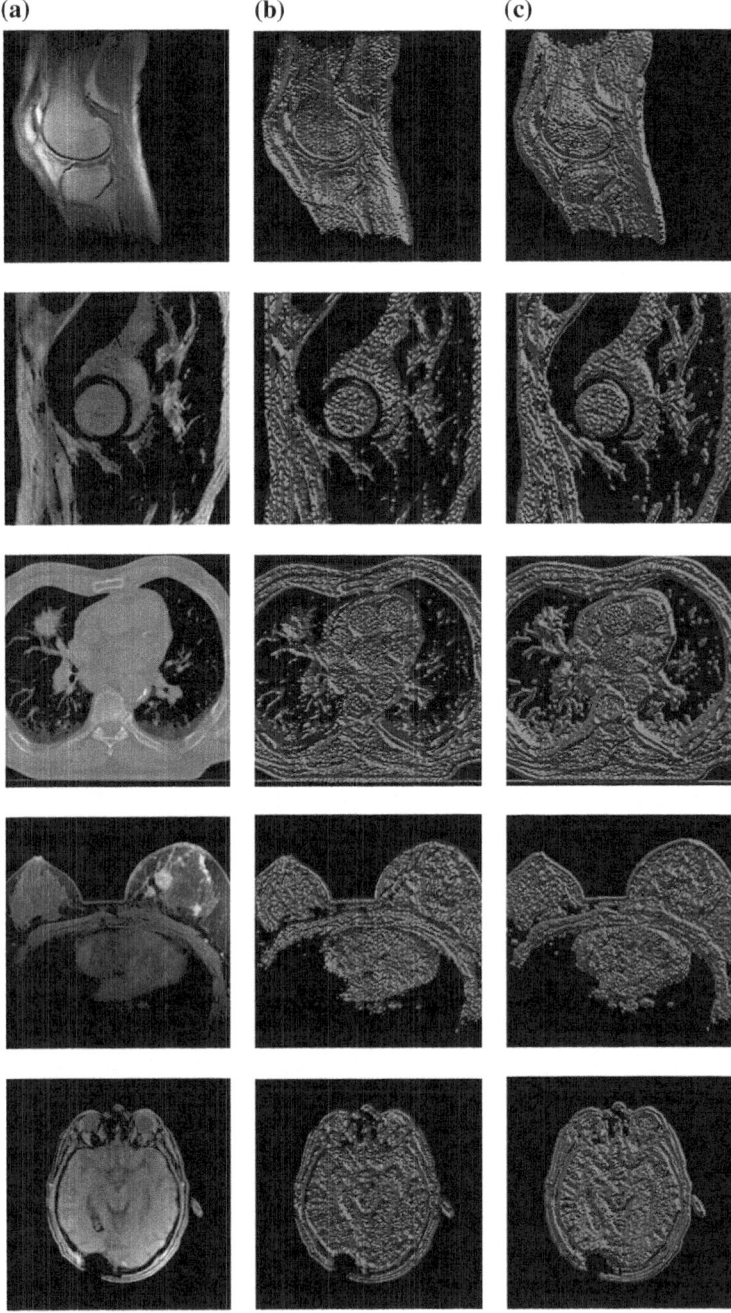

Fig. 5 **a** Original DICOM image **b** LDEP **c** LDLP

4 Conclusion and Future Work

In this paper, we presented a new and low-dimensional CT, MR image feature descriptor local diagonal Laplacian pattern and results are compared with LDEP. The proposed approach produces a 12-bit binary pattern and the number of iterations is reduced since, a second-order diagonal derivate has been introduced. Our future work involves application of LDLP in CT & MR image retrieval, image matching and image registration. Since, the computational cost and the dimension of the feature vector are drastically reduced LDLP is a better choice for image retrieval applications. This approach has not tested with the images contain noise. All the images acquired for testing purpose are of standard databases.

References

1. Dhawan, A.P.: Introduction. In: Medical Image Analysis. pp. 1–22, John Wiley and Sons Inc., New Jersy (2011)
2. Nixon, M.S., Aguado, A.S.: Low-level feature extraction (including edge detection). In: Feature Extraction and Image Processing for Computer Vision, 3rd edn., pp. 137–216. Academic Press, London (2012)
3. Krig, S.: Interest point detector and feature descriptor survey. In: Computer Vision Metrics: Survey, Taxonomy, and Analysis, pp. 217–282. Apress (2014)
4. Ojala, T., Pietikäinen, M., Maenpaa T.: Multiresolution gray-scale and rotation invariant texture classification with local binary patterns. IEEE Trans. Pattern Anal. Mach. Intell. **24**(5), 971–987 (2002)
5. Tan, X., Triggs, B.: Enhanced local texture feature sets for face recognition under difficult lighting conditions. IEEE Trans. Image Process. **19**(6), 1635–1650 (2010)
6. Pietikäinen, M., Hadid, A., Zhao, G., Ahonen, T.: Local binary patterns for still images. In: Computer Vision using Local Binary Patterns, pp. 13–47, Springer, London (2011)
7. Zhao, G., Pietikäinen, M.: Dynamic texture recognition using local binary patterns with an application to facial expressions. IEEE Trans. Pattern Anal. Mach. Intell. **29**(6), 915–928 (2007)
8. Hikkilä, M., Pietikänen, M., Scmid, C.: Description of interest regions with center-symmetric local binary patterns. In: 5th Indian Conference, ICVGIP 2006, pp. 58–69. Springer, Berlin, Heidelberg (2006)
9. Ren, J., Jiang, X., Yuan, J., Wang, G.: Optimizing LBP Structure for visual recognition using binary quadratic programming. IEEE Signal Process. Lett. **21**(11), 1346–1350 (2014)
10. Wang, K., Bichot, C.E., Zhu, C., Li, B.: Pixel to patch sampling structure and local neighboring intensity relationship patterns for texture classification. IEEE Signal Process. Lett. **20**(9), 853–856 (2013)
11. Murala, S., Wu, Q.M.J.: Local mesh patterns versus local binary patterns: biomedical image indexing and retrieval. IEEE J. Biomed. Health Inform. **18**(3), 929–938 (2014)
12. Murala, S., Wu, Q.M.J.: Local ternary co-occurrence patterns: a new feature descriptor for MRI and CT image retrieval. Neurocomputing **119**, 399–412 (2013)
13. Qi, X., Shen, L., Zhao, G., Li, Q., Pietikänen, M.: Globally rotation invariant multi-scale co-occurrence local binary. Pattern Image Vis. Comput. **43**, 16–26 (2015)
14. Murala, S., Maheswari, R.P., Balasubramanian, R.: Local tetra patterns: a new feature descriptor for content-based image retrieval. IEEE Trans. Image Process. **21**(5), 2874–2886 (2012)
15. Kaya, Y., Ertuğrul, Ö.F., Tekin, R.: Two novel local binary pattern descriptors for texture analysis. Appl. Soft Comput. **34**, 728–735 (2015)

16. Dubey, S.R., Singh, S.K., Singh, R.K.: Local wavelet pattern: a new feature descriptor for image retrieval in medical CT databases. IEEE Trans. Image Process. **24**(12), 5892–5903 (2015)
17. Dubey, S.R., Singh, S.K., Singh, R.K.: Local bit-plane decoded pattern: a novel feature descriptor for biomedical image retrieval. IEEE J. Biomed. Health Inform. **99** (2015)
18. Dubey, S.R., Singh, S.K., Singh, R.K.: Local diagonal extrema pattern: a new and efficient feature descriptor for CT image retrieval. IEEE Signal Process. Lett. **22**(9), 1215–1219 (2015)
19. Gupta, R., Patil, H., Mittal, A.: Robust order-based methods for feature description. In: IEEE Conference on Computer Vision and Pattern Recognition, pp. 334–341. San Francisco (2010)
20. Haralick, R.M., Shapiro, L.G.: Conditioning and labeling. In: Computer and Robot Vision, pp. 303–357, Addison-Wesley Longman Publishing Co. Inc., Boston (1992)
21. DICOM Library. http://www.dicomlibrary.com/
22. The Cancer Imaging Archive (TCIA) Public Access. https://wiki.cancerimagingarchive.net/display/Public/TCGA-PRAD
23. Digital Imaging and Communications in Medicine (NEMA) Database. ftp://medical.nema.org/medical/Dicom/Multiframe/
24. The Cancer Genome Atlas, National Cancer Institute, National Human Genome Research Institute. https://tcga-data.nci.nih.gov/tcga/tcgaHome2.jsp
25. The Cancer Imaging Archive (TCIA) Public Access. https://wiki.cancerimagingarchive.net/display/Public/RIDER+NEURO+MRI

A Novel Approach for Multimodal Biometric System Using Iris and PalmPrint

Yakshita Jain and Mamta Juneja

Abstract Biometric systems, which are being adopted as the most effective solution for security breaches, are pattern recognition systems which identify or verify the person based on their physical/behavioural traits. Multimodal biometric systems, being more reliable and accurate than unimodal systems, are gaining much popularity these days. This paper focuses on review of multimodal systems based on iris and palmprint. Also a new scheme is proposed for extraction of palmprint image from unconstrained background. Then, that image is fused with iris, and their combination is used for verification purpose. The approach mainly relies on IFCM technique for palmprint and RED algorithm for iris feature extraction. Score-level fusion is used for combining two modalities and utilizing hamming distance for generating matching scores for both the traits. Combination of iris and palmprint is a very powerful biometric trait because of the individual strengths and uniqueness of both the traits.

Keywords Biometric · Multimodal · Security · Iris · Palmprint
Intuitionistic · Background extraction

1 Introduction

Biometric systems are analytical systems that identify or verify an individual by analyzing his behavioural or physical characteristics. With the advancement in technology, it is getting difficult for traditional security methods like I-cards, badges and passwords to provide sufficient level of security and protect vital information from imposters. Even unimodal biometric systems sometimes fail to serve the matter. As unimodal biometric systems depend upon one biometric trait only,

Y. Jain (✉) · M. Juneja
UIET/CSE Department, PU Chandigarh, India
e-mail: yakshi.sliet@gmail.com

M. Juneja
e-mail: mamtajuneja@pu.ac.in

© Springer Nature Singapore Pte Ltd. 2018
K. Saeed et al. (eds.), *Progress in Advanced Computing and Intelligent Engineering*,
Advances in Intelligent Systems and Computing 563,
https://doi.org/10.1007/978-981-10-6872-0_8

they suffer from issues such as noisy or incorrect sensor data, dearth of individuality, high error rates, non-universality, spoofing attacks and lack of invariant representation [1, 2]. Considering any field today, from forensics to e-banking, issuing driving license to even entering to any office or country, security has become the most important aspect. As the threat of imposter breaching in the system increases, the methods of providing security must also get updated. So, people shifted from unimodal systems to multimodal systems as it uses two or more biometric modalities to complete their desired function. Multimodal systems have over-ruled many of the complications that unimodal systems suffered from.

The biometric modalities, on which the functioning of a biometric system mainly depends, can be listed among following two groups, i.e. behavioural and physical. The inherent, very stable and time-invariant type of traits of an individual are his physical traits, for example palmprint, footprint, iris, hand geometry, retina, fingerprint, height, hand vein, face and ears. Whereas the one depending on the habits or behaviour of the person is his behavioural traits, for example voice, signature, keystroke, walking speed, arm or leg movement and gait [3]. These biometric traits, when used in combination for multimodal biometric systems, can be fused at three different fusion levels [2, 4], those are:

I. Feature extraction level fusion: First possible level of fusing biometric modalities being used in the system is feature extraction level. As the raw data collected from sensors is the richest source of features/information and if fusion is done at this level, it gives the best results for the verification and identification process. But, this level of fusion is also the most difficult one as different sensors produces data in different form, they may or may not be compatible with each other for fusion of sensor data. Similarly, features extracted from different modalities can be in various forms, their compatibility must also be checked before their fusion.

II. Matching score-level fusion: Next level where fusion can be done is matching score level, where the scores generated by the matching classifiers for various feature vectors are fused instead of the feature vectors themselves. This method of fusion is the most used one till date, as this method is rich in information and easy to fuse as well. Matching scores of different feature vectors are generated using classifiers independently using their corresponding template stored in the database, and then, these scores values are fused to obtain a new matching score that can be further utilized by decision module for accepting or rejecting the individual's identity.

III. Decision-level fusion: Last possible level of fusing modalities is at decision level of the system. Decision for different modalities is taken independently depending on their matching scores. Then, these decisions are fused to take the final decision for the acception or rejection using schemes like majority voting. This level of fusion is the easiest one to imply but does not work well with real-time constraints. All these discussed level of fusions are demonstrated in Fig. 1 as well.

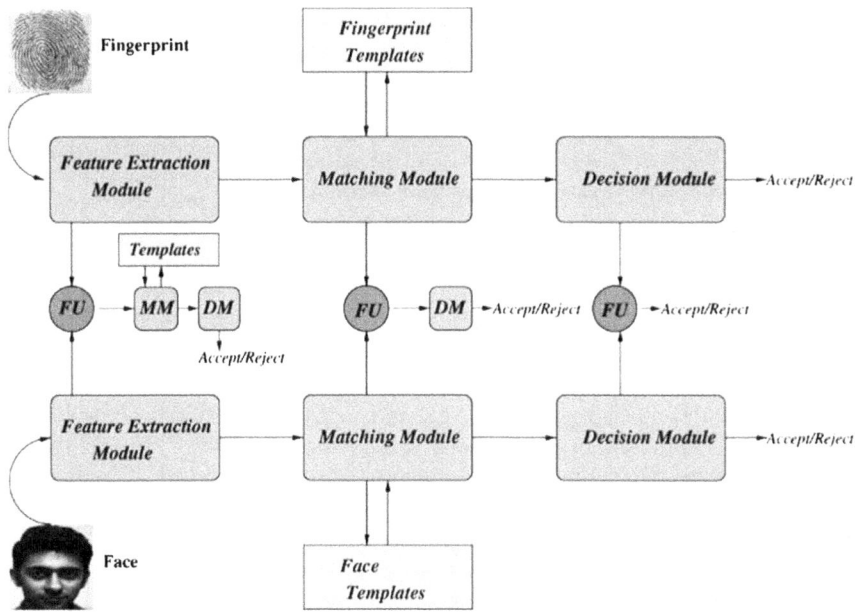

Fig. 1 Depicting various levels of fusion (FU: fusion module, MM: matching module, DM: decision module) [4]

In this paper, various techniques used in multimodal biometric systems based on iris, and palmprint are reviewed. A comparative analysis is made considering various aspects like level of fusion used, fusion technique used and various accuracy parameters. Then, a novel approach for extracting background from complex background hand images is presented. These extracted images of palm are then used along with iris for verification of the individuals.

Biometric systems work in two different modes depending on the need of the application. Those two modes are identification mode and verification mode [5, 6]. Identification mode is when the system compares the given biometric sample against all the templates present in the system's database to find out the unknown identity of the given trait. This mode of operation is very complex and time taking, but it is very helpful in negative recognition [7] in crucial areas like forensics and criminal cases. Whereas in verification mode, the system helps to confirm identity of the already known individual by comparing the given biometric trait against the template stored in the database along with that identity. This mode of operation takes less time as less number of comparisons is to be made. Applications like laptop or phone security system, attendance system, entry security in offices and e-banking are the examples of verification mode. The biometric system presented in this paper will work and get checked for results in future in verification mode.

2 Literature Review

The round-shaped flower like portion of human eye is called iris which is surrounded by pupil and sclera on both sides. Iris is among the most accurate biometric modalities being used since last decade. Many researchers like J. Daugman, Leonard Flom, Aran Safir and many more have worked and are still working great in developing new methods for iris recognition. J. Daugman [8] developed the most successful algorithm for iris recognition till date with an accuracy of 99.9% and very low FAR/FRR, i.e. 0.01/0.09. But this algorithm got commercial and hence very expensive, also it was very time consuming. Iris was first used as biometric trait successfully in 1987 by Flom and Safir [9]. After that many techniques like encoding iris code using 2-D Gabor filters [8], circular hough transform [10], RED (ridge energy detection) method [11, 12] and many more. Among these various algorithms, RED algorithm gained popularity after J. Daugman's presented algorithm. A major problem with iris recognition using all the above-discussed methods was its failure in unconstrained environment. But in 2009, Tan et al. [13] proposed a solution to this problem with a clustering-based algorithm for iris localization and integro-differential constellation pupil extraction. After that many other methods were presented to solve the same problem like 1-D and 2-D wavelet-based techniques [14], K-mean clustering and circular hough transform along with canny edge detector-based algorithm [15], Fuzzy c-mean clustering-based algorithm [16]. Among these, Fuzzy c-mean-based algorithm performed better as it considered membership functions for dividing the clusters.

On the other hand, palmprint is rather new in biometric field, but has many advantages over many other biometric traits. It can provide more information/features as compared to fingerprint. Sensors and hardware required for palmprint are cheaper than iris or retina like traits [17, 18]. Palmprint has so many features hidden in it which are grouped under five names, i.e. texture, line, geometric, point and statistical features. Combination of these features gives a very high accuracy rate in security systems. Some of the techniques developed to use these features include work done by Jain et al. [19] in 2001 who used prominent principle lines with feature points in palm region. 2-D Gabor filters were used for feature extraction in [20]. Sobel operator [21, 22], HMM (Hidden Markov Model) classifier [23], etc. were used for line features extraction. Techniques including PCA (principle component analysis) and ICA (independent component analysis) [24], DCT (discrete cosine transform) [25], Fourier transform [26], Scale-invariant feature transform for contactless images [27], Contourlet transform [28] and many more were used for extracting texture features of palmprint. The results of above-mentioned methods showed that texture features of palmprint give most accurate results among all five types of features.

Iris and palmprint both are very effective and reliable biometric traits but both have some limitations as well. Combining these two traits together can rule out their limitations and can develop a highly accurate and reliable security system. Some of the work done on the fusion of these two modalities includes the algorithm developed by Wu et al. [29] in 2007, which resulted in 0.012% MTR and 0.006% EER. Author used score-level fusion based on sum and product techniques. Another method using feature-level fusion based on wavelet packet transform technique developed by Hariprasath and Prabakar [30] in 2012 gave 93% accuracy rate. In the same year, R. Gayathri and Ramamoorthy [31] also used feature-level fusion using wavelet-based technique for extracting texture features generating an accuracy rate of 99.2% and FAR of 1.6%. After that, Kihal et al. [32] proved that quality of the image being used highly effects the results of the biometric system in 2014. They proved their point by working on three different datasets, performing all three levels of fusion on texture features of iris and palmprint. S.D. Thepade et al., in 2015, worked in transform domain [33] using Haar, Walsh and Kekre transform for extracting texture features and then performed score-level fusion proving that kekre transform works better with a GAR of 51.80 (approx.). Apurva et al. [34], on the other hand, worked in spatial domain using RED algorithm for iris and harris feature extraction algorithm for palmprint focusing on geometric features of palm. They used decision-level fusion for finalizing the results of their biometric system. Table 1 presents a comparative study of these algorithms discussed above.

Table 1 Comparision between fusion algorithms

Author	Dataset size (persons)	Fusion level	Fusion method	Parameters	Values
Wu et al. [29]	120	Score-level fusion	Sum, product, maximum, minimum strategies	MTR EER	0.012% 0.006%
Hariprasath and Prabakar [30]	30 (iris) 20 (palmprint)	Feature-level fusion	Wavelet Packet transform, concatenation	Accuracy	93.00%
R. Gayathri and Ramamoorthy [31]	125	Feature-level fusion	Wavelet-based technique	Accuracy FRR	99.2% 1.6%
Kihal et al. [32]	200	Feature fusion, score fusion, decision fusion	Concatenation, sum rule method, error fusion	GAR FAR[1] FAR[2]	100% 2.10^{-3} % 4.10^{-4} %
Thepade and Bhondave [33]	10	Score-level fusion	Mean square error method	GAR	50.20 (Walsh) 51.80 (Kekre) 50.20 (Haar)
Dhawale and Kale [34]	7	Decision-level fusion		RR	100% (iris) 100% (palmprint)

[1]FAR value for fusion of iris and CASIA palmprint database [32]
[2]FAR value for fusion of iris and PolyU palmprint database [32]

3 Proposed Work

A novel approach is presented in this paper for recognizing human's iris in an unconstrained environment. On the other hand, for palmprint segmentation, a new approach proposed for background extraction using IFCM (intuitionistic Fuzzy c-mean) algorithm. This will help in extracting human's hand image from any kind of unconstrained background making the system more suitable for real-time security applications. Figure 2 represents the flow diagram of major steps included in the proposed method. As it is a multimodal system, it will work on two modalities which are iris and palmprint. Both the traits are very unique and rich in feature information. But both are quite different kind of modalities and their feature sets are also very different so it will be very difficult and complex to fuse them at feature extraction level. The major steps that will be followed in the proposed method are:

I. Extract the red channel out for the hand image as red channel contains most of the important information of the image. So, it can be used individually for the background extraction purpose. This is done to reduce the overheads during background extraction process and making it fast. This is one of the pre-processing steps of the method.

II. IFCM algorithm is then applied to the extracted red channel of hand image for dividing the image into two clusters: one for background and one for human hand. This technique is chosen for clustering because it is assumed to give better results than Fuzzy c-mean algorithm for our purpose because it considers both the membership as well as non-memberships functions for creating the clusters in the image. The image will be divided into two clusters based on the intensity values, after that on the basis of the membership function of belongingness to each cluster, it is decided that which part of the

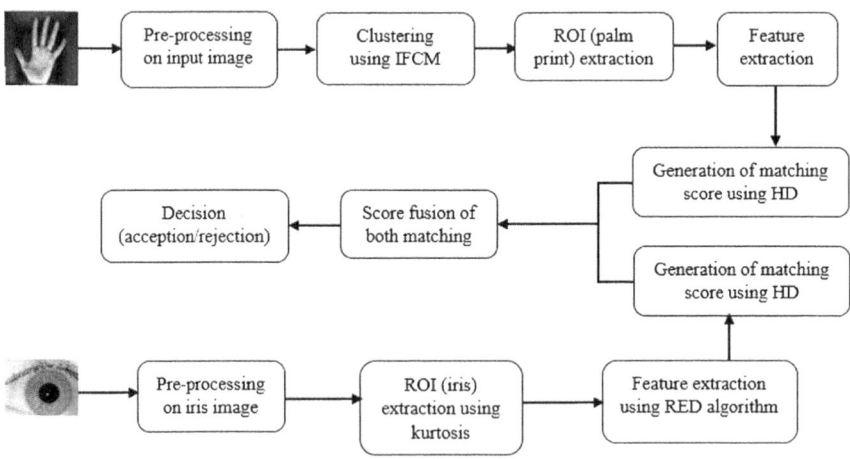

Fig. 2 Demonstrating the sequence of steps in proposed method

     image belongs to which cluster [16]. This technique performs better than Fuzzy c-mean algorithm as it rules out some of the difficulties faced by Fuzzy c-mean in creating clusters precisely by making use of non-membership functions as well.

III. ROI (palmprint) is then extracted out from clustered hand image using some morphological operations and enhancement algorithms. Palmprint region is full of different types features that are geometrical, statistical, line, point and textural. In this method, main focus will be kept on texture features and line features. Texture features are proved to give most accurate results among all five types of features, and line features are also found to be very unique such as principle lines. Extracted features will then be matched with the template stored in the dataset, and matching score will be generated using hamming distance method. These scores will then be stored.

IV. Iris image will then be operated on with various pre-processing steps like localization, normalization and then enhancement. These steps are very important for extracting an iris portion out correctly from an eye image. Iris segmentation is the step on which the recognition results of the system will depend upon. So, iris segmentation must be done accurately and carefully for best results. In our system, pre-processing will be done according to RED (Ridge energy detection) algorithm [11], i.e. some binary morphological operations for finding the centre and radius of pupil and then with the help of that and kurtosis (local statistics), finding the outer boundaries of iris. For this, image is converted to its polar coordinates with 21 possible centre of references around pupil. With this, the outer boundary will get oriented horizontally, and the best fit among these will determine the radius and centre of pupil. After this step, feature extraction can be performed easily.

V. RED algorithm will then again transform the iris image into polar coordinates, and then, features will be extracted using horizontal and vertical filtering and the extracted features/information will then be compared against the already retained template in the dataset using hamming distance method. These matching scores will again be stored.

VI. Both the generated matching scores will then be fused using sum rule and product rule, and then, final matching score will be generated. Depending on that matching score, final decision will be made for the verification process. If the results match successfully that means the person is verified to be the authentic person. Otherwise, he is not the authentic person and trying to be someone else or trying to hide his original identity.

4 Software and Datasets

We will be using Windows 10 with MATLAB 13a on an Intel core i5 processor. We will be using two different datasets of palmprint for experimenting the background extraction technique: Those will be COEP palmprint database and touchless palmprint database version 1.0 provided publically by IIT Delhi. This will be done to make sure that any kind of unconstrained background can be extracted using this technique. COEP palmprint database is publically available dataset maintained by College of Engineering, Pune consisting of palmprint samples of 167 different people with 8 different instances of same person. IIT Delhi touchless palmprint database is composed of left- and right-hand images from more than 230 subjects having 5 hand image instances from each of the hand. For palmprint feature extraction and matching process, IIT Delhi palmprint database will be used. For feature extraction of iris and its matching process, IITD Iris image database version 1.0 will be used. This dataset contains 2240 images acquired from 224 different users having 10 different instances of each user.

5 Conclusion

As we know, biometric systems are gaining very much importance for providing sufficient security to the vital information these days. Multimodal biometric system is the current trend in security systems. Iris and palmprint, the two traits considered in this paper are very efficient, unique and reliable biometric modalities. This paper provides a short review on both the modalities and their fusion, also a novel approach is presented for using palmprint and iris together for verifying the identity of any individual. In this approach, IFCM is used for extracting hand image from unconstrained background, morphological operations for extracting the ROI (palmprint) and RED algorithm for iris feature extraction. Score-level fusion is used for fusing both the modalities, and then, final decision is taken. As IFCM technique is not used till now for this purpose and is an improvement in Fuzzy c-mean algorithm so it is expected to give better results than previous work done in this area. This technique will be experimented for results on three databases, i.e. palmprint database provided by COEP, touchless palmprint database version 1.0 provided by IIT Delhi and Iris Image Database version 1.0 provided by IITD.

References

1. www.slideshare.net/piyushmittalin/multimodal-biometric-systems
2. Ross, A., Jain, A.K.: Multimodal biometrics: an overview. In: 2004 12th European Signal Processing Conference. IEEE (2004)
3. Sheena, S., Sheena, M.: A Study Of Multimodal Biometric System 93–97 (2014)

4. Ross, A., Jain, A.: Information fusion in biometrics. Pattern Recogn. Lett. **24**(13), 2115–2125 (2003)
5. Deshpande, S.D.: Review paper on introduction of various biometric areas. Adv. Comput. Res. **7**(1), 212 (2015)
6. Gupta, A., Mahajan, M.S.: An efficient iris recognition system using DCT transform based on feed forward neural networks (2015)
7. Arefin, M.M., Hamid, M.E.: A Comparative Study on Unimodal and Multimodal Biometric Recognition
8. Daugman, J.: How iris recognition works. IEEE Trans. Circuits Syst Video Technol **14**(1), 21–30 (2004)
9. Flom, L., Safir, A.: Iris recognition system, U. S. Patent 4641349 (1987)
10. Ma, L. et al.: Efficient iris recognition by characterizing key local variations. IEEE Trans. Image Process. **13**.6 (2004): 739–750
11. Ives, R.W. et al.: Iris recognition using the ridge energy direction (RED) algorithm. In: 2008 42nd Asilomar Conference on Signals, Systems and Computers. IEEE (2008)
12. Memane, M.M., Ganorkar, S.R.: RED algorithm based iris recognition. genetics 1 (2012): 2
13. Tan, T., He, Z., Sun, Z.: Efficient and robust segmentation of noisy iris images for non-cooperative iris recognition. Image Vis. Comput. **28**(2), 223–230 (2010)
14. Santos, G., Hoyle, E.: A fusion approach to unconstrained iris recognition. Pattern Recogn. Lett. **33**(8), 984–990 (2012)
15. Sahmoud, S.A., Abuhaiba, I.S.: Efficient iris segmentation method in unconstrained environments. Pattern Recogn. **46**(12), 3174–3185 (2013)
16. Kaur, N., Juneja, M.: A Novel Approach for Iris Recognition in Unconstrained Environment. J. Emerg. Technol. Web Intell. **6**(2), 243–246 (2014)
17. Chakraborty, S., Bhattacharya, I., Chatterjee, A.: A palmprint based biometric authentication system using dual tree complex wavelet transform. Measurement **46**(10), 4179–4188 (2013)
18. Patel, J.K., Dubey, S.K.: Deployment of palm recognition approach using image processing technique. IJCSI Int. J. Comput. Sci. Issues **10**(2) (2013)
19. Duta, N., Jain, A.K., Mardia, K.V.: Matching of palmprints. Pattern Recogn. Lett. **23**(4), 477–485 (2002)
20. Zhang, D et al.: Online palmprint identification. IEEE Trans. Pattern Anal. Mach. Intell. **25**(9), 1041–1050 (2003)
21. Han, C.-C. et al.: Personal authentication using palm-print features. Pattern Recogn. **36**(2), 371–381 (2003)
22. Wong, K.Y.E. et al.: Palmprint identification using Sobel operator. In: 2008 10th International Conference on Control, Automation, Robotics and Vision, ICARCV 2008. IEEE (2008)
23. Wu, X., Wang, K., Zhang, D.: HMMs based palmprint identification. In: Biometric Authentication, pp. 775–781. Springer, Berlin, Heidelberg (2004)
24. Connie, T. et al.: Palmprint recognition with PCA and ICA. In: Proceedings of Image and Vision Computing, New Zealand (2003)
25. Wong, K.Y.E., Sainarayanan, G., Chekima, A.: Palmprint identification using discrete cosine transform. In: World Engineering Congress (2007)
26. Li, W., Zhang, D., Zhuoqun, X.: Palmprint identification by Fourier transform. Int. J. Pattern Recognit Artif Intell. **16**(04), 417–432 (2002)
27. Morales, A., Ferrer, M., Kumar, A.: Improved palmprint authentication using contactless imaging. In: 2010 Fourth IEEE International Conference on Biometrics: Theory Applications and Systems (BTAS). IEEE (2010)
28. Butt, M. et al.: Palmprint identification using contourlet transform. In: 2nd IEEE International Conference on Biometrics: Theory, Applications and Systems (BTAS, 2008). IEEE (2008)
29. Wu, X. et al.: Fusion of palmprint and iris for personal authentication. In: Advanced Data Mining and Applications, pp. 466–475. Springer, Berlin, Heidelberg (2007)
30. Hariprasath, S., Prabakar, T.N.: Multimodal biometric recognition using iris feature extraction and palmprint features. In: 2012 International Conference on Advances in Engineering, Science and Management (ICAESM). IEEE (2012)

31. Gayathri, R., Ramamoorthy, P.: Feature level fusion of palmprint and iris. IJCSI Int. J. Comput. Sci. Issues **9**(4), 194–203 (2012)
32. Kihal, N., Chitroub, S., Meunier, J.: Fusion of iris and palmprint for multimodal biometric authentication. In: 2014 4th International Conference on Image Processing Theory, Tools and Applications (IPTA). IEEE (2014)
33. Thepade, S.D., Bhondave, R.K.: Bimodal biometric identification with Palmprint and Iris traits using fractional coefficients of Walsh, Haar and Kekre transforms. In: 2015 International Conference on Communication, Information & Computing Technology (ICCICT). IEEE (2015)
34. Dhawale, M.A.D., Kale, K.V.: Fusion of Iris and Palmprint Traits for Human Identification

Ultrasound Thyroid Image Segmentation, Feature Extraction, and Classification of Disease Using Feed Forward Back Propagation Network

U. Snekhalatha and V. Gomathy

Abstract The aim and objective of the study are to segment the ultrasound image of thyroid gland using PCA-based segmentation method and to extract the geometric and statistical features and to classify the disease using feed forward back propagation network. Thirty patients with thyroid disorder and thirty age- and sex-matched normal were included in this study. Ultrasound image of thyroid gland for all the patients was obtained. Automated segmentation algorithm using interclass variance analysis method was used for segmentation of thyroid gland. The geometrical features such as area and volume and statistical texture features were obtained from the segmented output image. Finally, the classification of thyroid disease was performed using feed forward back propagation network. The thyroid parameters correlated significantly with the feature extracted parameters for normal and abnormal cases. The feature extracted parameters show increased value for abnormal thyroid gland compared to normal. The developed automated image segmentation algorithm provides a quantitative analysis for estimation of area and volume of thyroid gland. The feed forward back propagation network provided the sensitivity of 85.71%, specificity of 95%, and accuracy of 91% in classification of abnormal thyroid gland and normal.

Keywords Thyroid gland · Ultrasound · Principle component analysis
Feature extraction · Multilevel wavelet transform

U. Snekhalatha (✉)
Department of Biomedical Engineering, SRM University,
Kattankulathur, Chennai 603203, Tamil Nadu, India
e-mail: sneha_samuma@yahoo.co.in

V. Gomathy
Department of Biomedical Engineering, Jerusalem College of Engineering,
Anna University, Chennai, Tamil Nadu, India
e-mail: gomathyvasu@gmail.com

© Springer Nature Singapore Pte Ltd. 2018
K. Saeed et al. (eds.), *Progress in Advanced Computing and Intelligent Engineering*,
Advances in Intelligent Systems and Computing 563,
https://doi.org/10.1007/978-981-10-6872-0_9

1 Introduction

The thyroid gland is a butterfly-shaped gland located in front of the neck and encases around the trachea. The thyroid hormone helps to regulate the body metabolism. The thyroid produces three active hormones namely: thyroxine (T3), triiodothyronine (T4), and thyroid stimulating hormone. These hormones assist in regulating the body metabolism, growth, and development. In worldwide, the most common causes of thyroid disorders are hyperthyroidism and hypothyroidism [1]. One-third of world population was affected by thyroid disorders [2]. In India, it has been estimated that 42 million people suffer from thyroid diseases [3]. Around 60% of individuals affected by thyroid disease are unaware of their condition. Women are five to eight times more likely affected than men to have thyroid problems [4].

The thyroid disease was diagnosed using numerous imaging modalities such as ultrasonography (USG), computed tomography (CT), magnetic resonance imaging (MRI), and optical coherence tomography. The US imaging modality is often preferred over other imaging techniques such as CT and MRI because of its mobility and cost-effectiveness. The thyroid ultrasound is used to detect the changes in appearance, size of organs, and abnormal masses [5].

Several researchers used various image segmentation methods such as local region-based active contour method, active contour models, variable background active contour model, and morphological operations to segment the thyroid gland [6–10]. The aim and objective of the study are to segment the ultrasound image of thyroid gland using principle component analysis (PCA)-based segmentation method and to extract the geometric and statistical features and to classify the disease using feed forward back propagation network.

2 Methodology

2.1 Patients

Thirty patients with thyroid disorder undergoing treatment at Kamatchi Hospital and research center, Chennai and 30 age- and sex-matched normal has participated in this study. The patients with throat cancer, throat inflammation, infection, cysts, diabetes, collagen disease, and heart diseases were excluded from the study. Patients with known thyroid disease and those with different degrees of thyroid dysfunction were included in the study. All the subjects who participated in this study had given the signed consent form.

2.2   Ultrasound Image Acquisition

The thyroid gland was scanned using the ultrasound real-time scanners (Logic 400 CL, GE Medical Systems, Korea) with 6–10 MHz linear array transducers which operated at 8 MHz frequency. The ultrasound examination was performed at neck region to diagnose the lump in the thyroid gland and the sonographer used minimal probe pressure after applying gel on the skin. All the subjects were requested to be in a supine position with their neck region extended by keeping the proper cushion under their shoulders. The ultrasound transducer is applied with the coupling gel and moved over the subject's neck region for acquiring the ultrasound image of the thyroid gland. The images were processed using MATLAB software version 8.5.1

2.3   Segmentation Algorithm and Feature Extraction

The ultrasound image of thyroid gland was segmented using the following algorithm:

   i. The principle component analysis (PCA) was used in preprocessing steps for dimensionality reduction.
   ii. The Gaussian filter was applied to remove the speckle noise.
   iii. The morphological closing operation is performed to remove unwanted redundancy.
   iv. The interclass variance analysis method was used for segmentation of thyroid gland.
   v. The area and volume were estimated from the segmented thyroid gland.
   vi. After segmentation, the feature extraction was performed using multiwavelet decomposition transform.
   vii. Three level of decomposition was performed. The low–high sub-band and high–low sub-band were used for further analysis.
   viii. The texture features such as energy, contrast, correlation, homogeneity, entropy, skewness, and kurtosis were extracted from the segmented thyroid gland.

The feature extraction is used to find the set of features which could clearly distinguish the normal and abnormal thyroid gland. The statistical features such as energy, contrast, correlation, homogeneity, and entropy were extracted from the segmented gray output image of the ultrasound. The energy provides the information concerning the randomness of the spatial distribution. It also quantifies the feature that measures the smoothness level in the image. Lower energy and reduced smoothness level are achieved in the uniformly distributed pixels, whereas the increased smoothness and energy level present in the case of non-uniform distributed region [11]. Contrast is measured as the average intensity level difference between the neighboring pixels. Correlation is defined as the linear dependencies of the intensity levels of the neighboring pixels. Homogeneity attains its maximum value, when the all pixels in the image are same. Entropy is defined as the statistical average of information that can be used to characterize the texture of the input image. The low-entropy images have little contrast, whereas the high-entropy images have high contrast level.

2.4 Feed Forward Back Propagation Network

The feed forward back propagation neural network contains the following layers: (i) input layer, (ii) one or more hidden layer, and (iii) output layer. It is trained by adjusting the weights to perform perfect classification using back propagation algorithm. The network used binary sigmoid activation function to scale the hidden layer and output layer. The input features are normalized between '0' and '1' before feeding to the network. The desired output is indicated as '0' for normal and '1' for abnormal. The classification process is categorized into four phases as follows: (a) training phase, (b) testing phase, (c) validation phase, (d) resultant phase. By using the input patterns, learning repetitions is performed with different training and validation data sets.

2.5 Statistical Analysis

The statistical analysis was accomplished by using SPSS software package version 19.0 (SPSS Inc., Chicago, USA). The correlations between thyroid parameters and feature extraction parameters in RA group were executed using the Pearson correlation. The student's t-test was performed to find the statistical difference between the normal and abnormal group.

3 Results and Discussion

The descriptive detail of thyroid gland parameters in normal and abnormal cases was indicated in Table 1. Among the three thyroid parameters, T3 exhibits a highly significant correlation between the thyroid parameters, geometric parameters, and feature extracted parameters. The geometrical features such as area and volume were increased significantly in abnormal cases compared to normal. By analyzing the statistical features, the parameter energy exhibits the highest percentage difference between the abnormal and normal thyroid gland. Hence, high energy in abnormal thyroid gland indicates the severity of the disease.

From the descriptive detail given in the Table 1, it was evident that mean T3 and T4 values were low, whereas mean Tsh value was high which indicates

Table 1 Descriptive details of thyroid gland parameters in normal and abnormal cases

Features	Normal (N = 30)	Abnormal (N = 30)	% difference	Statistical significance
i. Biochemical features				
T3 (ng/dl)	120 ± 15	65 ± 12	−84.61	0.0001
T4 (µg/dl)	5.27 ± 0.5	3.5 ± 0.2	−50.57	0.002
Tsh (µIU/l)	2.03 ± 0.8	8.5 ± 0.5	76.11	0.005
ii. Geometrical features				
Area	41.94 ± 4.5	77.4 ± 3.9	45.81	0.0012
Volume	75.8 ± 13.7	77.31 ± 16.1	1.95	0.05
iii. Statistical features				
Energy	0.13 ± 0.06	0.16 ± 0.09	18.75	0.001
Correlation	0.72 ± 0.09	0.78 ± 0.08	7.69	0.002
Homogeneity	0.4 ± 0.002	0.46 ± 0.006	13.04	0.004
Entropy	6.07 ± 0.5	6.36 ± 0.5	4.77	0.003

Table 2 Correlation between thyroid parameter, feature extracted parameters, and geometric parameters for the total population (N = 60)

Thyroid parameters	Feature extracted parameters					Geometric parameters	
	Energy	Contrast	Correlation	Homogeneity	Entropy	Area	Volume
T3	0.40	0.54	0.46	0.53	0.44	0.73	0.72
T4	0.57	0.71	0.92	0.66	0.57	0.35	0.35
Tsh	0.33	0.32	0.42	0.30	0.29	0.31	0.34

hypothyroidism in abnormal case compared to normal. The correlation between the biochemical parameters, features extraction parameters, and geometric parameters for the total population was indicated in Table 2.

Figure 1a depicts the input ultrasound image of the thyroid gland taken in the anterior view. Figure 1b indicates the Gaussian filtered and preprocessed image using principle component analysis method. Figure 1c illustrates the segmented image from the preprocessed image using morphological operations. Figure 1d indicates the final segmented image using interclass variance analysis method. From the obtained confusion matrix as given in Fig. 2, it was evident that the feed forward back propagation network provided the sensitivity of 85.71%, specificity of 95%, and accuracy of 92% in the classification of the abnormal thyroid gland and normal. The ROC curve plotted for various stages such as training, testing, and validation stage was given in Fig. 3.

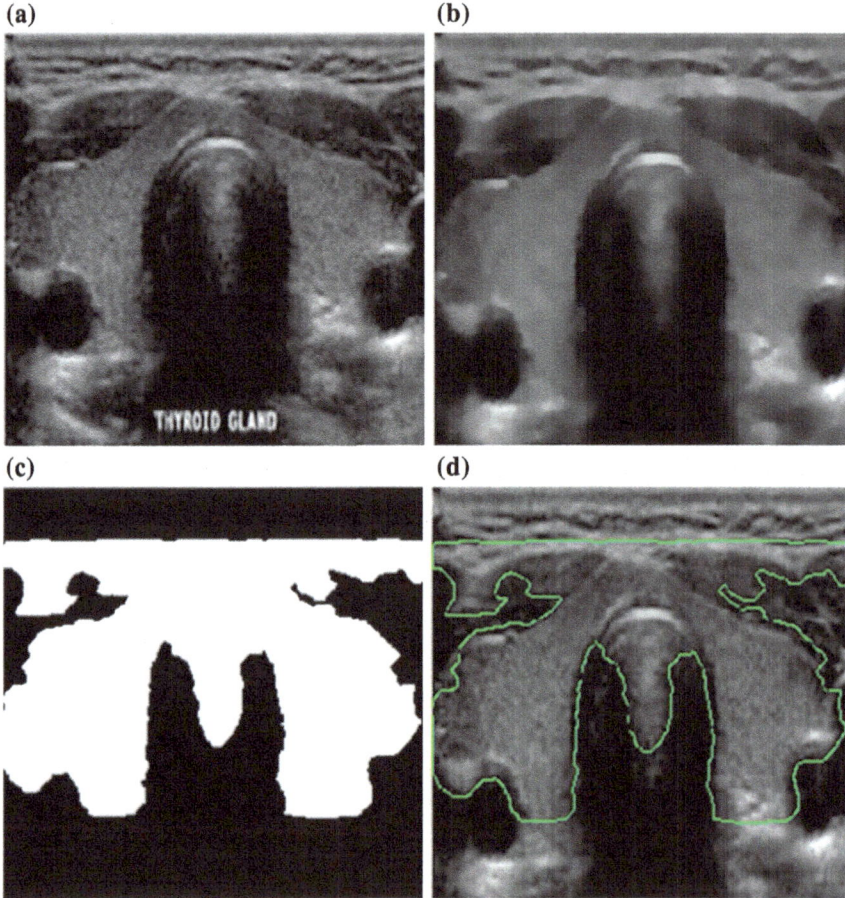

Fig. 1 **a** Input ultrasound image of thyroid gland, **b** the preprocessed image using principle component analysis (PCA), **c** the segmented image obtained using morphological operations, and **d** the final segmented image using interclass variance analysis method

Our study predicted that there was a significant association of thyroid parameters with the feature extracted parameters and geometric parameters. We found that reduced T3 and T4 values and increased TSH values in the abnormal group compared to the normal group which indicates the condition of hypothyroidism. It was examined that the geometrical features and statistical feature values in the abnormal group were increased compared to the normal group. Ertek et al. demonstrated in their studies that the serum zinc levels positively correlated with the thyroid volume in nodular goiter patients. Also, the serum zinc levels provided significant correlation with free T3 in subjects with the normal thyroid [12]. Milionis investigated that there is a significant positive correlation found between the thyroid parameters and body mass index (BMI) in women of euthyroid individuals

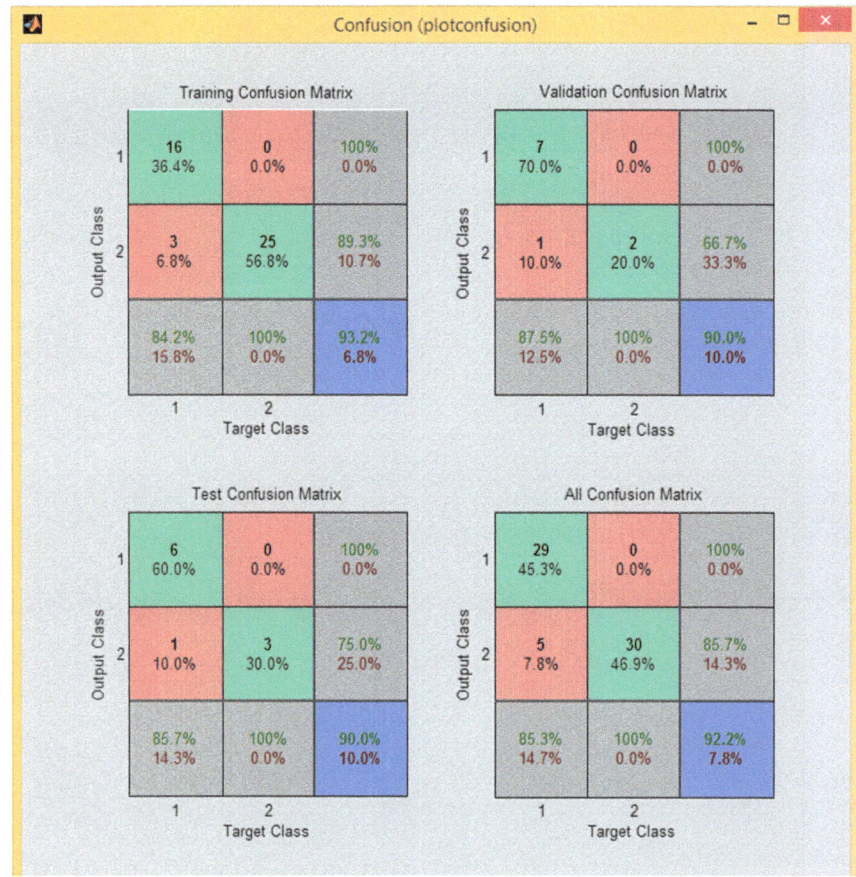

Fig. 2 Shows the confusion matrix for various stages such as training, testing, and validation stage

[13]. In contrast, other studies depicted that there is no relationship between the thyroid parameters and BMI in euthyroid persons or patients with subclinical hypothyroidism [14, 15]. In our study, the biochemical parameters significantly correlated with the feature extracted parameters and geometrical parameters.

Several researchers have determined and confirmed that the plausibility of automated methods was superior to the manual method in classifying the thyroid ultrasound image [16–19]. The described algorithm in this study involves automated segmentation, feature extraction, and neural network classification. The US is one of the vital imaging modality presently used for the assessment of thyroid disorders. But its image resolution and contrast was affected by multiplicative speckle noise which is the fundamental characteristics of US. Principle component analysis was used in the preprocessing stage to avoid speckle noise effectively. The

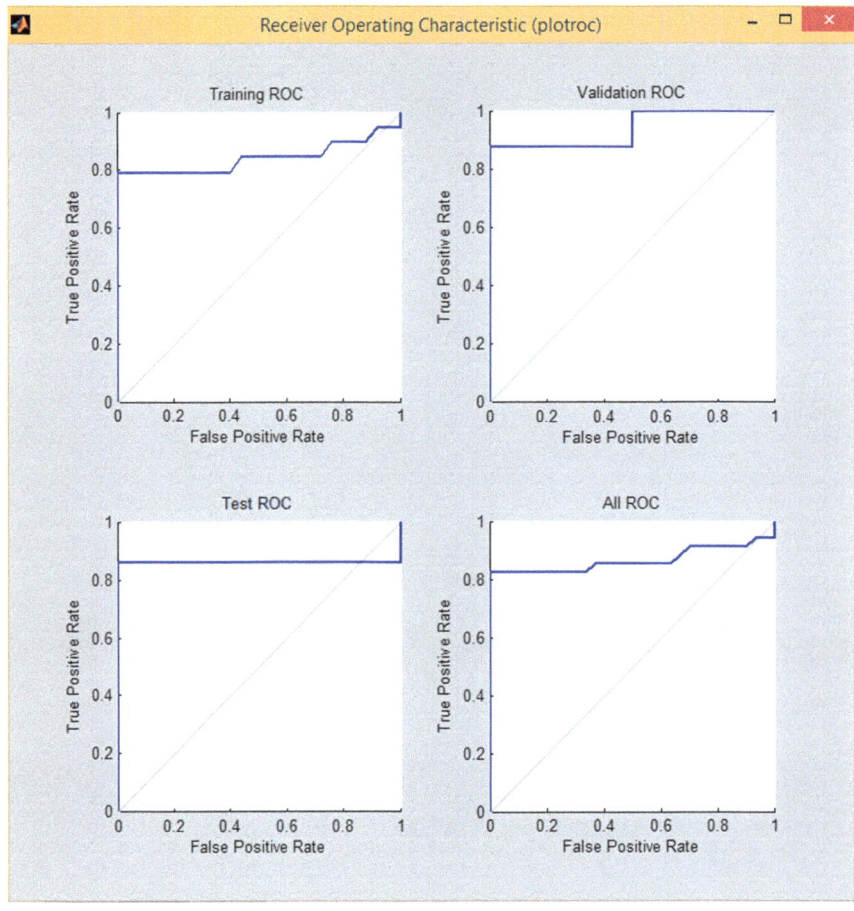

Fig. 3 Depicts the ROC curve plotted for various stages such as training, testing, and validation stage

ROI-based segmentation of thyroid gland using interclass variance analysis method provides better segmentation of gland region from the background compared to active contour models. The morphological operations employed in this process prevented the distortion from the background and smoothing the region. Singh et al. used Bayesian and KNN classifiers for classification of tumor in the thyroid gland and obtained the accuracy of 46.15% for Bayesian classifier and 38.46% for KNN classifier [6]. In our study, the feed forward BPN was used for classification of normal and abnormal thyroid gland which provided the accuracy of 92%. The errors caused due to manual method and the need for trained technician to analyze the US was eliminated by utilization of automated method.

4 Conclusion

In conclusion, the thyroid parameters correlated significantly with the feature extracted parameters for normal and abnormal cases. The feature extracted parameters show increased value for the abnormal thyroid gland compared to the normal. The developed automated image segmentation algorithm provides a quantitative analysis for estimation of area and volume of the thyroid gland. The feed forward back propagation network provided the sensitivity of 85.71%, specificity of 95%, and accuracy of 91% in the classification of the abnormal thyroid gland and normal.

Declaration Authors have obtained all ethical approvals from appropriate ethical committee for study on human subjects.

References

1. Vanderpump, M.P.J.: The epidemiology of thyroid disease. Br. Med. Bull. **99**, 39–51 (2011). https://doi.org/10.1093/bmb/ldr030
2. Zimmerman, M.B.: Iodine Deficiency. Endocr. Rev. **30**, 376–408 (2009)
3. Unnikrishnan, A.G., Menon, U.V.: Thyroid disorders in India: an epidemiological perspective. Indian J Endocrinol Metab. **15**, S78–S81 (2011). https://doi.org/10.4103/2230-8210.83329
4. http://www.thyroid.org/media-main/about-hypothyroidism/. Accessed 17 July 2015
5. Koundal, D., Gupta, S., Singh, S.: Computer aided diagnosis of thyroid nodule: a review. IJCSES. **30**, 67–83 (2012)
6. Singh, N., Jindal, A.: Ultra sonogram images for thyroid segmentation and texture classification is diagnosis of malignant (cancerous) and benign (non cancerous) nodules. IJEIT. **1**, 202–206 (2012)
7. Sahu, P.K., Bhawnani, D.K.: Thyroid segmentation and area measurement using active contour. IJEAT. **3**, 303–307 (2014)
8. Garg, H., Jindal, A.: Segmentation of thyroid gland in ultrasound image using neural network. In: Proceedings of Fourth International Conference on Computing, Communications and Networking Technologies. IEEE Explore. https://doi.org/10.1109/ICCCNT.2013.6726797
9. Iakovidis, D.K., Savelonas, M.A., Karkanis, S.A., Maroulis, D.E.: A genetically optimized level set approach to segmentation of thyroid ultrasound images. Appl. Intell. **27**, 193–203 (2007)
10. Kaur, J., Jindal, A.: Comparison of thyroid segmentation algorithms in ultrasound and scintigraphy images. Int. J. Comput. Appl. **50**(23), 24–27 (2012)
11. Salvatore, D., Reagle, D.: Theory and Problem of Statistics and Econometric, 2nd edn. Schaum outline series. Mcgraw-hill (2012)
12. Ertek, S., Cicero, A.F.G., Caglar, O., Erdogan, G.: Relationship between serum zinc levels, thyroid hormones and thyroid volume successful iodine supplementation. Hormones **9**, 263–268 (2010)
13. Milionis, A., Milionis, C.: Correlation between body mass index and thyroid function in euthyroid individuals in Greece. ISRN biomarkers 2013, Article ID 651494 (2013). http://dx.doi.org/10.1155/2013/651494

14. Manji, N., Boelaert, M.C., Sheppard, M.C., Holder, R.L., Gough, S.C., Franklyn, J.A.: Lack of association between serum TSH or free T4 and body mass index in euthyroid subjects. Clin. Endocrinol. **64**, 125–128 (2006)
15. Figueroa, B., Velez, H., Irizarry-Ramirez, M.: Association of thyroid–stimulating hormone levels and body mass index in overweight Hispanics in Puerto rico. Ethn. Dis. **18**, 151–154 (2008)
16. Chang, C.Y., Lei, Y.F., Tseng, C.H., Shih, S.R.: Thyroid segmentation and volume estimation in ultrasound images. IEEE Trans. Biomed. Eng. **57**, 1348–1357 (2010)
17. Mahmood, N.H., Rusli, A.H.: Segmentation and area measurement for thyroid ultrasound image. IJSER **2**, 1–8 (2011)
18. Lankton, S., Tannenbaum, A.: Localizing region based active contours. IEEE Trans. Image Process. **17**, 2029–2039 (2008)
19. Li, C., Xu, C., Gui, C., Fox.: Distance Regularized level set evolution and its application to image segmentation. IEEE Trans. Image Process. **19**, 3243–3254 (2010)

Recent Advancements in Detection of Cancer Using Various Soft Computing Techniques for MR Images

Ravindra Kr. Purwar and Varun Srivastava

Abstract Cancer is a lethal disease if not detected in an early stage. This paper presents an outline of different types of cancers and recent advancement in soft computing techniques for their detection. It focuses on how different image processing techniques are optimized using neural networks, fuzzy logic, and genetic algorithms to detect cancers.

Keywords Artificial neural network (ANN) · Cancer/tumor detection · Fuzzy logic · Genetic algorithms · Support vector machines (SVMs)

1 Introduction

Cancer has been a deadly disease from ages, and various image processing techniques like histograms, segmentation, and object detection are applied to identify cancer from mammograms [1]. Cancer is basically uncontrolled growth of cells in any organ. Thus, many researchers are also trying to identify abnormal mitosis situations so that cancer can be identified in an early stage itself [2]. However, with advancements in medical imaging techniques, better quality mammograms are generated, and more precise detection is made possible. Researchers these days are modifying these techniques further for faster and easier detection by combining these techniques with some soft computing techniques like neural network, SVMs, Principal Component Analysis (PCA), etc. [3].

R. K. Purwar (✉) · V. Srivastava
University School of Information and Communication Technology, GGSIPU,
New Delhi 110078, India
e-mail: ravindra@ipu.ac.in

V. Srivastava
e-mail: varun0621@gmail.com

© Springer Nature Singapore Pte Ltd. 2018   99
K. Saeed et al. (eds.), *Progress in Advanced Computing and Intelligent Engineering*,
Advances in Intelligent Systems and Computing 563,
https://doi.org/10.1007/978-981-10-6872-0_10

We discuss for various types of cancers, what all developments are made in recent years and where this research is heading to. Along with cancer, we also discuss anomalies that lead to cancer like rapid cell growth, abnormal mass deposition, etc. The paper is organized in the following sections. In Sect. 2, first we discuss the underlying technique for detection of any type of cancer. Thereby in following subsections different types of cancer detection techniques that are lately evolved are explained. Each subsection is concerned with a single type of cancer and its recent detection techniques. It is then followed by various results, discussions, and conclusion which discuss the use of different soft computing algorithms in cancer detection.

2   Materials and Methods

The various steps involved in detection of tumor involve preprocessing of the image to remove noise and enhance the cancer-affected area. Then, Segmentation allows us to extract features so that we can easily classify a portion having cancer by using various machine-learning classifiers. Figure 1 explains the process step-by-step.

2.1   *Generic Cancer Detection Techniques*

Cancer still is a deadly disease and is curable only if it gets detected in an early age [4]. Throughout the world, various doctors and researchers are working on various techniques for its early detection. Xu et al. [5] combined Particle Swarm Optimization (PSO) and ANN for reducing the dimensions that are used in identifying cancer.

The method implicated 80% classification accuracy. Ziaei et al. [6] used perceptron for cancer detection but the technique had limitations same as that of a perceptron in a classification network. Takhahashi et al. [7] combined ART, fuzzy, and sweep operator method for detection of leukemia and brain tumor. There have been attempts in identifying different types of cancers, e.g., Cho et al. [8] classified leukemia, colon, and lymphoma cancers using back propagation algorithm. Various fuzzy techniques have been employed for eliminating features recursively and thereby selecting the genes that are having tumor information [9].

Adali et al. [10] proposed that if we study micro-RNAs for early detection of cancer. The gene patterns in RNAs were studied, and back propagation algorithm was then used for detection. Since cancer shows its symptoms first in RNA and so it can be detected earlier using this method.

Weng et al. [11] also presented an interesting survey which showed that various analysis techniques that were used considered different datasets and different parameters for calculating the accuracy thereby compared the performances of various neural hybrid classifiers used for cancer prediction. They compared the techniques statistically and showed that ensemble classifiers are better than solo classifiers for most datasets.

Image Acquisition & Enhancement

- Obtain the MRI image dataset from an authentic source
- Enhance the mammograms so that cancer affected area can be easily identified.

Morphological Image Processing & Image segmentation

- Opening and closing operations are applied for boundary extraction
- Segmentation techniques are applied for extracting the affected area.

Object recognition and Feature Extraction

- Affected part is identified
- Various features like area, horizontal and vertical cell length etc. are identified to feed into machine learning systems for further classification.
- These features are then used for Image indexing in mammograms

Fig. 1 Steps in detection of cancer/tumor

Further, Belciug et al. [12] proposed neural networks with error correction using Bayesian paradigm. The techniques were then compared with other simple techniques that use neural network for classification of diseases like cancer, heart diseases, etc., and it was found to be better. Author considered feed forward multilayer neural networks along with Bayesian model, and Bayesian model was used for updating weights. They considered six medical datasets, and the error was found to be 0.05.

2.2 Breast Cancer Detection Techniques

Though image processing techniques are gaining popularity, human readable technique is more accurate [13]. Thereby, researchers have achieved progress in getting accurate and early results for breast cancer detection. Munklang et al. [14] used a fuzzy co-occurrence matrix and formulated a system of 14 features. He considered calcification, well-defined/circumscribed masses, architectural distortion, and speculated masses as few of the abnormalities. Fuzzy co-occurrence matrix shows better performance as compared to the gray-level co-occurrence matrix.

Gorkhan et al. [15] worked in the area of 3D template matching and ANNs for cancer and lesion detection in breast. Hassanien et al. [16] applied fuzzy type-II algorithms for preprocessing and then an adaptive ant-based segmentation technique for segmentation. Then, twenty statistical features were extracted and classified using perceptron neural classifier. Adaptive algorithm employed an advanced goodness function for segmentation. This technique was found to be better than already existing classical ant-based segmentation. Murat et al. [17] used a priori to remove the correlated features and thus producing a reduced dataset and a faster classification mechanism. For the classification among benign and malignant, Kullback Leiber classification method was used [18]. The method was used for breast cancers primarily.

Garro et al. [19] proposed techniques of classification of microarrays using neural networks in their research work. They also used ABC algorithm for reducing the dimensionality of the microarrays and thereby make the computational speed faster. This DNA microarray classification is the latest technique for the prediction of many diseases since it is very much accurate.

Bhardwaj et al. [20] introduced a combined approach for classification of cancer in which the neural architecture was obtained using genetic algorithm. This hybrid genetically optimized neural architecture was applied on Winconsin database of cancer, and the proposed algorithm reached from 98 to 100% accuracy for different datasets from the database. A Pixcal Refined Bandwidth algorithm was also proposed by Santra et al. [21] which scans the pixels to identify the microcalcification region. After that, a classification is done for normal/abnormal region.

J. Dheeba et al. [22] applied particle swarm optimization for extracting features and then wavelet neural networks for the identification of cancer. Sensitivity and specificity were used as parameters to judge the extent of correct classification. After preprocessing of the image, various kernels were chosen for extraction of features like level, edge, spot, wave, and ripple.

2.3 Liver Tumor Detection Techniques

Sharma and Kaur [23] combined PSO and Seeker optimization algorithm for image segmentation and edge detection in mammograms. SVMs, k-nearest neighbor, etc., have been used for identifying cancer when a dataset is given but for extracting the features in these datasets, we need various feature extraction algorithms of image processing. The connection between type-II diabetes and liver cancer was diagnosed and established by chi-square prediction model by Rau et al. [24]. Gender and age were also the factors that lead to the establishment of a connection between the two.

Huang et al. [25] used a cubic $5 * 5 * 5$ voxel to store features including texture, intensity, Sum and difference histogram, and trained ELM classifiers for further identification of liver cancer. The proposed technique used a one-class classifier and thus was extremely fast.

S.S. Kumar and Dr. Moni [26] used curvelet transform for the identification of liver cancer. First, a liver was segmented using adaptive threshold decision and morphological processing, and then, a neural network was trained to identify if the cancer is benign or malignant.

2.4 Lung Cancer Detection Techniques

Kuruvilla et al. [27] studied lung cancer in details and presented an algorithm where parameters like standard deviation, skewness kurtosis, fifth central moment, etc., were used, and the data was trained using feed forward back propagation network for classification of cancer. The traingdx function gave maximum accuracy. Feed forward back propagation networks are found to give better results as compared to feed forward neural networks. They also identified that skewness increased the accuracy by 5–8%.

Wu et al. [28] applied markers for detection of tumor in lungs. A technique m(RMR) KL was proposed by Korkmaz et al. [18] where the classification among benign and malignant cancers was made out. The approach used was found to be 98.3% correct.

Recent development shows the use of DNA microarrays in classification problems. ANNs and other soft computing techniques like radial basis function, etc., can be used to classify these microarrays which further can yield various characteristics about the sample in test. Likewise, Fernandes et al. [29] in their research focused on lung cancer which leads to the highest deaths in the world and considered 14 biomarkers to detect cancer. Biomarkers included age, gender, smoking habits, cooking with wood, etc.

2.5 Mass Detection Techniques

In case of cancer, since cells starts accumulating at a single area and so Mahersia et al. [30] developed intelligent systems based on Bayesian architecture and neural fuzzy model for such mass identification. In this case, first wavelet decomposition was applied followed by feature vector extraction. The pectoral muscles were removed using wavelet decomposition, and then, Bayesian rule-based back propagation algorithm was applied for further classification. These were found to be better than ANFIS used till date as a classifier.

Divyadarshini et al. [31] also worked in the detection of mammographic masses by first removing noise using a median filter and Contrast-Limited Adaptive histogram equalization and then using gray-level thresholding for segmentation.

2.6   Brain Tumor Detection Techniques

Various techniques have been used in past for detection of brain tumor and its classification as either benign or malignant. Simple watershed transformation and thereby extraction of features through various morphological processing techniques were applied in the past to detect it. The features can then be fed to a neural network for their further classification as the type of tumor in concern. Human brain is symmetrical about mid-sagittal, and its asymmetry may reflect the presence of the tumor.

Latest advancements in the detection of brain tumor in particular had shown the optimization of basic image processing algorithms using soft computing concepts. Mancas et al. [32] used watershed in an iterative format for segmentation of brain tumor and thereby highlighting the affected area. Iscan et al. [33] applied Zernike moments for the identification of brain tumor. First symmetry axis was determined using moments, and then, Zernike moments were applied separately to left and right hemisphere. Asymmetry denoted the presence of tumor. 2D CWT and ISNN both were applied by the author for the identification, and WAi and ND2 estimated the presence of the disease.

Arizmendi et al. [34] used wavelet transforms for identifying brain tumor and also reduced the features by using PCA. He used DWTs for image preprocessing and then applied ANN for the detection in mammograms.

Joshi et al. [35] used various feature extraction techniques on MRI images and thereby proposed a Neuro-fuzzy classifier for cancer detection.

Saha et al. [36] in his work used fast bounding box method for the identification of tumor. Their change detection process included a Bhattacharya coefficient which used histograms to speed up the process. Chan-Vese algorithm followed by FBB method yielded better results than Knowledge-based thresholding technique.

Nabizadeh et al. [37] identified the stroke-based lesion and tumor lesion from MRI images using histogram-based gravitational optimization algorithm with an accuracy of 91.5 and 88.1%. Here, histogram was made, and local maxima were identified. Upper cutoff border of the nth segment is then connected to lower cutoff border of n + 1th segment. Initial population and number of iterations were the convergence factor. Alomari et al. [38] thereby worked with circular algorithm to identify the rate of proliferation in brain cells for identification of tumor. He proposed a PRECAD system which had an accuracy of about 98.3%. The counting was based on circularity features, and accordingly, the proliferation rate was estimated. Color-based transformation, segmentation, and preprocessing being the other steps involved. A customized color modification is required for this approach. Four features, viz., use of dynamic iterative number, tolerance for degree of overlapping, irregular circular detection, and compatibility with high-resolution images were some of the key features of their approach.

Aslam et al. [39] improved Sobel edge detection algorithm for brain tumor segmentation. They proposed a closed region algorithm which finds a starting pixel and then expands for each region for identification. Eight neighbours in a 5 * 5 rectangular boundary are checked. Thereby, a closed contour search is developed. Vishnuvarthanan [40] further extended their approach and did dimensionality reduction and then applied SOM for centroid allocation.

Nazibadeh et al. [41] presented an integrated automated framework that is able to detect MR images and then segment the tumor implemented on T1-w, and FLAIR sequences separately were developed. After segmentation, gabor wavelets and statistical approach were used to extract features, and then, the refined features formed the basis of tumor identification.

Shanthakumar and Ganeshkumar [42] in their research work presented a neuro-fuzzy approach and modified the ANFIS classifier for brain tumor detection. An enhanced seed selection method was also proposed by them, and algorithm outperformed over others in factors like sensitivity, Similarity Index, OF, etc.

Fuzzy rule base is also used, and its membership function was then updated using the centroid found. Fuzzy K-mean algorithm is then applied for clustering of similar type of cells and thereby obtaining the segmented tumor region.

Table 1 Summary of various techniques used for cancer detection

Type of cancer	Recent advancements
1. Liver cancer	Watershed transform used
	Features like texture, intensity, sum, and difference histogram are used
	Combination of PSO and seeker optimization applied
	Connection with diabetes detected by chi-square model
2. Breast cancer	Particle swarm optimization is used
	Kullback Leiber classification proposed
	Classification of microarray used
	Fuzzy classifiers are applied
3. Brain tumor	Watershed transform used iteratively
	Fast bounding box method proposed
	Dimensionality reduction and then SOM applied for feature extraction
	Histogram-based gravitational optimization algorithm proposed for tumor lesion
	Sobel edge detection and circular algorithms for identifying cell proliferation
4. Lung cancer	Fourteen biomarkers to detect tumors which included age, smoking habits, etc.
	parameters like standard deviation, fifth central moment, etc., are used
	A technique m(RMR) KL is used

In the latest research approach, Chandra et al. [43] applied genetic algorithms for the detection of tumor. The initial population was obtained using K means which is then fed into genetic algorithm. A unique crossover method is proposed thereby to mix the characters which helps in identification further. Before applying GA, soft thresholding is done using wavelet transform, and the features extracted serve as input into genetic algorithm.

3 Results and Discussions

A lot of research has been carried out till date for detection of tumor but in past 3–4 years, many machine-learning algorithms have also been used for tumor detection. The milestones are presented in Table 1.

4 Conclusion

Various cancer detection methods are identified and categorized. Latest tools and techniques being used to optimization of analysis of images are also discussed. As we can see that a lot of research is going on for detection of different types of cancers, and thus, in future, following techniques can be explored:

1. Various image indexing methodologies for faster retrieval of images. 2. Image analysis using DNA microarrays for identifying different types of cancers. 3. Use of combined Neuro-DNA techniques to detect the presence of abnormal behavior by tissues.

References

1. Doi, K.: Current status and future potential of computer-aided diagnosis in medical imaging. Br. J. Radiol. **78**, 3–19 (2005)
2. Cirean, D.C.: Mitosis detection in breast cancer histology images with deep neural networks. In: Intervention MICCAI, pp. 411–418. Springer, Berlin, Heidelberg (2013)
3. Pentland, A., Picard, R.W., Sclaroff, S.: Tools for content based manipulation of image databases. In: Proceedings of SPIE: Storage and Retrieval of Image and Video Databases II 2185, pp. 34–47, Feb 1994
4. Amato, F., et al.: Artificial neural networks in medical diagnosis. J. Appl. Biomed. **11**(2), 47–58 (2013)
5. Xu, R., Cai, X., Wunsch II, D.C.: Gene expression data for DLBCL cancer survival prediction with a combination of machine learning technologies. In: Proceedings of the IEEE International Conference on Medicine and Biology, pp. 894–897 (2005)
6. Ziaei, L., Mehri, A.R., Salehi, M.: Application of artificial neural networks in cancer classification and diagnosis prediction of a subtype of Lymphoma based on gene expression profile. J. Res. Med. Sci. **11**(1), 13–17 (2006)

7. Takahashi, H., Murase, Y., Kobayashi, T.: Hiroyuki Honda New cancer diagnosis modeling using boosting and projective adaptive resonance theory with improved reliable index. Biochem. Eng. J. **33**, 100109 (2007)
8. Cho, S.-B., Won, H.-H.: Cancer classification using ensemble of neural networks with multiple significant gene subsets. Appl. Intell. **26**, 243–250 (2007). Springer Science and Business Media
9. Agrawal, S., Agrawal, J.: Neural network techniques for cancer prediction: a survey. In: 19th International Conference on Knowledge Based and Intelligent Information and Engineering Systems, Elseveir Procedia Computer Science, vol. 60, pp. 769–774 (2015)
10. Adala, T., Ekerolu, B.: Analysis of micrornas by neural network for early detection of cancer. Proc. Technol. **1**(449), 452 (2012)
11. Weng, C.-H., Huang, T.C.-K., Han, R.-P.: Disease prediction with different types of neural network classifiers. Telemat. Inform. **33**, 277292 (2016)
12. Belciug, S., Gorunescu, F.: Error-correction learning for artificial neural networks using the Bayesian paradigm. Appl. Autom. Med. Diagn. J. Biomed. Inform. **52**, 329337 (2014)
13. Ganesan, K., et al.: Computer-aided breast cancer detection using mammograms: a review. IEEE Rev. Biomed. Eng. **6**, 77–98 (2013)
14. Munklang, Y., Auephanwiriyakul, S., Theera-Umpon, N.: Examination of mammogram image classification using fuzzy co-occurrence matrix. Int. J. Tomogr. Simul. **28**(3), 82–95 (2015)
15. Ertasa, G., Gulcura, H.O., Osmanb, O., Ucanc, O.N., Tunacd, M., Dursund, M.: Breast MR segmentation and lesion detection with cellular neural networks and 3D template matching. Comput. Biol. Med. **38**, 116–126 (2008). Elseveir
16. Hassanien, A.E., Moftah, H.M., Azar A.T.: MRI breast cancer diagnosis hybrid approach using adaptive ant-based segmentation and multilayer perceptron neural networks classifier. Appl. Soft Comput. **14**, 6271 (2014). Elseveir
17. Murat Karabatak, M.: Cevdet Ince An expert system for detection of breast cancer based on association rules and neural network Elseveir. Expert Syst. Appl. **36**, 34653469 (2009)
18. Korkmaz, S.A., Korkmaz, M.F.: A new method based cancer detection in mammogram textures by finding feature weights and using KullbackLeibler measure with kernel estimation Optik, **126**, 25762583 (2015)
19. Beatriz, A., Garro, K.R., Vzquez, R.A.: Classification of DNA microarrays using artificial neural networks and ABC algorithm. Appl. Soft Comput. **38**, 548560 (2016). Elseveir
20. Bhardwaj, A.: Aruna Tiwari breast cancer diagnosis using genetically optimized neural network model. Expert Syst. Appl. **42**, 46114620 (2015)
21. Santra, A.K., Singh, W.J., Arul, S.D.: Detection of microcalcifications using Pixcals refined bandwidth algorithm in digitized mammograms. Int. J. Tomogr. Simul. **16**(W11), 81–90 (2011)
22. Dheeba, J., Albert Singh, N., Tamil Selvi, S.: Computer-aided detection of breast cancer on mammograms: a swarm intelligence optimized wavelet neural network approach. J. Biomed. Inform. **49**, 4552 (2014)
23. Akanksha, S., Parminder, K.: Optimized liver tumor detection and segmentation using neural network. Proc. Int. J. Recent Technol. Eng. (IJRTE) **2**(5), 7–10 (2013)
24. Rau, H.-H., Hsu, C.-Y., Lin, Y.-A., Atique, S., Fuad, A., Wei, L.-M., Hsu, M.-H.: Development of a web-based liver cancer prediction model for type II diabetes patients by using an artificial neural network. Comput. Methods Prog. Biomed. **125**, 5865 (2016)
25. Huang, W., Li, N., Lin, Z., Huang, G.B., Zong, W., Zhou, J., Duan, Y.: Liver tumor detection and segmentation using kernel-based extreme learning machine. In: 2013 35th Annual International Conference of the IEEE Engineering in Medicine and Biology Society (EMBC), pp. 3662–3665 (2013)
26. Kumar, S.S., Moni, R.S.: Diagnosis of liver tumor from CT images using curvelet transform. Int. J. Comput. Sci. Eng. **2**(4), 1173–1178 (2010)
27. Jinsa Kuruvilla, K.: Gunavathi Lung cancer classification using neural networks for CT images. Comput. Methods Prog. Biomed. **113**, 202209 (2014)
28. Yongjun, W., Yiming, W., Wang, J., Yan, Z., Qu, L., Xiang, B., Zhang, Y.: An optimal tumor marker group-coupled artificial neural network for diagnosis of lung cancer, Elseveir. Expert Syst. Appl. **38**, 1132911334 (2011)

29. Flores-Fernndez, J.M., Herrera-Lpez, E.J., Snchez-Llamas, F., Rojas-Calvillo, A., Cabrera-Galeana, P.A., Leal-Pacheco, G., Gonzlez-Palomar, M.G., Femat, R., Martnez-Velzquez, M.: Development of an optimized multi-biomarker panel for the detection of lung cancer based on principal component analysis and artificial neural network modeling. Expert Syst. Appl. **39**, 1085110856 (2012)
30. Mahersia, H., Boulehmi, H., Hamrouni, K.: Development of intelligent systems based on Bayesian Regularization network and neuro-fuzzy models for mass detection in mammograms: a comparative analysis. Comput. Methods Prog. Biomed., 46–62 (2015)
31. Divyadarshini, K., Vanithamani, R.: Segmentation of mammographic masses using gray level thresholding. Int. J. Tomogr. Simul. **28**(1), 22–32 (2015)
32. Mancas, M., Gosselin, B., Macq, B.: Fast and automatic tumoral area localisation using symmetry. In: ICASSP 2005 Proceedings. ICASSP 2005 (du 18/03/2005 au 23/03/2005), pp. 725–728 (2005)
33. Iscan, Z., Dokur, Z.: Tamer lmez Tumor detection by using Zernike moments on segmented magnetic resonance brain images. Expert Syst. Appl. **37**, 25402549 (2010)
34. Arizmendi, C., Vellido, A.: Enrique Romero classification of human brain tumours from MRS data using discrete wavelet transform and bayesian neural networks Elseveir. Expert Syst. Appl. **39**, 52235232 (2012)
35. Joshi, D.M., Rana, N.K., Misra, V.M.: Classification of brain cancer using artificial neural network. In: Proceedings of International Conference on Electronic Computer Technology (ICECT), pp. 112–116 (2010)
36. Saha, B.N., Ray, N., Greiner, R., Murtha, A., Zhang, H.: Quick detection of brain tumors and edemas: a bounding box method using Symmetry. Comput. Med. Imag. Graph. **36**, 95–107 (2012)
37. Nabizadeh, N., John, N.: Clinton Wright Histogram-based gravitational optimization algorithm on single MR modality for automatic brain lesion detection and segmentation. Expert Syst. Appl. **41**, 78207836 (2014)
38. Alomaria, Y.M., Abdullahb, S.N.H.S., Zinc, R.R.M., Omar, K.: Iterative randomized irregular circular algorithm for proliferation rate estimation in brain tumor Ki-67 histology images. Expert Syst. Appl. **48**, 111129 (2016)
39. Aslam, A., Khan, E., Beg, M.M.S.: Improved edge detection algorithm for brain tumor segmentation. Proc. Comput. Sci. **58**, 430–437 (2015)
40. Vishnuvarthanan, G., et al.: An unsupervised learning method with a clustering approach for tumor identification and tissue segmentation in magnetic resonance brain images. Appl. Soft Comput. **38**, 190–212 (2016)
41. Nabizadeh, N.: Miroslav Kubat brain tumors detection and segmentation in MR images: Gabor wavelet vs. statistical features. Comput. Electr. Eng. **45**, 286301 (2015)
42. Shanthakumar, P., Ganeshkumar, P.: Performance analysis of classifier for brain tumor detection and diagnosis. Comput. Electr. Eng. **45**, 302311 (2015)
43. Chandra, G.R., Rao, K.R.H.: Tumor detection in brain using genetic algorithm. Proc. Comput. Sci. **79**, 449–457 (2016)

Insight on Symmetrical Posture Detection of Human Using SIFT

Smita Kulkarni and Sangeeta Jadhav

Abstract It is proposed to develop a system for systematic evaluation process of human posture recognition in video sequences which is essential for many solicitations. Video frame has variant information such as different posture, gauge and brightness. This paper implements the algorithm Scale-Invariant Feature Transform (SIFT) to detect and recognize humans posture, in which image is invariant to scrabbling and rendition. This paper explores the application of the SIFT approach in the framework of human posture detection and can deal with the circumstantial noise disputes. The proposed detector is rotation invariant and achieves satisfactory human posture detection. Results are demonstrated based on the video frames, voice broadcast of weight lifting, footage from the movie and a cricket match, etc. To appraise the performance of the suggested algorithm, experiments have been conducted by employing an ANN classifier on a database.

Keywords Human posture · SIFT features · ANN

1 Introduction

The human action recognition (HAR) has evolved considerably in recent past years with an interest in the various analyses related to human action recognition, due to its versatile nature in diversified fields related to computing- and gaming-related activities. This paper relates to a very complex aim of designing and implementing a reliable and dependable system capable of recognizing various human actions which has many significant and influential applications which are automated surveillance systems, human computer interface, smart home, health care management, etc. This

S. Kulkarni (✉)
MIT Academy of Engineering, Alandi(D), Pune, India
e-mail: sskulkarni@entc.maepune.ac.in

S. Jadhav
Army Institute of Technology, Dighi, Pune, India
e-mail: djsangeeta@rediffmail.com

© Springer Nature Singapore Pte Ltd. 2018
K. Saeed et al. (eds.), *Progress in Advanced Computing and Intelligent Engineering*,
Advances in Intelligent Systems and Computing 563,
https://doi.org/10.1007/978-981-10-6872-0_11

paper addresses the problem of human activity recognition to a very genuine level of authenticity to advanced level of Industrial as well as Home automation. The purpose of an algorithm is to recognize actions which include various postures and activities related to sports and general daily-life activities from the input sequences from video. In this paper, we have used data sets from UCF [1] which include various sports-related clips, collected from channels which are broadcasted on BBC and ESPN. The paper confined towards the most important postures in human activity such as standing and bending which relates to multiple postures that can be programmed to the system and achieve very accurate results for various sensitive applications where accuracy in results is of upmost importance than recognition of humans. Several approaches are available in human activity recognition [2, 3]. To recognize motion patterns of human action, local spatio-temporal features [4] are used. Image frame features of spatial and the temporal relations progress recognition. To reduce features dimensionality of human image FS method [5]. Compressed video-based motion information is described for humane action [6]. By converting videos, frames into signal image by binary motion [7] human action flow can identify. The extraction of features which are appropriate for studying or recognizing activity is critical to detection of actions in video sequences. To collect interest points from video frames as suitable features for human action recognition implemented [8], it is suggested to use complex standard database. SIFT algorithm finds detection of feature point and feature description. This enriched SIFT algorithm can successfully avoid the noise points, structural distinct matching points, so this can largely improve the accuracy of image matching for gender [9], face [10] detection.

2 Proposed Work

The proposed algorithm is composed of following stages in Fig. 1 Proposed method focuses on choosing the best feature set for posture recognition. To detect human presence in video frames preprocessing and segmentation are implemented. Classifier is used to test the frame to recognize human posture.

2.1 *Preprocessing*

The preprocessing increases performance of video processing by removing noise and insignificant features. In this section, introduced adaptive median filter [11] for preprocessing is important for enhancing video frame features.

2.2 *Segmentation and Background Subtraction*

In background subtraction, a model is developed which covers simply the background sight which is stationary starved of any forefront objects. Object segmentation is defined as the given object or image is divided into its different region or

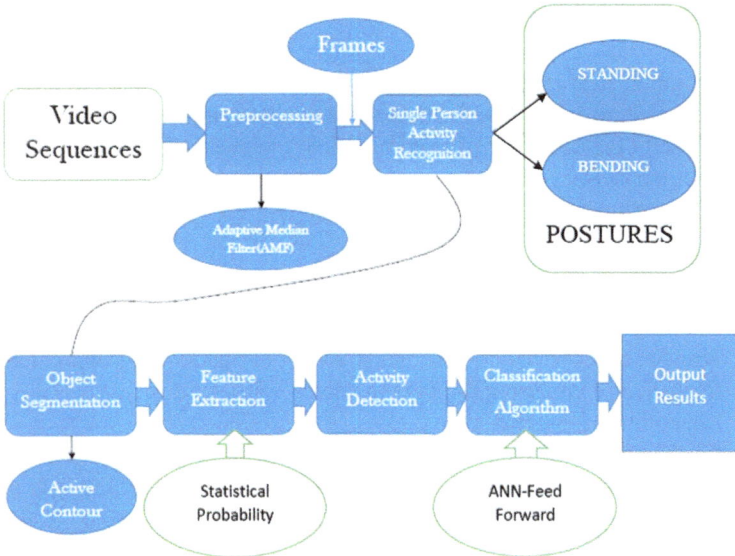

Fig. 1 Flow of posture detection

object, and it is done if the image pixel qualities/properties differ from the pixels of an image and also analyse it. Segmentation observes various types of objects within a given image. It also used to finds different regions of connected pixels having different properties. It also determines the boundaries between more than one region and avoids unwanted regions. Active contours [12] are used to decide region of human presence in the image frame.

2.3 Feature Extraction

The extraction of features should be appropriate for recognizing activity to detection of actions in video sequences. Visual features handle the following method for action detection.

2.3.1 Method I—Statistical Probability Parameter

Here we reduced the dimensionality of the data, so a high-level representation is built where a set of features is obtained and it is based on region and contour, and area and eccentricity are features of the region. Correlation, homogeneity and energy are features for contour. These features provide an approximate value of the time series data, and this is used to calculate many statistical features and measure different variable properties. To signify important features from video image frames, statistical probability parameters are extracted [13].

2.3.2 Method II—SIFT

In a video input, it is challenging to extract features from frames to identify the symmetrical posture due to illumination, scale, rotation and affine. To extract feature SIFT proposed [14] as extended features. SIFT transformed unknown image into collection of local visual descriptors with invariant properties. The objective of SIFT is to identify features in frames scale space that is invariant to translation, scaling, rotation and small distortion.

To implement SIFT [14], the following are the steps:

Step 1   Scale-Space Extrema Detection
         This phase takes effort to recognize locations and scales that are detectable from diverse visions of the same posture.
Step 2   Keypoint Localization
         Keypoints are selected based on the processes of their constancy for invariance to transformation. These locations are superior from removal point of low disparity [14].
Step 3   Orientation Assignment
         An orientation is then allotted to each key point established on local image frame features. Processes are accomplished to the allocated orientation, scale and location for each posture feature, providing invariance to these transformations [14].
Step 4   Keypoint Descriptor
         In this step, calculate a descriptor to make it invariant to remaining disparities.

2.4 Classifier: ANN (Feed Forward, Back Propagation)

ANN is established as model classifiers for recognition of symmetrical human posture. This technique is created by training for two symmetrical postures bending and standing. For training, feature vector is created by two methods, statistical probability and extend the feature from SIFT as input layer neuron. The model represents each posture class as bending and standing as output layer neuron. The testing of the model is done for these two classes.

3 Result and Discussion

Used video database of UCF [1] investigation results is described in this section. To execute the recognition of symmetrical human posture bending and standing, MATLAB software development tool is used.

3.1 Video Database

UCH [1] sport database with a lifting action is used. The database includes a total of 150 sequences with the resolution of 740×480. Frame rate of video is 10 fps. Through these videos, the proposed experiment is performed on database to recognize standing and bending postures.

3.2 Preprocessing

To extract proper features from video frames, preprocessing and segmentation are performed as shown in Fig. 2.

3.3 Features Extraction—SIFT

Figure 3 shows standing and bending images with keypoint vector using the SIFT algorithm [14].

Bending   Standing

Fig. 2 Preprocessing and segmentation

Fig. 3 Posture with SIFT features

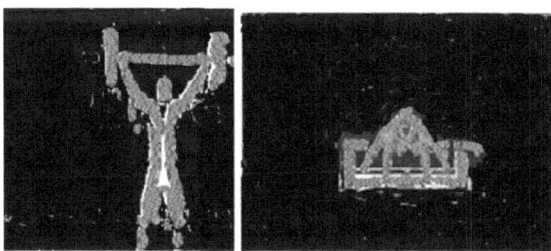

Table 1 Percentage accuracy, percentage sensitivity and percentage specificity

	Method I—statistical	Method II—SIFT
Accuracy (%)	86.32	93.26

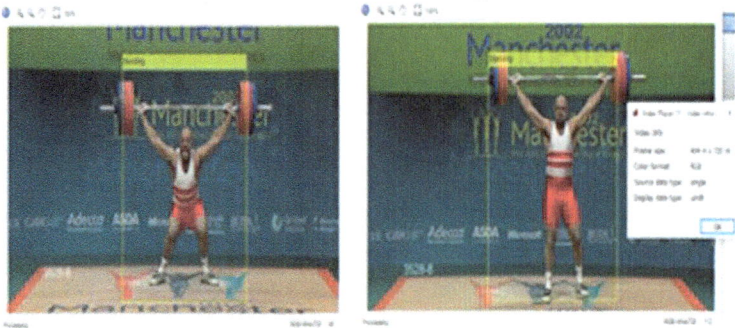

Fig. 4 Symmetrical posture bending and standing recognition

3.4 Classification Accuracy

Table 1 shows how classification results for two methods in which feature extraction by SIFT can improve the recognition rate of posture.

3.5 Symmetrical Posture Recognition

As shown in Fig. 4, classifier recognizes human symmetrical posture bending and standing with correct, accurate, using extended features from SIFT [14].

4 Conclusion

This paper has analysed the performance of recognition in human action for symmetrical posture bending and standing. In human activity recognition, we took the advantage of depth information in the feature description because of its insensitiveness towards illumination changes. In this paper, to extract features of human posture of UCF database statistical probability parameter, SIFT was implemented. For image frame classification, the ANN (feed forward, back propagation) was used. The experimental result shows the accuracy of the ANN on the test set image frames. From the achieved investigation, best result is marked using the SIFT method with accuracy of 93.26% was achieved.

References

1. http://crcv.ucf.edu/data/UCF_Sports_Action.php
2. Kanchan Gaikwad, M.S., Narawade, V.: HMM classifier for human activity recognition in computer science & engineering. Int. J. (CSEIJ) **2**(4), 27–36 (2012)
3. Wang, L., Cheng, L., Thi, T.H., Zhang, J.: Human action recognition from boosted pose estimation. In: IEEE International Conference on Digital Image Computing. Techniques and Applications, pp. 308–313 (2010)
4. Schuldt, C., Laptev, I., Caputo, B.: Recognizing human actions: a local SVM approach. In: IEEE Proceedings of the 17th International Conference on Pattern Recognition, pp. 32–36 (2004)
5. Gonzlez, S., Sedano, J., Villar, J.R., Corchado, E., Herrero, I., Baruque, B.: Features and models for human activity recognition. Neurocomputing **167**, 52–60 (2015)
6. Venkatesh, B.R., Ramakrishna, K.R.: Recognition of human actions using motion history information extracted from the compressed video. Image Vis. Comput. 597–607 (2004)
7. Dobhal, T., Sitole, V., Thomas, G., Navada, G.: Human activity recognition using binary motion image and deep learning. Proc. Comput. Sci. **58**, 178–185 (2015)
8. Moussa, M.M., Hamayed, E., Fayek, M.B., El Nemr, H.A.: An enhanced method for human action recognition. J. Adv. Res. **6**, 163169 (2015)
9. Zahedi, M., Yousefi, S.: Gender recognition based on sift features. Int. J. Artif. Intell. Appl. (IJAIA) **2**(3), 87–94 (2011)
10. Aly, M.: Face Recognition using SIFT Features. http://www.vision.caltech.edu
11. Sharma, N., Pal, U., Blumenstein, M.: Recent advances in video based document processing: a review. 10th IAPR International Workshop on Document Analysis Systems (DAS). IEEE (2012)
12. Kass, M., Witkin, A., Terzopoulos, D.: Snakes: active contour models. Int. J. Comput. Vis. **1**(4), 321–331 (1988)
13. Holalu, S.S., Arumugam, K.: Breast tissue classification using statistical feature extraction of mammograms. Med. Imaging Inf. Sci. **23**(3), 105–107 (2006)
14. Lowe, D.G.: Distinctive image features from scale-invariant keypoints. Int. J. Comput. **2**(60), 91–110 (2004)

Complete Forward and Reverse Text Transcription of Odia Braille and Hindi Braille

Vinod Jha and K. Parvathi

Abstract Braille translation software translates electronic documents written in different writing system of the world into Braille code. From these files, a refreshable Braille display can present tactile Braille or a Braille embosser can produce a hard copy on special paper. Few products are available which are economically costly and they mainly focus on national and international languages. There is a need for such products for regional languages of India also. Several researches are carried out on different languages to convert it into Braille. Odia Braille is one of the regional language in which moderate research has been carried out. This paper is focused on translation of electronic document of Odia into Braille as well as Braille to text. The generated Braille code is compatible with Braille embosser of any printer. The mapping table of Unicode of Odia letters and Braille code has been developed for translation. The translation algorithm is also tested on the national language Hindi for its effectiveness. The inverse translation of Braille to text also verified and found the results are satisfied.

Keywords Braille · Odia · Hindi · UTF-8

1 Introduction

There are 285 million blind and visually impaired people in the world as stated by WHO in 2014. These people also contribute to the society in some or the other way but they face stiff challenge to communicate with the rest of the world because they use a different kind of language called "Braille" for their study and communication. Braille is a system of reading and writing for visually challenged people developed by Frenchman Louis Braille. Every character or symbol in Braille is made up of six

V. Jha (✉) · K. Parvathi
School of Electronics Engineering, KIIT University, Bhubaneswar, India
e-mail: v.jha85@gmail.com

K. Parvathi
e-mail: kparvati16@gmail.com

© Springer Nature Singapore Pte Ltd. 2018   117
K. Saeed et al. (eds.), *Progress in Advanced Computing and Intelligent Engineering*,
Advances in Intelligent Systems and Computing 563,
https://doi.org/10.1007/978-981-10-6872-0_12

dots arranged in 3 × 2 matrix pattern. One or more dots are raised simultaneously to represent different characters of any language.

1.1 Introduction to Braille

All the measures provided in Fig. 2 are in inches. Braille is a system of writing that uses patterns of raised dots to inscribe characters on paper. It therefore allows visually—impaired people to read and write using touch rather than vision. A Braille cell for Indian scripts comprises of six dots. By using one or more dots at a time 64 letters (symbols) in any language can be represented. Braille dots are embossed on a particular type of sheet, and it is read by sensing the embossment over the Braille sheet by using fingers. Although the thickness of the sheet is also standardized, it varies place to place. Thickness of the Braille sheet is directly proportional to its life. Less the thickness, less the number of times it could be read. A standard Braille sheet consists of 40 lines with maximum 25 characters each.

1.2 Related Works

There is a need of system which can do the text transcription from Braille to other global languages. Many researches have been carried out in national and international languages for unidirectional or bidirectional conversion of texts into Braille. There is open source as well as paid software for the text transcription of English Braille. There are systems available that not only convert scanned Braille sheets into universal languages but read them also by incorporating text to speech converters. Process, which converts a Braille sheet into natural languages, is called optical Braille recognition or OBR. The OBR system comprises of two main steps [1, 2]: first is segmenting the Braille cells [3] and second is converting the Braille cell into corresponding natural language character. The first step may be accomplished by image preprocessing, denoising, morphological operation, and segmentation. One of the major problems which come up in segmentation of a scanned sheet is image skewing [4], which can be minimized by image rotation, Hough transformation and linear regression analysis. When images are taken using camera then along with image skewing, non-uniform illumination also creates problem while segmenting.

The second step comprises of dot extraction, dot pattern extraction, and character mapping. Accuracy of dot extraction depends completely on the success of segmentation of cells and later on segmentation of dots in a cell. Since the sizes of the cells are standardized, so if the segmentation of the cell is successful then chances of error in dot extraction are very less. However, presence of salt and pepper noise reduces the dot extraction success rate sharply.

Lots of works have been done for transcription of English [1], Arabian [5], Tamil [6], Bengali, Hindi, Kannada, etc., languages into Braille.

S. R. Rupanagudi et al. (2014) proposed a work for transcription of Kannada Braille for converting the obtained text into speech using Verilog HDL language on a Xilinx Spartan 3e series of FPGA [7].

A versatile tool has been created by Shiva Kumar and Ramakrishnan [6] to convert Tamil documents into RTF, XML, or BRF (Braille) file format. It is developed in JAVA using eclipse SWT and runs on Linux, Windows and MAC operating system. The system also suggests the possible words in case of error.

In [8], Jie Li et al. used Haar feature extraction and provided it in SVM for Braille character recognition. The algorithm uses the scanned Braille sheets and preprocesses for any geometrical corrections after converting it to gray scale. The sliding window crops each image into subimage whose Haar feature is being detected and passed on to the SVM to detect the dots thus converting the gray scale to binary image. The simple search technique is applied to the binary image to get the English characters from the Braille sheet. The Spider SVM, a MATLAB implementation followed with radial basis function makes the paper more compatible. Thus, the processing time is easily reduced here by using C implementation of SVM. This is one of the rare uses of classifier for Braille recognition.

There are similar works for Bengali [9] in which Bengali Braille is decoded into Bengali text and Gujarati [10] in which the authors discussed various works scanned Gujarati Braille sheets are converted to respective natural language. In [11], the author presented a tool which accesses, edits, and creates Devanagari Unicode text using digital Braille typewriter.

Although these papers does not discuss about the reverse transcription, still lots of work have been done in Bengali, Tamil, Gujarati, and Kannada languages. But there is scarcity of such researches for many other regional languages like Odia. In 2015, bijet Samal et al. [12, 13] published their work related to transcription of Odia letters into Braille letters and vice versa. The need of strong database and lack of full text conversion were the issues left to work upon. Moreover, almost all researches conducted so far emphasized either on text to Braille conversion or Braille to text conversion. This paper illustrates a bidirectional text transcription of printed Odia text into Braille text. The same process has been applied to Hindi document also for verification. This method can further be modified to work with scanned Braille documents as well as scanned Odia or Hindi documents. Main advantages of this method are its independence from any database and its ability to do both forward and reverse text transcription.

This method takes the text documents as input and reads it letter wise and map the letters into corresponding Braille dot patterns. Corresponding to the dot patterns, Braille cells are generated as an 27×18 image and are saved in a standard Braille sheet format. Using the obtained file a refreshable Braille display can present tactile Braille or a Braille embosser can produce a hard copy on special paper. All the work has been done on MATLAB platform with Nirmala UI Semilight font for display of all the characters.

2 Proposed Method

2.1 Text to Braille Conversion

Reading text file with UTF-8 as default encoding scheme
Conversion of UTF-8 code of each letter into its Unicode(hex code)
Mapping of the hex code to corresponding braille dot pattern from mapping table
Creation of the braille cell for the dot pattern
Arranging the braille cell in standard structure of 40 lines with maximum 25 cell
Storing the braille pages in .j2c and .jpg formats

2.2 Braille Sheet to Text Conversion

Segmentation of each braille cell rowwise
recognition of dot pattern of each braille cell
Mapping of the dot pattern to the corresponding hindi or oriya unicode
Converting Unicode to corresponding UTF-8 code
Storing the characters corresponding to UTF-8 into a text file in a standard format

2.3 UTF-8 Coding and Decoding Scheme

UTF-8 is a universal character set (UCS) coding scheme [14] which encompasses most of the world's writing systems. UTF stands for universal transformation format which uses 8-bit blocks to represent a character. One byte UTF-8 codes are equivalent to ASCII codes whereas representation of characters from non-English languages may require 2, 3, or 4 bytes depending upon there hex code in the Unicode table.

To represent a character in UTF-8 coding scheme, the hex code of it in the Unicode table, is converted to binary.

If its binary comprises of 1 bit to 7 bits like $a_1a_2a_3a_4a_5a_6a_7$ then its UTF code would be "$1a_1a_2a_3a_4a_5a_6a_7$" and only one byte would suffice. All the English letters and frequently used punctuation marks fall in this category.

If it's binary comprises of 8 bits to 11 bits like $a_1a_2a_3a_4a_5a_6a_7a_8a_9a_{10}a_{11}$, then its UTF-8 code would be "$10a_1a_2a_3a_4a_5a_6$ $110a_7a_8a_9a_{10}a_{11}$" and it would require two bytes of space. If its binary comprises of 12 bits to 16 bits like $a_1a_2a_3a_4a_5a_6a_7a_8a_9a_{10}a_{11}a_{12}a_{13}a_{14}a_{15}a_{16}$, then its UTF-8 code will be "$110a_1a_2a_3a_4a_5$, $10a_6a_7a_8a_9a_{10}a_{11}$, $110a_{12}a_{13}a_{14}a_{15}a_{16}$" and it will require three bytes of space. All the Indian languages fall in this particular category and hence they require 3 bytes for every character representation.

If it's binary comprises of 17 bits to 21 bits like $a_1a_2a_3a_4a_5a_6a_7a_8a_9a_{10}a_{11}$ $a_{12}a_{13}a_{14}a_{15}a_{16}a_{17}a_{18}a_{19}a_{20}a_{21}$, then its UTF-8 code will be "$110a_1a_2a_3a_4a_5$, $10a_6a_7a_8a_9a_{10}a_{11}$, $110a_{12}a_{13}a_{14}a_{15}a_{16}$, $110a_{17}a_{18}a_{19}a_{20}a_{21}$" and it will require four bytes of space. UTF-8 coding is required in the conversion of Braille to text.

While decoding UTF-8 code, the number of bytes of the code is found out and arranged in the order given above. Lastly binary code is converted to hex code (Unicode) of the character. The UTF-8 decoding is required in the conversion of text to Braille.

2.4 Mapping Table

2.4.1 Oriya Mapping Table [15, 16]

See Table 1.

2.4.2 Hindi Mapping Table

See Table 2.

V. Jha and K. Parvathi

Table 1 Odia Braille mapping table

Letter	Hex code	Braille dot pattern	Letter	Hex code	Braille dot pattern
ଅ	BO5	100000	ଧ	B27	011011
ଆ, ା	B06, B3E	010110	ନ	B28	110110
ଇ, ̂	B07, B3F	011000	ପ	B2A	111010
ଈ, ୀ	B08, B40	000110	ଫ	B2B	001110
ଉ, ୁ	B09, B41	100011	ବ	B2C	101000
ଊ, ୂ	B0A, 42	101101	ଭ	B2D	010100
ଏ, େ	B0F,B47	001001	ମ	B2E	110010
ଐ, ୈ	B10, B48	010010	ଯ	B2F	110111
ଓ, ୋ	B14, B4C	011001	ଲ	B32	101010
କ	B15	100010	ଵ	B35	101011
ଖ	B16	010001	ଶ	B36	110001
ଗ	B17	111100	ଷ	B37	111011
ଘ	B18	101001	ସ	B38	011010
ଙ	B19	010011	ହ	B39	101100
ଚ	B1A	110000		B01	000010
ଛ	B1B	100001	○	B02	000101
ଜ	B1C	011100	ଃ	B03	000001
ଝ	B1D	000111		94D	010000
ଞ	B1E	001100	̖	B3D	001000
ଟ	B1F	011111	ଡ଼	B5C	111101
ଠ	B20	011101	,	2C	001000
ଡ	B21	111001	;	3B	001010
ଢ	B22	111111	:	3A	001100
ଣ	B23	010111	SPACE	20	000000
ଥ	B24	011110	?	3F	001011
ଦ	B25	110101	!	21	001110
ଥ	B26	110100	|	7C	001101

Table 2 Hindi Braille mapping table

Letter	Hex code	Braille dot pattern	Letter	Hex code	Braille dot pattern
अ	905	100000	त	924	011110
आ, ाा	906, 93E	010110	थ	925	110101
इ, ि	907, 93F	011000	द	926	110100
ई, ी	908, 940	000110	ध	927	011011
उ, ु	909, 941	100011	न	928	110110
ऊ, ू	90A, 942	101101	प	92A	111010
ए, े	90F, 947	001001	फ	92B	001110
ऐ, ै	910, 948	010010	ब	92C	101000
ओ, ाॉ	912, 94A	110011	भ	92D	010100
ओ, ो	913, 94B	100110	म	92E	110010
औ, ौ	914, 94C	011001	य	92F	110111
क	915	100010	र	930	101110
ख	916	010001	ल	932	101010
ग	917	111100	व	935	101011
घ	918	101001	श	936	110001
ङ	919	010011	ष	937	111011
च	91A	110000	स	938	011010
छ	91B	100001	ह	939	101100
ज	91C	011100	ँ	901	000010
झ	91D	000111	ं	902	000101
ञ	91E	001100	ः	903	000001
ट	91F	011111	्ृ	94D	010000
ठ	920	011101	s	93D	001000
ड	921	111001	ड़	95C	111101
ढ	922	111111			
ण	923	010111			

2.5 Braille Cell Generation

In this paper, the raised dots are read as '1' and other as '0'. The Braille cell shown in Fig. 1 is read row-wise and is read as '100100'. The default spatial resolution of a computer screen is 96 dpi. Size of Braille cell, Braille dots, cell spacing, and line spacing for a Braille embosser is fixed according to the tactile resolution of the person's fingertips as shown in Fig. 2. The rectangular Braille cell is divided in six equal squares. Depending upon the specifications of Braille cell and spatial resolution of computer screens, it is calculated as shown below that each square occupies a space of approximately 9×9 pixels.

$$a = 0.057, \quad a + 2b = 0.092, \quad 2a + 4b + c = 0.245, \quad a = 0.057 \equiv 5.472 \, pixels,$$
$$b = 0.0175 \equiv 1.68 \, pixels, \quad c = 0.061 \equiv 5.856 \, pixels.$$

So a Braille cell is made of 27×18 pixels and adjacent cells have a gap of approximately six pixels. One raised Braille dot (represented as '1' in this paper) may be represented as a 9×9 binary image as shown in Fig. 4. The non-raised dot

Fig. 1 A Braille cell

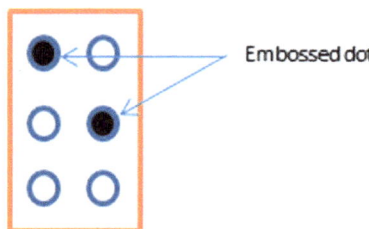

Fig. 2 Braille specification

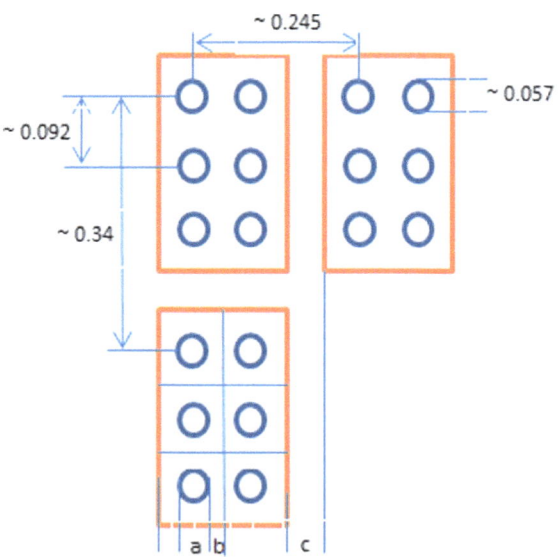

Fig. 3 A Braille cell

Fig. 4 Image matrix for
raised dot

1	1	1	1	1	1	1	1	1
1	1	1	0	0	0	1	1	1
1	1	0	0	0	0	0	1	1
1	0	0	0	0	0	0	0	1
1	0	0	0	0	0	0	0	1
1	0	0	0	0	0	0	0	1
1	1	0	0	0	0	0	1	1
1	1	1	0	0	0	1	1	1
1	1	1	1	1	1	1	1	1

Fig. 5 Image matrix for
non-raised dot

1	1	1	1	1	1	1	1	1
1	1	1	1	1	1	1	1	1
1	1	1	1	1	1	1	1	1
1	1	1	1	1	1	1	1	1
1	1	1	1	1	1	1	1	1
1	1	1	1	1	1	1	1	1
1	1	1	1	1	1	1	1	1
1	1	1	1	1	1	1	1	1
1	1	1	1	1	1	1	1	1

(represented as '0') can be represented as a 9×9 binary image with all pixel values equal to '1' as shown in Fig. 5. A Braille cell hence generated for dot pattern say, '101011' appears like Fig. 3.

2.6 Segmentation of Braille Sheet

As the size of the Braille sheet is standardized, segmentation, and classification of dot in '0' or '1' is done by calculating area of the dots.

3   Results

3.1 Odia Braille

3.1.1   Odia Input Text

ରଚନ: ତୁଳସୀ ଦାସ ଦଶ।ହା

ଶ୍ରୀ ଗୁରୁ ଚରଣ ସରୋଜ ରଜ ନିଜମନ ମୁକୁର ସୁଧାରି ।ବରଣଉଁ ରଘୁବର ବିମଳଯଶ ଜୋ ଦାୟକ ଫଳଚାରି ॥ ବୁଦ୍ଧିହୀନ ତନୁଜାନିକେ ସୁମିରଉଁ ପବନ କୁମାର ।ବଲ ବୁଦ୍ଧି ବିଦ୍ୟା ଦହେ ମୋହି ହରହୁ କଲେଶ ବିକାର ॥ ଧ୍ୟାନମ ଗଓଷ୍ପଦୀକୃତ ବାରାଣିଂ ମଶକୀକୃତ ରାକ୍ଷସମ । ରାମାୟଣ ମହାମାଲା ରତ୍ନଂ କଂଦେ ଅନିଲାତ୍ମଜମ ॥ ସ୍ତର ସ୍ତର ରଘୁନାଥ କୀର୍ତନଂ ତତ୍ର ତତ୍ର କୃତମସ୍ତକାଂଜଲିମ ।ଭାଷ୍ପବାରି ପରିପୂର୍ଣ ଲୋଚନଂ ମାରୁତିଂ ନମତ ରାକ୍ଷସାଂତକମ ॥ ଚେଁପାଇ ଜୟ ହନୁମାନ ଜ୍ଞାନ ଗୁଣ ସାଗର ।ଜୟ କପୀଶ ତିହୁ ଲୋକ ଉଜାଗର ॥ 1 ॥ ରାମଦୁତ ଅତୁଳିତ ବଲଧାମା ।ଅଂଜନି ପୁତ୍ର ପବନସୁତ ନାମା ॥ 2 ॥ ମହାବୀର ବିକ୍ରମ ବଜରଙ୍ଗୀ ।କୁମତି ନିବାର ସୁମତି କେ ସଙ୍ଗୀ ॥3 ॥ କଂଚନ ବରଣ ବିରାଜ ସୁବେଶୋ ।କାନନ କୁଂଡଲ କୁଂଚିତ କେଶୋ ॥ 4 ॥ ହାଥବଜ୍ର ଓଁ ଧ୍ବଜା ବିରାଜେ ।କାଂଥେ ମୁଂଜ ଜନେଉ ସାଜେଁ ॥ 5॥ ଶଂକର ସୁବନ କେସେରୀ ନନ୍ଦନ ।ତେଜେ ପ୍ରତାପ ମହାଜଗ ବନ୍ଦନ ॥ 6 ॥ ବିଦ୍ୟାବାନ ଗୁଣୀ ଅତି ଚାତୁର ।ରାମ କାଜ କରିବେ କୋ ଆତୁର ॥ 7 ॥ ପ୍ରଭୁ ଚରିତ୍ର ସୁନିବେ କୋ ରସିୟା ।ରାମଲଖନ ସୀତା ମନ ବସିୟା ॥ 8॥ ସୁକ୍ଷ୍ମ ରୂପଧରି ସିୟହି ଦିଖାବା ।ବିକଟ ରୂପଧରି ଲଂକ ଜରାବା ॥ 9 ॥

3.1.2   Generated Braille Sheets

See Fig. 6.

Fig. 6 Obtained Odia Braille
sheet

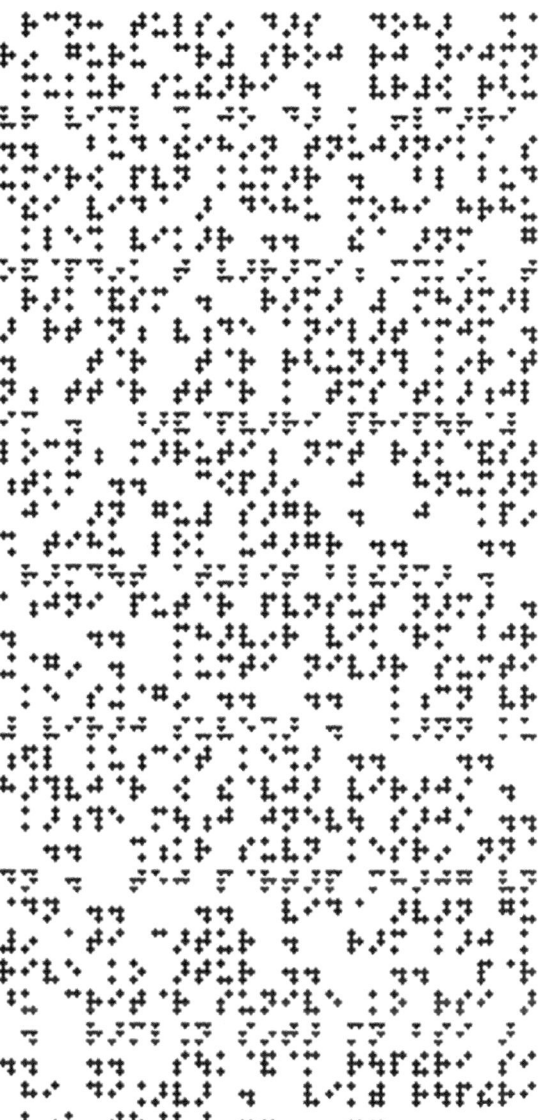

3.1.3   Braille to Odia Back Conversion

ରଚନନ୍ତ ତୁଳସୀ ଦାସ ଦେ।ହା ଶୁରୀ ଗୁରୁ ଚରଣ ସର।ଜ ରଜ ନିଜମନ ମୁକୁର ସୁଧାରି | ବରଣଉଁ ରଘୁବର ବିମଲ ଶ କେ। ଦା କ ଫଲଚାରି || ବୁଦ୍ଧିହୀନ ତନୁଜାନିକେ ସୁମିର।ଉଁ ପବନ କୁମାର | ବଲ ବୁଦ୍ଧି ବିଦ୍ ଆ ଦଯୁ ମେ।ହି ହରହୁ କଲେଶ ବିକାର || ଥ୍ ଆନମ ଗେ।ଷ୍ଟପ୍ରଦୀକ ତ ବାରାଶିଂ ମଗଲୀକ ତ ରାକ୍ଷସଯମ | ରାମୀ ଶ ମହାମାଲା ରତ୍ନଂ ଚଂଦ୍ର ଅନିଲାତ୍ମଜମ || ତର ତ୍ର ରଘୁନାଥ କୀର୍ତନଂ ତତ୍ର ତତ୍ର କ ତମସ୍ୟତକାଞ୍କଲିମ | ଭାଷ୍ପବାରି ପରିପୁର୍ଣ ଲେ।ଚନଂ ମାରୁତିଂ ନମତ ରାକ୍ଷସାଂତକମ || ତେ।ପାଇ ଜ ହନୁମାନ ଅ ଆନ ଗୁଣ ସାଗର | କ କପୀଶ ତିହୁ ଲେ।କ ଉଜାଗର || || ରାମଦୂତ ଅତୁଲିତ ବଲଧାମା | ଅଂଜନି ପୁତ୍ର ପବନସୁତ ନାମା | || ମହାବୀର ବିକ୍ରମ ବଜରଙ୍ଗୀ | ମତି ନିବାର ସୁମତି କ ସଙ୍ଗୀ | || କାଂଚନ ବରଣ ବିରାଜ ସୁବଶୋ | କାନନ କୁଂଡଲ କୁଂଚିତ କଶୋ || ହାଥବଜ୍ର ଓଂ ଧ୍ବଜା ବିରାଜଉଁ | କାଂଥେ ମୃଂଜ ଜନବେ ସାଜଉଁ || ଶଂକର ସୁବନ କସେରୀ ନନ୍ଦନ | ତଜେ ପ୍ରତାପ ମହାଜଗ ବନ୍ଦନ || || ବିଦ୍ ଆବାନ ଗୁଣୀ ଅତି ଚାତୁର | ରାମ କାଜ କରିବେ କେ। ଆତୁର || || ପ୍ରଭୁ ଚରିତ୍ର ସୁନିବେ କେ। ରସି ଆ | ରାମଲଖନ ସୀତା ମନ ବସି ଆ || ସୁକ୍ଷ୍ମ ରୂପଧରି ସି ହି ଦିଖାବା | ବିକଟ ରୂପଧରି ଲଂକ ଜରାବା || ହୀ ବିଚ୍ଛେନ୍ଦ୍ରୀ ପାଲ ସଡ ମିଲନର ଚଲୀ ଗୀ | ଅରୁଣିମା ନେ ପାଲ କୀ ନିଗରାନୀ ମେଂ ଟ୍ରେନିଂଗ ଗୁରୁ କୀ| କୀ ମୁସ୍ୱୀବତେଂ ଆଈ ଲକିନ ଉନ୍ହେ।ଂନେ ହାର ନହୀଂ ମାନୀ ଓଂର ଧୀରେ ଧୀରେ ପର୍ବତାର।ହଣ କୀ ଟ୍ରେନିଂଗ ପୁରୀ କୀ|

3.2   Hindi Braille

3.2.1   Input Hindi Text

कुछ भी करें, लोग आलोचना करते ही हैं; इसलिए उस ओर ध्यान न देते हुए स्वयंको जो योग्य लगता है, वैसा आचरण करना ही हितदायक है ।

एक बार एक पिता-पुत्र एक घोडा लेकर जा रहे थे । पुत्रने पितासे कहा ''आप घोडेपर बैठें, मैं पैदल चलता हूं ।'' पिता घोडेपर बैठ गए । मार्गसे जाते समय लोग कहने लगे, ''बाप निर्दयी है । पुत्रको धूपमें चला रहा है तथा स्वयं आरामसे घोडेपर बैठा है । यह सुनकर पिताने पुत्रको घोडेपर बैठाया तथा स्वयं पैदल चलने लगे । आगे जो लोग मिले, वे बोले, ''देखो पुत्र कितना निर्लज्ज है ! स्वयं युवा होकर भी घोडेपर बैठा है तथा पिताको पैदल चला रहा है ।'' यह सुनकर दोनों घोडेपर बैठ गए । आगे जानेपर लोग बोले, ''ये दोनों ही भैंसेके समान हैं तथा छोटेसे घोडेपर बैठे हैं । घोडा इनके वजनसे दब जाएगा ।'' यह सुनकर दोनों पैदल चलने लगे । कुछ अंतर चलनेपर लोगोंका बोलना सुनाई दिया, ''कितने मूर्ख हैं ये दोनों ? साथमें घोडा है, फिर भी पैदल ही चल रहे हैं ।''

तात्पर्य : कुछ भी करें, लोग आलोचना ही करते हैं; इसलिए लोगोंको क्या अच्छा लगता है, इस ओर ध्यान देनेकी अपेक्षा ईश्वरको क्या अच्छा लगता है, इस ओर ध्यान दीजिए । सर्व संसारको प्रसन्न करना कठिन है, ईश्वरको प्रसन्न करना सरल है ।

3.2.2   Generated Braille Sheets

Generated Braille sheets are shown in Figs. 7 and 8.

Fig. 7 Obtained Hindi
Braille sheet 1

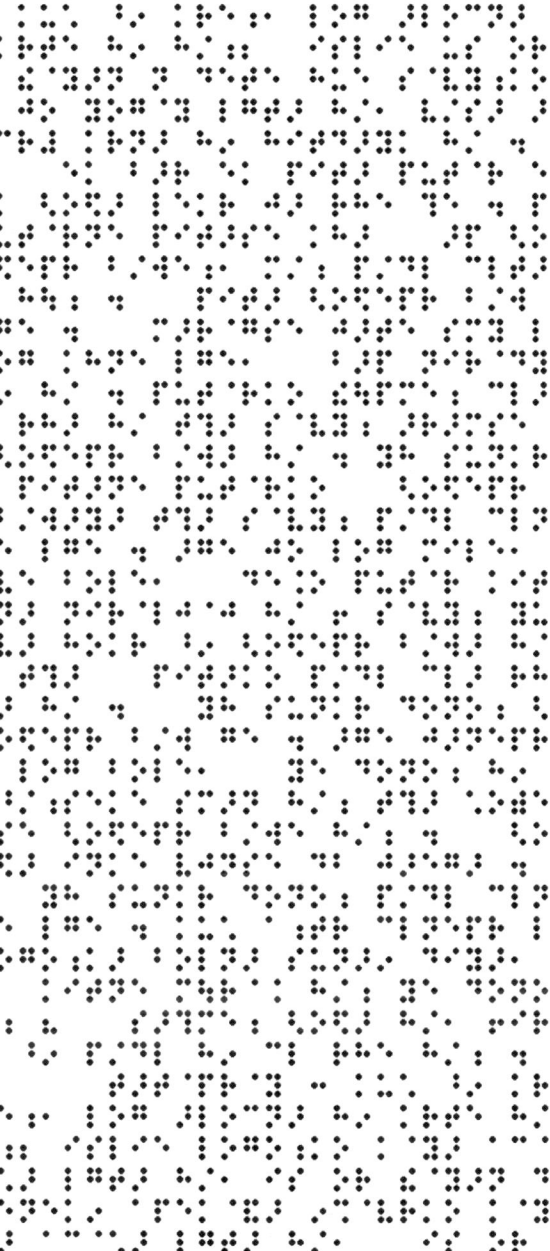

Fig. 8 Obtained Hindi
Braille sheet 2

3.2.3   Braille to Hindi Back Conversion

कुछ भी करें, लोग आलोचना करते ही हैं; इसलिए उस ओर ध्यान न देते हुए स्वयंको जो योग्य लगता है, वैसा आचरण करना ही हितदायक है | एक बार एक पिता पुत्र एक घोडा लेकर जा रहे थे | पुत्रने पितासे कहा   आप घोडेपर बैठें, मैं पैदल चलता हूं | पिता घोडेपर बैठ गे | मार्गसे जाते समय लोग कहने लगे, बाप निर्दयी है | पुत्रको धूपमें चला रहा है तथा स्वयं आरामसे घोडेपर बैठा है | यह सुनकर पिताने पुत्रको   घोडेपर बैठाया तथा स्वयं पैदल चलने लगे | आगे जो लोग मिले, वे बोले, देखो पुत्र कितना निर्लज्ज है फ स्वयं युवा होकर भी घोडेपर बैठा है तथा   पिताको पैदल चला रहा है | यह सुनकर दोनों घोडेपर बैठ गे | आगे जानेपर लोग बोले, ये दोनों ही भैंसेके समान हैं तथा छोटेसे घोडेपर बैठे हैं | घोडा इनके वजनसे दब जाएगा | यह सुनकर दोनों पैदल चलने लगे | कुछ अंतर चलनेपर लोगोंका बोलना सुनाई दिया, कितने मूर्ख हैं ये दोनों ? साथमें घोडा है, फिर भी पैदल ही चल रहे हैं |

तात्पर्य अ कुछ भी करें, लोग आलोचना ही करते हैं; इसलिए लोगोंको क्या अच्छा लगता है, इस ओर ध्यान देनेकी अपेक्षा ईश्वरको क्या अच्छा लगता है, इस ओर ध्यान दीजिए | सर्व संसारको प्रसन्न करना कठिन है, ईश्वरको प्रसन्न करना सरल है

4   Discussion and Conclusion

The results are more than 99% accurate, based on the number of characters successfully decoded. The main reasons for errors have been the ability of Braille to represent only 64 symbols uniquely. For that reason in Indian languages vowels and Matras (vowel signs) have same Braille representation. To differentiate between them while decoding in this paper, it is presumed that if a vowel comes just after a consonant then it represents Matra, not a vowel. This is one kind of error which is difficult to get away with. For simplicity, this paper does not include numerals in its mapping table, whose Braille again matches with certain letter's Braille representation. Some punctuation marks also have common Braille representations, hence resulting in error. The main advantage of this method is that it does not rely upon any database. While decoding the Braille sheet into texts, matching of the Braille letters is done on the basis of number of dots present in the detected Braille. It means if number of dots is four then only four dot Braille patterns are searched and matched. This reduces lots of time while decoding Braille sheet.

Further this method is very useful to integrate all the regional languages conversion into Braille. In future, this method will be used to develop software for bidirectional translation of printed Bharti Braille.

References

1. Padmavathi, S., Manojna, K.S.S., Sphoorthy Reddy, S., Meenakshy, D.: Conversion of Braille text in English Hindi and Tamil languages. In: IJCSEA, vol. 3, no. 3 (2013)
2. Isayed, S., Tahboub, R.: A review of optical Braille recognition. In: The 2nd World Symposium on Web Applications and Networking, At Sousse, Tunisia, vol. 1 (2015)
3. Al-Salman, et al.: A novel approach for Braille images segmentation. In: International Conference on Multimedia Computing and Systems (ICMCS), pp. 190–195 (2012)
4. Shreekanth, T., Udayashankara, V.: A review on software algorithms for optical recognition of embossed Braille characters. Int. J. Comput. Appl. **81**(3), 0975–8887 (2013)
5. Al-Shamma, S.D., Fathi, S.: Arabic Braille recognition and transcription into text and voice. In: 5th Cairo International Biomedical Engineering Conference, Cairo, Egypt, pp. 16–18, December 2010
6. Shiva Kumar, H.R., Ramakrishnan, A.G.: Tool that converted 200 Tamil books for use by blind students
7. Rupanagudi, S.R., Huddar, S., Bhat, V.G., Patil, S.S., Bhaskar, M.K.: Novel methodology for Kannada Braille to speech translation using image processing on FPGA. In: Advances in Electrical Engineering (ICAEE), International Conference (2014)
8. Li, J., et al.: Optical Braille recognition with haar wavelet features and support-vector machine. In: International Conference on Computer, Mechatronics, Control and Electronic Engineering (CMCE) (2010)
9. Halder, S., Hasnat, A., Khatun, A., Bhattacharjee, D., Nasipuri, M.: Development of a Bangla character recognition (BCR) system for generation of Bengali text from Braille notation. In: IJITEE, vol. 3, issue-1, (2013). ISSN: 2278–3075
10. Jariwala, N.B., Patel, B.: Conversion of Gujarati text into Braille: a review. In: IJIAC, vol. 4, issue 1 (2015)
11. Mishra, D., Pandey, B.: Editing unicode devanagari text through digital Braille typewriter. In: 22nd International Congress of Vedanta, December 2015
12. Samal, B.M., Parvathi, K., Das, J.K.: A bidirectional text transcription of Braille for Odia, Hindi, Telugu and English via image processing on FPGA. Int. J. Res. Eng. Technol. **04**(07) (2015)
13. Parvathi, K., Samal, B.M., Das, J.K.: Odia Braille: text transcription via image processing. In: Futuristic Trends on Computational Analysis and Knowledge Management (ABLAZE), IEEE Conference, pp. 138–143, February 2015
14. Yergeau, F.: UTF-8, a transformation format of ISO 10646. Network Working Group, November 2003
15. http://www.fileformat.info/info/charset/UTF-8/list.htm
16. World Braille Usage, 3rd edn. Perkins, National Library Service for Blind and Physically Handicapped, Library of congress, UNESCO, Washington, D.C. (2013)

GPU Based Bag of Feature for Fast Activity Detection in Video

Vikas Tripathi, Durgaprasad Gangodkar, Samin Badoni and Sagar Singh Bisht

Abstract Classification of an image on the basis of independent patches selected from image has been effectively used due to its performance. These Image blocks information is collectively represented by Bag of features. Bag of features consist four major parts, block sampling, descriptor generation from collection, characterization of distributions extracted from descriptor and classification of images. In this paper we present GPU based bag of feature generation for effective activity detection in video. Proposed framework focuses on parallel implementation strategy to reduce time taken by traditional bag of visual feature approaches. In bag of visual feature approach clustering takes significant amount of time. We propose GPU based implementation of centroid calculation and dataset generation by distance calculation to reduce time taken by clustering. Feature extraction from video dataset is performed by using MHI with energy and further HOG descriptor. We have used two datasets UT interaction and ATM to validate our results. Our experiments demonstrate that calculations are reduced to almost ten times without affecting accuracy.

Keywords Activity analysis · GPU · UT interaction · Clustering
Bag of feature

1 Introduction

Bag of visual words is showing great improvement in search results from large datasets of varying images [1]. Clustering techniques like K-mean algorithm is applied to generated visual word from extracted descriptor values [2]. Two major criteria have been identified to extend BOF. A first criterion is enhancement in

V. Tripathi (✉)
Uttarakhand Technical University, Sudhowala, India
e-mail: vikastripathi.be@gmail.com

V. Tripathi · D. Gangodkar · S. Badoni · S. S. Bisht
Graphic Era University, Dehradun, India

© Springer Nature Singapore Pte Ltd. 2018
K. Saeed et al. (eds.), *Progress in Advanced Computing and Intelligent Engineering*,
Advances in Intelligent Systems and Computing 563,
https://doi.org/10.1007/978-981-10-6872-0_13

133

processing time taken to generate visual features or frequency vectors. Other criterion is to enhance the discriminative supremacy of the visual words. Bag of Features perform different steps on large datasets for image classification. These steps are feature extraction, quantization and classification. Bag of features uses clustering technique in its quantization step which works as a descriptor in BOF. When video datasets size increased then BOF takes large amount of time to process all videos. Major component of computational time is used by clustering algorithm to calculate centroids and distance. In this paper we present framework to parallelize the clustering part of BOF so that the time taken by it decreases and it becomes feasible. Although the BOF uses K means clustering algorithm to do so we have used a similar approach in our method. We propose the parallel method of clustering that will make the BOF processing faster, this will be time efficient.

2 Literature Review

Data mining applications have increased significantly hence repository of informative data sets is also becoming extremely large. This poses a challenge for real time analysis and scalability of the current clustering algorithm [3]. Recent research to enhance speed of clustering algorithms has demonstrated that parallel implementations of these algorithms can produce immense benefits [4]. Dhillon and Modha [5] have proposed A parallel k-means clustering algorithm. Stoffel and Belkoniene [1] described in their paper the parallel approach of K means algorithm over 32 PC's on an Ethernet and shown a near linear speedup for large data sets. Scalability of the parallel k-means algorithm has also been demonstrated by others [6, 7]. The motive behind this paper is the parallelization of clustering technique to decrease the time taken by the big datasets to perform various calculations. To analyse performance of bag of feature there is need of feature generators. Several researchers have shown that for better efficiency strong feature descriptor is required. Laptev et al. [8] proposed that histogram of oriented Gradient (HOG) based descriptor and histogram of optical flow (HOF) descriptors can provide better results when used together. Dollar et al. [9] compared a variety of restricted space-time descriptors in local patches based on normalization of pixel values, intensity gradient of pixels, and variation in window size of optical flow. Zhen et al. [2] paper about action recognition by spatiotemporal oriented energies. Tripathi et al. [10] proposed a security framework for ATM premises using MHI and Hu moments. They have shown that MHI can be very useful to extract motion from images. Dalal and Triggs [11] proposed new descriptor histogram of gradient (HOG) for pedestrian detection in static images. We have used the concept MHI with oriented Energy to represent information of image and HOG is used as efficient descriptor to convert video frames into quantitative values. This data set used in BOF to represent performance of our parallel approach.

3   Methodology

The proposed methodology/system uses parallel computation technique in two parts firstly in clustering applied on features extracted from series of videos and second in Distance Calculation. As shown in Fig. 1 video frames are combined and MHI image is generated for every five frames. Energy images are generated from each MHI frame. In next step features are extracted by applying HOG on energy images, further GPU based algorithms are applied for centroid calculation, distance calculation and grouping of clusters. The features extracted are treated as the datasets on which clustering is applied and the resultant is used as an input for Random Forests [12] method.

3.1   Feature Extraction

The MHI is a view-based temporal template method suitable for human activity recognition and motion analysis. The MHI is a simple and robust binary image where pixel intensity is a function of the recency of motion in a video sequence.

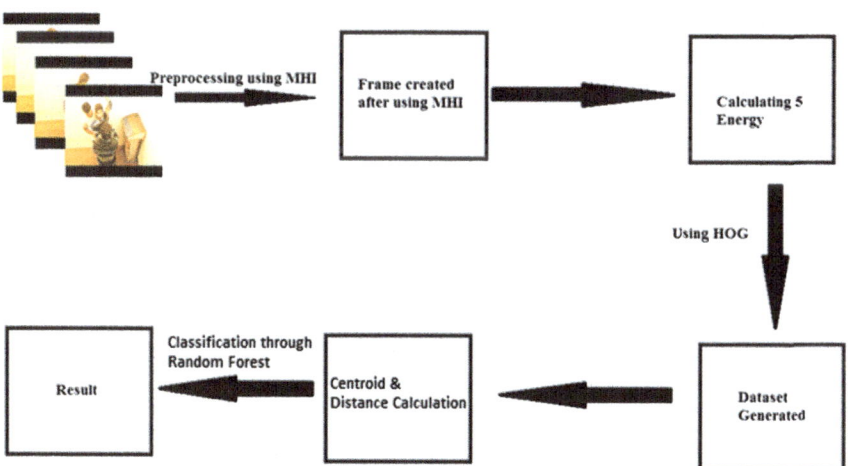

Fig. 1 Functional architecture for the proposed framework

The template matching approach first converts an image sequence into a static shape pattern, and then compares it to pre-stored action prototypes during recognition. Once MHI image is calculated further oriented energy generated. Distributions of spatiotemporal oriented energy are used to model behaviour. We have calculated five oriented energy images left, right, top, bottom, and centre. HOG is applied on individual image to generate descriptor.

The HOG is feature descriptor used in computer vision and image processing for the purpose of object detection. The technique counts occurrences of gradient orientation in localized portions of an image-detection window, or a region of interest (ROI). HOG is used to effectively depict the shape and look of local object within an image by breaking down of ardency gradients and edge directions. The descriptor used by HOG has few upper hands over the other descriptors. As it works on local cells, it is unvarying to geometric and photometric transfigurations, apart for object orientation.

3.2   Centroid and Distance Calculation on GPU

Calculating centroid is one of the most important part of the clustering technique. Equation 1 shows centroid calculation formula.

$$J = \sum_{j=1}^{k} \sum_{i=1}^{n} \left(x_i^{(j)} - c_j \right)^2 \tag{1}$$

Centroids of different groups are independent of each other. If we calculate these centroids parallel, computation remains the same but the computation time decreases, which is the objective of this paper. Centroid is calculated parallel by dividing the datasets into groups. The datasets are then distributed column-wise to different threads from each group parallel.

In our method we calculated centroid through parallel technique as shown in Algorithm 1 with the help of thread concept. As centroids for different groups is different so we can perform parallel technique in this, for this we calculated the mean of all the members in a group and the mean is calculated column wise. Each of these means are send to different threads and these threads perform the calculations parallel which takes very less time as compared to the native method. Once all the calculations are done by the threads, summation of these mean are taken into account for the centroid to be calculated. The next step is to calculate distance which is done using the Euclidean distance formula shown in Eq. 2, where p and q are points between which we are calculating the distance. In this case, these are members and centroids.

```
Algorithm 1: Centroid Calculation
Input: Descriptor generated from HOG
Output: Centroids
```

```
m=0;

limit= train _data_size_row / number_of_cluster;

label 1: LOOP e =0

              sum=0;

              label 2 : LOOP z=0

                            sum+=data[m][i];

                            z++; m++;

                        EXIT LABEL2 WHEN

                        z<limit&&m<train_data_size_row;

                 centroid[e][i]=sum/limit;

              EXIT LABEL1 WHEN e< number_of_cluster   ;

end LOOP label 1;
```

$$\mathbf{d}(\mathbf{p}, \mathbf{q}) = \mathbf{d}(\mathbf{q}, \mathbf{p}) = \sqrt{(q_1 - p_1)^2 + (q_2 - p_2)^2 + \cdots + (q_n - p_n)^2} \qquad (2)$$

As shown in Algorithm 2 distance is calculated between every element of the dataset and the centroid. As the dataset is very large, it takes a lot of time to compute this distance for each and every element. Also, distance for every element is different and independent from others, so we can parallelize this step. We have used parallel approach for calculating the distance as well, same technique of threading is used earlier in centroid calculation.

```
Algorithm 2: Distance Calculation
I/P:Centroid,Train/Test
O/P: New dataset
```

```
count_of_thread =blockIdx.x*blockDim.x+threadIdx.x;
temp= 0.0;
IF count_of_thread < size_of_train/test_dataset THEN
   label 1 : LOOP   j=0
                    sum=0.0;
                    label 2 : LOOP z=0
temp=absolute_of(data_train/test_set[count_of_thread][z]-
centroid[j][z]);
sum=temp*temp+sum;J++;
EXIT LABEL 2 WHEN j<number_of_cluster;

new_dataset[count_of_thread][j]=square_root_of (sum);
                                        z++;
            EXIT LABEL 1 WHEN z<number_of_feature ;
END IF;
```

In this we used threads to calculate the difference between the members of the group and the centroid of their respective groups. After the calculation of all the centroids and distances, the final step is to check whether the members are in correct group or not. To group the members, check the calculated distances. The member that is closest to a group is added to that group. The final result is then obtained which can be used for further processing of different methods like calculation of accuracy using Random forest classification technique.

4   Experimental Results and Analysis

The system has been tested on a computer having 5th generation Intel core i5, 2.20 GHz processor with 2 GB RAM, 1 GB Graphic card with 96 cores on a video of 320 × 240 resolution. The GPU is used for all the parallel calculations that we have performed i.e., centroid and distance calculation which is done using threads, as the datasets are very large and we needed parallel technique to perform the calculation fast so we distributed the dataset accordingly to the threads and all the calculations were carried out simultaneously. The system has been tested against different classes such as catch, clap, climb, drive, dribble and many more. After the feature extraction by the HOG, the dataset generated is divided in the ratio 7:3 of which 7 is train file and 3 is test file. We have tested our approach on two dataset UT interaction [13] and ATM [10]. Tables 1 and 2 show complete detailing

Table 1 Classes description of ATM dataset [10]

Class	Train_m	Test_m
Single	1459	343
Single abnormal	1561	116
Multiple	1786	446
Multiple abnormal	1349	203
Total	6155	1108

Table 2 Class description of UT interaction dataset [13]

Class	Train	Test
Handshake	262	25
Hug	275	36
kick	177	18
Point	218	27
Punch	154	20
Push	202	22
Total	1288	148

regarding classes and corresponding frames per class used for training and testing. 480 represents feature per frame extracted by feature descriptor.

As shown in Table 3 time comparison of normal code, normal code with K means and our code, it effectively demonstrate that the time taken by our method for both UT interaction and ATM is very less as compared to the other two methods. Table 4 shows the detailed description of TP (true positive), FP (false positive), Precision, Recall, ROC, Accuracy of single processor implementation, normal code with parallel K means and our approach for UT interaction and ATM. It shows that our approach is not only reducing time efficiently but also maintaining accuracy as well.

Table 3 Time comparison of all the codes in UT and ATM

Dataset	Time taken in sec (centroid and distance calculation in CPU)	Time taken in sec (centroid in CPU and distance calculation in GPU)	Time taken in sec (centroid and distance calculation in GPU)
ATM	175.952	67.86000001	15.39699984
UT	33.1319997	14.30099988	4.782999992

Table 4 TP, FP, accuracy of all codes in UT and ATM datasets (CT-centroid, DT-distance)

Dataset		TP rate	FP rate	Precision	Recall	ROC	Accuracy (%)
ATM	CT and DT CPU	0.903	0.043	0.899	0.903	0.989	90.2527
	CPU-CT and GPU-DT	0.912	0.036	0.914	0.914	0.989	91.4182
	CT and DT GPU	0.906	0.04	0.904	0.906	0.989	90.6052
UT	CT and DT CPU	0.399	0.111	0.461	0.399	0.755	39.8649
	CPU-CT and GPU-DT	0.449	0.105	0.538	0.449	0.761	44.898
	CT and DT GPU	0.442	0.103	0.504	0.442	0.752	44.2177

5 Conclusion

This paper presents the parallel techniques based on GPU used to parallelize centroid calculation which is core part of any clustering technique for real time activity detection with MHI energy based feature descriptor. Further Distance Calculation from centroid to dataset also performed using GPU. The need of parallel clustering and Distance Calculation is because clustering and Distance Calculation takes a lot of time which make algorithm very slow. Reduction in time take to process achieved by using thread and the multiple core of GPU, hence parallel clustering and Distance Calculation can be used in real time scenario as compared to the normal clustering and Distance Calculation. To validate our approach we have performed analysis on two datasets ATM and UT interaction. Results calculated on datasets ATM and UT shows that average time taken is 10 Times faster and accuracy approximately remain same as compared to CPU based approach. In future optimization can be performed on algorithms to enhance accuracy and utilization of GPU cores. Other feature descriptor can be used to enhance accuracy.

References

1. Kilian, S., Belkoniene, A.: Parallel k/h-Means Clustering for Large Data Sets: Euro-Par'99 Parallel Processing, pp. 1451–1454. Springer, Berlin (1999)
2. Xiantong, Z., Shao, L., Li, X.: Action recognition by spatio-temporal oriented energies. Inf. Sci. **281**, 295–309 (2014)
3. Ganti, V., Gehrke, J., Ramakrishnan, R.: Mining very large databases. Computer **32**(8), 38–45 (1999)
4. Judd, D., McKinley, P., Jain, A.: Large-scale parallel data clustering. In: Proceedings of the International Conference on Pattern Recognition, pp. 488–493 (1996)

5. Dhillon, I.S., Modha, DS.: A data-clustering algorithm on distributed memory multiprocessors: large-scale parallel data mining. In: Lecture Notes in Artificial Intelligence, vol. 1759, pp. 245–260. Springer, Berlin (2000)
6. Nagesh, H., Goil, S., Choudhary, A.: A scalable parallel subspace clustering algorithm for massive data sets. In: Proceedings International Conference on Parallel Processing. IEEE Computer Society, pp. 477–484 (2000)
7. Ng, M.K., Zhexue, H.: A parallel k-prototypes algorithm for clustering large data sets in data mining. Intell. Data Eng. Learn. **3**, 263–290 (1999)
8. Laptev, I., Marszałek, M., Schmid, C., Rozenfeld, B.: Learning realistic human actions from movies. In: Proceedings of the IEEE Conference on Computer Vision and Pattern Recognition, pp. 1–8 (2008)
9. Dollar, P., Wojek, C., Schiele, B., Perona, P.: Pedestrian detection: an evaluation of the state of the art. IEEE Trans. Pattern Anal. Mach. Intell. **34**(4), 743–761 (2012)
10. Tripathi, V., Gangodkar, D., Latta, V., Mittal, A.: Robust abnormal event recognition via motion and shape analysis at ATM installations. J. Electr. Comput. Eng. (2015)
11. Dalal, N., Triggs, B.: Histograms of oriented gradients for human detection. In: Computer Vision and Pattern Recognition, CVPR, vol. 1, pp. 886–893 (2005)
12. Breiman, L.: Random Forests: Machine Learning, vol. 45, no. 1, pp. 5–32 (2001)
13. Michael, R.S., Aggarwal, J.K.: UT-interaction dataset, ICPR contest on semantic description of human activities (SDHA). In: IEEE International Conference on Pattern Recognition Workshops, vol. 2 (2010)

A New Approach Using Discrepancy Theory for MR Image Segmentation

Abir Hudait, Nitish Pandey, Lalit Vashistha, M. N. Das
and Amitava Sen

Abstract This article introduces an approximate rehabilitation of unidentifiable edge in medical images by settling demarcation line based on experimental evidence, by using the essence of classical discrepancy theoretic approach. The main intriguing aspect of this method is distributing the quality points over the ambiguous area by the virtue of jittered sampling method for validating the points in a deterministic way, after finding the decisive point. Before this textual enhancement on unidentifiable edge, a known image classification model is introduced by using wavelet transformation to classify the brain MRI in three categories as normal, benign, and malignant. Here, we are taking timestamp one snapshot of an image, not to consider the aliasing and anti-aliasing effect. We are concentrated on the edges those are not detected even after applying edge detection method.

Keywords Wavelet transform · Principal Component Analysis (PCA)
Kernel SVM · Fuzzy edge detection · Decisive points · Lebesgue measure
Convex set

A. Hudait (✉) · N. Pandey · L. Vashistha · M. N. Das · A. Sen
School of Computer Engineering, KIIT University, Bhubaneswar 751024, India
e-mail: abirhudait@gmail.com

N. Pandey
e-mail: nitish5808@gmail.com

L. Vashistha
e-mail: lalitkvashishtha@gmail.com

M. N. Das
e-mail: mndas_prof@kiit.ac.in

A. Sen
e-mail: amitavasen@hotmail.com

© Springer Nature Singapore Pte Ltd. 2018   143
K. Saeed et al. (eds.), *Progress in Advanced Computing and Intelligent Engineering*,
Advances in Intelligent Systems and Computing 563,
https://doi.org/10.1007/978-981-10-6872-0_14

1   Introduction

Discrepancy theory [1] provides a mathematical basis to quantify the scalar measure that is the deviation between real and expected evidence. Now with this measure, we can easily identify image or image segment having edge discontinuity. So to reduce the discontinuity, initially, we have taken some points randomly on unidentifiable edges; they are called as quality points, and finally, they are validated, and based on that, those edges are reformed by accumulating those quality points. These concepts are applied after the evaluation of edges of brain MRI through fuzzy inference system (FIS). Before all these computations, at the initial stage, a well-renounced automatic brain MRI classifier is used to classify brain MRI into three classes as normal, benign (non-cancerous), and malignant (cancerous). For the classification, wavelet transformation (daubechies wavelet (db4)) is used. Then, PCA has been introduced to reduce the curse of dimensionality of extracted feature set, and finally, model is trained with kernel SVM and also by concatenating K-fold stratified cross validation approach. The aim of the entire proposed approach is to reduce the intervention of some invasive approach in medical domain like biopsy, spinal tap, or lumbar puncture method.

In medical domain, our proposed approximation-based approach for ambiguous area or unidentifiable edges of FMRI images provides some extra atom of information under discrepancy measure which may put forward a new, useful way of image understandings that allow investigators to inspect their data in as flexible, scientifically informative, and convenient manner as is possible.

The rest of the chapters are structured as follows: Sect. 2 consists of survey on existing contribution, Sect. 3 provides the detail about the input dataset, Sect. 4 mainly focuses on brain MRI classification, in Sect. 5 a bird eye view on FIS for edge detection, Sect. 6 concerns about the detail discussion of mathematical foundation of discrepancy theory and the implementation detail of full framework, in Sect. 7 a method is introduced for validation of newly form unidentifiable edges, and finally highlights the conclusion and future work.

2   Related Work

In the domain of image processing, among all alluring works on identification of unidentifiable edge of an image, the work by Shoji and Tsumoto [2], where they proposed an approach for the recognition of unidentifiable edge of an MRI image using Rough representation of a region of interest in medical images. Jenkinson and Stephen [3] determined a global optimization method for robust affine brain images. On the main aspect, 'discrepancy computation' is presented in the article by William et al. [1], discussed fundamental concept about discrepancy theory and its classification, application, and complexity. Devid and Devid [4] provided the way for the computation of the discrepancy. Devid et al. [5] computed discrepancy with the application to super sampling patterns. So after assessing all the contributed work, our aim is to fill the deficiency by detecting the unidentifiable edges as accurately as possible.

3   Input Dataset

The datasets consist of T2-weighted MR brain images in axial plane and on average 256×256 in-plane resolution, from Harvard Medical School, OASIS dataset; also, some data are collected locally from Kalinga Institute of Medical School (KIMS).

Arbitrarily, 25 normal brain MRIs, one type of benign tumor brain (49), and six types of abnormal (malignant) brain MRI ($150 = 6$ types of diseases $\times$ 25 images/ diseases), in total 224 images are taken for our experiment.

4   Image Classification

Before going into the details to our proposed model, a known framework is used using discrete wavelet decomposition (db4) to classify images into normal brain image, benign (non-cancerous), and malignant (cancerous) brain image [6, 7]. The phases are as follows:

4.1   Feature Extraction

The fundamentals of DWT are presented as follows: Suppose $z(t)$ is a square integrable function, then the continuous WT of $z(t)$ with respect to a given wavelet $\Psi(t)$ is represented as

$$WT_\Psi(p, q) = \int_{-\infty}^{\infty} z(t)\Psi_{p,q}(t)dt. \tag{1}$$

$$\Psi_{p,q}(t) = \frac{1}{p}\Psi\left(\frac{t-p}{q}\right). \tag{2}$$

Here, the wavelet $\Psi_{p,q}(t)$ is evaluated from the mother wavelet $\Psi(t)$ by translation and dilation: p is the dilation factor and q the translation parameter. Equation (1) can be discretized by keeping p and q to a discrete lattice ($p = 2^q$ and $p > 0$) to give the DWT, which can be represented as follows.

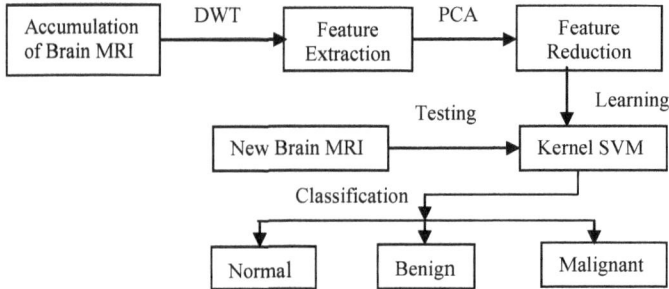

Fig. 1 Stepwise depiction of brain image classification model

$$ca_{k,l}(m) = DS\left[\sum_m z(m)u_k * (m - 2^k l)\right]. \qquad (3)$$

$$cd_{k,l}(m) = DS\left[\sum_m z(m)v_k * (m - 2^k l)\right]. \qquad (4)$$

Here, $ca_{k,l}$ and $cd_{k,l}$ allude to the approximation components and the detail components of an image, respectively. $u(m)$ and $v(m)$ indicate for the low-pass filter and high-pass filter, respectively. k and l refer the wavelet scale and translation factors, respectively. DS operator means the down-sampling. Equation (3) is the basal of wavelet decomposition. It decomposes signal $z(m)$ into two signals, the approximation coefficients $ca(m)$ and the detail components $cd(m)$.

For 2D images, the DWT is integrated to each dimension individually. As an outcome, there are four sub-band (LL, LH, HH, and HL) images at every scale. The sub-band LL is used for next 2D DWT. The LL sub-band and the LH, HL, and HH sub-bands can be considered as the approximation component and the detailed components of the image, respectively. In this algorithm, level-4 decomposition via daubechies wavelet was utilized to extract features (Figs. 1, 2, 3, and 4).

4.2 Principal Component Analysis

PCA is used to lessen the curse of dimension of a feature set. It can be represented as the orthogonal projection of the data onto a lower dimensional linear space, such that the variance of the projected data is maximized. If E is the matrix consisting of eigenvectors of the covariance matrix as the row vectors formed by transforming a data vector x, we get

$$y = E(x - m_x). \qquad (5)$$

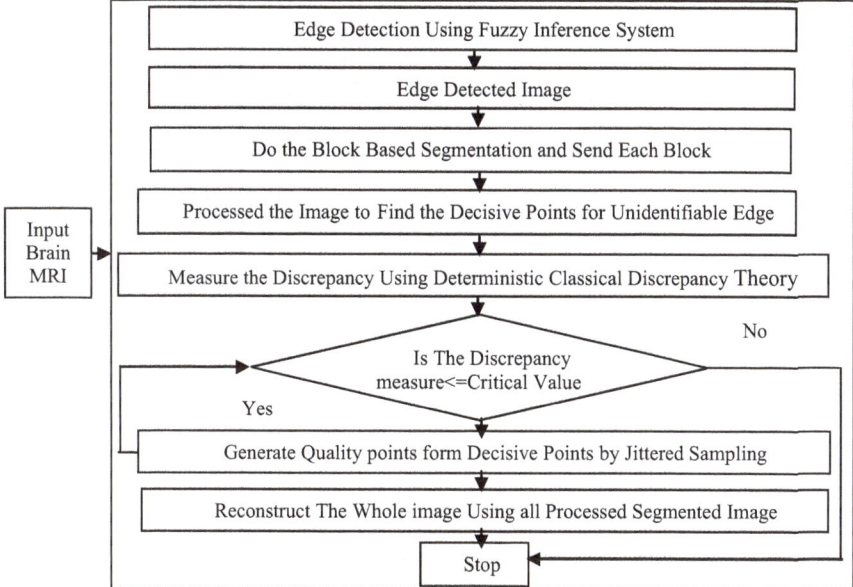

Fig. 2 Overall representation of the proposed approach

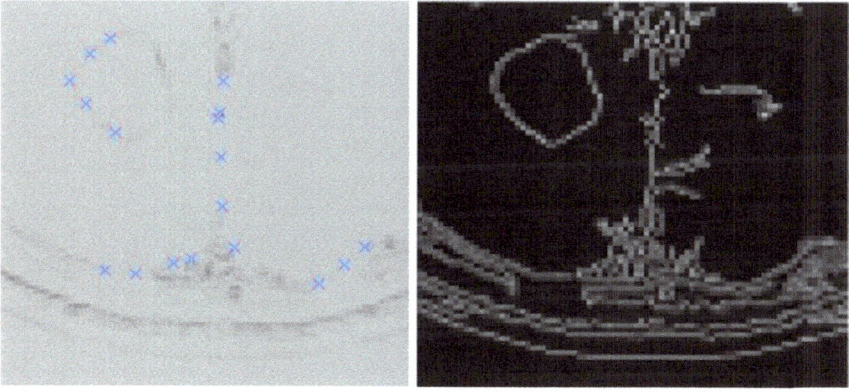

Fig. 3 Sample representation of quality points for a segmented part and image after enhancement

4.3   Kernel SVM and K-Fold Cross Validation

In the following method, the nonlinear kernel is related to the transform $\phi(x_i)$ by the equation $k(x_i, x_j) = \phi(x_i)\phi(x_j)$. The value ω is also in the transformed space, with $\omega = \sum_i \alpha_i \eta_i \phi(x_i)$. Dot products with ω for classification can be computed by

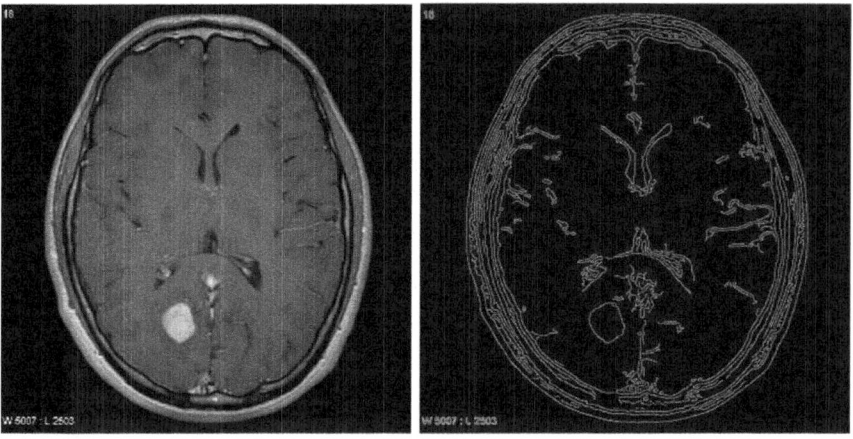

Fig. 4 Original and final output of the sample image, respectively

$\omega\phi(x) = \sum_i \alpha_i \eta_i k(x, x_i)$. To circumvent over fitting, cross validation method is initiated, where we pragmatically calculated K as 7 through the hit and miss technique (Tables 1, 2, 3, and 4).

Table 1 Estimation of different attributes

Attributes	Img1	Img2	Img3	Img4	Img5	Img6
Contrast	0.2564	0.2819	0.2249	0.2032	0.2313	0.2919
Correlation	0.1302	0.0939	0.0991	0.1125	0.1072	0.1440
Energy	0.7567	0.7551	0.7690	0.7553	0.7418	0.7724
Homogeneity	0.9330	0.9297	0.9365	0.9331	0.9297	0.9343
Mean	0.0046	0.0047	0.0020	0.0019	0.0042	0.0057
Standard deviation	0.0896	0.0896	0.0020	0.0897	0.0897	0.0896
Entropy	3.3877	3.2516	3.5181	3.6549	3.5516	3.2348
RMS	0.0898	0.0898	0.0898	0.0898	0.0898	0.0898
Variance	0.0079	0.0079	0.0080	0.0079	0.0080	0.0080
Smoothness	0.9453	0.9463	0.8849	0.8782	0.9403	0.9554
Kurtosis	7.1747	8.4626	6.7672	5.8116	6.0614	10.580
Skewness	0.5570	0.7868	0.4412	0.3407	0.5104	1.1383
IDM	0.8125	−1.4012	0.5461	1.0010	0.3130	1.2394

Table 2 Output of image classification for experimental image inputs

	Img1	Img2	Img3	Img4	Img5	Img6
Type of tumor	Normal	Normal	Benign	Benign	Malignant	Malignant

Table 3 Approximate accuracy measure with different types of kernel

RBF	Linear	Polygonal	Quadratic	Type of tumor
97.76	91.07	95.53	94.19	Normal
				Abnormal
96.48	90.45	94.91	92.96	Benign
				Malignant

Table 4 Estimated coefficients

	Estimate	SE	tStat	pValue
b1	14.17	2.1134	6.7051	0.00027614
b2	30	NaN	NaN	NaN

5 Fuzzy Inference System (FIS) and Comparisons Among Different Edge Detection Technique

With the fundamental knowledge of fuzzy logic, fuzzy inference system helps us to highlight the edges of an image. For detecting an edge of our experimental image, we have used a very well-known edge detection technique that is edge detection using fuzzy inference system. For FIS system, we have used Gaussian MF as input and used triangular MF to represent the output.

5.1 Traditional Fuzzy Edge Detector Versus Fuzzy Edge Detector

In case of fuzzy edge detector, it does not need parameter setting as for classical edge detector such as threshold and σ, and great computational complexity, even high-level noise does not affect the detection [8, 9]. Even the edge thickness can easily be changed by adding new rules or changing output parameters. So the system has dynamic structure adopted by changing rules.

6 Discrepancy Theoretic Approach

Suppose, d $>= 2$ be an integer and it is fixed throughout the problem. Here, in Euclidean space R^d, where d defines the dimensions, a domain is considered and designated by X. X is containing a set of unit Lebesgue measure [1]. Suppose that c is a set of measurable subsets of X, endowed with an integral geometric measure, and normalized so that the total measure is equal to unity. Suppose further that Pt is a collection of p points in X. For every subset C $\in c$ of X, let

$$Z[Pt; C] = |(Pt \cap C)|. \tag{6}$$

represent the number of points of Pt that fall into C. We have selected the random quality points from the set $(Pt \cap C)$ on the discontinuous region of our experimented image. We consider it as definite point calculation of Pt in C, with respect to the expectation $N\mu(C)$. On the basis of discrepancy of Pt in C, we signify the dissimilarity as

$$(DISP[Pt; C] = Z[Pt; C] - N\mu(C)). \tag{7}$$

However, for upper bound analysis, take the corresponding L2-norm.

$$\|DISP[Pt]\|_2 = (\int_A |DISP[Pt; C]|^2 dC)^{\frac{1}{2}}. \tag{8}$$

as well as the corresponding Le-norms where $2 < e < 1$. Now, the lower bound result can be represented of the form $\|DISP[Pt]\|_2 > f(n)$ for all sets Pt of n points in X, and upper bound result is of the form $\|DISP[Pt]\|_2 < g(n)$ for some sets Pt of n points in X.

Therefore, our main concern is to build such a point set Pt or to validate that those point sets exist. So, the major duty is to construct lower and upper bounds where the order of magnitude for the functions $f(n)$ and $g(n)$ is identical.

6.1 Large Discrepancy Approach

Let consider a Euclidean space R^d with the dimension d. We take as our domain X the unit square $[0, 1]^d$, treated as a torus for simplicity. Let O $[0, 1]^d$ be a compact and convex set that meets a practical criterion

$$O(\gamma, \varsigma, \chi) = \gamma(\varsigma O) + \chi. \tag{9}$$

Now if we put a ball with maximum radius inside O, then that radius is defined by $\delta(O)$, and n is the total number of points in the point sets Pt under consideration. Let Γ denotes the group of all orthogonal transformations in Rd, normalized so that the total measure is equal to unity. For any contraction $\gamma \in [0, 1]$, orthogonal transformation $\varsigma \in \Gamma$ and translation $\chi \in [0, 1]^d$, we consider the similar copy of O. We then consider the collection

$$c = \left\{ O(\gamma, \varsigma, \chi) : \gamma[0, 1]^d, \ \varsigma \in \Gamma, \ \chi \in [0, 1]^d \right\}. \tag{10}$$

$$\|DISP[Pt]\|_2 = \left(\int\limits_{[0,1]^d} \int\limits_{\varsigma} \int\limits_0^1 |DISP[Pt; O(\gamma, \varsigma, \chi)]|^2 d\gamma d\varsigma d\chi \right)^{\frac{1}{2}}. \tag{11}$$

of every identical photocopies of O, where the integral geometric measure is given by a natural combination of the standard Lebesgue measures of γ and χ and the measure of Γ. Additionally, for any set Pt of n points in $[0, 1]^d$, we have the L2-norm

$$\|DISP[Pt]\|_\infty = \underset{\substack{\gamma \in [0,1]^d \\ \varsigma \in \Gamma \\ \chi \in [0,1]^d}}{SUP} |DISP[Pt; O(\gamma, \varsigma, \chi)]|. \tag{12}$$

We have used this property of discrepancy significantly in order to change the point count over unidentifiable edge by the orthogonal transformation and convex set constraints [1].

6.2 Implementation of Theoretic Representation on MR Image

In this stage, the edge detected image is taken and tried to segment it first region wise. Then we have considered each of the segmented image to check in which segment discrepancy has occurred. Based on that, finally, we have got a few segment those are actually having the discontinuous edge and then we send them to next step for enhancement. Thus, the computational time for processing of image segment is reduced.

We have considered each of the individual image segment those are having discrepancy. Now during processing, we have tried to traverse five adjacent pixels in each neighboring side of a particular pixel. After traversing the neighbors, if we get another pixel after all 5×5 neighbors, having pixel value less than equal to average threshold value of the image, then we saved that as new quality points. Otherwise, if points are occurring within that constrained 5×5 neighboring pixel, we will not consider them for quality points, though they are contributing for edge continuity.

So, after finding the quality points, consider each single quality point and consider a d dimensional Euclidean space R^d [1]. Let us assume that the number of points is a perfect square, so that $n = N^2$ for some natural number N. We may then choose to split the unit square $[0, 1]^2$ in the real method into a merge of $n = N^2$ small squares, each of which length of each side is N^{-1}, and then distributes deterministically in the middle of every individual small square. Now some of the generated points will be overlapped with some existing points, which will make discontinuous edge more concrete.

Let $C = O(\gamma, \varsigma, \chi)$ where $\gamma \in [0, 1]$, $\varsigma \in \Gamma$, $\chi \in [0, 1]^2$ is an identical photocopy of a compact and convex set O, provided that the O is a fixed set. Now, we calculate the discrepancy $DISP$ [Pt; C]. Let Pt represent the set of the $n = N^2$ little squares T of side length N^{-1}. Therefore, based on the additive property of the discrepancy function, it provides

$$DISP[Pt; C] = \sum_{T \in \xi} DISP[Pt; T \cap C]. \tag{13}$$

Now, we formulate an easy consideration that, $DISP[Pt; C] = DISP[Pt; T \cap C]$, if $T \subseteq C$ or $T \cap C = \phi$, The identity (13) then becomes

$$DISP[Pt; C] = \sum_{\substack{T \in \xi \\ T \cap \partial C \neq \phi}} DISP[Pt; T \cap C]. \tag{14}$$

where ∂C represents the periphery of C. Ultimately, notice that each of the following two $0 \leq Z[Pt; T \cap C] \leq 1$ and $0 \leq N\mu (T \cap C) \leq 1$, so that $|DISP[Pt; T \cap C]| \leq 1$ and it obeys from (14) and the triangle inequality that

$$|DISP[Pt; C]| \leq |\{T \in \xi : T \cap \partial C \neq \phi\}| \ll N = n^{\frac{1}{2}}. \tag{15}$$

Here, we build a significant analysis that the term $\#\{T \in \xi : T \cap \partial C \neq \phi\}$ in (15) and the length of the boundary curve ∂O of O are having a complicated association, note that the set C is an identical photocopy of the appointed compact and convex set O.

6.3 Sampling Method for the Verification of Quality Points on Unidentifiable Edge

The basic task in sample point generation is, given some desired no. of samples, N, to find N sample point on the unit square [7]. Jittered sampling is one way [6], where the unit square is partitioned into a set of equal area rectangles, and a point is chosen uniformly from each rectangle

```
for i = 0 to nx-1
    for j = 0 to ny-1
        k = inx + j
        xk = randfrom (i/nx, (i+1)/nx)
        yk = randfrom (i/ny, (i+1)/ny)
```

where, as in regular sampling, $n_x n_y = N$.

After all the processing of those selected image, finally, we have merged them in order to get the full image with some extra token of information, better visibility, and understanding, which might helpful for the experts from medical domain.

7 Validation of Newly Formed Edge

After the formation of newly generated edge segment, the quality points are validated. A nonlinear regression model with sigmoid function as a kernel is applied to analyze the trend of the newly generated point.

After validation of quality points, we simply join them. For some set of sequence of quality points, they form an arc of certain circle; so the formation of edge is quite easy. For some point set, they are closer to the part of certain circle to which most number of quality points belongs; in that case, we have tried to form a zone of contention to form those certain arc. In the following image, we have shown the validation for one segment after the total processing for Img6. Nonlinear regression model can be represented as

$$y = b1 * (1 - \exp(-b2 * x)).$$ (16)

Number of observations: 8, Error degrees of freedom: 7
Root Mean Squared Error: 5.98
R-Squared: 1.11e-16, Adjusted R-Squared 1.11e-16
F-statistic versus zero model: 45, p-value = 0.000276.

8 Future Work and Conclusion

Our work can be extended to further classification of each different category of tumor, find exact location of tumor and find the volume of the tumor and also computation discrepancy using probabilistic approach.

Our algorithm has tried to compute the exact deficiency, but in many cases have slow running times because of the underlying combinatorial complexity. Which definition of discrepancy is better and whether there are any sample distributions with low projected discrepancies for several directions. Finding the decisive points and generating the quality points for distribution properly still be a problem, which can be enhanced by using better different algorithmic approach. Also, some popular super sampling patterns have discrepancies with better asymptotic behavior than random sampling. So, these super sampling methods might be more effective for the reduction of asymptotic complexity.

References

1. William, C., Anand, S., Giancarlo, T.: A Panorama of Discrepancy. Springer Cham, Heidelberg (2014). ISBN 978-3-319-04695-2
2. Shoji, H., Tsumoto, S.: Rough representation of a region of interest in medical images. Elsevier Int. J. Approximate Reasoning **40**(2005), 23–34 (2005)
3. Jenkinson, M., Smith, S.: A global optimisation method for robust affine registration of brain images. Med. Image Anal. **5**,143–156 (2001). Elsevier
4. Devid, D., Devid, E.: Computing the discrepancy. In: 9th Annual Computation Geometry (1993). ACM 0-89791.583.6/93/0005/0047
5. Devid, D.P., Devid, E., Don, M.P.: Computing the Discrepancy with the Application to Super Sampling Patterns, p. 08544. Princeton University, Princeton, NJ (1983)
6. Zhang, Y., Wu, L.: An MRI brain image classifier via principal component analysis and kernel support vector machine. Prog. Electromagnet. Res. **130**, 369–388 (2012)
7. John, P.: Brain tumor classification using wavelet and texture based neural network. Int. J. Sci. Eng. Res. **3**(10) (2012). ISSN 2229-5518
8. Boopathi Kumar, E., Sundaresan, M.: Fuzzy inference system based edge detection using fuzzy membership functions. Int. J. Comput. Appl. **112**(4), 0975–8887 (2015)
9. Anjali, D., Satnam, S.: Fuzzy inference system based edge detection in images. Int. J. Core Eng. Manage. (IJCEM) **1**(10) (2015). ISSN: 2348 9510
10. Müller, H., Michoux, N., Bandon, D., Geissbuhler, A.: A review of content-based image retrieval systems in medical applications—clinical benefits and future directions. Int. J. Med. Inform. **78**(2009), 638 (2009)
11. Peters, G., Lingras, P., Sl̦ezak, D., Yao, Y.: Rough Sets: Selected Methods and Applications in Management and Engineering. Springer, London (2012). ISSN 1610-3947

Error Detection and Correction Using Parity and Pixel Values of Image

Narander Kumar and Jaishree

Abstract The problem have been noticed of error in the images while sending, the noise affect in the image so that the result is, the distortion of the secret message embedded on the image in case of watermarking technique. We have seen that the LSB technique is very prone to the noise, so in order to remove this problem, this paper proposes a method for error detection and correction on images entitled as error detection and correction using parity and pixel values; in this, we are adding two extra information that is parity bit and pixel value in order to correct and detect the error bit.

Keywords Parity · Pixel value · LSB · Error detection · Error correction

1 Introduction

Interference and defects in the environment and in the communication network medium can cause the random distortion of the data bit. Data can be interrupted by the network defects. So in order to correct and detect the error in the data on receiving end, the technique of error detecting is introduced by which the error present in the data can be detected and corrected. Various error-correcting and error-detecting techniques are there which help to find out the error and also to correct it. The data are encoded with the extra information which helps to find out about the original data; these extra information are called the redundancy. So many error-detecting techniques are introduced depending on what types of error are there.

N. Kumar (✉) · Jaishree
Babasaheb Bhimrao Ambedkar University (A Central University), Lucknow, India
e-mail: nk_iet@yahoo.co.in

Jaishree
e-mail: jaishreebansal99@gmail.com

© Springer Nature Singapore Pte Ltd. 2018
K. Saeed et al. (eds.), *Progress in Advanced Computing and Intelligent Engineering*,
Advances in Intelligent Systems and Computing 563,
https://doi.org/10.1007/978-981-10-6872-0_15

Now as we are talking about the methods, we have two strategies error-detecting codes and error-correcting codes. We have two types of error that are one is single error and other is burst error.

There are various error-detecting techniques as follows:

1. Simple parity check
2. 2-D parity check
3. Check sum
4. Cyclic redundancy check

Now, we have error-correcting codes techniques as backward error correction and forward error correction.

In this paper, we are working on the method which can be useful for error detecting as well as error correcting on any image matrix. We are taking two information matrices to store the parity bit and pixel value, and these can work as redundancy bit which can be used for further method of detecting and correcting data bit. We will discuss the technique with the help of a working example.

2 Related Work

Many theories have been given by the researchers for finding the error of any image matrix in order to correct the error. Codes are developed in [1] by which random errors can be detected and corrected as well as also detect unidirectional error with fewer random errors. Modification of BCH and LDPC concatenation is proposed in [2] on Chen et al., and comparison between previous and new modification is done which results in low error floor. Index coding scheme and error-correcting methods has been studied and analyzed in [3]. The discussion on lee type error vector and 5-error-correcting algorithm is presented in [4]. The comparison of decoding performance between binary LDPC and non-binary LDPC codes for IEEE 802.11n for probability domain and log domain decoding algorithms are discussed [5]. Light weight channel codes are being traced with real error with respect to IEEE.802.15.4-2006 in [6] also with timing constraints of industrial wireless communication. Suggestion of Reed–Solomon (12, 7) block is also mentioned. Testing for multiple gross error situations is done in [7] of many gross error scenario which is having different redundancy level. Testing of error in transmission line is done in [8] firstly in order to check if measurement set has gross error. In [9], the difference of both error bound in gap matric and positive real balanced truncation is discussed. The method of storing code words in buffer structure and also using error control blocks at receiver as well as on transmitter is discussed in [10]. All the single bit errors are being detected and corrected in this. Convergence of result and predefined accuracy has achieved through refining the mesh described in [11]. From the above review of work, we have concluded that so many theories have been given in order to detect and correct error, but pure results are not achieved; if achieved, then the method is becoming more complex and time taking without giving an optimal solution. So in order to try to solve these problems, a simpler methodology is being proposed which gives better and optimal results.

3 Proposed Methodology

Aiming to correct and also to detect the error in the image matrix, first, an image should be converted into binary form then the matrix of that binary image should be generated. As we know, by the defect in the network communication medium, the bits can be changed. We here taking an image of which a pixel matrix is obtained (Fig. 1).

This image is again converted into binary matrix so that the error-detection and error-correction technique would be applied on that image. So to convert it into the binary matrix, we are using command of MATLAB, i.e.

```
%Read the image
im = imread('peppers.png')
%Convert it to bw
im = im2bw(im)
```

So the resultant matrix obtained from this is binary as we can see the matrix with 0 and 1 values only (Fig. 2).

This image matrix can also be converted in binary form. So in order to retrieve original data, we are adding two information matrix along with the image matrix at the sender side; the first matrix is the parity matrix; the second is the pixel value, and the third is the own original image matrix. The first and second matrix are of

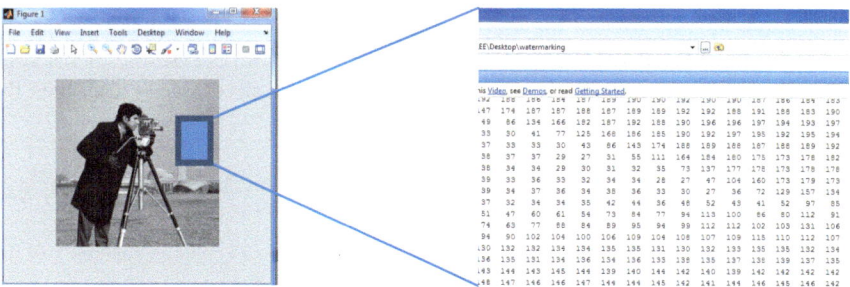

Fig. 1 Matrix form for the image

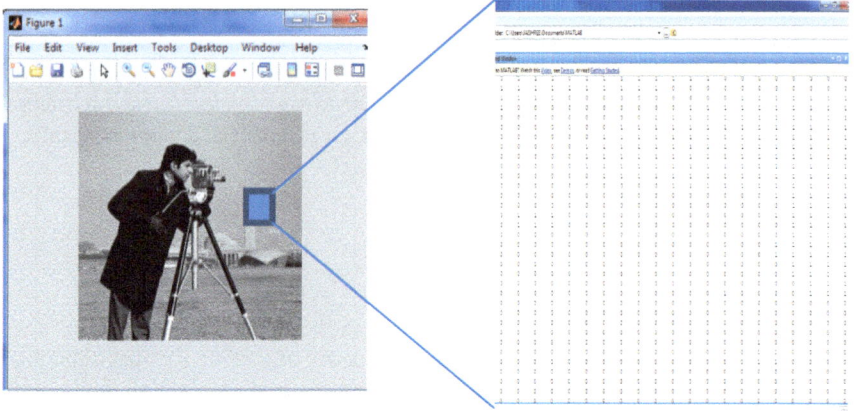

Fig. 2 Binary matrix for image

Index value	8	4	2	1
Data bit	0	1	0	0

Fig. 3 Allotment of index value

Index value	16	8	4	2	1
Data bit	1	0	1	1	0

Fig. 4 Column index value of 5 × 5 matrix

size [m × 1] and [1 × n], respectively. Here, the m and n are the number of rows and columns. The parity which is inserted in first matrix is on the basis of number of 1's; if the number of 1 is even, then the parity would be 0 else the parity would be 1. Now moving toward the second matrix, the value is inserted as (Fig. 3).

The values assign to the rows and column are in the form of power (2, k); here, k is the bit position, and it is for row and column; if the matrix is of 5 × 5, then the values of index would be (Fig. 4).

Similarly in case of row indexes in Fig. 5 as:

Then, we calculate the pixel value and parity; the pixel value can be calculated as (Fig. 6).

$$16 * 0 + 8 * 1 + 4 * 0 + 2 * 1 + 1 * 1 = 11$$

Index value   data bit

Index value	data bit
16	0
8	1
4	1
2	0
1	1

Fig. 5 Row index value of 5 × 5 matrix

Index value	16	8	4	2	1
Data bit	0	1	0	1	1

Fig. 6 Index value of 5 × 5 matrix

So the pixel value of first row is 11; similarly, we can find the pixel value of each row and each column and can be stored for further use.

Now when the altered matrix is obtained with random or unidirectional error, then again the pixel value of rows and column of new altered matrix is calculated, and then by subtracting the pixel value of both the matrices, therefore the resultant error bits is calculated and corrected. Then, after the correction of the error, the parity index is compared for surety of obtaining correct matrix.

4 Proposed Algorithm

1. Take Image A[4] [4] as input// here we taking example of 4 × 4 matrix
2. Create Transform Image B[6] [6] matrix by adding parity and pixel value of row and column.

```
For(i=0;  i<=3;  i++)
{    row=0 ;  col=0;  k=3;
          Row_value=0; col_value=0;
          For(j=0;  j<=3;  j++)
                  {        Image B[i][j]= Image A[i][j];
                  Row= row+ Image A[i][j];
                  Col= col+Image A [j][i];
                  Row_value= row_value  +  Image  A[i][j]   *
          pow(2,k);
                  Col_value=  col_value   +   ImageA[j][i]*
          pow(2,k);
                        k--;
                  }
Image B[i][4]= row;
Image B[4][i]= col;
Image B[i][5]= row_value;
Image B[5][i]= col_value;
```

3. Now send Image B[6] [6] over network, due to network let the change is made in Image B[4] [4] part noted as image A'

   ```
   Image A' [4] [4]
   ```

4. Create Image B' [6] [6] as

   ```
       For(i=0;  i<=3;  i++)
       K=3; row_value=0; col_value=0
   }
       For(j=0;  j<=3;  j++)
             Image B'[i][j]= Image A'[i][j]
             Row_value=     row_value+     Image     A'[i][j]
       *pow(2,k);
             Col_value=  col_value  +  Image  A'[j][i]  *pow
       (2,k);
             K-- ;
   }
          Image B'[i][4]= Image B[i][4];
       Image B'[4][i]= Image B[4][i] ;
       Image B'[i][5]= row_value;
       Image B'[5][i]= col_value ;
   ```

5. Compare Image B[i] [5] with Image B' [i] [5] and Image B[5] [i] with Image B [5] [i]

```
For(i=0; i<=3; i++)
{        if Image B'[i][5] > Image B[i][5])
         Difference_row[i]=  Image  B'[i][5]  –  Image
    B[i][5];
Else
         Difference_row[i]=  Image  B[i][5]  –  Image
    B'[i][5];
         If ( Image B'[5][i]> Image B[5][i])
         Difference_col [i]= Image  B'[5][i]  –  Image
    B[5][i];
Else
         Difference_col[i]=  Image  B[5][i]  –  Image
    B'[5][i];
}
```

6. Mark the bits from Difference_row[i] &Difference_col[i] in Image A' [4] [4] matrix.
7. Compliment the marked bits.
8. Check Image A and Image A' and also compare with the parity of each row and column.

5 Working Example

The techniques will be simply understand by an example of 4 × 4 matrix; we will solve the errors by going step by step through the technique in order to find the error and to correct it.

Step 1. Input matrix (Fig. 7).
Step 2. Insert 2-D Parity (Fig. 8).
Step 3. Calculate pixel value and insert it into the matrix (Fig. 9).
Step 4. Now after the effect of noise when the matrix is altered, let the error is in (8,8) (8,4) (4,2), and (1,1) so the matrix obtain after alteration will be and the pixel value according to the new bit will be as follows (Fig. 10).
Step 5. Subtract the pixel value of original matrix from altered matrix. Take the module of the result obtain (Figs. 11 and 12).

Fig. 7 Input 4 × 4 matrix

1	0	1	0
0	0	1	1
1	1	0	0
0	0	0	1

Parity

0	1	0	1	0
0	0	0	1	1
0	1	1	0	0
1	0	0	0	1
	0	1	0	0

Parity

Fig. 8 Input matrix with parity bit

Parity Pixel value

1	0	1	0	0	10
0	0	1	1	0	3
1	1	0	0	0	12
0	0	0	1	1	1
0	1	0	0		
10	2	12	5		

Parity (row 5), pixel value (row 6)

Fig. 9 Input matrix with pixel value and parity bit

Now if we shade the bit positions which are obtained, we can see that indicates the error position (Fig. 13).

Step 6. Now, take the complement of each bit position which are obtained and compare with the parity; we can obtain the original matrix.

We have applied the above technique on various matrices 4 × 4, 5 × 5 and 6 × 6; so we have obtained the optimal and correct solution in every size matrices. The technique improves many problems of different methods which are used for error detection and correction. The main advantage of this technique is that it provides optimal and reliable solution in every case, and also with the help of parity matrix, we recheck the result and make sure if the solution we are getting is correct. The other advantage is that it is easy to implement and easy to understand as compared to other techniques and also we do not compromise with the results because it give correct results.

				Parity	pixel value
0	1	1	0	0	16
0	0	0	1	0	1
1	1	0	0	0	12
0	0	0	0	1	0
Parity	0	1	0	0	
Pixel value	2	10	8	4	

Fig. 10 Matrix with error bits

Pix. Val. ori. row	Pix. Val. Alt. row	Bit positn.
10	6	4th
3	1	2th
12	12	0
1	0	1st

Fig. 11 Calculate row value

Fig. 12 Calculate column value

Pix. Val. ori. col	Pix. Val. Alt. col	Bit positn.
10	2	8th
2	10	8th
12	8	4th
5	4	1st

Fig. 13 Matrix after finding errors

0	1	1	0
0	0	0	1
1	1	0	0
0	0	0	0

6 Conclusions and Future Work

From the above work, we concluded that this technique can be used for error correction and also detection for the noisy image matrices; in order to do that, we have to first convert the image into binary form and then we can apply this technique. This technique uses two extra information matrices that are parity matrix and pixel value matrix; with the help of these, the error can be detected and also corrected. This technique requires less computation and better performance which are the necessary conditions for every technique. In future, this technique can also be used with the watermarking technique or in secure transmission method.

References

1. Pradhan, D.K.: A new class of error-correcting/detecting codes for fault-tolerant computer applications. IEEE Trans. Comput. (1980)
2. Shieh, S.-L.: Concatenated BCH and LDPC coding scheme with iterative decoding algorithm for flash memory. IEEE Commun. Lett. 327–330 (2015)
3. Dau, S.H., Skachek, V., Chee, Y.M.: Index coding and error correction. In: IEEE International Symposium on Information Theory Proceedings (ISIT), pp. 1787–1791 (2011)
4. Lahtonen, J.: Decoding the 6-error-correcting Z4-linear Calderbank-McGuire code. In: Proceedings IEEE International Symposium on Information Theory (2000)
5. Venkateshwari, J., Anbuselvi, M.: Decoding performance of binary and non-binary LDPC codes for IEEE 802.11n standard. In: Recent Trends In Information Technology (ICRTIT), pp. 292–296 (2012)
6. Barac, F., Gidlund, M., Zhang, T.: Channel coding and interleaving in Industrial WSN. In: Abiding to Timing Constraints and Bit Error Nature Measurements and Networking Proceedings (M&N) pp. 46–51 (2013)
7. Bretas, N., Bretas, A., Martins, A.: Convergence property of the measurement gross error correction in power system state estimation, using geometrical background. In: IEEE PES General Meeting | Conference & Exposition (2014)
8. Bretas, N.G., Castillo, M.R.M., London, J.B.A.: The innovation concept for parameter error identification and correction using the composed measurements errors in power system state estimation. In: IEEE Power and Energy Society General Meeting, pp. 1–7 (2012)
9. Guiver, C., Opmeer, M.R.: A Counter example to positive realness preserving model reduction with norm error bounds. IEEE Trans. Circuits Syst. I 1410–1411 (2011)
10. Raheemi, B.: Error correction on 64/66 bit encoded links. In: Canadian Conference on Electrical and Computer Engineering, pp. 412–416 (2005)
11. Al-Qedra, M.., Okhmatovski, V.: Novel error-control methodology for finite difference and finite element based electrostatic green's function computation in inhomogeneous substrates. In: Radio Science Meeting (Joint with AP-S Symposium), USNC-URSI, p. 237 (2015)

Computer-Aided Diagnosis of Type 2 Diabetes Mellitus Using Thermogram of Open Mouth

Priyam Singh, Amrita Basundhara Pal, M. Anburajan and J. S. Kumar

Abstract Type 2 diabetes mellitus (DM) is one of the most chronic diseases worldwide. Presently, there are no specific symptoms for this disorder. Hence, an early diagnosis is very critical to prevent the long-lasting complication caused by the disease. In this paper, the efficiency of thermogram was investigated for the evaluation of this disease when compared with biochemical standard, i.e., glycated hemoglobin (HbA_{1c}). The various gray-level co-occurrence matrix (GLCM) and statistical features were extracted from the thermograms of open mouth with the help of MATLAB software. The best result obtained from classification using artificial neural network had an accuracy level of 86.7%, sensitivity of 80%, and specificity of 93.3%. The computer-aided diagnosis (CAD) model was developed for the evaluation of type 2 DM from thermogram. This study proved that the thermogram of open mouth can be an efficient, faster, and powerful diagnostic tool for type 2 DM when compared with the HbA_{1c}.

Keywords Type 2 diabetic mellitus · Thermogram · GLCM
Artificial neural network · Computer-aided diagnosis

P. Singh (✉) · A. B. Pal · M. Anburajan
Department of Biomedical Engineering, SRM University,
Chennai, Tamil Nadu, India
e-mail: priyam26may@gmail.com

A. B. Pal
e-mail: basundhara89@gmail.com

M. Anburajan
e-mail: hod.biomedi@ktr.srmuniv.ac.in

J. S. Kumar
Department of Diabetology, SRM Medical College and Hospital,
Kattankulathur, Chennai 603203, Tamil Nadu, India
e-mail: Kumar.j.s@ktr.srmuniv.ac.in

© Springer Nature Singapore Pte Ltd. 2018
K. Saeed et al. (eds.), *Progress in Advanced Computing and Intelligent Engineering*,
Advances in Intelligent Systems and Computing 563,
https://doi.org/10.1007/978-981-10-6872-0_16

1 Introduction

Type 2 diabetes mellitus (DM) is one of the most chronic and long-term metabolic disorders that is described by high blood sugar level and insulin resistance [1]. As the disease progresses, cells fail to respond to insulin properly and result in the relative lack of insulin in the body. There are no specific symptoms of this disease so it is possible that the individual unaware of the disease for an extended period of time which could be hazardous to their health. The long-term complication of this disease comprises cardiovascular symptoms, amputations, peripheral neuropathy with foot ulcer, neuropathy leading to renal failure, autonomic neuropathy causing gastrointestinal, and retinopathy with potential vision loss, genitourinary and sexual dysfunction. It is one of the most chronic diseases in the world [2]. It was predicted that in India, DM may affect up to 79.4 million individuals by the year 2030. Out of which, ninety percent cases are suffering from type 2 DM [3]. Hence, the early diagnosis is very vital to detect type 2 DM. Thermal imaging is one of the methodologies for type 2 DM which are considered to serve as a second diagnostic tool after various medical tests [4]. It is a noninvasive diagnostic technique that identifies the variation of temperature on the surface of the human skin and widely used in the medical field. It is painless non-risky, non-irradiant, economical, quick, non-contact, and user-friendly imaging method [5, 6]. In this paper, we present an efficient method for diagnosis of type 2 DM from thermogram of the open mouth. Thermogram alone will not be sufficient for the healthcare professionals to make a diagnosis; therefore, the use of analytical methods like biostatistical technique and classification through artificial neural network (ANN) will be used to analyze and achieve a higher level of accuracy rate in diagnosis when compared with clinical examination [7, 8]. This might aid the medical practitioner in identifying high-risk type 2 DM patient at a low cost and proceeding with further diagnosis.

2 Materials and Methods

2.1 Study Population

In this research study, thirty thermograms of open mouth of both type 2 DM and normal subjects were collected from the outpatient department of diabetology at SRM Medical College Hospital and Research Center, Kattankulathur, Chennai, Tami Nadu, India, during the free medical camp conducted for the disease on 25 October, 2015. The ethical committee of SRM Medical College Hospital and Research Center, Kattankulathur, Chennai, approved this study, and consent form was signed by each patient before taking the thermograms. Subject with known type 2 DM for quite a long time while (over a year) without the other difficulties like cardio- vascular disease, neuropathy, nephropathy, retinopathy, and foot damage was included in the study.

While, the participants with known both type 1 DM as well as type 2 DM with any one of the above-mentioned diseases were excluded from the study.

2.2 Measurements

In each subject, three measurements were carried out. The first measurement was demographic which include the detailed information about age, body height, and body weight of the subjects. The second measurement was biochemical blood test which includes the HbA_{1c}. According the criteria set by the American Diabetes Association (ADA), if the measured HbA_{1c} was greater than or equal to 6.5% (7.7 mmol L^{-1}), then the subject was characterized as having DM. In each subject, HbA_{1c} was measured as per the ADA criteria and the subjects were grouped into two groups, in which fifteen were known type 2 DM and remaining were age- and sex-matched normal subjects. Therefore, Group-I: healthy control subjects (n = 15, mean ± SD age = 43.1 ± 6.7 years, women: men = 10:5) and Group-II: type 2 DM patients (n = 15, mean ± SD age = 68.7 ± 8.6 years, women: men = 10:5). The third measurement included the acquisition of thermograms. It was done in the controlled room temperature of 26 ± 10 °C (approximately) and partially darkened during the test. The subjects were asked to come with overnight fasting and requested to sit and relax for 10 min in the controlled room temperature. The subjects especially in the case of female were requested to remove any accessories which may alter the skin temperature. The standard thermogram of center part of open mouth from anterior to posterior in upright sitting position of the subject was acquired using a handheld thermal camera (A305sc, FLIR) from the distance of 0.9 m with a resolution of (320 × 240) pixels. The thermogram was obtained in the JPEG format of IRON palette and its thermal sensitivity of reported as 0.95 mk.

2.3 Thermal Image Processing

The image processing was further divided into two parts. First is the average skin surface temperature (SST) measurement of region of interest (ROI) in thermogram, which was done by the software supplied by manufacturer of the camera (FLIR software version 1.2). It is a powerful and easy to use thermal analysis software package for fast speed data recording, image analysis, and data sharing. After uploading all the images in the software, a tool named 'analyze' was selected. Then, in the thermogram, ROI has to be defined. In this study, the ROI was the center part of open mouth view, which mainly comprised the middle part of tongue. For selecting the ROI, a rectangular box was chosen and it was kept in the center part of open mouth carefully excluding tooth and lips of each thermogram. Now, in each thermogram, the chosen rectangular analyze box of almost same area was placed appropriately, and the following measurements were made: (i) average SST (°C); (ii) area of ROI (mm); and (iii) average SST per unit area of ROI (°C/mm). The inspection report was exported into excel sheet for further analysis. The thermal image processing was done in MATLAB R2103a programming language version (8.1.0.430).

The algorithm is summarized as follows:

Step-1: Thermogram of open mouth of a subject was given as an input image;
Step-2: Conversion of input RGB image into grayscale image;
Step-3: Cropping and resizing of the ROI (center part of open mouth view) manually (128 × 128) pixels.
Step-4: Removal of noise from the resized image using adaptive filter.
Step-5: Perform GLCM feature extraction [(i) Autocorrelation; (ii) Correlation, (iii) Variance; (iv) Dissimilarity; (v) Cluster shade; (vi) Contrast; (vii) Energy; (viii) Entropy; and (ix) Homogeneity] and Statistical analysis [(i) Mean; (ii) Standard deviation; (iii) Range; (iv) Kurtosis; and (v) Skewness]. Thus, a total number of fourteen features were extracted from each thermogram of the open mouth.
Step-6: Perform feature selection by linear regression (LR).
Step-7: Perform classification using back propagation (BP) based ANN.
Step-8: Plot receiver operating curve (ROC) from the obtained classification result.
Step-9: Develop computer aided diagnosis (CAD) model for the evaluation of type 2 DM with good accuracy.

2.4 Feature Selection

Feature selection is one of the critical steps in data analysis as it generally increases the accuracy level of the system in the field of pattern recognition. In this paperwork, linear regression (LR) method used the feature selection [9] as it is usually computes the whole process more effectively and computationally. The LR correlation analysis was done between all the extracted features from thermogram of open mouth to HbA$_{1c}$ (%), as standard using IBM SPSS software 10.0 (SPSS Inc., Chicago, USA) [10]. It was found that only the following five features, out of fourteen features extracted from thermogram of open mouth viz. (i) energy; (ii) entropy; (iii) contrast; (iv) correlation; and (v) homogeneity, were correlated statistically significant with HbA$_{1c}$ (%). Therefore, the above-mentioned features of thermogram were chosen as inputs to be given in the classifier for further analysis.

2.5 Classification of Type 2 DM Using Artificial Neural Network (ANN)

ANN was used for pattern recognition, and its data processing system and inter-connected framework were based on the function of the human cerebrum [10]. In this paperwork, for classification, back propagation (BP) algorithm was applied in ANN. There are three interconnected layers viz. input, output, and hidden layers. According to the various inputs given to BP algorithm, the desired output was obtained [11, 12]. In the input layer, dataset was given to the network through an input buffer, whereas the output layer carried the output response to a given input [13]. The hidden layers are very crucial parameters in the network. In this paper-work, there were five input layers, one output layer and one hidden layer. The various operations done for the training of the system after the training of dataset is over; it is stopped without further training is done. Testing dataset has no impact on training and also provides a completely independent measure of network performance during and after training.

2.6 CAD Model for Evaluation of Type 2 DM

The CAD is a kind of method which is nowadays used to help doctors and other medical professionals for the better interpretation of medical images. The CAD model was developed using graphical user interface (GUI). The GUI is a built-in program of MATLAB that permits users to interact with electronic gadgets through graphical symbols and visual indicators. The activities in a GUI are usually performed through direct manipulation of the graphic-17 call components in addition to PCs. A CAD model was developed for the evaluation of type 2 DM with good accuracy from the thermogram of open mouth.

3 Experimental Result

As we mentioned earlier, the aim of this study was to investigate the efficiency of the thermogram with biostatistical methods and ANN in the diagnosis of type 2 DM when compared with HbA_{1c} (%), a biochemical standard. The standard IR thermograms of open mouth of an individual from both groups have given below in Fig. 2. The ROI has been selected manually with care, so that the tongue region alone has been selected with almost uniform area in each subject. The results of various procedures are given in Tables (Fig. 1).

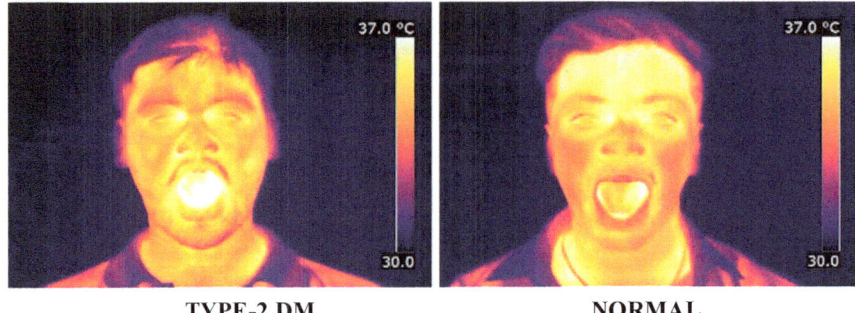

TYPE-2 DM   **NORMAL**

Fig. 1 Region of interest (ROI), i.e., center view of open mouth of thermograms of both type 2 DM and normal subjects

Table 1 Statistical correlation between various extracted features from open mouth thermogram and HbA_{1c}

Features extracted from thermogram of open mouth (n = 30)	Biochemical variable: HbA_{1c} (%)	
	Linear regression	P value
Energy	−0.371	0.044
Entropy	0.402	0.020
Homogeneity	0.383	0.037
Correlation	−0.421	0.021
Contrast	−0.364	0.048
Average SST/area of open mouth	0.286	0.027

3.1 Statistical Correlation Analysis

In all 30 subjects studied (both type 2 DM patients and normal subjects), the Pearson's correlation was performed using SPSS software. The result was tabulated in 3.1 which showed that there were statistically significant positive correlations between HbA1c (%) with the following extracted features of thermogram of open mouth namely: (i) entropy (r = 0.402, p = 0.020); (ii) homogeneity (r = 0.383, p = 0.037). Also, HbA_{1c} was statistically significant and negatively correlated with the following extracted features from thermogram of open mouth, namely: (i) energy (r = −0.371, p = 0.044); (ii) contrast (r = −0.364, p = 0.048); and (iii) correlation (r = −0.421, p = 0.021). The average SST per unit area of open mouth correlated statistically significant with HbA_{1c} (r = 0.286, p = 0.027) Table 1.

3.2 Neural Network Analysis

In the evaluation of type 2 DM, the selected features from thermogram of open mouth (total subjects = 30, both type 2 DM patients and normal subjects) were trained as well as tested using BP-based ANN classifier.

The architecture of BP-based ANN was given in Fig. 2. From the Table 2, it can be easily concluded that the total number of samples given for the training was 30 while only 25 samples were selected for the testing procedure. The number of input neurons totally depended on the number of features selected for the further classification, and in this study, it was five; similarly for hidden layer, we opted one because it was calculated according the numbers of input neurons given for the classification. The performance of the system is 0.36 which represents the good performance level of the system. From the Table 3, it can be easily concluded that the accuracy of the classification study for thermal image was found to be 86.7% and sensitivity was found to be 80% with the specificity of 93.3%. Figure 3 is the confusion matrix obtained after the classification, and the above score rate was tabulated according the confusion matrix of the classification. The ROC curve was plotted for the same result (Fig. 2) between the true positive rate on y-axis and false positive rate on x-axis, values obtained from the confusion matrix after the classification. It showed the good performance level of the proposed system.

3.3 Computer-Aided Diagnosis (CAD)

In all 30 subjects studied (both type 2 DM patients and normal subjects), it can easily see that the CAD model from the Fig. 4 was developed for the evaluation of type 2 DM from the thermograms of open mouth. The CAD model was made for the best classification output obtained from the matricial neural network for the better accuracy in evaluation of the disease. For this study, there was one edit text in which title for the CAD model has been given. There were three axes and three pop up menu has been selected. Three axis were used to display original image, gray image, and region of interest while their respective titles were shown into pop up menu. There were two push buttons; first one is used for browsing while the

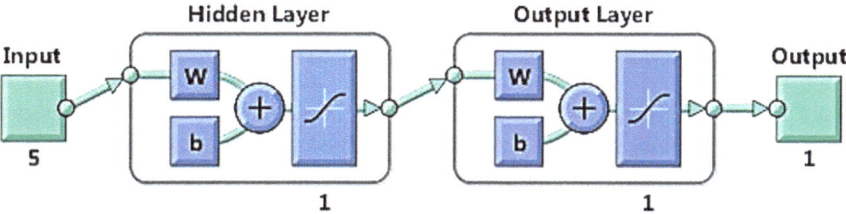

Fig. 2 An architecture of artificial neural network (ANN) for classification of type 2 DM using selected features of thermogram of open mouth

Table 2 BP-based ANN using selected extracted features from open mouth thermogram in the evaluation of type 2 DM

Artificial Neural Network (ANN)	Total subjects
No. of training samples	30
No. of testing samples	25
No. of input neurons	5
No. of hidden neuron	1
No. of output neuron	1
Performance	0.36

Table 3 Performance of BP-based ANN classifier in the evaluation of type 2 DM (n = 30)

Result	Number
True negative	14
False negative	3
False positive	1
True positive	12
Accuracy	86.70%
Sensitivity	80%
Specificity	93.30%
Positive predictive value	82.40%
Negative predictive value	92.30%

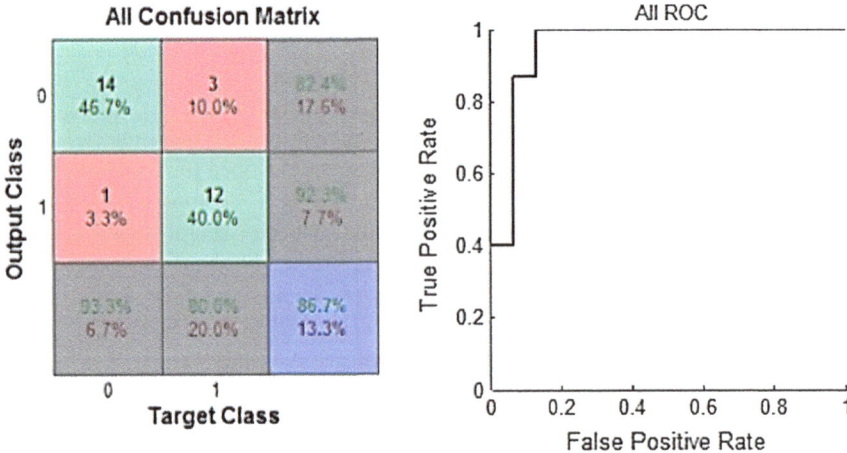

Fig. 3 Confusion matrix of BP-based ANN classifier using selected features of open mouth thermogram in the evaluation of type 2 DM and ROC plotted on the best result of classification

second one for diagnosis. The images were browsed from the database and appeared as 'original image' on the first axis after that it was automatically converted into gray image. From the gray image, ROI has been chosen very carefully and after that it immediately classified the given input image either normal or type 2 DM accordingly for the study. The final evaluated result has been showed into the static box Fig. 4.

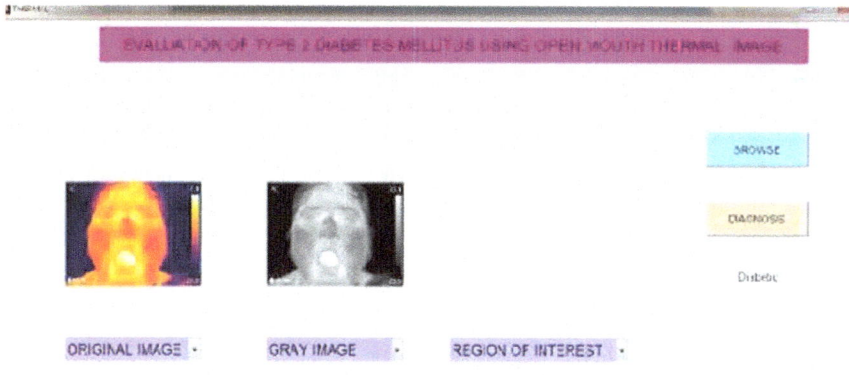

Fig. 4 Developed CAD in the evaluation of type 2 DM using thermogram of open mouth

4 Discussion

The thermogram is widely used in the medical field for detection of various health problems and diseases [2, 5, 7]. In this paper, we investigated the usefulness of the thermogram in the diagnosis of type 2 DM. There were no specific symptoms and causes listed for this disease [3]. The some of the precautions are regular routine checkup, control on diet and no alcoholic consumption, and regular exercised can be beneficial, but it is not sufficient. Therefore, in this study, we tried to find out the suitable, quick, low cost, and patient-friendly method for diagnosis of type 2 DM to low down the mortality rate in India due to this disease. Type 2 DM is one of the most chronic diseases worldwide, and the present method of diagnosis is very painful, invasive as well as costly for many people, and it is not available everywhere [1]. There were many research works that have been done on the various body parts (mainly face, retina, palm-region, and feet). In this study, we had chosen the central part of open mouth, which mainly comprised the tongue region. Because the tongue region diagnosis is one of the most crucial diagnostic methods in the clinical as well as biomedicine field as it is very easy and fast to observe any abnormal changes in the tongue [14]. In some cases like breast cancer, thermal imaging has the potential to detect the disease, approximately ten years before the standard method, i.e., mammography. Even though thermal imaging has its unique merits, medical practitioner is not ready to accept the thermography alone as a reliable diagnostic tool, because of inconsistencies in the result of diagnosis obtained from the thermogram [8]. Therefore, in this paper, we incorporated the biostatistical tool and ANN with thermal imaging for better accuracy and result. The BP-based ANN is a supervised learning neural network model which is effectively used in the most clinical applications and research work all over the world [15]. The slower speed of classification is one of the disadvantages of using this method, but its performance and accuracy level are better than other algorithms [19]. Therefore, we used this method for the classification of subject having type 2 DM [16]. The ROC curve implied the good performance and accuracy level of the system in the diagnosis of type 2 DM when compared with HbA_{1c} [17, 18]. Also, a CAD model was developed to evaluate type 2 DM with good accuracy using BP-based ANN classifier from selected extracted features of open mouth thermogram. This can be considered as one of the low cost and faster screening procedure to identify patient, who is prone to have high risk of type 2 DM.

5   Conclusion

Early detection of type 2 DM played a crucial role in the decrease of mortality rate in India. In this paper, we presented an effective and practical approach for medical diagnosing of type 2 DM using thermograms of open mouth. The feature extraction and selection were done to check the efficiency. The classification was done by using back propagation algorithm of ANN which is resulted with accuracy score of 86.7% and sensitivity of 80% with the specificity of 93.3%. So, it can be say that the thermograms of open mouth can be an effective and faster medical diagnosing tool for type 2 DM.

References

1. Devi, M.N., Balamurugan, A., Reshma Kris, M.: Developing a modified logistic regression model for diabetes mellitus and identifying the important factors of type II Dm. Indian J. Sci. Technol. **9** (2016)
2. Kalaiselvi, C., Nasira, G.: Prediction of heart diseases and cancer in diabetic patients using data mining techniques. Indian J. Sci. Technol. **8** (2015)
3. Tama, B., Rodiyatul, F.S., Hermansyah, H.: An early detection method of type-2 diabetes mellitus in public hospital. TELKOMNIKA (Telecommun. Comput. Electron. Control). **9**, 287 (2011)
4. Duarte, A., Carrão, L., Espanha, M., Viana, T., Freitas, D., Bártolo, P., Faria, P., Almeida, H.: Segmentation algorithms for thermograms. Procedia Technol. **16**, 1560–1569 (2014)
5. Devulder, J., Dumoulin, K., De Laat, M., Rolly, G.: Infra-red thermographic evaluation of spinal cord electrostimulation in patients with chronic pain after failed back surgery. Br. J. Neurosurg. **10**, 379–384 (1996)
6. Acharya, U., Ng, E., Sree, S., Chua, C., Chattopadhyay, S.: Higher order spectra analysis of breast thermograms for the automated identification of breast cancer. Expert Systs. **31**, 37–47 (2012)
7. Dayakshini, D., Kamath, S., Prasad, K., Rajagopal, K.: Segmentation of breast thermogram images for the detection of breast cancer—a projection profile approach. J. Image Graph. **3** (2015)
8. El-Melegy, M.: Random sampler m-estimator algorithm with sequential probability ratio test for robust function approximation via feed-forward neural Networks. IEEE Trans. Neural Netw. Learn. Syst. **24**, 1074–1085 (2013)

9. Ng, E., Kee, E.: Advanced integrated technique in breast cancer thermography. J. Med. Eng. Technol. **32**, 103–114 (2008)
10. Pal, S., De, R., Basak, J.: Unsupervised feature evaluation: a neuro-fuzzy approach. IEEE Trans. Neural Netw. **11**, 366–376 (2000)
11. Chon, T., Park, Y., Moon, K., Cha, E.: Patternizing communities by using an artificial neural network. Ecol. Model. **90**, 69–78 (1996)
12. Gevrey, M., Dimopoulos, I., Lek, S.: Review and comparison of methods to study the contribution of variables in artificial neural network models. Ecol. Model. **160**, 249–264 (2003)
13. Watta, P., Sudjianto, A.: Pattern Recognition and Neural Networks [Book Reviews]. IEEE Trans. Neural Netw. **8**, 815–816 (1997)
14. Zhang, B., Kumar, B., Zhang, D.: Detecting diabetes mellitus and nonproliferative diabetic retinopathy using tongue color, texture, and geometry features. IEEE Trans. Biomed. Eng. **61**, 491–501 (2014)
15. Tiwari, A., Dubey, A., Patel, A.: Comparative study of short term load forecasting using multilayer feed forward neural network with back propagation learning and radial basis functional neural network. SAMRIDDHI J. Phys. Sci. Eng. Technol. **7** (2015)
16. De Jesus, O., Hagan, M.: Backpropagation algorithms for a broad class of dynamic networks. IEEE Trans. Neural Netw. **18**, 14–27 (2007)
17. Huang, C., Wu, Y., Hwang, C., Jong, Y., Chao, C., Chen, W., Wu, Y., Yang, W.: The application of infrared thermography in evaluation of patients at high risk for lower extremity peripheral arterial disease. J. Vasc. Surg. **54**, 1074–1080 (2011)
18. Misnikova, I., Dreval, A., Kovaleva, Y., Gubkina, V., Odnosum, A.: Significance of HbA1c targets based on an individual approach to the treatment of patients with type 2 diabetes mellitus. DM. **17**, 4 (2014)
19. Niyati Gupta, N.: Accuracy, sensitivity and specificity measurement of various classification techniques on healthcare data. IOSR-JCE. **11**, 70–73 (2013)

3D Surface Measurement through Easy-Snap Phase Shift Fringe Projection

A. Harshavardhan, T. Venugopal and Suresh Babu

Abstract Smaller objects reconstruction using three-dimensional techniques is one among the challenging tasks from the decade. Researchers in graphical designing and professionals of photography are continuously working on the reconstruction of 3D object techniques to meet the demand of real-time applications of almost all in every walk of real life. Reconstruction of 3D objects has a major role in the reverse engineering applications too. The major challenges in successful 3D object reconstruction are high computational costs and lack of accuracy. Fringe projection has come into view as a propitious 3D reconstruction mechanism with low computational cost for high precision and resolutions. It makes use of digital projection, structured light systems, and phase analysis on fringed images. Its performance is shown as acceptable in the research analysis carried out on the implementation of it and its insensitiveness to ambient light. An overview of some of the fringe projection techniques are presented in this paper and also propose a new simple fringe projection system, which can yield the more accurate and acceptable results with different objects.

Keywords Digital fringe projection · 3D reconstruction · Fringe analysis
Phase shifting · Structured light systems

A. Harshavardhan (✉)
Department of CSE, JNTUH, Hyderabad, India
e-mail: harshavgse@gmail.com

A. Harshavardhan
Department of CSE, SREC, Wgl, Warangal, India

T. Venugopal
Department of CSE, JNTU, Sultanpur, Medak, India
e-mail: t_vgopal@rediffmail.com

S. Babu
Department of CS, KDC, Warangal, India
e-mail: sureshd123@gmail.com

© Springer Nature Singapore Pte Ltd. 2018   179
K. Saeed et al. (eds.), *Progress in Advanced Computing and Intelligent Engineering*,
Advances in Intelligent Systems and Computing 563,
https://doi.org/10.1007/978-981-10-6872-0_17

1 Introduction

Measuring techniques objects using 3D three-dimensional approaches have been attracted intensive research in every walk of our real-time applications over the past decade, e.g., time of flight, stereo vision, and structured light mechanisms [1]. Every mechanism has its own unique features and can be implemented into various applications [1]. Adopting phase shift fringe projections number of techniques for 3D object measurement is developed and available today commercially. Fringe projection techniques are broadly categorized into either multi-shot or single-shot mechanisms. Multi-shot mechanisms are employed often due its high reliability and accuracy in results, especially on the stationary objects, where application does not have acquisition time constraints. On the other hand, single-shot mechanisms are employed on moving objects for 3D object measurement. Fourier transformation method a current single-shot mechanism does not provide accuracy in measuring of objects. The rainbow 3D cameras possess the ability to snapshot and achieve the measurement of objects with color-coded single patterns with the limitations of color cross talk and distortion of color. Phase shifting mechanism is a digital fringe projection multi-shot mechanism with high resolutions. Measuring of highly dynamic objects is a challenging task because the number of fringe patterns needed for the measurement of objects, minimum of three patterns. Using of supersonic cameras and the pattern binary projections with digital light processing projectors are the latest advancements to attain the speed of kHz for the measurement of 3D objects.

2 Overview of 3D Shape Measurement Techniques

As in below Fig. 1, 3D shape measurement mechanisms are broadly categorized as contact-based and non-contact-based mechanisms [1]. Non-contact mechanisms again classified as active and passive approaches, active approaches are further divided into TOF or structured light mechanisms based on the way of reconstruction of objects. Based on the time of flight of the light beam, TOF mechanisms reconstruct the objects shape. The reconstruction of the shape of objects is based on triangular relationships in structured light mechanisms and is classified as FPP and non-FPP based on fringe patterns employed [1]. The FPP mechanisms based on number of fringe patterns employed they are further categorized into single-shot and multi-shot mechanisms.

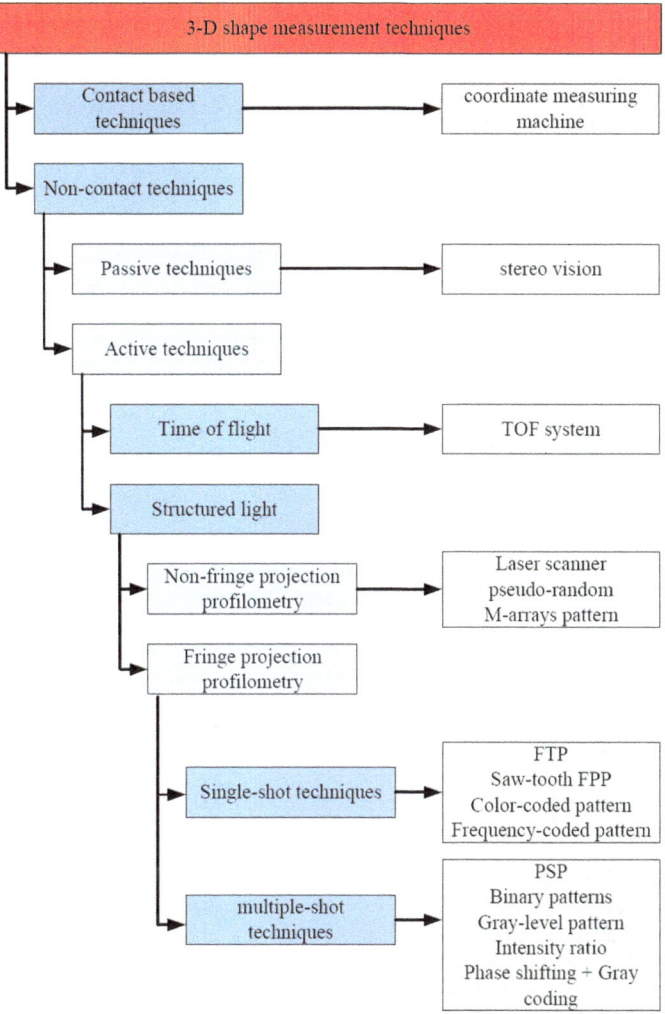

Fig. 1 Taxonomy of 3D shape measurement techniques

2.1 Fringe Projection Profilometry System

As in below Fig. 2, a fringe projection profilometry system consists a projection, image acquisition, and processing or analyzing unit. Fringe projection mechanism for measuring the shape of an object involves: (1) structured pattern projection on the surface of the object; (2) fringe pattern image capturing which is modulated in phase with the height distribution of the object; (3) calculation of phase modulation by analyzing the image with the fringe analysis; (4) calculating the continuous phase distribution using an appropriate phase unwrapping algorithm which is

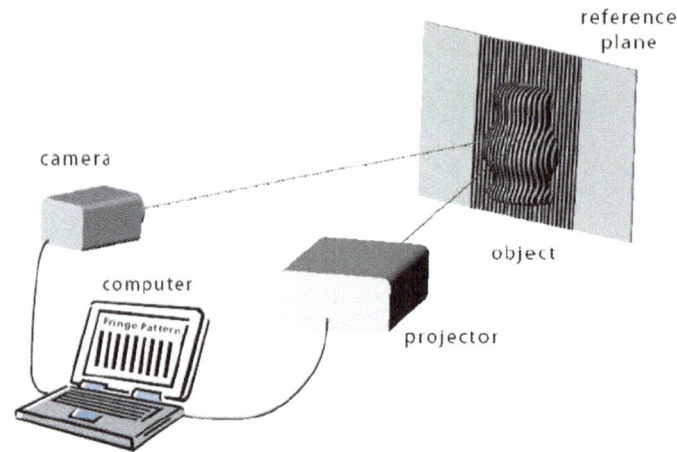

Fig. 2 Fringe projection profilometry system

Fig. 3 Fringe projection
phases

Patterns of fringe projection
Image acquisition
Fringe analysis and phase map obtaining
Unwrapping of phase
Conversion of phase to height

similar to variations of the object height; and (5) system calibration to map unwrapped distributed phase to the real-world environment [1]. Figure 2 gives the clear image of major phases involved for measuring the 3D objects employing the fringe projection mechanisms (Fig. 3).

3   Fringe Projection Mechanisms

3.1   Single Frame Digital Fringe Projection Profilometry

Projecting the linear fringes over the plane or curved object surfaces is a digital fringe projection mechanism which helps to implement HT method easily on curved surfaces along with the straight forward manner [2]. Only one frame is needed for the evaluation surface shape of the objects, sinusoidal fringes are focused onto the object with a portable digital light processing device connected to a computer in this mechanism. Images with high quality, brightness, contrast, and spatial repeatability are projected using a Digital Micromirror Device seamlessly. Currently, linear interference fringe patterns with gray scales are generated on MATLAB and projected on the objects. The systems' sensitivity depends on the axis angle of the camera and projectors'; with the increase in angle, the sensitivity increases. Accordingly, if the height of the object to be evaluated it needs to set a suitable angle, and the received fringes are buffered for fringe analysis.

3.2   General Solution for High Dynamic Range Three-Dimensional Shape Measurement using the Fringe Projection Technique

For measuring the objects with high reflectivity or with shiny specular surface, there exist three mechanisms. The reflectivity's of the object measured is divided into groups depending on the histogram distribution, and for each groups, optimal exposure time is predicted so that it is possible to handle the bright and dark areas on the measured surface without any compromise for the high reflectivity surfaces [3]. At calculated exposure times, capturing of phase-shifted images is carried out, and a phase-shifted image with composite characteristic is extracted from the optimally exposed pixels in the raw fringe images. Two orthogonal polarizers came into picture placed separately before the camera and projector into the high reflectivity mechanism for the shiny specular surface. With the combination of the above two mechanisms, a third mechanism is developed in which altering of angles among the polarizers of the second mechanism is adopted. The reflective object mechanism increases the measurement accuracy with LRR on diffuse objects. On the other hand, the objects with low specular reflection are measured adopting the second mechanism. The third mechanism can inspect the surfaces with strong specular reflections very accurately.

3.3 On Axis Fringe Projection

In this mechanism, the angle of projection and direction is zero, but it is highly sensitive because of projection divergence which leads to the change in the frequency fringes on z-axis [4]. Due to the zero angles, it reduces the shading problem considerably, and general calibration is enough to get better results. In this, the phase differences are negligible in comparison with standard fringe projecting mechanism because of divergent light used in projection. Its signal-to-noise ratio is very less and highly sensitive to the noise produced by camera.

3.4 Velocity Sensing Approach

A sinusoidal pattern of fringes are projected onto the moving objects, and observes the fringes on the object by a CCD camera placed at various different angles. Fringes are blurred due to the long exposure time of the camera; these fringes provide extra information to calibrate the depth and to identify the velocity vector [5]. Periodically, multi-shot measurements are not needed to address changes in 3D positions. As a result, no image registration is required, and it makes possible to inspect several objects simultaneously.

3.5 Multi-Polarization Fringe Projection Imaging Algorithm

In MPFP, the light beam is linearly polarized prior and projected on the object, than the fringes are captured after the multi-polarization reflection [6]. A pixilated polarized camera of four states attached to sensor. Different object surfaces modulate different polarized fringes on reflection, which leads to dissimilar measurements in multi-polarized channels.

3.6 Snapshot Phase Shift Fringe Projection

It uses polarized light illumination and polarization camera. Single source of light is split into two beams: one among of it is left and the other is right circularly polarized for illuminating the object simultaneously [7]. A polarization camera of four-channel focal plane division is used to capture the light reflected from the surface of the objects. Reconstruction of a 3D object surface is carried out by extracting a snapshot image with a phase shift of $\pi/2$ form the four images. It is the foremost mechanism of snapshot for surface imaging of 3D objects with a high potential of ultrafast imaging of 3D objects.

In the process of reconstruction of the 3D shape of the objects in contrast to the single-shot mechanisms, PSP, binary pattern profilometry, and the intensity ratio mechanism need more than one fringe pattern. Multi-shot are robust when compared to single-shot due to the use of more information and can achieve accurate result. At the same time, the time consuming is very high to acquire fringe patterns and gives the accurate results, the object must be kept stationary during the multiple fringe pattern projection. As the object moves errors will be a raised and can be nullified.

4   Easy-Snap Phase Shift Fringe Projection

In this paper, we propose a unique easy-snap phase shift fringe projection for measuring 3D surfaces by the illumination of polarized light and polarization camera. The source light initially splits into two beams, one among the two is left circularly polarized and the other beam is again split into two beams, one of the beams is right circularly polarized, the other is linearly polarized, and all the beams are superimposed for illuminating the objects simultaneously. For capturing of the reflected light from the surface of the objects, polarization camera of four-channel division-of-focal-plane (DoFP) is used. From the easy-snap image, four varied images of the object are extracted by a phase shift of $\pi/2$ and analyzed for the reconstruction of 3D surface of the objects. It can be more insensitive to the objects in motion and with a potential of rapid 3D surface imaging (Fig. 4).

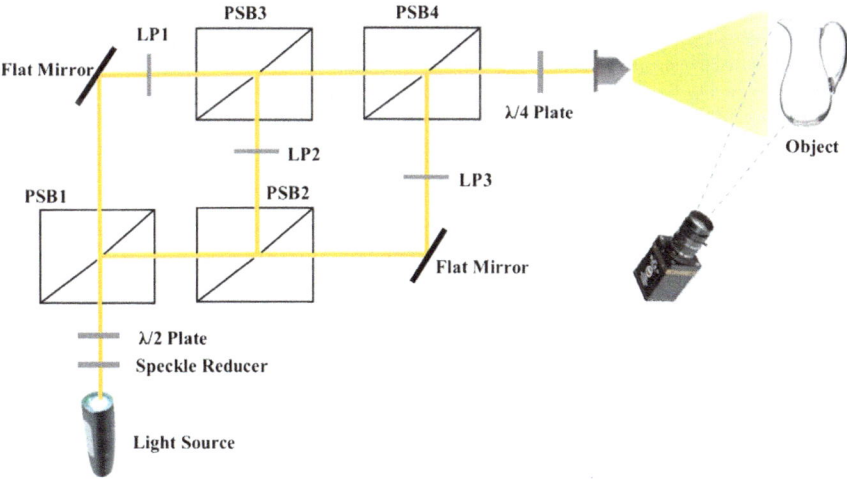

Fig. 4 Easy-snap phase shift fringe projection experimental setup

5 Conclusion

For reconstructing the 3D objects, fringe projection is a powerful mechanism. In this paper, various 3D objects shape measurement mechanisms such as single-shot and multi-shot mechanisms were analyzed using phase shift method for fringe pattern analysis. In this paper, we propose an easy-snap phase shift fringe projection which is for the measurement of 3D surface shape of the objects. In this mechanism, single light source is twice split into two beams and polarized for illuminating the objects simultaneously. A polarization camera of four-channel division-of-focal-plane (DoFP) is used. From the easy-snap image, a four varied images of the object are extracted by a phase shift of $\pi/2$ and analyzed for the reconstruction of 3D surface of the objects. By employing fast polarization cameras built from fast CCD or CMOS cameras with pixilated polarizer arrays, using this method, we can measure the objects moving or varying shape objects with high speed, which can be the foremost mechanism of single-shot fringe projection for surface imaging of 3D objects with a high potential for rapid imaging of 3D objects.

References

1. Lu, L.: Improving the accuracy performance of fringe projection profilometry for the 3D shape measurement of objects in motion, Doctor of Philosophy thesis, School of Electrical, Computer and Telecommunications Engineering, University of Wollongong (2015)
2. Paul Kumar, U., Somasundaram, U., Kothiyal, M.P., Mohan, N.K.: Single frame digital fringe projection profilometry for 3-D surface shape Measurement. Elsevier Optik **124**, 166–169 (2013)
3. Feng, S., Zhang, Y., Chen, Q., Zuo, C., Li, R., Shen, G.: General solution for high dynamic range three-dimensional shape measurement using the fringe projection technique. Elsevier Opt Lasers Eng. **59**, 56–71 (2014)
4. Sicardi-Segade, A., Estrada, J.C., Martínez-García, A., Garnica, G.: On axis fringe projection: a new method for shape measurement. Elsevier Opt Lasers Eng **69**, 29–34 (2015)
5. Sun, W.-H, Co, W.-T.: A real-time, full-field, and low-cost velocity sensing approach for linear motion using fringe projection techniques. Elsevier Opt Lasers Eng **81**, 11–20 (2016)
6. Salahieh, B., Chen, Z., Rodriguez, J.J., Liang, R.: Multipolarization fringe projection imaging for high dynamic range objects. Opt Express **22**(8) (2014)
7. Chen, Z., Wang, X., Liang, R.: Snapshot phase shift fringe projection 3D surface measurement. Opt Express **23**(2) (2015)

Off-Line Handwritten Odia Character Recognition Using DWT and PCA

Abhisek Sethy, Prashanta Kumar Patra and Deepak Ranjan Nayak

Abstract In this paper, we propose a new approach for Odia handwritten character recognition based on discrete wavelet transform (DWT) and principal component analysis (PCA). Statistical feature descriptors like mean, standard deviation, energy have been computed from each sub-band of the second level DWT and are served as the primary features. To find the most significant features, PCA is applied. Subsequently, back-propagation neural network (BPNN) is harnessed to perform the classification task. The proposed method is validated on a standard Odia dataset, containing 150 samples from each of the 47 categories. The simulation results offer a recognition rate of 94.8%.

Keywords Discrete wavelet transform (DWT) · Optical character recognition (OCR) · Principal component analysis (PCA)

1 Introduction

Optical character recognition (OCR) systems have received much more attention over the last decades because of its wide range of applications in bank, postal, and industries. These systems have been designed not only for printed characters but also for handwritten characters. However, recognizing the handwritten characters has become a difficult task for researchers due to the variations in writing styles of

A. Sethy (✉) · P. K. Patra
Department of Computer Science and Engineering,
College of Engineering and Techonoloy, Bhubaneswar, India
e-mail: abhisek052@gmail.com

P. K. Patra
e-mail: principalcet@cet.edu.in

D. R. Nayak
Department of Computer Science and Engineering,
National Institute of Techonoloy, Rourkela, India
e-mail: depakranjannayak@gmail.com

© Springer Nature Singapore Pte Ltd. 2018
K. Saeed et al. (eds.), *Progress in Advanced Computing and Intelligent Engineering*,
Advances in Intelligent Systems and Computing 563,
https://doi.org/10.1007/978-981-10-6872-0_18

187

human beings and remains an open problem. In recent years, many OCR systems have been proposed by diverse researchers over different languages like Japanese, Chinese, and Arabic [1]. This paper aims at developing an OCR system based on handwritten Odia characters. Odia is the one of ancient and famous regional language in the eastern India and mostly spoken in the state of Odisha, and Kolkata. Recognition of Odia characters has become a tiresome and demanding task. In handwritten Odia characters, a lot of ambiguities can occur as most of the characters are similar in shape and structure and have same orientations. Therefore, it is required to design a robust OCR system that can correctly discriminate the characters. Odia language consists of 49 characters (14 vowels and 35 consonants) and some special conjunct consonants. During past years, different authors have made an attempt for analysis on Odia scripts [2]. Feature extraction stage plays an important role for better recognition. In this paper, we have applied discrete wavelet transform (DWT) on the character images for feature extraction. The coefficients of the level-2 decomposition are computed, and then statistical features like mean, median, min, max, standard deviation, mean absolute deviation, median absolute deviation, Euclidean distance and energy have been calculated for each sub-band. These values are considered as the key feature values for each character image. Thereafter, principal component analysis (PCA) has been employed to obtain the more important features, and subsequently, BPNN is utilized to classify the characters. Simulation results on a standard dataset offer 94.8% accuracy.

The remainder of this paper is structured as follows. We summarize the related works in Sect. 2. We present different methodologies adopted in the proposed system in Sect. 3. We report the simulation results in Sect. 4, and at last we draw conclusions and outline the future scope in Sect. 5.

2 Related Works

In the last two decades, numerous works have been introduced over printed and handwritten Odia scripts. Orientation, angular rotation of character images are reported by Patra et al. in [3]. The authors have used Fourier-Modified Direct Mellin Transform (FMDMT) and Zernike moments over variant and invariant scaled character to get the features. Their experiments were conducted on 49 Odia character images, and the classification accuracy of 99.98% is achieved through probabilistic neural network classifier. In [4], Pal et al. have developed an offline Odia handwritten character recognition system based on curvature feature. PCA was used to reduce 1176 dimensional feature vector to 392, and finally they got 94.6% accuracy by using modified quadratic classifier [5]. Chaudhuri et al. [6] suggested an OCR system for printed Odia script where each character is recognized by a combination of stroke, run-number based feature. In addition, water reservoir-based features were also used. Later on, a binary stage recognition system for handwritten characters is introduced by Padhi and Senapati [7]. They have calculated average zone centroid distance and also reported the mismatch among several characters for achieving high recognition rate. They used two artificial neural networks (ANNs) for the characters of similar groups and for each individual one. Wakabayashi et al. [8] have given a comparative analysis of similar shaped characters of Odia with respect to other languages like Arabic/Persian, Devnagari, English, Bangla, Tamil, Kannada, Telugu. They have introduced Fisher ratio (F-ratio) approach for character along with gradient feature for feature extraction. Dash et al. [9] have utilized PCA for dimensional reduction for feature vector. The unconstrained numerals were taken for classification. A new method called Kirsch edge operator is used which is an edge detection algorithm. They have implemented Modified Quadratic Discriminate Function (MQDF) and Discriminative Learning Quadratic Discriminate Function (DLQDF) as the classifier and achieved 98 and 98.5% recognition rate using MQDF and DLQDF, respectively. Various authors independent writing is classified by Chand et al. in [10]. They have used support vector machine and got promising results. Kumar et al. in [11] have proposed the ant-miner algorithm (AMA) for offline handwritten Oriya character recognition. It is an extension of ant colony optimization algorithm. They have taken the matrix space analysis method and feature analysis method for analyzing the handwritten images. Stroke prevention algorithms are also established by Pujari et al. in [12]. They have proposed a new parallel thinning algorithm to preserve significant features.

The literature study reveals that works in this area are still limited. Most of them are implemented on a smaller dataset. Hence, there is a scope to enhance the recognition rate further on a larger database.

3 Proposed Method and Materials

In this section, we portray the proposed system which aims at recognizing the Odia characters and present the materials used on it. The general steps of the proposed approach are depicted in Fig. 1. which mainly consists of the following steps: We first preprocess the input data and then use DWT to extract features; we use PCA for feature reduction, and subsequently we employ BPNN for classification.

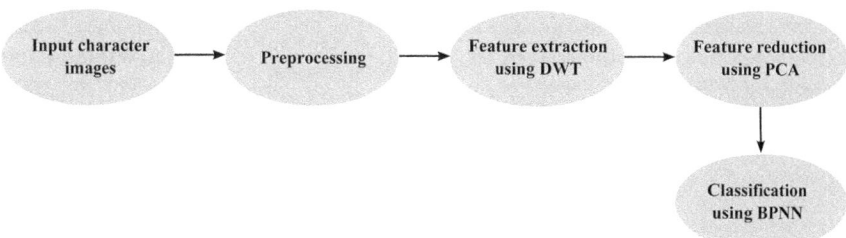

Fig. 1 Proposed model for Odia character recognition

Fig. 2 Samples of 47 Odia characters

3.1 Materials

A standard database, called NIT Rourkela Odia database, has been used to validate the proposed system, which was designed by Mishra et al. in [13]. The database consists of handwritten Odia characters and numerals of different users. Around 15040 images were reported there. But in this paper, we use only 47 Odia characters which are numbered from 1 to 47 as shown in Fig. 2. We have chosen 150 numbers of samples from each character, and hence a total of $47 \times 150 (=7050)$ images have been considered for simulation.

3.2 Preprocessing

Preprocessing is one of the most essential steps of any recognition system. It helps to maintain the originality of the character image by removing the unwanted things from the image. Here, we first resize the input grayscale image into 81×81. Additionally, we employ min-max normalization on the character matrix of the respective images in order to obtain a normalized data set; finally, a morphological operation called dilation is utilized.

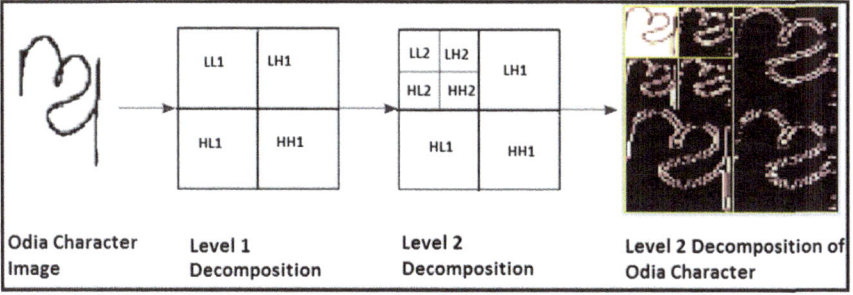

Fig. 3 Sample Odia character and its wavelet decomposition at second level

3.3 Feature Extraction Using DWT

A good feature extractor leads to a better recognition rate [14]. This paper utilizes the coefficients of level-2 decomposition of DWT for feature extraction. DWT [5] is usually an extension of Fourier transformation which analyzes the signal at different scales or resolutions and, therefore, has become a popular method for feature extraction. It is designed with a low-pass and high-pass filters along with down samplers. Whenever DWT is applied to an image, it produces sub-band images like LL, LH, HL, and HH for each individual level. Among these LH, HL, HH represents detail components in the horizontal, vertical, diagonal directions and LL are the approximation component which is further used in the second level of decomposition [15]. In this work, we have considered Haar wavelet which is orthogonal in nature. It helps to achieve high recognition rates.

Figure 3 depicts wavelet coefficients of a sample Odia character and its level-2 decomposition. This work considers the coefficient of all sub-bands at level-2 decomposition to extract the feature. The key features like mean, median, min, max, standard deviation, mean absolute deviation, median absolute deviation, Euclidean distance and energy have been computed from all the sub-bands. Therefore, we get a feature vector of length 63 (9 features from 7 sub-bands) for each character image. Eventually, a feature matrix of size 7050 × 63 is constructed. The obtained feature matrix is then passed to PCA to reduce the dimension further.

3.4 Feature Reduction Using PCA

The presence of insignificant features leads to high computation overhead, more storage memory and sometimes reduce the performance of the classifier. Hence, it is necessary to find the most significant features from the original feature set. In this work, we have used PCA to reduce the dimension of the feature. PCA is the most well-known approach that has been broadly harnessed for dimensionality reduction and data visualization in many applications [16]. It projects the input data onto a lower dimensional linear space, termed as the principal subspace with an aim of maximizing the variance of the projected data. The main motivation of using PCA in this paper is to lessen the dimensionality of the features which results in a more accurate and efficient classifier.

3.5 Classification Using BPNN

Neural networks have gained popularity in classification problems because of its several advantages over probabilistic-based classifiers. It can be defined as a massively parallel processor which can learn through examples [17]. In this paper, we have used a back-propagation neural network having sigmoid activation in the hidden layer and linear activation in the output layer, to classify characters into a set of target categories. For training, we used scale conjugate gradient technique as it produces faster convergence than gradient descent approach. The input layer of BPNN consists of 9 neurons as nine features are selected by the PCA. It may be noted that the number of neurons in hidden and output layer is set to 25 and 47, respectively. The performance of BPNN was measured by mean square error (MSE) and is defined as

$$MSE = \frac{1}{n} \sum_{n} (T - O)^2 \tag{1}$$

where T is the target output, O is the actual output, and n is the total number of training data.

3.6 Implementation

The overall pseudocode of the proposed system is presented in Algorithm 1. It is divided into two phases: offline and online phase. In offline phase, we train the network based on the features extracted in the extraction and reduction step; however in online phase, we can predict a class label for an unknown sample.

4 Simulation Results and Discussions

All the methods of the proposed OCR system were simulated using MATLAB 2014a. After preprocessing, in feature extraction step we got a feature vector of length 63 for each image. Then, PCA is used to reduce the dimension to 9 as 9 principal components (PCs) can able to preserve more than 80% of the total variance. In addition, it has been found that with only nine PCs, the system earns highest accuracy. Then,

Table 1 Recognition rate achieved using the proposed system

Sl no	N_{cc}	N_{mc}	Recognition rate (%)	Sl No	N_{cc}	N_{mc}	Recognition rate (%)
1	145	5	96.6	25	144	6	96.0
2	142	8	94.6	26	146	4	97.3
3	142	8	94.6	27	140	10	93.3
4	140	10	93.3	28	141	9	94
5	140	10	93.3	29	143	7	95.3
6	143	7	95.3	30	144	6	96.0
7	141	9	94.0	31	144	6	96.0
8	141	9	94.0	32	144	6	96.0
9	141	9	94.0	33	140	10	93.3
10	141	9	94.0	34	140	10	93.3
11	139	11	92.6	35	140	10	93.3
12	140	10	93.3	36	141	9	94.0
13	143	7	95.3	37	142	8	94.6
14	143	7	95.3	38	143	7	95.3
15	143	7	95.3	39	142	8	94.6
16	144	6	96.0	40	142	8	94.6
17	146	4	97.3	41	142	8	94.6
18	146	4	97.3	42	142	8	94.6
19	142	8	97.3	43	142	8	94.6
20	146	4	97.3	44	142	8	94.6
21	140	10	93.3	45	142	8	94.6
22	141	9	94.0	46	144	6	96.0
23	142	8	94.6	47	144	6	96.0
24	143	7	95.3	Overall recognition rate = 94.8%			

Algorithm 1 Pesudocode of the proposed system

Offline learning:

1: The input images are pre-processed and are decomposed by DWT.
2: Calculate the statistical features from all the sub-bands of 2nd level DWT.
3: PCA is carried out and principal component (PC) coefficient matrix is generated.
4: The reduced set of features along with its corresponding class labels are used to train the BPNN classifier.
5: Report the performance.

Online prediction:

1: Users presented a query image to be classified
2: DWT is performed on the query image and then statistical features are calculated from all seven the sub-bands
3: PC score is obtained by multiplying feature vector into PC coefficient matrix
4: The PC score is given input to the previously trained BPNN to predict the class label

---

a reduced feature matrix is obtained. The input dataset is divided into 70% training and 30% testing samples. Thereafter, we design a BPNN network $9 \times 25 \times 47$ to perform classification. The overall classification results of the proposed system are listed in Table 1, where the 47 characters are numbered from 1 to 47. N_{cc} denotes the number of times the characters correctly classified, and N_{mc} indicates the number of times the characters are miss-classified. From the table, it has been observed that the overall recognition rate is 94.8%.

5   Conclusion and Future Scope

This paper presents an automatic OCR system for Odia characters and achieved 94.8% accuracy on a benchmark dataset. In the preprocessing step, we perform operations like normalization and dilation. Two-dimensional DWT has been used for feature extraction from character images followed by PCA for feature reduction. Finally, the reduce set of features is fed to the BPNN classier. For feature extraction, DWT is decomposed up to two levels; however, features from high levels of decompositions may be considered. Other feature selection techniques like filter-based techniques or evolutionary-based approaches can be applied to find the most significant features. Different combination of classifiers can also be taken into account for character recognition in the future.

References

1. Plamondon, R., Srihari, S.N.: On-line and off-line handwritten recognition: a comprehensive survey. IEEE Trans. PAMI **22**, 62–84 (2000)
2. Pal, U., Jayadevan, R., Sharma, N.: Handwriting recognition in indian regional scripts: a survey of offline techniques. ACM Trans. Asian Lang. Inf. Process. **11**(1), 1–35 (2012)
3. Patra, P.K., Nayak, M., Nayak, S.K., Gabbak, N.K.: Probabilistic neural network for pattern classification. In: International Joint Conference on Neural Networks, pp. 1200–1205 (2002)
4. Pal, U., Wakabayashi, T., Kimura, F.: A system for off-line oriya handwritten character recognition using curvature feature. In: 10th International Conference on Information Technology, pp. 227–229 (2005)
5. Pratt, W.K.: Digital Image Processing. Willey (2007)
6. Chaudhuri, B.B., Pal, U., Mitra, M.: Automatic recognition of printed oriya script. In: Sixth International Conference on Document Analysis and Recognition, pp. 795–799 (2001)
7. Padhi, D., Senapati, D.: Zone centroid distance and standard deviation based feature matrix for odia handwritten character recognition. In: International Conference on Frontiers of Intelligent Computing Theory and Applications (FICTA), vol. 199, pp. 649–658 (2005)
8. Wakabayashi, T., Pal, U., Kimura, F., Miyake, Y.: F-ratio based weighted feature extraction for similar shape character recognition. In: 10th International Conference on Document Analysis and Recognition (2009)
9. Dash, S.K., Puhan, N.B., Panda, G.: A hybrid feature and discriminate classifier for high accuracy handwritten odia numeral recognition. IEEE Region 10 Symposium, pp. 531–535 (2014)
10. Chand, S., Frank, K., Pal, U.: Text independent writer identification for oriya script. In: IAPR International Workshop on Document Analysis Systems (2012)

11. Kumar, B., Kumar, N., Palai, C., Jena, P.K., Chattopadhyay, S.: Optical character recognition using ant miner algorithm: a case study on oriya character recognition. Int. J. Comput. Appl. **61**(3), 0975–8887 (2013)
12. Pujari, A.K., Mitra, C., Mishra, S.: A new parallel thinning algorithm with stroke correction for odia characters. In: Advanced Computing, Networking and Informatics, vol. 1, pp. 413–419. Springer (2014)
13. Mishra, T.K., Majhi, B., Sa, P.K., Panda, S.: Model based odia numeral recognition using fuzzy aggregated features. Front. Comput. Sci. Springer, pp. 916–922 (2014)
14. Kumar, G., Bhatia, P.K.: A Detailed review of feature extraction in image processing systems. In: Fourth International Conference on Advanced Computing & Communication Technologies (2014)
15. Nayak, D.R., Dash, R., Majhi, B.: Brain MR image classification using two-dimensional discrete transform and adaboost with random forests. Neurocomputing **177**, 188–197 (2016)
16. Bishop, C.M.: Pattern Recognition and Machine Learning. Springer, New York (2006)
17. Haykin, S.: Neural Networks: A Comprehensive Foundation. Prentice Hall (1999)

Lesion Volume Estimation from TBI–MRI

O. V. Sanjay Sarma, Martha Betancur, Ramana Pidaparti
and L. Karumbaiah

Abstract Traumatic brain injury (TBI) is a major problem affecting millions of people around the world every year. Usually, TBI results from any direct or indirect physical impact, sudden jerks, or blunt impacts to the head, leading to damage to the brain. Current research in TBI is focused on analyzing the biological and behavioral states of patients prone to such injuries. This paper presents a technique applied on MRI images in estimation of lesion volumes in brain tissues of traumatic brain-injured laboratory rats that were subjected to controlled cortical impacts. The lesion region in the brain tissue is estimated using segmentation of the brain, diffusion, and the damage regions. After the segmentation, the area of the damaged portion is estimated across each slice of MRI and the combined volume of damage is estimated through 3D reconstruction.

Keywords Traumatic brain injury · Controlled cortical impact (CCI)
Magnetic resonance imaging (MRI) · Image segmentation

1 Introduction

Traumatic brain injury (TBI) is the damage to the brain caused by external impacts such as injuries resulting from vehicle accidents, accidents in sports, sudden jerks, blunt impacts. Over 2 million people around the world are severely affected by TBI

O. V. Sanjay Sarma (✉) · R. Pidaparti
College of Engineering, University of Georgia, Athens, GA, USA
e-mail: sanjaysarmaov@uga.edu

R. Pidaparti
e-mail: rmparti@uga.edu

M. Betancur · L. Karumbaiah
Regenerative Bioscience Center, The University of Georgia, Athens, GA, USA
e-mail: marthbet@uga.edu

L. Karumbaiah
e-mail: lohitash@uga.edu

© Springer Nature Singapore Pte Ltd. 2018   197
K. Saeed et al. (eds.), *Progress in Advanced Computing and Intelligent Engineering*,
Advances in Intelligent Systems and Computing 563,
https://doi.org/10.1007/978-981-10-6872-0_19

every year. A TBI can lead to death or a temporary or permanent impairment in behavioral, cognitive, and emotional activities. Current research in this field is focused on associating behavioral patterns of patients with physical injuries, analyzing tissue recovery, investigating how they compensate for memory loss, and also growing stem cells in the lesion regions [1–3]. In the current work, experimental animals, laboratory rats, were subjected to control cortical impact (CCI), where physical impacts are made on the brains, thereby damaging the brain tissues. These TBI rats are scanned periodically using magnetic resonance imaging (MRI), and the damage volumes were estimated through segmentation and reconstruction of the brain. These estimations can further help analyzing the recovery in rats and also identify other behavioral parameters associated with the damage [4].

The advent of magnetic resonance imaging (MRI) technology makes it possible to observe the direct physical condition of the patient (in vivo). It is becoming increasingly popular for its ability to detect disease without exposing patients to hazardous radiation like in X-Ray and CT imaging. Two types of mechanisms, the spin–lattice T_1 and spin–spin T_2, are used in the imaging process in generating two different contrasting tissue images [1]. The obtained images are analyzed, in separating the various regions of interest. We employed an elimination strategy, where the region of interest was separated and image metrics were obtained for area and volume of the damaged region.

The lesion volumes for estimating the damage were computed from the MRI images scanned at Bio-Imaging Research Center (BIRC), UGA. BIRC at University of Georgia has a 7T Varian System and is capable of performing various imaging methods. The facility uses image analysis tools like NUTS, jMRUI, Image J (NIH), and FSL for data and image analysis. The current paper details the strategies employed in automating the process of determining the damage volume from the MRI data of TBI rats.

The experimentation and surgical procedures followed are discussed in Sect. 2, followed by details of the segmentation process used to differentiate the regions of interest and the estimation of physical parameters. Procedures involved in the 3D reconstruction of the brain are discussed in a subsequent section.

2 Experimentation

2.1 Subjects

Male Sprague–Dawley rats (200–250 g, Harlan Laboratories) received a craniotomy operation followed by a direct injury to the cortex. All procedures involving animals strictly followed the guidelines set forth in the Guide for the Care and Use of Laboratory Animals (US Department of Health and Human Services, Pub no. 85-23, 1985) and were approved by the University of Georgia Institutional Animal Care and Use Committee.

2.2 Controlled Cortical Impact on Rats

In order to study the effect of TBI, the experimental laboratory rats were subjected to repeatable direct injury to the left fronto-parietal cortex via a controlled cortical impactor (CCI) [5, 6]. Briefly, each rat was placed under anesthesia using 5% isoflurane for induction, and its head was then quickly shaved using an electric shaver. The animal received a continuous flow of 2% isoflurane and oxygen via a fitted anesthesia mask after transferring to a stereotaxic frame (David Kopf Instruments, CA). The animal's body temperature was kept warm using a controlled heating blanket, and the breathing rate was monitored throughout the procedure. The shaven head surface was sterilized, and a single incision was made slightly to the left of the sagittal suture. The skin and muscle tissue were retracted, and the skull was cleaned with 3% hydrogen peroxide followed by saline as previously described [7, 8]. A 5-mm craniotomy was made using a 5-mm (diameter) trephine bur on an electrical drill, at 0.5 mm from coronal suture, and 0.5 mm from sagittal suture on the left parietal bone. The bone disk was removed, and the area was cleaned with a thin piece of gel foam (PFIZER, New York, NY) saturated with saline. The dura was evaluated for any ruptures and for any bleeding. The brain was covered with the gel foam, while the CCI was quickly set up for impact. Prior to starting the surgery, the velocity (2.25 m/s) and dwell time (250 ms) were calibrated and tested. The 3-mm tip was sterilized and screwed onto the pistol of the CCI. The tip was then lowered, and the gel foam was moved to the side, to allow the tip to come in direct contact with the dura. The tip was retracted, and the gel foam was placed back on the open brain area. The retracted tip was lowered 2 mm, in order to cause a 2 mm deep injury. The gel foam was quickly moved to the side, and the CCI was fired. After the impact, the gel foam was placed over the bleeding brain, and the extent of hematoma and herniation of the brain was recorded. The speed of the pistol was calculated and recorded into each animal's surgical sheet, a couple of minutes after the bleeding stopped; additional gel foam was used to cover the craniotomy followed by sealing with UV curing dental cement. The skin was sutured back after the hemostats were removed. The animal was given a subcutaneous injection with buprenorphine HCl and allowed to recover under a heat lamp in a new cage before it was returned to the animal room.

2.3 Brain Tissue Collection

Four weeks post-TBI, the animals were anesthetized using ketamine (50 mg/kg)/ xylazine for gross anatomy analysis. The brains were carefully removed, and each extracted brain was post-fixed overnight in 4% PFA. The next day, each brain was placed in 30% sucrose and imaged under a dissecting microscope in order to capture the entire brain superficial plain. Each brain was placed on a brain matrix, and a coronal incision was made down the center of the injury. Images of the coronal view were also taken to access the extent of the gap or area voided of brain tissue.

2.4 Scanning in MRI

Four weeks after receiving a brain injury, each rat MR received an MRI scan as previously described [9, 10]. Briefly, each animal was kept under 2% isoflurane anesthesia in oxygen-enriched air via a built in nose cone and placed directly on the animal tube coil assembly. The animal's head was securely positioned directly under the 20-mm dual-tuned surface coil, and the animal tube assembly was placed inside the 210-mm horizontal bore of the small animal 7-T MRI system (Varian Magnex). Each animal received coronal scans of the head, with each slice having a thickness of 1 mm and zero distance between each slice. The entire injured tissue was imaged and with the addition of two extra slides added to the number of slices needed to capture the entire injury, one at the rostral end and one at the caudal end. T1-weighted (T1WI) and T2-weighted (T2WI) images were collected for injury volume analysis. For T1-weighted MRIs, animals were scanned using (TR) = 0.50 s, echo time (TE) = 14 ms, a matrix size of 256 × 256 with a field of view (FOV) = 40 mm × 40 mm were chosen. The total acquisition time with two averages per phase—encode step was 4 min 17 s per rat. While, T2 images had 2.0 s, 32 ms, and 40 mm × 40 mm for TR, TE, FOV, respectively, with a 256 × 256 matrix. A total acquisition time of 4 min 20 per rat was fixed on the rats with 4 averages per phase encoding step.

2.5 Quantification of Cortical Tissue Lost from Brain Ex Vivo Images

Brain images taken of the superior view of the injured brains, along with coronal images of the injured left fronto-parietal cortex, were analyzed using Image J (NIH) to calculate the approximate total tissue loss. The images collected from the superior view of the injured brains were used to calculate the loss of tissue by manually outlining the large area of the cortex where cortical tissue was missing. The coronal view images were used to determine the total depth or thickness of the gap left by the missing tissue. The volume of cortical tissue missing was estimated by multiplying the total superior area by the depth of the injury. The results of the quantifications are illustrated in the graph below, along with sample images of the ex vivo brain superior view (A) and coronal view (B).

MRI, a noninvasive medical imaging technique, uses magnetic fields and radio waves, where a powerful magnetic field is made to oscillate at a high frequency, generating oscillations in hydrogen molecules (protons) in water. These oscillating hydrogen atoms generate radio frequency pulses which are transformed to images. Hence, tissues with higher water content hold their energy for a longer time, extending the duration of emission of radio waves. MRI images are used for identifying different anatomical structures or pathologies in tissue via weighted image contrast. After excitation, the differences in tissue content determine the contrast, or different gray levels, obtained by the independent processes of T1 (spin–lattice) and T2 (spin–spin) relaxation as tissue returns to its equilibrium state. The T1-weighted images are obtained when MR signal is measured after magnetization is allowed to recover through repetition time (TR). Cerebral cortex assessment, general morphological information identification, and post-contrast imaging can be done through this image T2-weighting. On the other hand, the image magnetization in T2-weighted imaging is allowed to decay prior to MR signal measurement for a varying echo time (TE). This method is useful for detecting edema, white matter lesions, hypoxia, inflammation and assessing hypoxic ischemic changes in brain injury. Additionally, chronic consequences of brain injury such as diffuse atrophy and gliosis have been identified using T2 MRIs [11]. T2 MRIs tend to be the most common MRIs performed to assess brain injuries because of the broad spectrum of different tissue injuries that can be observed using T2 MRIs. While the T2 MRI images allow for an overall qualitative observation of the injury, the need for a method to accurately and reliably quantify the volume of injured tissue continues to be unmet. In order to address this issue, we developed a method to efficiently quantify hyper-intensity in T2 MRIs. Figure 1 shows the scanned portions of the brain, and Fig. 2 shows coronal brain images of a T2 MRI scan obtained from one of the rats. MATLAB was used for converting the image formats and in separating layers of MRI images. The regions of interest are separated applying the designed algorithm. As shown in Fig. 2, the diffusion (white) and damaged (blue) region areas were computed in every image and all the images were combined for estimating the volume of damage. The images of different layers of MRI images are presented in Fig. 2 (Fig. 3).

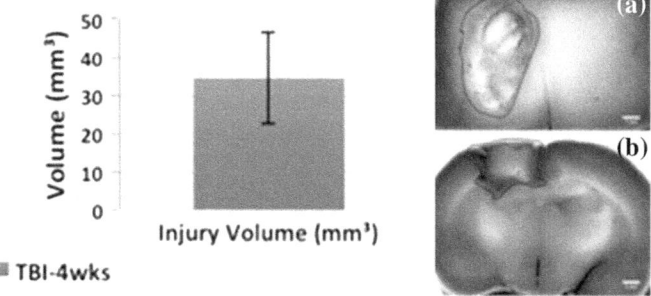

Fig. 1 Coronal view images used to determine the total depth or thickness of the gap left by the missing tissue

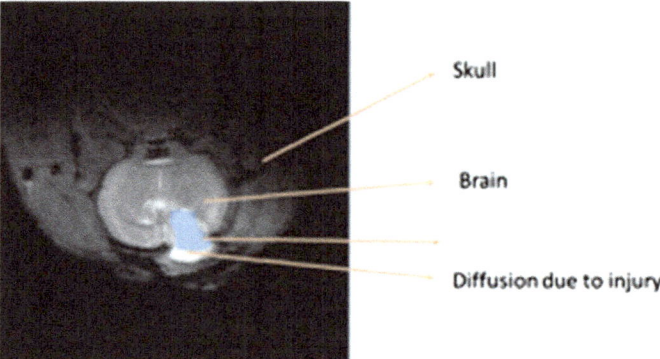

Fig. 2 Region of interest in MRI images of TBI rats

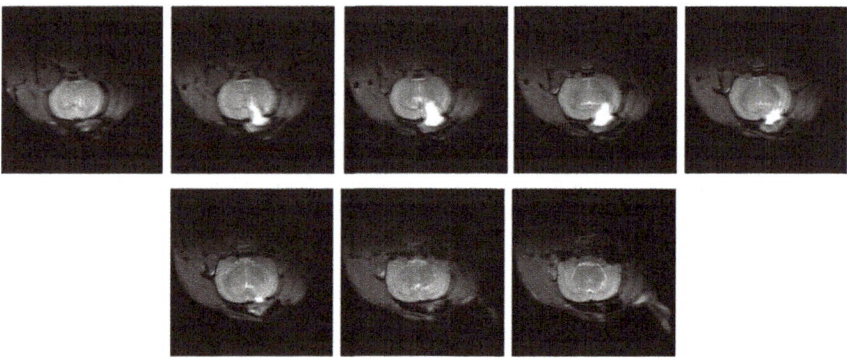

Fig. 3 MRI scanning of TBI rats

3 Implementation and Results

This section describes the procedures followed in separating the region of interest from the rest of the image. First, the brain region is separated from the skull and muscle portion surrounding it. It must be noted that a healthy brain is more or less evenly hydrated and represents a narrow band of intensities in the image. The skull and other muscle portions, on the other hand, contain lesser water content and hence display lower intensities. The algorithms implemented were taken from [12].

Also, the diffusion portion which is largely composed of water shows very high intensity in the images, and hence, it is easy separating diffusion portion from the healthy portion of the brain. In order to compute the damage area, the diffusion inside the brain region is computed, discussed later on in this section.

The separated regions are summed up for the number of pixels, and their corresponding areas are computed. The overall algorithm flowchart is presented in Fig. 4.

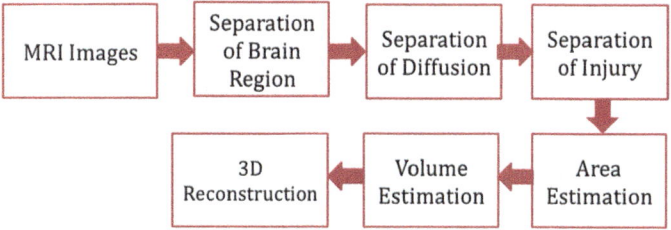

Fig. 4 Algorithm overview

3.1 Segmenting Brain Region

Each of the MRI images separated is subject to adaptive threshold technique. The threshold value obtained usually ensures that the brain region and the skull region are separated, as the number of pixels in the brain region is higher.

The separated brain region, however, contains unwanted regions falling within the threshold limits. Hence, the image is labeled using connected component labeling (CCL). This allows the separation of the regions which are large in size. The detailed segmentation of the image is presented in the flowchart in Fig. 5.

3.2 Segmenting Diffusion

The images within the brain region are further segmented and processed for separating the diffusion region. The diffusion portion is usually at a higher intensity, and its threshold generally falls above 200. Hence, the regions with intensities higher than the threshold are separated, and the largest connected portions are selected through CCL. The segmentation procedure is shown in Fig. 6.

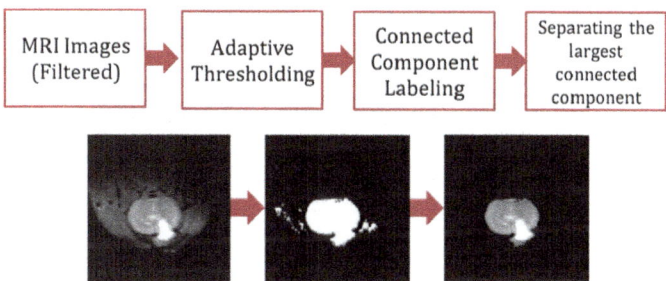

Fig. 5 Separation of brain and skull in MRIs

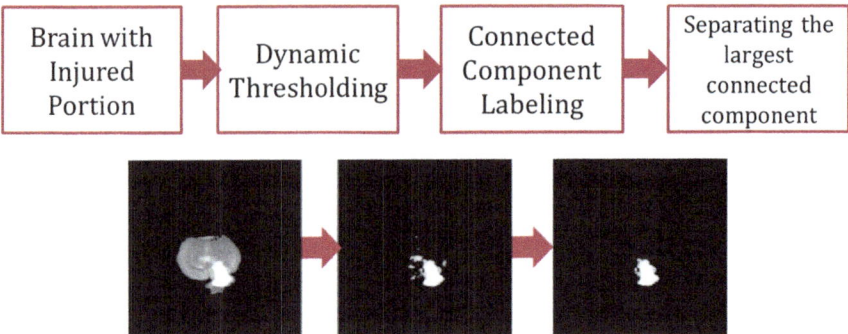

Fig. 6 Segmentation of diffusion region

3.3   Injury Identification

The separated brain and diffusion segments of the image are raster scanned in parallel. In every row, the number of high pixels (1s) in the brain segment should always be greater than the diffusion segment high pixels. Otherwise, the pixels in that row are made zero. This gives the approximate diffusion region falling inside the brain region, which is nothing but the damaged portion. The flowchart for the algorithm is presented in Fig. 7.

3.4   Area and Volume Computation

Each of the images in the MRI scan image constitutes 256×256 pixels corresponding to 4 mm $\times$ 4 mm of tissue. The area constituted by n pixels is given by

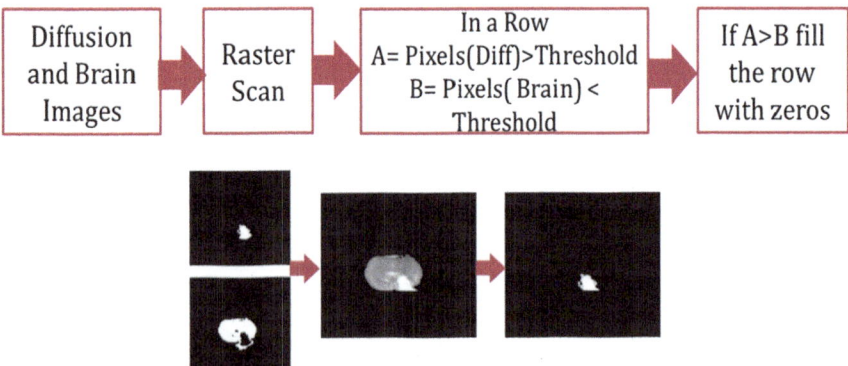

Fig. 7 Separation of injury

Fig. 8 2D reconstruction

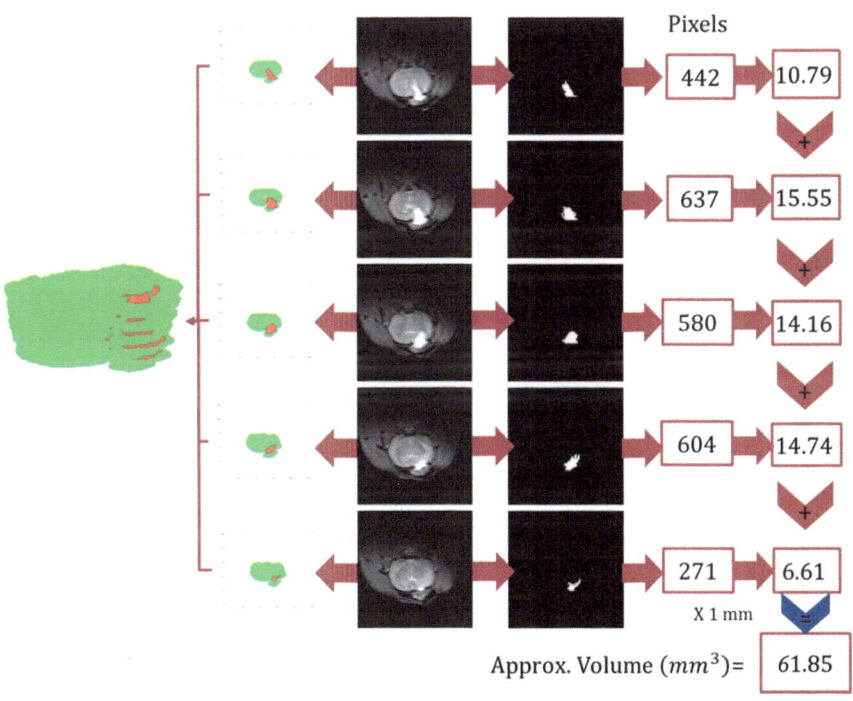

Fig. 9 2D, 3D reconstruction and area and volume estimation form a sample MRI of a TBI rat

$$Area = n \times \frac{1600}{256 \times 256} = n \times 0.02441 \text{ mm}^2 \tag{1}$$

where n is the number of pixels of interest. The process is repeated on all the layers of MRI images. All the areas are summed up and multiplied with the layer's thickness for computing approximate volume. The segmented data obtained after applying the algorithm is reconstructed on to a 2D image as shown in Fig. 8. The volume estimation and the 3D reconstructed model for a sample MRI are presented in Fig. 9.

$$Volume = \sum_{i=1}^{L} n_i \times 0.02441 \times S \text{ mm}^3 \tag{2}$$

where

i Number of layer/MRI image
L Total number of MRI images
n_i Number of pixels of interest in ith layer
S Thickness of the layer

4 Conclusions

We presented an algorithm for segmenting brain diffusion and injury in T2 images. We prototyped it for preliminary analysis of rat TBI–MRI images and estimated the area and volume of damage. Further, 2D and 3D reconstructions of the brain were done for visualization. Since the method presented here is limited to the intensity differences in the T2 MRIs which indicate a broad spectrum of injury at the edge of the cortex, additional noise created by the craniotomy and brain herniation needs to be accounted for in order to eliminate non-cortical areas. However, this quantification method can be applied to scans of more central areas of the brain and/or to other types of scans designed to identify more specific changes in tissue.

References

1. Kim, E.S., et al.: Human umbilical cord blood-derived mesenchymal stem cell transplantation attenuates severe brain injury by permanent middle cerebral artery occlusion in newborn rats. Pediatric Res. **72**(3), 277–284 (2012)
2. Byun, J.S., et al.: Engraftment of human mesenchymal stem cells in a rat photothrombotic cerebral infarction model: comparison of intra-arterial and intravenous infusion using MRI and histological analysis. J. Korean

3. Wei, X.-E., et al.: Dynamics of rabbit brain edema in focal lesion and perilesion area after traumatic brain injury: a MRI study. J. Neurotrauma **29**(14), 2413–2420 (2012)
4. Turtzo, L.Christine, et al.: The evolution of traumatic brain injury in a rat focal contusion model. NMR Biomed. **26**(4), 468–479 (2013)
5. Dixon, E.C., Clifton, G.L., Lighthall, J.W., Yaghmai, A.A., Hayes, R.L.: A controlled cortical impact model of traumatic brain injury in the rat. J. Neurosci. Methods **39**, 253–262 (1991)
6. Shear, D.A., Tate, M.C., Archer, D.R., Hoffman, S.W., Hulce, V.D., Laplaca, M.C., et al.: Neural progenitor cell transplants promote long-term functional recovery after traumatic brain injury. Brain Res. **1026**, 11–22 (2004)
7. Karumbaiah, L., Norman, S.E., Rajan, N.B., Anand, S., Saxena, T., Betancur, M., et al.: The upregulation of specific interleukin (IL) receptor antagonists and paradoxical enhancement of neuronal apoptosis due to electrode induced strain and brain micromotion. Biomaterials **33**, 5983–5996 (2012)
8. Saxena, T., Karumbaiah, L., Gaupp, E.A., Patkar, R., Patil, K., Betancur, M., et al.: The impact of chronic blood-brain barrier breach on intracortical electrode function. Biomaterials **34**, 4703–4713 (2013)
9. Turtzo, L.C., Budde, M.D., Gold, E.M., Lewis, B.K., Janes, L., Yarnell, A., et al.: The evolution of traumatic brain injury in a rat focal contusion model. NMR Biomed. **26**, 468–479 (2013)
10. Kim, E.S., Ahn, S.Y., Im, G.H., Sung, D.K., Park, Y.R., Choi, S.H., et al.: Human umbilical cord blood-derived mesenchymal stem cell transplantation attenuates severe brain injury by permanent middle cerebral artery occlusion in newborn rats. Pediatr. Res. **72**, 277–284 (2012)
11. Huang, B.Y., Castillo, M.: Hypoxic-ischemic brain injury: imaging findings from birth to adulthood. RadioGraphics **28**, 417–439 (2008)
12. Haidekker, M.A.: Shape Analysis. Advanced Biomedical Image Analysis, pp. 276–309. Wiley (2010)

Part II
Authentication Methods, Cryptography and Security Analysis

Analytical Assessment of Security Mechanisms of Cloud Environment

Bijeta Seth and Surjeet Dalal

Abstract Cloud computing is the budding paradigm nowadays in the world of computer. It provides a variety of services for the users through the Internet and is highly cost-efficient and flexible. However, despite all its advantages, security and privacy have evolved to be of significant concern in Cloud computing. Online computing is preferred by consumers and businesses only if their data are assured to remain private and secure. This paper aims to study several key concepts, namely Cloud characteristics, delivery models, deployment models, and security issues. The document also discusses the work done on Cloud security and privacy issues.

Keywords Clouds · Cloud computing · Issues · Security · Privacy
Security tools

1 Introduction

The computer has evolved from the bigger and expensive first-generation vacuum tube-based computers to smaller and more affordable personal computers and now to a new revolution called Cloud computing system, abbreviated as Cloud in the year 2008. Cloud computing provides a variety of services for the users through the Internet, enabling an application to be run from anywhere in the world without the need to run personal computer every time.

The article is ordered as follows: Sect. 1 discusses Cloud characteristics, delivery models, and service offerings. Section 2 mentions Cloud platform and technologies. Section 3 mentions security in Clouds, discussing challenges and issues. Section 4 describes real-world case studies, and Sect. 5 highlights the survey done in a table. Conclusion comprises the last part of the paper.

B. Seth · S. Dalal (✉)
Department of Computer Science & Engineering, SRM University,
Sonipat, Haryana, India
e-mail: profsurjeetdalal@gmail.com

© Springer Nature Singapore Pte Ltd. 2018
K. Saeed et al. (eds.), *Progress in Advanced Computing and Intelligent Engineering*,
Advances in Intelligent Systems and Computing 563,
https://doi.org/10.1007/978-981-10-6872-0_20

211

1.1 Cloud Characteristics

Cloud computing has subsequent crucial characteristics as described by The Cloud Security Alliance [1]:

- **Self-service as per requirement**—A Cloud client is capable to use the Cloud-based services on demand from anywhere.
- **Broad network access**—Constant high-speed interconnection is required to provide services to users via tools like PCs, mobile phones.
- **Resource pooling**—Various resources can be used by multiple customers according to their needs which becomes the biggest advantage of the Cloud.
- **Rapid elasticity**—Customers can purchase and release the unlimited available capabilities of Cloud computing in any extent at any time.
- **Measured service**—The consumers pay for the services according to their usage (Fig. 1).

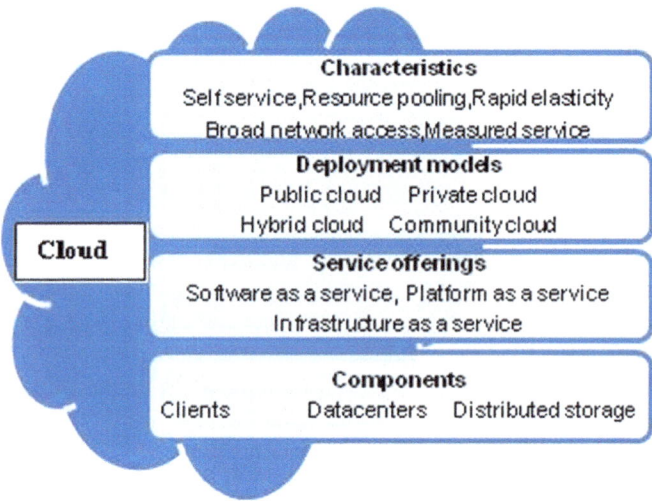

Fig. 1 Cloud computing paradigm [2]

1.2 Delivery Models

Cloud services can be provided as four basic Cloud delivery models which are as follows:

Public Cloud

Public Cloud provides the interface between the unrestricted customers and the owner group (third party), for example, Amazon Cloud service. Multiple enterprises can work on the infrastructure provided, at the same time. Public SaaS offerings are Salesforce Chatter, Gmail Dropbox. Public PaaS offerings are like Google App Engine. Public IaaS offerings are AWS and Windows Azure. It is more cost-effective, highly reliable and flexible, and location Independent. But they are less secure and customizable [3].

Private Cloud

Private Cloud affords the services merely for an organization in an exclusive manner, for example, CIO/G6. It provides high security and more control in comparison to public clouds. But such models have restricted area of operation and have a high price with limited scalability [4].

Community Cloud

It provides the services for the specific groups instead of whole public groups. They all work together for common concerns, for example, Government or G-Cloud. Cost effectiveness and more security are the advantages of using community clouds.

Hybrid Cloud

It is formed by combining any of the public, private, or community Clouds, for example, CIO/G6/APC + Amazon EC2. Enhanced scalability, security, and flexibility are the advantages of this model. But it faces networking issues and security compliance.

1.3 Cloud Service Offerings

The fundamental type of Cloud service offerings is as follows:

Software as a Service (SaaS) It allows the client to have an application for lease from Cloud service provider instead of buying, installing, and running software, for example, Gmail Docs.

Platform as a Service (PaaS)

It provides a stage to the users upon which applications can be prepared and executed in the Cloud, for example, Windows Azure.

Infrastructure as a Service (IaaS)

Users can access resources according to their requirements from huge pools installed in data centers, for example, Elastic Cloud Compute [5, 6].

2 Cloud Computing Platforms and Technologies

Below we discuss different platforms and frameworks in Cloud computing:

2.1 Amazon Web Services (AWS)

AWS is a group of Web services that work in cooperation to deliver Cloud services. It permits users to store and replicate data across geographical regions.

2.2 Google App Engine

Google App Engine is an Internet-based collection of applications which uses distributed file system (DFS) to collect data. It provides single sign-on (SSO) service to integrate with LDAP. Google protects its operating system with the means of proprietary software by checking binary modifications [7].

2.3 Microsoft Azure

Applications in Azure are structured around the notion of roles, which recognize and embody a distribution unit for an application mainly Web role, worker role, and virtual machine role [8].

2.4 Hadoop Apache

Hadoop is an open-source structure suitable for processing big data sets on commodity hardware. Hadoop is an implementation of MapReduce, which provides two fundamental operations for data processing: map and reduce.

2.5 Force.com and Salesforce.com

Salesforce provides Force.com for building business applications and uses Stateful packet inspection firewall. For authentication purposes, LDAP is used. Unknown address connection requests are denied [9].

3 Security in Cloud Environment

This section mentions the challenges and issues associated with Cloud computing.

3.1 Challenges

Following are the main challenges that occur in adoption of Clouds:

- **Outsourcing**—Privacy violations can occur as the customers actually lose control on their data and tasks.
- **Multi-tenancy**—New vulnerabilities and security issues can occur because of the shared nature of Clouds between multiple customers.
- **Massive data and intense computation**—Traditional security mechanisms cannot be applied to Clouds due to large computation or communication overhead.
- **Heterogeneity**: Integration problems arise between diverse Cloud providers using different security and privacy methods.
- **Service level agreement**: A negotiation mechanism between provider and consumer of services needs to be established.

Security is regarded as the dominant barrier amongst the nine challenges in accordance to the survey done by IDC in August 2008, existing in Clouds as shown in Table 1.

Table 1 Challenges/issues in Clouds [10]

S.No.	Challenge/issue	% age
1.	Security	74.6
2.	Performance	63.1
3.	Availability	60.1
4.	Hard to integrate with in-house IT	61.1
5.	Not enough ability to customize	55.8
6.	Worried on demand will cost more	50.4
7.	Bringing back in-house may be difficult	50.0
8.	Regulatory requirements prohibit Cloud	49.2
9.	Not enough major suppliers yet	44.3

3.2 Cloud Computing Security Issues

There are following security issues as given below:

- **Trust**: The Cloud service provider is required to provide sufficient security policy to reduce the risk of data loss or data manipulation.
- **Confidentiality**: The confidentiality can be breached as sharing or storage of information on remote servers is done in Cloud computing which is accessed through the Internet.
- **Privacy**: It refers to the willingness of a user to control the disclosure of private information. An illegal admittance to user's sensitive data may bring security issues.
- **Integrity**: It is to guarantee the precision and uniformity of data. Therefore, the Cloud service provider should provide security against insider attacks on data.
- **Reliability and availability**: Trustworthiness of Cloud service provider decreases when a user's data get leaked.
- **Authentication and authorization**: To prevent unauthorized access, software is required outside the organization's firewall.
- **Data loss**: Removal or modification of data lacking any backup could lead to data loss.
- **Easy accessibility of Cloud**: Cloud services can be utilized by anyone by a straightforward enlistment show. This opens an opportunity to get to administrations for the cunning personalities.

4 Case Studies

Many real-world scenarios where Cloud computing was compromised by attacks and their feasible prevention methods are listed below in Table 2.

5 Related Work

One of the most difficult aims in Cloud computing is to provide security and protecting data privacy. But due to its shared nature, it becomes difficult to prevent threats in Cloud computing, so information can be leaked by unauthorized access. This section presents an outline of existing review articles allied to security and privacy. In Table 3, Cloud refers to Cloud computing.

Table 2 Case studies

Type of attack	Definition	Example	Solution
XML signature wrapping attack	Wrapping attack inserts a fake element into the signature and then makes a Web service request	In 2011, Dr. Jorg Schwenk discovered a cryptographic hole in Amazon EC2 and S3 services	A proposed solution is to use a redundant bit called STAMP bit in signature in the SOAP message
Malware injection	Hacker attempts to insert malicious code by inserting code, scripts, etc., into a system	In May 2009, four public Web sites were set offline for the BEP in which hackers introduced undetectable iFrame HTML code that redirected guests to a Ukrainian Web site	Web browsers like Firefox should install NoScript and set Plugins The FAT table can be used to determine the validity and integrity of the new instance
Social engineering attack	It depends on human interaction, thereby breaking normal security procedures	In August 2012, hackers completely destroyed Mat Honan's digital life by deleting data from his iPad, iPod, and MscBook by exploiting Amazon and AppleID Account of the victim	Apple forced its customers to use Apple's online "iForgot" system to provide stronger authentication. Various account settings like credit card, e-mail addresses cannot be altered on phone by Amazon customer service head
Account hijacking	It compromises confidentiality, integrity, and availability by stealing credentials of accounts	In July 2012, UGNazi, entered CloudFare's personal Gmail account by exploiting Google's e-mail and password recovery mechanism	CloudFare has stopped sending password reset and transactional messages for security purpose

Table 3 Summary of related work

Authors	Year	Topics discussed	The approach used
Passent [11]	2015	Security issues, Cloud models, security measures	Focused on discussion on security functions and comparison being made on some Cloud models. Different security measures were also discussed
Thilakanathan [12]	2014	Privacy issues, attacks, secure data sharing, ABE, PRE	Described how proper key management provides secure and confidential data sharing and association using Cloud and discussed ABE and PRE techniques
Yuan et al. [13]	2014	Neural network, back propagation, Cloud computing, privacy preserving	Discussed neural networks and preservation of privacy was done with multilayer back propagation neural networks with homomorphic encryption for a multiparty system
Chen and Zao [14]	2012	Cloud security, threats, defense strategy	Analyzed data security and privacy protection issues and their solutions
Aguiar et al. [15]	2013	Access, virtualization, availability, storage computation	Provided a broad overview of literature covering security aspects in Cloud, attacks and protection mechanisms, improving privacy and integrity of data in Cloud
Zhou et al. [16]	2011	Cloud, security, defense strategy, privacy	Discussed five goals required to be achieved for security and legal and multilocation issues in privacy
Bohli et al. [17]	2013	Multicloud architecture, application partitioning, tier division, data separation, secure multiparty computation	Mentioned four major multicloud approaches, with its drawbacks, compliance with legal obligations, feasibility
Pearson [18]	2013	Privacy, trust, compliance, access, software virtualization	Discussed why and how security, trust and privacy issues occur in Cloud
Dillon et al. [19]	2010	Cloud adoption challenges, Cloud interoperability issues, SOA	Discussed grid computing, Service-oriented computing and their relationship with Cloud, several challenges and interoperability issues were discussed
Tari [20]	2010	Cloud, security, threats, requirement analysis, protecting user data	Identified new threats and vulnerabilities, mentioned how to protect outsourced computation

(continued)

Table 3 (continued)

Authors	Year	Topics discussed	The approach used
Takabi et al. [21]	2010	Security and privacy implications, challenges, and approaches	Discussed the unique issues exacerbating security and privacy challenges, different approaches to address the challenges and explored the future work for Cloud system
Diogo [22]	2014	Clouds, Cloud computing, issues, security	Discussed security issues, analyzed literature together with studies from the academia and industry and some real-life examples

6   Conclusion and Future Work

Cloud computing is the most recent development in online computing. Because storage and computing services are provided in Clouds at very low cost, Cloud computing is becoming very popular. The article provided a broad outline of literature covering security aspects in Cloud computing. Our research indicates that security and privacy are the major issues that are needed to be countered. This document has addressed several security approaches to overcome the issues in security in Cloud computing. Various real-world examples illustrating attacks in Cloud computing were discussed. For future enhancements, efforts are being made to build up a Multicloud architecture as an efficient scheme that can provide security and eliminate the disadvantages of single Cloud.

References

1. Xiao, Z., Xiao, Y.: Security and privacy in cloud computing. IEEE Commun. Surv. Tutor. **15**, 843 (2013)
2. Abuhussein, A., Bedi, H., Shiva, S.: Evaluating security and privacy in cloud computing services: a stakeholder's perspective. In: The Seventh International Conference for Internet Technology and Secured Transactions, ICITST-2012. IEEE (2012). ISSN 978-1-908320-08/7
3. Subashini, S., Kavitha, V.: A survey of security issues in service delivery models of cloud computing. Netw. Comput. Appl. **34**(1), 1–11 (2011). https://doi.org/10.1016/j.jnca.2010.07.006
4. Ahuja S.P., Komathukattil, D.: A survey of the state of cloud security. Netw. Commun. Technol. **1**(2), 66–75 (2012). https://doi.org/10.5539/nct.v1n2p66
5. Banafar, H., Sharma, S.: Secure cloud environment using hidden markov model and rule based generation. Int. J. Comput. Sci. Inf. Technol. **5**(3), 4808–4817 (2014)
6. Saroj, S., Chauhan, S., Sharma, A., Vats, S.: Threshold cryptography based data security in cloud computing. In: IEEE International Conference on Computational Intelligence and Communication Technology, pp. 207–209. IEEE (2015). https://doi.org/10.1109/CICT.2015.149

7. Gonzalez, N., Miers, C., Redigolo, F., Simplicio, M., Pourzandi, M.: A quantitative analysis of current security concerns and solutions for cloud computing. J. Cloud Comput. Adv. Syst. Appl. **1**(11), 1–11 (2012). https://doi.org/10.1186/2192

8. Joseph, NM, Daniel, E., Vasanthi, A.: Survey on privacy-preserving methods for storage in cloud computing. In: Amrita International Conference of Women in Computing AICWIC, vol. 4, 1–4 Jan 2013

9. Murray, A., Begna, G., Blackstone, J., Patterson, W.: Cloud service security and application vulnerability. In: IEEE SoutheastCon, pp. 1–8. IEEE (2015). https://doi.org/10.1109/SECON.2015.7132979

10. Rittinghouse, J.W., Ransome, J.F.: Cloud Computing: Implementation, Management, and Security, p. 154. CRC Press (2009). ISBN 9781439806807

11. El-Kafrawy, P.M., Abdo, A.A.: Security issues over some cloud models. In: International Conference on Communication, Management and Information Technology (ICCMIT), ScienceDirect, Procedia Computer Science, vol. 65, pp. 853–858. Elsevier (2015). https://doi.org/10.1016/j.procs.2015.09.041

12. Thilakanathan, D., Chen, S., Nepal, S., Calvo, R.: Secure data sharing in the cloud. In: Security, Privacy and Trust in Cloud Systems, Book Part 1, pp. 45–72. Springer, Berlin, Heidelberg (2014). https://doi.org/10.1007/978-3-642-38586-5_2

13. Yuan, J., Yu, S.: Privacy preserving back-propagation neural network learning made practical with cloud computing. IEEE Trans. Parallel Distrib. Syst. **25**(1), 212–221 (2014). https://doi.org/10.1109/TPDS.2013.18

14. Chen, D., Zhao, H.: Data security and privacy protection issues in cloud computing. Int. Conf. Comput. Sci. Electron. Eng. **1**, 647–651 (2012). https://doi.org/10.1109/ICCSEE.2012.193IEEE

15. Aguiar, E., Zhang, Y., Blanton, M.: An Overview of Issues and Recent Developments in Cloud Computing and Storage Security, pp. 1–31. Springer, Berlin (2013)

16. Zhou, M., Zhang, R., Xie, W., Qia, W., Zhou, A.: Security and privacy in cloud computing: a survey. In: Sixth International Conference on Semantics, Knowledge and Grids, pp. 149–150 (2008). https://doi.org/10.1109/SKG.2010.19 (2010)

17. Bohli, J.M., Gruschka, N., Jensen, M., Lo Iacono, L.: Security and privacy-enhancing multicloud architectures. IEEE Trans. Dependable Secure Comput. **10**(4), 212–214 (2013). https://doi.org/10.1109/TDSC.2013.6IEEE

18. Pearson, S.: Privacy, security and trust in cloud computing. In: Privacy and Security for Cloud Computing, pp. 3–42. Springer (2013). https://doi.org/10.1007/978-1-4471-4189-1_1

19. Dillon, T.: Cloud computing issues and challenges. In: 24th IEEE International Conference on Advanced Information Networking and Application, AINA, pp. 27–33. Perth, Australia, IEEE, Apr 2010. ISSN: 978-0-7695-4018-4

20. Tari, Z.: Security and privacy in cloud computing. IEEE Cloud Comput. RMIT Univ. **1**(1), 54–57 (2014). https://doi.org/10.1109/MCC.2014.20May

21. Takabi, H., Joshi, J., Ahn G.-J.: Security and privacy challenges in cloud computing environments. IEEE Comput. Reliab. Soc. **8**(6), 24–31 (2010). https://doi.org/10.1109/MSP.2010.186

22. Diogo, A.B.F., Liliana, F.B.S., Gomes, J.V., Freire, M.M., Inacio, P.R.M.: Security issues in cloud environments: a survey. Int. J. Inf. Secur. **13**(2), 113–170 (2014). https://doi.org/10.1007/s10207-013-0208-7

BRRC: A Hybrid Approach Using Block Cipher and Stream Cipher

Binay Kumar, Muzzammil Hussain and Vijay Kumar

Abstract Design of cryptographic algorithms has ever been meant for making messages confidential strong enough for cryptanalysts. This has resulted into different categories of encryption/decryption algorithms while they have their own pros and cons. The cryptanalysts have played roles in observing confidential messages, identifying used algorithms to encrypt messages, and they have been successful in exploiting the weak points, if found any, in the algorithms' designs. But Hybrid cryptographic algorithms, which bring mixture of randomness in the ciphered text, are ones of some obstacles for them. It is more difficult to break the hybrid ciphered text than that of block or stream cipher text. The current work presents design of a hybrid algorithm that makes uses of content replication removal in input data, block cipher and stream cipher cryptographic algorithms.

Keywords Cryptography · Cipher break · Design of cryptographic algorithms Replication removal · Block cipher · Stream cipher

1 Introduction

The art and science of keeping messages secure is cryptography, and it is practiced by cryptographers [1]. Ensuring confidentiality/security of messages becomes more important when the messages are allowed to travel through communication channels in digital devices' networks (e.g. computer networks). The assurance is given by fundamentals of cryptography, e.g. algorithms for encryption (encoding the

B. Kumar (✉) · M. Hussain · V. Kumar
Department of Computer Science and Engineering, Central University
of Rajasthan, Kishangarh, Ajmer, Rajasthan, India
e-mail: 2014MTCSE008@curaj.ac.in

M. Hussain
e-mail: mhussain@curaj.ac.in

V. Kumar
e-mail: vijaykumar.mtech12@curaj.ac.in

© Springer Nature Singapore Pte Ltd. 2018   221
K. Saeed et al. (eds.), *Progress in Advanced Computing and Intelligent Engineering*,
Advances in Intelligent Systems and Computing 563,
https://doi.org/10.1007/978-981-10-6872-0_21

messages) and decryption (decoding the encoded version of messages). In addition to the algorithms for encryption and decryption cryptographers use other elements too, e.g. keys, initial values. According to William Stalling, the security of encrypted data is entirely dependent on two things: the strength of the cryptographic algorithm and the secrecy of the key [2].

Based on the keys used for cryptography fundamentally, there are two categories of algorithms that perform cryptographic acts; symmetric and asymmetric. For symmetric key cryptography, there have been again more categories of algorithms, e.g. block cipher [2], stream cipher [2]. Block cipher, e.g. AES, DES, IDEA, divides a given file/text into fixed sized of chunks and encrypt them one by one. Stream cipher encrypts the file/text character by character, e.g. AC4. All cryptographic algorithms say CAi have their own strength and weakness in terms of persisting the confidentiality of encrypted messages say C despite of attempts made to reveal the hidden message from C.

There exists the art and science of breaking ciphertext (encrypted message); that is, seeing through the disguise called cryptanalysis [1]. Cryptanalysis does not refer only to reveal hidden message in ciphertext maliciously but to also as a test on how easy/difficult it is to compromise the algorithms and the keys used in encrypting a message.

Ciphertext-Only Attack is one of the general methods described by Christopher [3] that is used by cryptanalysts to reveal the hidden message from a cipher one. Also there are tools available that analyse ciphertext, e.g. bfcrypt [4], which extracts important information about the encrypted file. There have been works [5, 6], which distinguish block ciphertext and stream ciphertext when unknown ciphered message is given. The information like size of block ciphered data as a multiple of some number say N, random patterns in the data, replication of data chunks in a plain file etc. have been used for doing cryptanalysis. Although for the replication of some data chunks in a plain file, it is also possible that block cipher algorithms encrypt every replicated chunks differently when specific modes, e.g. CBC are applied. However, not all the time it is assured that all block cipher algorithms support such modes. Once algorithms used to encrypt text/files is identified, cryptanalyst gets deep and finally break the cipher to get the original message. There have been noticed security attacks on stream ciphertext [7].

To make the algorithms CAi more proven in terms of their cipher strength, two or more of them are combined to produce a new one called Hybrid cipher [8, 9]. Although a hybrid cryptography has been meant mostly as combined uses of symmetric key and asymmetric key encryption/decryption algorithms. However, we present design of a hybrid cryptography that makes use of only symmetric key algorithms in the respect of future concerns that may be possible because of the attempts made to distinguish block cipher and stream ciphertext. The proposed work is meant also for removal of replicated chunks of data in a message and bringing mixture of random patterns of encrypted results. The work is organised as follows.

Section 2 is the proposed work whose implementation we present in Sect. 3. Thereafter, we analyse the algorithm design in Sect. 4. Finally, our work is concluded in Sect. 5, and some future work added in Sect. 6.

2 Proposed Work

The proposed work is meant for a hybrid cryptography mechanism, which combines block cipher and stream cipher algorithms on the basis of content redundancy found in a plain file/text. This also means to bringing mixture of random patterns of encryption, randomising the size of encrypted file so that the produced cipher contents should be enough resistant against ciphertext only attack. The work is processed in two phases. The first phase removes redundant units of contents in a file/text, and it creates a stream of references for the replicated units corresponding to their unique ones. This makes output file/text containing unique blocks of text and references. Resultantly, size of file/text is reduced and also it produces a middle level cryptographic output before the cipher algorithms are applied. So the paper title goes meaningful as it refers to "Block Replication Removal Cryptography". For understanding the BRRC mechanism, let there be a file containing some text such that there are some text chunks TCi with size N. Of the TCi, TC7, TC8 and TC10 are replication of TC1, and TC4 and TC6 are replication of TC3. Thus, the output file is figured as TC1 + TC2 + TC3 + TC4 + TC5 + upto TCn + (R_TC7 + R_TC8 + R_TC9) + (R_TC4 + R_TC6). Here, TC1 + ⋯ + TCn are arranged unique blocks and (R_TC7 + R_TC8 + R_TC9) + (R_TC4 + R_TC6) are streamed references.

Note-"+" denotes a concatenation of two blocks or references.

In the second phase, block cipher and stream cipher techniques are applied for encrypting arranged unique blocks and streamed references, respectively, then the produced output is converted to base64 encoding. This makes the final output whose length is not as same as its input file/text. It makes possibly infeasible to find block size used by block cipher algorithm, and so guessing a key size also becomes difficult. We analyse it later in a section.

The proposed work is consisted of removal of replicated content units in a file/text, uses of block cipher and stream cipher for encrypting unique units of contents and stream of references, respectively, concatenating results of block ciphers and stream ciphers, and finally converting the combined results into base64 encoding. We describe the constituents in detail as following along with some definitions.

2.1 Definitions

Text Block A chunk of text in a file/text can be used by a block cipher algorithm as a block of data. Also it is a chunk of text in a file/text whose replication is checked in the file/text. We consider that size of a text block say Bi used by block cipher algorithms is same as that of a text block say Bj used in this proposed work. i.e. size_of(Bi) = size_of(Bj).

Unique Block A text block say Ui in all Bi of a file/text such that Ui is not equal to more than one Bi.

Replicated Block A text block say Ri in all Bi of a file/text such that Ri is equal to more than one Bi.

Reference This is a cardinality of a unique block Uj "j" in all Ui such that Uj is equal to a replicated block Ri in a file/text. We say that a replicated block Ri has reference j in respect to a unique block Uj such that Ri is equal to Uj.

2.2 Removal of Replicated Blocks

We consider that a file/text can have redundant blocks of contents. Removal of redundant text blocks refers to finding unique blocks in a given file/text Fi and replacing replicated blocks with numbers referring to the corresponding unique blocks. The removal process is described in the following algorithm.

Algorithm1: Block Replication Removal

```
1. Ai= A block cipher algorithm.
2. Bs= Size of a block in Ai.
3. Fs= Size of a given file/text in Fi
4. Bn= Ceil (Fs/Bs ). // Rounded up as a posotive integer
5. Lu= A list to hold unique blocks.
6. Li= Some lists to hold references for replicated blocks.

7. do
--- Bi= Read (Bs, Fi) // Read text of size Bs from Fi
--- If length(Bi) <> Bs then
----- Do padding // Complete size of Bi as Bs by random text
--- Endif
--- If (Bi found in Lu) <> true then
----- Add Bi in Lu.
--- else
----- Ib= Cardinal of Bi in Fi.
----- Il= Cardinal of Bi in Lu.
----- Lr= Li where ( i==Il )
----- Add Ib in Lr
----- Li=Lr // updates the Li with references to replicated
        blocks in correspondence to unique block Bi
---- Endif
---- count++

while ( count <= Bn) // Repeat for all blocks in file/text
```

2.3   Use of Block Cipher

All the unique blocks are ciphered using block cipher algorithms using following
procedure.
   Algorithm 2: Encrypting unique text blocks.

```
1. Ca= A block cipher algorithm // refers to Ai in Algorithm 1
2. Ki= Key for Ca
3. count= 0
4. Bl= Length of Lu // Referring algorithm1
5. SBc="" // A string of block ciphered text
6. while ( count < Bl )

--- Bi= Lu [count] // An element from list of unique blocks
--- Ci= Encrypt( Bi, Ca, Ki) // Encrypt Bi using Ca and Ki
--- SBc= SBc + Ci // concatenate
--- count++
6. return SBc
```

2.4   Use of Stream Cipher

We use stream cipher algorithm to encrypt references associated to replicated blocks
in correspondence to their unique blocks. The procedure is described as follows.
   Algorithm 3: Encrypting streamed references.

```
1. Cb= A stream cipher algorithm // Referring algorithm2
2. Kj= Ki // Key for Cb = Key for Ca
3. Rn= Number of lists that hold references to replicated
   blocks
4. Li= Li ( Step 6 Algorithm 1 )
5. i=0
6. Bl= Bl // Refers step4 in Algorithm 2, sizeof(Bl) = Say N
   Bytes
7. C=""
8. La [ ]= List of alphabets
9. Ls [ ]= List of symbols.
10. while ( i < Rn ) // till all the lists Li visited

--- index=0
--- C= C + Random ( Ls [ ] ) // Add a random symbol to mark
```

```
      start of elements of a list
----- while (Li [index] <> null ) // till Li has elements
------- C= C + Random ( La [ ] ) + Li [index] // La[random]
      separates two Li[index]
------- index++
----- Endwhile
--- i++
Endwhile

11. SSc= Encrypt(C, Cb, Kj) + Encrypt( Bl, Cb, Ki) //
    concatenate encrypted number of unique blocks Bl
12. return SSc
```

2.5 Base64 Conversion

Results returned Sc, SSc are concatenated, and the concatenated string is converted into base64 encoding.

2.6 Proposed Process

The complete process of this approach is described in following figures. Figure 1 gives the proposed steps whereas Figure 2 explains the block replication removal and creation of references.

2.7 Decryption

Decryption of encrypted text is performed by taking reverse operation of the algorithms given above. Because final output contains number of blocks as a final string FS in fixed number of bytes, so the first step in decryption is to extract this value N following stream cipher algorithm. Now, the encrypted content is divided into two chunks. First refers to list of unique blocks whose total size in encrypted content is 0 to $N*Block\_size - 1$, and second is the stream of references with the length $N*Block\_size$ to $X - 1$, here $X = size\_of(encrypted\ content) - size\_of(FS)$. Now, the block cipher and stream cipher help in order to get the middle level cryptographic output. Reverse of Algorithm 1 does accordingly and finally original content is got back.

Fig. 1 BRRC process

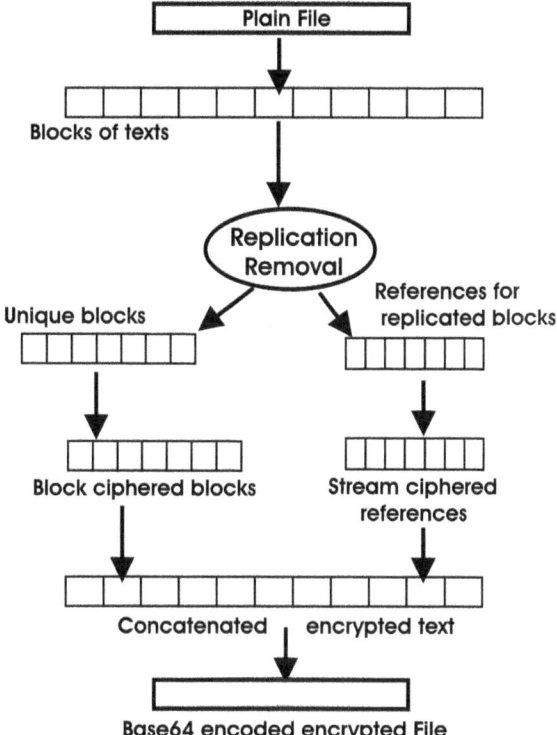

Fig. 2 Block replication removal

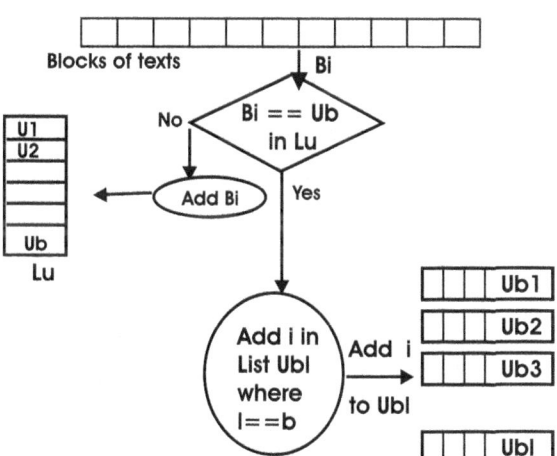

Unique Blocks = U1, U2, .. Ub in Lu

References = Elements of Ub1, Ub2, ... Ubl

3 Implementation

We implement the proposed cryptographic design for text files with following parameters.

Block cipher algorithm = AES 128
Stream cipher algorithm = ARC4
Block size = 16 bytes
Key length = 128 bits
String size for storing number of unique blocks as final string = 3 bytes

The implementation is done in python using Pycrypto library. The computer configuration consists of intel i5 quad core CPU, 2 GB RAM and installed operating system on the computer is Ubuntu 14.04 LTS.

Following message is the input plain text (Fig. 3).

The message is processed using Algorithm 1 and obtained following results. Red underlined text is the number of unique blocks, and blue underlined text is meant to stream of references that is produced following Algorithm 1 (Fig. 4).

Algorithms 2 and 3 are applied for enciphering the different parts of transformed message (Fig. 5).

Finally, we test it by taking reverse of Algorithms 2 and 3. This gives different decrypted parts of the transformed message, which are shown following (Fig. 6).

Fig. 3 Plain text

Fig. 4 Transformed message

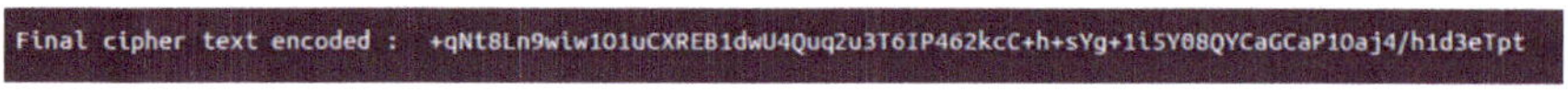

Fig. 5 Encrypted text in Base64 encoding

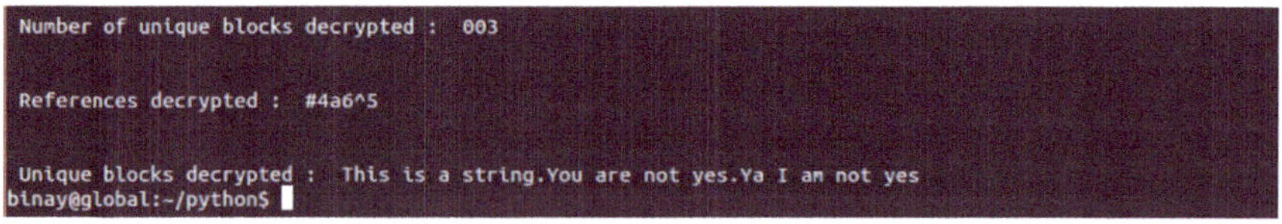

Fig. 6 Decrypted transformed message

4   Analysis and Discussion

Design of the proposed cryptography is meant for replication removal in input file/text, random size of the output file, random patterns of ciphertext in output file. Following subsections analyse these terms one by one.

4.1   Replication Removal

In experiments done by Dutch T. Meyer [10], it has been found that replication strategy named block-level deduplication [11] can remove 68% of data in computer storage. Here, a block refers to a chunk of data of size say N bytes/kilobytes. Now, it is easy to conclude that a file can have replicated chunks of data. Also other authors have found that amount of deduplicate data increase when block granularity is decreased; although it is true that IO performance decrease in such cases [12]. Some authors have suggested to use 4 KB as size of a data block taking into consideration of other system parameters.

On the another hand, recent block cipher algorithms like AES, DES, IDEA use 16, 32, 64, 128 bytes as a size of one data block. If we compare the maximum size taken in block cipher algorithms to the size suggested by authors of deduplicaion technologies, we find the ratio of the blocks in the two algorithm as 1:32. It is impressive to know that the required file storage space can be decreased much and much if we apply deduplication during encryption. However, a proven magnitude of deduplication in individual files is a subject of experiments, which we keep for our future work. But we have concluded that replication of data chunks can compress the input file with a large ratio.

4.2   Size of Output File

Previous subsection concludes that size of the output file can not be equal to as its input file, if we don't talk about some exceptions with probability that a file may not have any replicated units of content. It is not mandatory that replication can save always a fixed fraction of the input file size as it depends on the randomness of data in the file. So it is easy to conclude that one can not estimate the input file size if output file size is given. Again to say that replication removal and adding stream of references randomise size of the output file. So, it becomes difficult for cryptanalysts to guess the block size taken in encrypting the file content using block ciphers.

4.3 *Random Patterns in Cipher Text*

To distinguish between block cipher and stream ciphertext, the machine learning methods used by authors [5] conclude that random patterns of these two ciphertext can be identified with some errors. In other words, it has not been found that all the chunks of ciphertext can be identified to its cipher category.

The output generated from our work consists encrypted data of both cipher mechanisms. The random size of the output file raises questions for cryptanalysts on choosing a block size. If we randomize the positions of block cipher data, encrypted references, encrypted number of unique blocks to be arranged in the final output then it will be more difficult for the cryptanalysts. Because complexity in making decision for the boundary of block cipher and stream cipher text will increase. If it is done randomly, then errors reported by authors [5] say that not all the data blocks will be identified whether it is block cipher or stream cipher. Ultimately, it becomes much difficult to apply ciphertext-only attack on our proposed cryptography design until key is compromised, attacked or revealed.

5 Conclusion

Replication of data chunks in input for encryption, patterns of random characters in encrypted output, size of output and block lengths of data chunks have helped cryptanalyst in forging the encrypted data. Machine learning and other tools, techniques have reduced the human effort to do computation against encryption processes. In such scenarios, hybrid algorithms can put ambiguity in encrypted data, and cryptanalysts can face more difficulties in order to identify the correct algorithms, keys used in ciphertext. We have proposed design of a cryptographic algorithm, which makes uses of replication removal in input data, compressing it, use block cipher and stream cipher algorithms for producing symmetric hybrid cryptographic algorithms. We have designed, implemented, tested and analysed our work from different aspects. It has been found that size of output is random and unequal to the input size. It will be hard for machine-learning mechanisms to identify the block ciphered text and the stream ciphered text. Hence, identifying algorithms and keys for ciphering text will also be more difficult.

6 Future Work

Current work faces deficiency of some experiments such as finding average ratio of content replication in individual files, although the file contents may be random. Executing the algorithm designs with various combination of existing block and cipher algorithms, testing patterns of randomness in them, feeding these patterns in machine learning methods that distinguish block and stream cipher data are some more required experiments. We expect these experiments to be done in our future work.

References

1. Menezes, A.J., Van Oorschot, P.C., Vanstone, S.A.: Handbook of Applied Cryptography. CRC Press (1996)
2. Stallings, W.: Cryptography and Network Security, 4th edn. Pearson Education India (2006)
3. Swenson, C.: Modern Cryptanalysis: Techniques for Advanced Code Breaking. Wiley (2008)
4. https://github.com/fwhacking/bfcrypt
5. Saxena, G., Karnik, H., Agrawal, M.: Classification of ciphers using machine learning. Master's thesis, Department of Computer Science and Engineering, Indian Institute of Technology, Kanpur (2008)
6. Kant, S.: Classification models for symmetric key cryptosystem identification. Def. Sci. J. **62**(1), 38 (2012)
7. Klein, A.: Attacks on the RC4 stream cipher. Des. Codes Cryptogr. **48**(3), 269–286 (2008)
8. Shemaili, M.B., et al.: A new lightweight hybrid cryptographic algorithm for the internet of things. In: 2012 International Conference for Internet Technology and Secured Transactions. IEEE (2012)
9. Bhatele, K., Sinhal, A., Pathak, M.: A novel approach to the design of a new hybrid security protocol architecture (2012)
10. Meyer, D.T., Bolosky, W.J.: A study of practical deduplication. ACM Trans. Storage (TOS) **7**(4), 14 (2012)
11. He, Q., Li, Z., Zhang, X.: Data deduplication techniques. In: 2010 International Conference on Future Information Technology and Management Engineering (FITME), vol. 1. IEEE (2010)
12. Zhao, X., et al.: Liquid: a scalable deduplication file system for virtual machine images. IEEE Trans. Parallel Distrib. Syst. **25**(5), 1257–1266 (2014)

A Novel and Efficient Perceptual Image Encryption Based on Knight Moves and Genetic Operations

Jalesh Kumar and S. Nirmala

Abstract To create pursuit in the multimedia information along with protection, perceptual security techniques are gaining more attention. A novel approach for perceptual encryption of images based on moves of knight pieces is proposed. The novelty of the approach lies in simulating unpredictable key sequence generation based on crossover operation and knight piece moves. Three stages are carried out in the proposed method. In the first stage, the initial population of a genetic process is selected based on the amount of perceptual information. In the second stage, random keys are generated based on pseudorandom number generator for crossover operation. Two different cases are considered in the proposed approach for crossover operation. Crossover of one-point and two-point operations is applied to the random positions of the knight piece. In the third stage, movement of knight pieces is carried out iteratively for fixed number of times, and crossover operation is performed according to one and two points on selected pixels. Structural similarity index, PSNR, entropy, and correlation coefficient are analyzed to measure the performance of the proposed approach. The comparative study with existing perceptual method reveals that the proposed work provides better perceptual security for images.

Keywords Knight piece · Crossover · Perceptual encryption
Digital image

1 Introduction

Most of the multimedia information is voluminous in size, and to secure such information in reduced time is a challenging task. There is always a trade-off between the amount of security and speed of computation in security-providing

Jalesh Kumar (✉) · S. Nirmala
Department of CSE, JNNCE, Shivamogga 577204, India
e-mail: jalesh_k@yahoo.com

S. Nirmala
e-mail: nir_shiv_2002@yahoo.co.in

© Springer Nature Singapore Pte Ltd. 2018
K. Saeed et al. (eds.), *Progress in Advanced Computing and Intelligent Engineering*,
Advances in Intelligent Systems and Computing 563,
https://doi.org/10.1007/978-981-10-6872-0_22

233

techniques. This composition could be overcome by providing security at different levels. Security levels vary from predictability to unpredictability of the information contents. The blurred or degraded quality of multimedia information attains predictability of the information which constitutes a low level of security. In a high level of security, information contents are completely scrambled and there are no residuals which reveal the contents of original information. So the amount of degrading the perceivability of the information leads to different levels of security.

Transmitting the document images, satellite images, medical images or instant sharing of images acquired by mobile phones needs the security techniques which provide the protection at different levels. Several techniques are found in the literature to secure the images either partially or fully. Further, to attract about the information without giving the details of information, scalable security techniques are needed. Such perceptual encryption technique is gaining importance in today's multimedia environment.

Perceptual encryption is the technique where end user perceives the content in partial but not in original form. For example, videos or images will be released in a degraded version of the information to create interest to the customers. Customers are able to perceive the contents and can communicate for getting the original version. To degrade the image contents, scalable encryption approach based on alpha rooting is proposed in [1]. The logarithmic Michelson contrast measure (AME) and the logarithmic AME by entropy are applied as parameters for enhancing the image. By controlling the enhancement parameters, degradation is achieved which results in the partial encryption. The enhancement parameters are used to protect the confidential information in an image. Limitation of the method is the protection of the information that depends only on the controlling parameter. In [2], a perceptual encryption technique is explored to degrade the quality of the image. Degradation process is carried out using transposition of pixels on the basis of kernel generated from geometric objects. Using geometric objects, different swapping patterns are provided and security is provided at different levels. The process is time-consuming. A new method of cyclic bit manipulation for partial encryption is discussed in [3]. But the cyclic bit manipulation method is unable to provide variable security. Yang et al. [4] presented shuffling and reversible histogram spreading for perceptual image encryption. Local details are protected completely to adopt perceptual security of the information. The approach consumes more time to protect the information.

Versatile techniques are available in the literature to generate key sequences to protect the information. The behavior of knight pieces in the cryptography applications is discussed in [5]. A slip encryption-filter template matrix based on moves of the knight is used in [6]. The full encryption is carried out, and computation time is more. For encryption of the image, pixels are transposed between red, green, and blue channels on the basis of rules for movement of knight piece and chaos-based pseudorandom bit generator [7]. Separating color channels and process them separately requires additional computation time. Genetic operations along with moves of knight pieces to encrypt the document images are proposed in [8]. The computation time is depending on the number of pixels and moves of knight piece. Some of the evolutionary methods are attaining significance in securing the information contents. The evolutionary approach to security is described in [9]. In [10], mutation and crossover operation are discussed to secure the information in an image. In [11], discrete Fourier transform is used for image encryption. Magnitude and phase are modified using crossover operation. The process of decryption results in a lossy image.

From the survey, it is evident that the past perceptual encryption techniques are computationally complex and do not address security at different levels. It is observed from the survey that movements of knight pieces could be used for the creation of unpredictable sequences. It is also perceived that genetic process is gaining importance in encryption of the information contents. In this work, perceptual encryption technique is proposed to secure information contents of the image. The novelties of the work are use of moves of knight piece and crossover operation in generating the unpredictable key sequence.

2 Behavior of Knight Piece on Chessboard

Chess is a popular game played between two people which consists 8 × 8 board and six different types of movable objects such as rook, knight, bishop, queen, king, and piece. Each chess piece is having following behavior.

(a) Rook—it moves straight from present position
(b) Bishop—it moves along the diagonal direction
(c) Knight—it moves straight and side both the directions together

Compared to all the pieces on the chessboard, moves of knight have specific behavior. Only knight piece can jump on other piece and be having many possible next moves. From the current position, knight piece occupies next position at most eight different ways. Consider 64 different knight pieces occupy in each position of 8 × 8 matrix. Among that, one knight piece is selected to process iteratively. Selected knight piece is having eight different next possible positions. Among these positions, one position is chosen randomly. The process could be continued depending on fixed number of iterations. Generally, for knight pieces, position in a corner, a number of next possible positions are two. For other positions, a number of next possible moves are three, four, six, or eight. Each selection of position creates a different path for knight piece. This behavior of knight piece is utilized to generate the unpredictable key sequence in securing the information.

3 Proposed Method

In the proposed method, perceptual encryption of image is carried out based on moves of knight pieces and crossover operations of the genetic algorithm. Figure 1 shows the block diagram of the proposed method. All knight pieces process on the

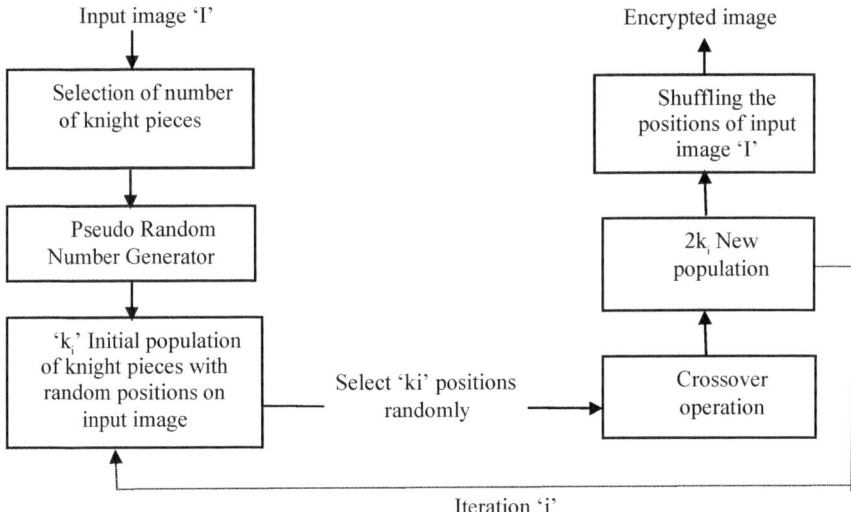

Fig. 1 Block diagram of the proposed method

image simultaneously. After every move of knight pieces, pixels from the corresponding positions are selected for crossover operation. The encryption process is repeated for fixed number of iterations.

Stage 1: Selection of Number of Knight Pieces

Using pseudorandom number generator, nonoverlapping initial positions of the knight pieces are generated. Pixels in the selected positions act as initial population.

A number of initial positions 'k' of knight pieces are generated based on the amount of perception of information

$$k = z * (M \times N) \tag{1}$$

where 'M' is the width and 'N' is the height of the input image, 'z' is variable; a value of 'z' ranges between 0 and 1. '0' represents there is no knight piece on the chessboard. '1' indicates a number of knight pieces initialized are equal to a size of the input image.

Stage 2: Crossover Operations of Genetic Process on Selected Pair of Pixels

Genetic process is a generation of new individuals based on old population based on selection, crossover, and mutation operation. Crossover operation works similar to transposition technique and mutation like substitution operation [1]. In this work, one-point and two-point operations are used for the encryption process.

(i) One-point crossover operation

Based on the single crossover point chosen, new individuals are generated from the parents. For example, at crossover point 2 in parents 'A' and ' B', new individuals 'A$_1$' and 'B$_1$' are created as shown in Fig. 2.

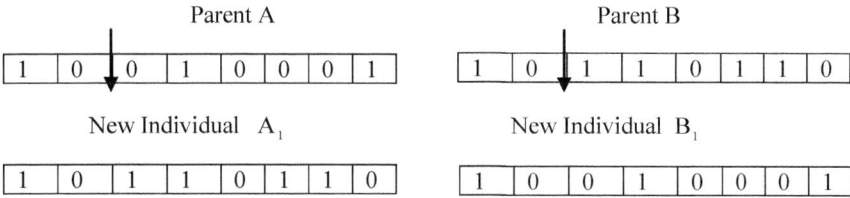

Fig. 2 One-point crossover operation

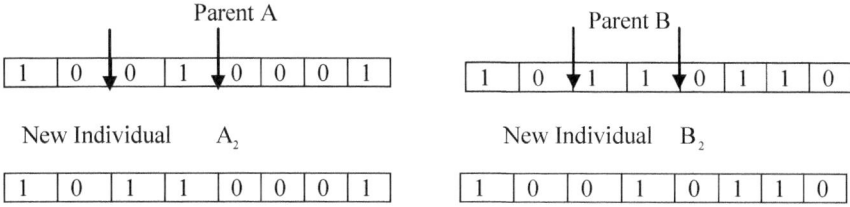

Fig. 3 Two-point crossover operation

(ii) Two-point crossover operation

From the parents, two points are selected to generate new individuals. For example, two points are selected from parents 'A' and 'B' shown below, and after crossover operation, new individuals 'A₂' and 'B₂' created are shown in Fig. 3.

Stage 3: Perceptual Encryption Process

Two cases are considered in the proposed approach for the perceptual encryption process.

Case 1: Shuffling process with one-point crossover operation
Case 2: Shuffling process with two-point crossover operation

According to moves of knight pieces in stage 1, two positions are selected for cross operation. Crossover points are randomly generated using key stream generator for one-point and two-point crossover operations, respectively. Operations are repeated for fixed number of iterations. Crossover operation is carried out on different combination of pixels. Thus, pixel values change a multiple number of times on the same position.

The reverse of the encryption process results in decryption. Crossover operation and key generation process should be reversed to get the original image.

4 Experimental Results and Discussions

Standard image database USC [12] is used for testing the proposed approach. Corpus contains the sample images of different size, colors, and textures. The results are shown for values of 'z' (in Eq. 1) equal to 0.01, 0.02, 0.03, 0.04, 0.05, and 0.1. The average number of pixels participating in the encryption operation is computed, and values are recorded in Table 1 for all images in the corpus. Experiments are conducted by varying the number of initial knight pieces. Further, moves of each knight piece are varied from 10 to 100. From the values in Table 1, it is observed that the percentage of pixels participating in encryption is more than 99% when a number of knight pieces are set to more than 4% of total size of the image and moves of each knight piece are 100. Further, the percentage of pixels participated in encryption is more than 50%, when knight piece move is fixed to 50 and a number of initial knight pieces are more than or equal to 1% of a size of the input image. Hence, increase in the initial number of knight piece or increase in a number of moves of each knight piece both achieves good encryption. However, processing time also increases with increase in the amount of encryption.

It is observed from Table 1 that as 'z' value reduces to zero perceptibility of the information in image increases and increase in a value of 'z' near to one reduces the perceptibility, additionally, results in more degradation. It is also evident from Table 1 that 50 moves and initial knight pieces equal to 1% of a size of the input image are sufficient to achieve 51% of degradation of information contents. Hence, 49% of information contents are perceivable from the image. In the proposed method, the value of 'z' is chosen depending on the different levels of security. Perceivability above 24% indicates a low level of security, between the range 15–24% perceivable information provides a medium level of security, and below 15% perceivability indicates security at higher levels. As a decrease in perceivability of information, a level of security increases.

The result obtained with varying number of initial knight pieces is shown in Figs. 4 and 5 for a number of knight moves equal to 50. Figure 4 shows the result obtained from case 1. Figure 5 shows the result obtained from the case 2. The increase in a number of initial knight pieces results in the increase of unpredictability of contents of a source image. In both the cases, images are degraded and

Table 1 Average values for the performance metrics of the images in the corpus (for both cases 1 and 2)

Variable 'z'	Number of pixels participating in encryption process (%)									
	Number of moves of each knight piece									
	10	20	30	40	**50**	60	70	80	90	100
0.01	16.55	28.22	37.30	44.49	**51.88**	57.9	63.27	68.12	72.16	73.20
0.02	28.92	48.23	60.68	68.48	**75.98**	81.96	86.59	88.42	91.00	92.30
0.03	37.34	61.45	75.12	82.27	**87.57**	91.97	94.14	95.55	96.95	97.80
0.04	45.89	71.24	82.84	88.77	**93.11**	96.19	97.13	98.06	98.93	99.42
0.05	53.45	77.25	87.78	92.76	**95.74**	97.83	98.35	99.10	99.38	99.78
0.1	78	88.23	93.23	98.10	**98.90**	99.12	99.31	99.45	99.60	99.92

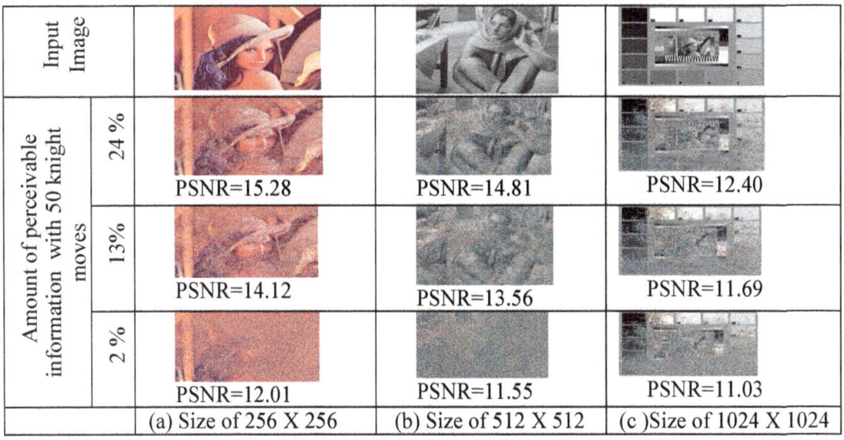

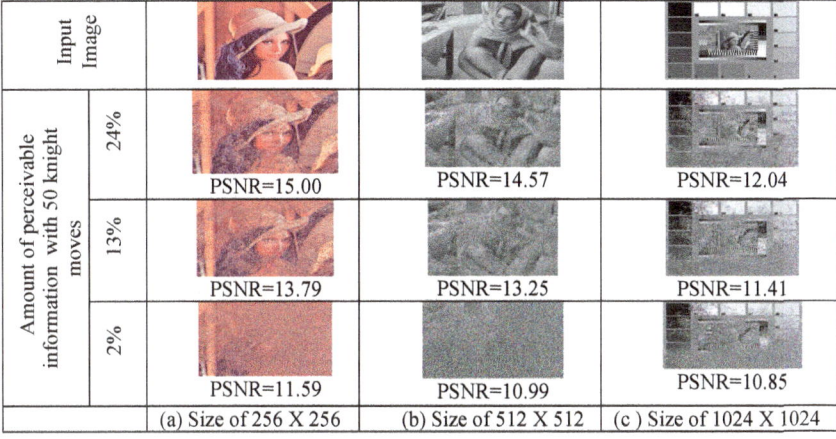

Fig. 4 Results of proposed approach for varying amount of perceivable information (case 1)

Fig. 5 Results of proposed approach for varying amount of perceivable information (case 2)

unable to reveal the complete information. Compared to case 1, in case 2 shows the increased amount of degradation due to the selection of two random points for crossover operation.

5 Security Analysis

Statistical measures are used to analyze the encrypted images. The performance of the method is evaluated using structural similarity index, entropy, PSNR, and correlation coefficient values.

5.1 Structural Similarity Index

A similarity between two images is measured in terms of structural similarity index [13]. The value of SSIM is in the range [0 1]. The value '0' indicates that there is no correlation between two images. Value '1' indicates two images are similar.

5.2 Entropy

Information entropy value is used to measure the amount of randomness. It is calculated by [2]

$$\text{Entropy} = \sum Q(i) \log_2 \frac{1}{Q(i)} \tag{2}$$

5.3 Correlation Coefficient

A correlation coefficient is a measure of correlation between two adjacent pixels in an image [2]. Correlation coefficient value '1' indicates adjacent pixels are highly correlated. '0' indicates there is no correlation between the adjacent pixels. As degradation in the image increases, correlation coefficient value reaches near to zero.

5.4 Peak Signal-to-Noise Ratio

Reduction in the quality of images is measured in terms of peak signal-to-noise ratio (PSNR). The PSNR can be computed by Lian [14]

$$PSNR = 10 \log_{10} \frac{W^2}{MSE} \tag{3}$$

where mean square error is calculated by

$$MSE = \frac{1}{n} \sum_{i=0}^{n-1} (c_i - p_i)^2 \tag{4}$$

In the above equation, 'W' is the gray intensity of the pixel.

SSIM, correlation coefficient, entropy, and PSNR values are evaluated for the performance analysis. These parameters are computed between input images and

Table 2 Average values for the performance metrics of all the encrypted images in the corpus

Amount of perceivable information with 50 knight moves (%)	Proposed method	Performance metrics			Correlation coefficient		
		SSIM	PSNR	Entropy	Horizontal	Vertical	Diagonal
51	Case 1	0.389	15.37	7.77	0.239	0.156	0.279
	Case 2	0.387	15.01	7.78	0.236	0.201	0.185
24	Case 1	0.206	14.28	7.78	0.195	0.133	0.216
	Case 2	0.203	14.01	7.81	0.159	0.156	0.111
13	Case 1	0.139	12.34	7.80	0.167	0.084	0.151
	Case 2	0.135	12.00	7.83	0.136	0.118	0.072
7	Case 1	0.101	11.65	7.81	0.121	0.079	0.138
	Case 2	0.098	10.95	7.84	0.105	0.131	0.090
5	Case 1	0.085	10.6	7.82	0.114	0.071	0.108
	Case 2	0.081	10.02	7.86	0.115	0.122	0.071
2	Case 1	0.063	9.05	7.86	0.091	0.051	0.082
	Case 2	0.060	8.55	7.90	0.025	−0.0013	0.017

corresponding encrypted images. Results are tabulated in Table 2 for both the cases 1 and 2. It is evident from values in Table 2 that as the initial knight piece increases (perceivable information reduces), the quality of the image degrades more and perceptual security improves. As amount of perceivable information reduces, the value of entropy also increases. Correlation coefficient values of adjacent pixels in diagonal, vertical, and horizontal direction are recorded in Table 2. It is observed from the values in Table 2 that as perceivability decreases, the value of correlation coefficient decreases. It is observed from the PSNR values that quality of the image also decreases. It is also observed from the values tabulated in Table 2 that the amount of randomness is more in case 2 compared to case 1.

6 Comparative Study

The result obtained in the proposed method is compared with the method [2]. In method [2], the geometrical objects are used as the kernel to permute the pixels to achieve perceptual encryption. Different level of degradation is provided by adjusting the kernel objects. The result obtained for the input image in Fig. 5a, b in case 2 is compared with method [2] in terms of PSNR and correlation coefficient. Images obtained by encryption using kernel formed by four objects in method [2] are compared with encrypted image with perceivable information about 2%.

Table 3 Comparison of case 2 with method [2]

Metrics		Figure 5a		Figure 5b	
		Method [2]	Proposed method	Method [2]	Proposed method
Correlation coefficient	Horizontal	0.0459	0.025	0.0602	0.036
	Vertical	−0.0261	−0.0013	0.0702	0.0155
	Diagonal	−0.0146	0.017	−0.0223	0.0155
PSNR		11.988	11.59	11.197	10.99

From the Table 3, it is evident that the proposed method outperforms the method [2] of perceptual encryption. Depending on the amount of perceivability, degradation has been carried out in the proposed approach with reduced computation complexity.

7 Conclusion

In the proposed approach, movement of the knight pieces is considered for crossover operation of the genetic process to achieve perceptual encryption. Two different cases of crossover operation are considered for encryption. Two-point crossover with knight piece moves achieves more degradation in information contents of the image in comparison with one-point crossover. In the proposed approach, the initial seeds of pseudorandom number generator, the number of initial knight pieces, and moves of each are considered as key sequences for the encryption process. Further, crossover operation on the different combination of pixels with a multiple numbers of times enhances the unpredictability in the key sequence. It has been observed from the experimental results that depending on the amount of perceivability, different levels of perceptual security are achieved. Further, increase in the number of moves of knight or number of initial knight piece better security is achieved.

References

1. Wharton, E.J., Panetta, K.A., Againg, S.S.: Scalable encryption using alpha rooting. In: SPIE Defense and Security Symposium, International Society for Optics and Photonics, pp. 69820G–69820G (2008)
2. Jagadeesh, P., Nagabhushan, P., Kumar, R.P.: A novel perceptual image encryption scheme using geometric objects based kernel. Int. J. Comput. Sci. Inf. Technol. 5(4), 165 (2013)
3. Parameshachari, B.D., Sunjiv Soyjaudah, K.M., Sumithra Devi, K.A.: Image quality assessment for partial encryption using -modified cyclic bit manipulation. Int. J. Innov. Technol. Explor. Eng. (2013)

4. Yang, B., Busch, C., Niu, X.: Perceptual image encryption via reversible histogram spreading. In: Proceedings of 6th International Symposium on Image and Signal Processing and Analysis, 2009. ISPA 2009, pp. 471–476. IEEE (2009)
5. Philip, A.: A generalized pseudo knight's tour algorithm for encryption of an image. IEEE Potentials 10–16 (2013)
6. Delei, J., Sen, B., Wenming, D.: An image encryption algorithm based on knight's tour and slip encryption-filter. In: International Conference on Computer Science and Software Engineering (2008)
7. Diaconu, A.-V., Costea, A., Costea, M.-A.: Color image scrambling technique based on transposition of pixels between RGB channels using knight's moving rules and digital chaotic map. Math. Probl. Eng, **2014**, Article ID 932875, 15 (2014). https://doi.org/10.1155/2014/932875
8. Jalesh Kumar, Nirmala, S.: A new light weight encryption approach to secure the contents of image. In: Advances in Computing, Communications and Informatics (ICACCI, 2014), pp. 1309–1315 (2014)
9. Picek, S., Golub, M.: On evolutionary computation methods in cryptography. In: Proceedings of the Information Systems Security, MIPRO 2011, 23–27 May, pp. 1496–1501
10. Husainy, M.: Image encryption using genetic algorithm. Inf. Technol. J. **5**(3), 516–519 (2006)
11. Abuhaiba, I.S.I., Hassan, M.A.S.: Image encryption using differential evolution approach in frequency domain. Int. J. Singal Image Process. **2**(1) (2011)
12. http://sipi.usc.edu/database/miscellaneous
13. Wang, Z., Bovik, A.C., Sheikh, H.R., Simoncelli, E.P.: Image quality assessment: from error visibility to structural similarity. IEEE Trans. Image Process. **13**(4), 600–612 (2004)
14. Lian, S.: Multimedia Content Encryption Techniques and Applications. CRC Press (2012)

A Flow Marking Based Anti-spoofing Mechanism (FMAS) Using SDN Approach

Saifuddin Ahmed and Nabajyoti Medhi

Abstract Security has become a potential issue among the Internet users all over the globe. IP spoofing is one of the major security threats to the Internet. Spoofing is a technique in which an attacker does forgery of ID's of a legitimate host in the network. Spoofing is a serious problem to the present Internet world. It creates serious security problems such as packet authenticity, launching DDOS attacks, etc. In this paper, we propose a flow marking based anti-spoofing (FMAS) mechanism for wide area networks (WAN) which uses marking of source IP address field of a flow for encoding. We have tested our anti-spoofing mechanism using software-defined networking (SDN) approach by dynamically marking network traffic flows for enhanced security.

Keywords Software-Defined Networking (SDN) · OpenFlow
Flow marking based anti-spoofing (FMAS) · Denial of Service (DOS)
Wide Area Network (WAN)

1 Introduction

IP spoofing is an IP address forgery or host file hijacking technique in which an attacker masquerades as a legitimate host to conceal its identity. An attacker may spoof the MAC address of a legitimate host also, and then the attacker may launch DOS or DDOS attack [1]. So far, many anti-spoofing mechanisms have been proposed, but none of them is able to give a better effective solution. Most of the anti-spoofing techniques have used the concept of IP traceback [2]. IP traceback technique basically finds out the origin of the attack node and the path traverses by the anonymous DDOS attack with spoofed address [3, 4]. But using only the basic

S. Ahmed (✉) · N. Medhi
Department of Computer Science & Engineering, NIT Meghalaya, Shillong, India
e-mail: saif783335@gmail.com

N. Medhi
e-mail: nabajyoti.medhi@nitm.ac.in

© Springer Nature Singapore Pte Ltd. 2018
K. Saeed et al. (eds.), *Progress in Advanced Computing and Intelligent Engineering*,
Advances in Intelligent Systems and Computing 563,
https://doi.org/10.1007/978-981-10-6872-0_23

IP traceback technique, it is difficult to locate the real attack host [5]. So far, various procedures have been used for IP traceback technique like ICMP traceback message, link testing, packet marking, logging, etc. [6]. There are different types of IP traceback techniques that have been used for anti-spoofing mechanisms—in probabilistic packet marking (PPM) approach of IP traceback, the routers or the network-forwarding devices mark the packets probabilistically, and the destination router recreates the path by observing such marked packets [2, 7–9]. According to Belenky and Ansari [10], mark spoofing still may happen in PPM. So they proposed the concept of deterministic packet marking (DPM). DPM mainly focuses to mark all the packets across the network by some kinds of ingress router. According to Jin and Yang [11] and Belenky and Ansari [12], basic DPM may fail to handle some special kinds of DDOS attacks—such as multiple attack nodes use the same spoofed address. We have implemented our anti-spoofing mechanism (FMAS) with SDN because it provides a better security perspective solution than the traditional Internet network. In SDN approach, we can implement high-level rules; we can set up new policies easily by writing some codes; or by implementing our own logic in the control plane as per our reliability. By the proper use of IDS (intrusion detection system) with SDN controller, SDN approach can provide better networking in terms of security [13]. IDS is a device or software application that looks after the network or system activities for malicious activities. IDS analyzes the traffic pattern, whether it is a kind of malicious traffic or not [14–16]. But in reality to analyze all the traffic patterns, whether those are malicious or not, is almost practically impossible or it would be too much time consuming. So rather than detecting the malicious packets and then discarding, it is assumed to be a better approach to detect the malicious source node or attack node.

In our approach, we proposed a mechanism to trace the host node and then conclude whether it is an attacked node or not. In our anti-spoofing mechanism, FMAS, we have used a somewhat similar concept like MPLS (Multiprotocol Label Switching). MPLS is a mechanism in which data packets are assigned with some kinds of labels. Packets are being sent based on those labels. All the switches along the path of the packets just perform packet forwarding based those labels rather than look for IP header. MPLS gives some valuable applications to the traffic engineering [17].

In this paper, we propose a method which enables marking of a flow by encoding the source IP address field of an outgoing flow.

2 Related Work

Foroushani and Heywood [18] proposed a deterministic flow marking approach (DFM) for IP traceback. This method uses ECDSA (Elliptic Curve Digital Signature Algorithm) crypto-system algorithm for authentication purpose of the marked identification data.

Kwon et al. [19] proposed a path-based anti-spoofing mechanism called BGP-BASE (Border Gateway Protocol—Based Anti-Spoofing Extension) which is well compatible to work in software-defined networking. BASE uses BGP update message for marking value distribution.

Jing et al. [5] proposed a Distributed Log-based Scheme (DLS). It is basically a combination of PPM (Probabilistic Packet Marking) and logging technique. In DLS, they use message authentication code (MAC) technique for data integrity in case of marking and log transmission.

According to Foroushani and Heywood [18], deterministic methods have comparatively more processing overheads than probabilistic marking methods of IP tracebacking.

The proposed approach of Saurabh and Sairam [20] is a combination of linear packet marking and a randomized Remainder Packet Marking (RPM) schemes, which requires less number of packets for IP traceback. Both methods use ID and TTL field of IP packet for marking purpose.

According to the proposed method by Jin et al. [21], the across-domain deterministic packet marking for IP traceback which basically traces the malicious traffic in a network and then filters it is assumed to divide the network into two kinds of ASs and routers.

According to [18, 19] only tracking the malicious traffic rather than the malicious source and then filters them is not an efficient and reliable concept in scenarios like huge number of DDOS attack from different sources.

Jin and Yang [11] have suggested a Deterministic Packet Marking approach based on Redundant Decomposition for IP Traceback (DPM-RD). This approach uses one marking and a recovery algorithm, and it basically focuses in decreasing the false positive ratio of recovery.

According to Stoica and Zhang [22], only about 0.25% or less of packets are fragmented. So the use of ID field of IP packet to store the marking value will not affect much. Therefore, many researchers in their anti-spoofing mechanisms have used ID field to store marking values [18, 19, 23, 24]. But we know that disabling the basic functionality of ID field of IP packet is the major drawback in scenarios like packet reassembly.

To avoid the above-mentioned problems of the aforementioned anti-spoofing mechanisms and to fill up some new requirements, we have proposed Flow Marking based Anti-Spoofing mechanism (FMAS) with the following objectives:

- The proposed flow marking based anti-spoofing mechanism (FMAS) uses a centralized OpenFlow controller for flow marking which improves the overall network visibility.
- In our FMAS mechanism, the packets are being marked flow-wise instead of being marked packet-wise randomly.
- Instead of marking any field of the IP packet header such as Identification (ID) or TTL field, etc., we choose to mark source IP address field of an IP packet for sending traffic to the wide area network. For a normal communication of a

packet across the Internet, the source IP address packet is an easy target for spoofing. We can choose this field to store the marking value assigned dynamically to ensure integrity of the marking value. In this process, we should be able to successfully decode the actual source IP at the destination side.

The marking or encoding of packets as well as the validation of the marking values and filtering of the spoofed packets are being done based on some functions computed at the SDN controller in FMAS mechanism.

3 FMAS Mechanism

In FMAS approach, we have used source IP address-based marking procedure that is done specially for wide area network (WAN). In this process, the packet marking procedure is done by dynamically marking source IP addresses of the IP packets of a flow.

This paper represents the scenario where the packets are being analyzed at the egress switches, based on pre-defined marking values. For better manageability in FMAS mechanism, either a single OpenFlow controller is used or multiple controllers are used with similar policies for encoding and decoding of a flow at source and destination, respectively. In FMAS, two kinds of switches are considered—the nearest switch or router from the source node as the ingress router or switch and the same of the destination node as egress router or switch. In this paper, we use the term switch to represent both switch and routers. The marking/encoding of packets is done at the ingress switch based on source IP address field, and the egress switch is basically used for the comparison of the marking values of the packets. Both ingress and egress switches contain flow matching table and specific rules defined by the OpenFlow controller to authenticate the correct marking at both the ends.

The marking value assigned to the packets changes dynamically after some interval of time. So it reduces the guessing probability of marking values by the attackers.

FMAS is basically an encryption and decryption technique to authenticate and validate the marking values at the ingress and egress switches. FMAS algorithm is explained in the next section.

3.1 FMAS Flow Encoding-Decoding Algorithm

In the OpenFlow controller connected to the ingress switch, when a Packet-In message reaches the controller, it instructs the switch with Flow-mod messages to maintain some actions

Input: Time-stamp: t, randomly taken integer: n, duration: Δt, random operation: op, list of authenticated user ids, random functions: f(n), h(t)

begin

    Initialize t = 1

    Procedure *Controller-Encode*()

        Evaluate g(n)

        Evaluate h(t)

        f(n, t) = concatenated g(n) and ceil(h(t))

        M(f(n,t)) = f(n,t) op Src-IP

        Return M(f(n, t))

    end Procedure

    Procedure *Controller-Decode*(Src-IP)

        Evaluate g(n)

        Evaluate h(t)

        f(n, t) = concatenated g(n) and ceil(h(t))

        Src-IP = M(f(n,t)) op f(n,t)

        if Src-IP ϵ (list) **then**

            Action: Forward to the destination port

        else

            Action: Drop

        end if

    end Procedure

end

We use a function f(n, t) which gives a 32-bit number irrespective of choosing any random value n and t by the controller. In the marking procedure, we use this 32-bit number as the marking value in place source IP address.

$$f(n, t) = g(n) + h(t)$$

Here, g(n): This is a dynamic random function that depends on n. This random function g(n) and n are chosen in a manner such that the integer value of the function g(n) can be represented in 16 bits.

h(t): This is a time variant function in terms of t, where h(t) is chosen in such a way so that the ceil(h(t)) can result in an integer value, which can be represented in 16 bits.

n: randomly taken integer from a list of numbers maintained in the controller.

The function g(n) changes with every new selection of the n value. But the function h(t) remains unchanged for a certain duration of Δt and then it changes after some interval as defined by the controller.

The above function is used by the controllers at both the ends.

3.2   FMAS Ingress Switch Algorithm

Both the ingress and egress switch contain flow matching table. Once an ingress switch receives packets from any specific flow, for a duration of time, it calls the

Controller-Encode() function to perform the marking on the source IP and calculate the marking value by using the function defined by the controller.

m_i = Formulated marking value by the Ingress switch algorithm.

In our proposed method, when a packet reaches the ingress switch, it does

begin
         Procedure INGRESS ALGO
              for a duration Δt **do**
                    for all the flow **do**
                          Controller-Encode()
                          $m_i = M(f(n, t))$
                          Action: nw-src = m_i
                    end for
              end for
         end Procedure
end

3.3 FMAS Egress Switch Algorithm

When a packet reaches the egress switch, it compares the Src-IP address with its calculated marking value. Then the controller decides whether to discard or send the packet to the destination.

m_i = Formulated marking value by the egress switch.

When packet of a flow reaches to the egress switch, it does

begin
         Procedure EGRESS ALGO
              for a duration Δt **do**
                    for each incoming new flow at the Egress switch **do**
                          Controller-Decode(Src-IP)
                    end for
              end for
         end Procedure
end

3.4 Packet Encoding Scenario

Let us take a scenario where the source host wants to send its packets to the destination host. The ingress switch sends the packet to the controller if it is a packet of a new flow. The controller defines the particular function to be used by the ingress switch for the purpose of marking. Otherwise, for an existing flow, the ingress switch formulates marking value for the packet using the already defined function from the controller. We use source IP address field of the IP packet to store this marking value, and these fields keep unchanged until the packet reaches the destination or egress switch. The communication between source and destination occurs based on some dynamic source IP addresses of packets which are generally vulnerable to attack for the purpose of spoofing.

3.5 Marking Validation and Egress Switch Filtering

The marking value validation and packet filtering are done at the egress switch by calling the controller-Decode () function at the controller.

When egress switch receives a packet, at first, it checks the Src-IP address against its flow table. If table miss happens, then the egress switch simply drops the packet. But if it is present (table hit), the egress switch compares the marking value of the packet, i.e., Src-IP to its calculated marking value and if it matches then sends the packet to the destination host.

Now, suppose an attacker host spoofs the address of any legitimate host and sends the packet to the egress switch, concealing its own identity. When the attack host spoofs the source node address and sends the packet to the egress switch using that address for the purpose of sending to the destination host, the egress switch compares the marking value of the packet to its calculated marking value and then it filters the packet, the attacker host address is recorded in the controller as well.

4 Experimental Result

We implemented FMAS mechanism by taking a set of simple functions and some small random integers to test its functionality. We tested our model inside mininet and using with POX controller. In mininet, a three-layer tree topology is used as shown in Fig. 1. A pair hosts in the topology works as legitimate hosts, whereas another pair works as malicious hosts. A separate pair of hosts is taken as the target hosts. All the four source hosts send TCP traffic using Iperf to the target host. We compare our model with a network model without any anti-spoofing mechanism. Traffic is generated from the legitimate as well as malicious hosts toward the target hosts.

We have tested our FMAS approach with the help of three experiments.

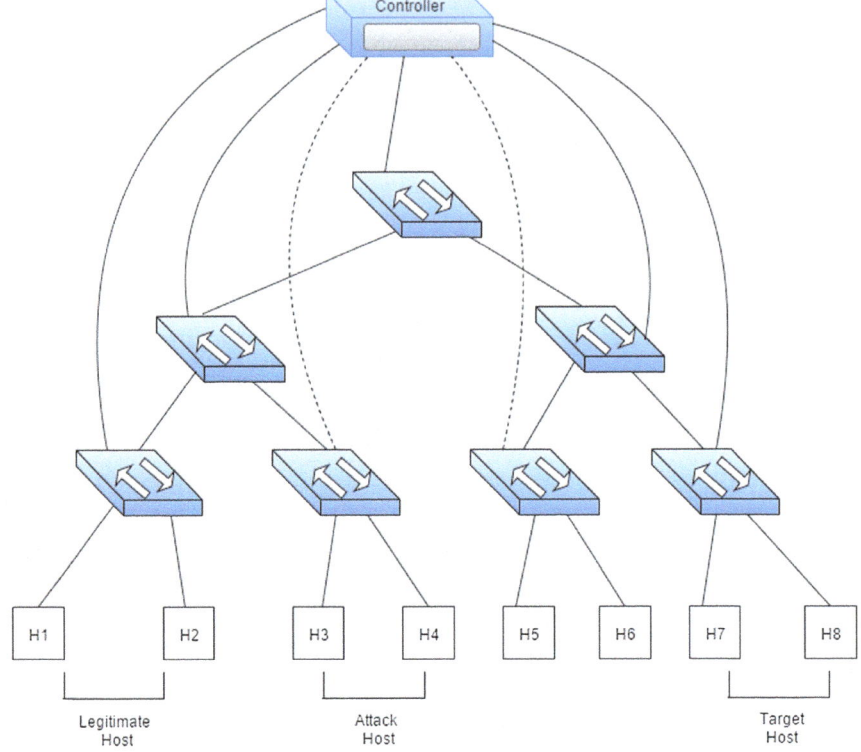

Fig. 1 A basic spoofing attack scenario in a tree topology under a single controller

4.1 Experiment No: 01

We measure the throughput of legitimate traffic in both the cases. From Fig. 2a, b, we observe that FMAS allows the legitimate traffic to pass faster with higher throughput as compared to the network model without any anti-spoofing mechanism. This happens due to the dropping of malicious traffic which in turn provides a better throughput for legitimate traffic.

4.2 Experiment No: 02

In this experiment, we observe the dropping ratio of legitimate traffic flows in the presence of malicious traffic flows.

We further experimented by increasing the number of legitimate and malicious flows in the same topology with FMAS enabled in the switches. Network traffic is generated using D-ITG traffic generator [25]. The number of flows passing between

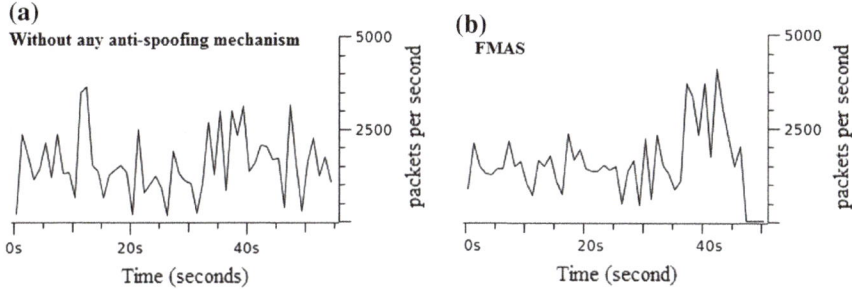

Fig. 2 a Throughput of legitimate traffic without using any anti-spoofing mechanism. **b** Throughput of legitimate traffic with FMAS

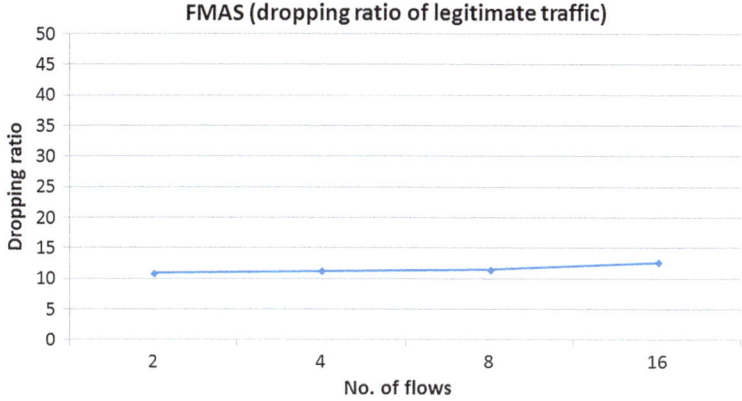

Fig. 3 Dropping ratio of legitimate traffic

each pair of hosts is doubled, and the dropping ratio is observed as shown in Fig. 3. It is observed that as the number of flows increases in equal proportion for both legitimate as well as the malicious traffic, the dropping ratio of legitimate flows increases very slowly. In FMAS, we observed around 12.7% dropping ratio of the legitimate traffic.

4.3   Experiment No: 03

In this experiment, we tested the dropping ratio of malicious traffic using our anti-spoofing mechanism. We have observed that dropping ratio of malicious traffic range lies in the range from 62 to 99%. The dropping ratio increases for decreasing volume of malicious traffic in comparison to the legitimate traffic. The maximum ratio is attained when the volume of malicious traffic is very low in the network.

5 Conclusion

In this paper, we implemented a flow marking based anti-spoofing mechanism (FMAS) which can be very effective in the scenario like IP address spoofing. It can prevent spoofing-based attacks by means of marking source IP field of a particular flow. With a centralized control using SDN, FMAS mechanism becomes flexible and manageable as new rules can be added dynamically at source and destination routers or switches for enhancing security. In FMAS, dropping of the legitimate packets is very much negligible due to its robust and dynamic encoding and decoding at the switches. This work can be extended further to a real-time WAN scenario where distributed controllers can be used for centralized synchronization of encoding and decoding at the source and destinations.

References

1. Vijayakumar, R., Selvakumar, K., Kulothungan, K., Kannan, A.: Prevention of multiple spoofing attacks with dynamic MAC address allocation for wireless networks. International Conference on Communication and Signal Processing (ICCSP), 3–5 Apr. IEEE, India (2014)
2. Gong, C., Sarac, K.: Toward a more practical marking scheme for IP traceback. In: 3rd International Conference on Broadband Communications, Networks and Systems, 2006. BROADNETS 2006. IEEE (2006)
3. Oe, M., Kadobayashi, Y., Yamaguchi, S.: An implementation of a hierarchical IP traceback architecture. In: 2003 Symposium on Applications and the Internet Workshops, 2003. Proceedings. IEEE (2003)
4. Saurabh, S. Sairam, A.S.: Linear and remainder packet marking for fast IP traceback. In: 2012 Fourth International Conference on Communication Systems and Networks (COMSNETS). IEEE (2012)
5. Jing, Y., Tu, P., Wang, X., Zhang, G.: Distributed-log based scheme for IP traceback. In: The Fifth International Conference on Computer and Information Technology, 2005. CIT 2005. IEEE (2005)
6. Choi, K.H., Dai, H.K.: A marking scheme using Huffman codes for IP traceback. In: 7th International Symposium on Parallel Architectures, Algorithms and Networks, 2004. Proceedings. IEEE (2004)
7. Goodrich, M.T.: Efficient packet marking for large-scale IP traceback. In: Proceedings of the 9th ACM Conference on Computer and Communications Security. ACM (2002)
8. Perrig, A., Yaar, A., Song, D.: FIT: fast Internet traceback. In: 24th Annual Joint Conference of the IEEE Computer and Communications Societies, INFOCOM 2005. Proceedings IEEE, vol. 2. IEEE (2005)
9. Izaddoost, A., Othman, M., Rasid, M. F. A.: Accurate ICMP traceback model under DoS/DDoS attack. In: International Conference on Advanced Computing and Communications, 2007. ADCOM 2007. IEEE (2007)
10. Belenky, A., Ansari, N.: IP traceback with deterministic packet marking. IEEE Commun. Lett. 7(4), 162–164 (2003)
11. Jin, G., Yang, J.: Deterministic packet marking based on redundant decomposition for IP traceback. IEEE Commun. Lett. 10(3), 204–206 (2006)
12. Belenky A., Ansari, N.: Tracing multiple attackers with deterministic packet marking (DPM). In: 2003 IEEE Pacific Rim Conference on Communications, Computers and Signal Processing, 2003. PACRIM, vol. 1. IEEE (2003)

13. Dabbagh, M., Hamdaouri, B., Guizani, M., Rayes, A.: Software-defined networking security: pros and cons. IEEE Commun. Mag.—Commun. Stand. Suppl. (2015)
14. Min, M.M., Hla, K.H.S.: Security on software life cycle using intrusion detection system. In: 6th Asia-Pacific Symposium on Information and Telecommunication Technologies, 2005. APSITT 2005 Proceedings. IEEE (2005)
15. Vokorokos, L., Chovanec, I.M., Látka, I.O., Kleinová, I.A.: Security of distributed intrusion detection system based on multisensor fusion. In: 6th International Symposium on Applied Machine Intelligence and Informatics, 2008. SAMI 2008, pp. 19–24. IEEE (2008)
16. Salah, S., Abdulhak, S.A., Sug, H., Kang, D., Lee, H.: Performance analysis of intrusion detection systems for smartphone security enhancements. In: 2011 International Conference on Mobile IT Convergence (ICMIC), pp. 15–19. IEEE (2011)
17. Awduche, D.O.: MPLS and traffic engineering in IP Network. IEEE Commun. Mag. 37(12), 42–47 (1999)
18. Foroushani, V.A., Heywood, A.N.Z.: Deterministic and authenticated flow marking for IP traceback. In: IEEE 27th International Conference on Advanced Information Networking and Applications (2013)
19. Kwon, J., Seo, D., Kwon, M., Lee, M., Perrig, A., Kim, H.: An incrementally deployable anti-spoofing mechanism for software defined network. Korea Zurich 8092, Switzerland
20. Saurabh, S., Sairam, A.S.: Linear and remainder packet marking for fast IP traceback. In: Fourth International Conference on Communication Systems and Networks (COMSNETS). IEEE (2012)
21. Jin, G., Yang, J., Wei, W., Dong, Y.: Across-domain deterministic packet marking for IP traceback. In: Second International Conference on Communication and Networking in China, 2007, CHINACOM'07, pp. 382–386. IEEE (2007)
22. Stoica, I., Zhang, H.: Providing Guaranteed Services Without per Flow Management, vol. 29, no. 4. ACM (1999)
23. Savage, S., Wetherall, D., Karlin, A., Anderson, T.: Practical network support for IP traceback. ACM SIGCOMM Comput. Commun. Rev. 30(4) (2000)
24. Yaar, A., Perrig, A., Song, D.: Pi: A path identification mechanism to defend against DDoS attacks. In: 2003 Symposium on Security and Privacy, 2003. Proceedings. IEEE (2003)
25. Avallone, S., Guadagno, S., Emma, D., Pescap, A., Ventre, G.: D-ITG distributed internet traffic generator. In:. First International Conference on the Quantitative Evaluation of Systems, 2004 Proceedings. QEST 2004, pp. 316–317. IEEE (2004)

Cheating Immune Visual Cryptographic Scheme with Reduced Pixel Expansion

Kanakkath Praveen and M. Sethumadhavan

Abstract One of the drawbacks in visual cryptography is cheating attacks, where the malicious adversaries can cheat the honest participant by submitting fake shares during reconstruction phase. Cheating immune visual cryptographic schemes are used for mitigating cheating attacks in visual cryptography. There are two types of cheating immune schemes: One is share authentication-based schemes, and the other is blind authentication-based schemes. For the existing blind authentication-based schemes, the pixel expansion value will increase in the order of $O(n)$. In this paper, a blind authentication-based cheating immune visual cryptographic scheme is proposed by modifying the existing scheme based on uniform codes where the pixel expansion value will increase in the order of $O(\log n)$.

Keywords Secret sharing · Visual cryptography · Cheating prevention
Cheating immune · Blind authentication

1 Introduction

Visual cryptography is an unconditionally secure secret splitting technique used to generate n shares from a secret image (SI). During the distribution phase, these shares are given to each of the n participants, and the secret image will be visible only during the reconstruction phase when sufficient participants combine their shares. In visual cryptographic scheme (VCS) for reconstruction, the Boolean operators OR, AND, NOT, XOR are used instead of complicated computation as in conventional cryptography. The quality of a VCS is quantified using pixel expansion m and contrast $\alpha.m$. A pixel in SI is converted to m sub pixels in all

K. Praveen (✉) · M. Sethumadhavan
TIFAC-CORE in Cyber Security, Amrita School of Engineering Coimbatore,
Amrita Vishwa Vidyapeetham, Amrita University, Coimbatore, India
e-mail: k_praveen@cb.amrita.edu

M. Sethumadhavan
e-mail: m_sethu@cb.amrita.edu

© Springer Nature Singapore Pte Ltd. 2018   257
K. Saeed et al. (eds.), *Progress in Advanced Computing and Intelligent Engineering*,
Advances in Intelligent Systems and Computing 563,
https://doi.org/10.1007/978-981-10-6872-0_24

shares. In the reconstructed image, gray levels of black and white pixel differ by α. m. The participants in the qualified (resp. forbidden) set can (resp. cannot) reconstruct the secret image. VCS are of different types namely $(2, n)$, (k, n), and general access structure.VCS can also be classified into deterministic and probabilistic scheme depending upon the reconstruction of the secret. In 1994, Naor and Adi Shamir [1] developed deterministic VCS's, where OR (stacking) operation is used for reconstruction of secret image. The constructions of conventional VCS cannot resist against cheating attacks. So attacks (deterministic white to black (DWtB), deterministic black to white (DBtW) and region based) are possible against honest participants or victims by the malicious adversaries (collusive cheaters, malicious participant, and malicious outsider), by submitting fake shares during reconstruction process. There are two types of cheating immune visual cryptographic scheme (CIVCS): One is share authentication (SA)-based CIVCS, and the other is blind authentication (BA)-based CIVCS. In SA, apart from the shares of participants, extra information generated by the dealer or the participant is needed to verify cheating but in BA the shares are constructed in such a manner that the cheaters are not able to identify the structure of other participant's share. Two SA-based CIVCS are proposed by Yang and Laih [3] in 1999. The verification of shares is done with (resp. without) the help of an online trusted third party in the first (resp. second) CIVCS. The collusive cheating attack in (k, n)—VCS and its two mitigation techniques was developed by Horng et al. [4] in 2006. The first one is a SA technique, where each participant needed to carry extra verification transparencies, while the second technique is a $(2, n)$ scheme based on BA, where $(n + l)$ shares are generated in the sharing phase but only randomly selected n shares are used for distribution which makes cheaters hard to accomplish a successful attack. But, second technique protects only black pixels from cheating while white pixels are vulnerable to attack. Tsai et al. [5] in 2007 proposed a BA-based CIVCS using genetic algorithm by creating multiple homogeneous secret images. Here, the probability of successful cheating is highly decreased compared to [4] second scheme. Hu and Tzeg [6] in 2007 identified that cheating attack is possible by malicious participant or a malicious outsider if the VCS is not following perfect black criteria. The authors showed attacks on [3] first cheating method and on [4] scheme. The paper [6] also proposes SA-based CIVCS with reduced pixel expansion than [3] second method. De Prisco et al. [7] in 2010 proposed an (n, n) BA-based CIVCS and two $(2, n)$ BA-based CIVCS. The first $(2, n)$ scheme is a simple scheme but with a weakness that white pixels are not protected. The second scheme protects both white and black pixels by making use of larger m. Tsai et al. [8] in 2010 showed that the genetic algorithm-based CIVCS given in Tsai et al. [5] decodes the secret share incorrectly. Liu et al. [9] in 2011 proposed a SA-based CIVCS by disclosing t secret pixels to participants during the share distribution phase. The scheme is applicable to every VCS by verifying that the positions of randomly choosing t pixels in the secret are following the same color or not in the reconstructed image. Wang et al. [10] in 2011 proposed a SA-based tagged VCS for cheating prevention. Chen et al. [11] in 2011 showed a new variant of attack called Region cheating attack (RCA) which cheat human visual system (HVS), when for a region in the secret image if there is a white pixel surrounded by lot of black pixels or a black pixel surrounded by lot of white pixels. The paper also shows that DWtB attack and RCA are possible in the second construction of [7] scheme even though the pixel expansion is high. Chen et al. [12] in 2011 also proposed a BA-based (2, n) scheme which is immune to RCA attack but still vulnerable to DWtB attack. Both Chen et al. [12] and Liu et al. [9] showed that scheme proposed by Hu and Tzeg [6] is not a CIVCS. Chen et al. [13] in 2012 suggested an improvement to [6] CIVCS. Chen et al. in 2012 [13] (resp. 2013 [14]) proposed SA-based (2, n) CIVCS, with (resp. without) extra verification transparency for each participant.

Here, this paper proposes a novel (2, n) CIVCS. The preliminaries for VCS are given in Sect. 2. In Sect. 3, the background for BA-based CIVCS and collusive cheating attacks on VCS is discussed. In Sect. 4, the novel construction of (2, n) CIVCS based on OR operation is explained.

2 Preliminaries

Let $PA = \{pa_1, pa_2, pa_3, ..., pa_n\}$ be the participant set and 2^{PA} is the cardinality of power set of PA. Then, the share S_i, $1 \leq i \leq n$ of SI is distributed to each pa_i. Let us denote Γ_{Qual} as a collection of qualified sets and Γ_{Forb} as a collection of forbidden sets, where $\Gamma_{\text{Qual}} \cup \Gamma_{\text{Forb}} = 2^{PA}$ and $\Gamma_{\text{Qual}} \cap \Gamma_{\text{Forb}} = \varphi$, then $\Gamma = (PA, \Gamma_{\text{Qual}}, \Gamma_{\text{Forb}})$ is called the access structure of VCS. Any set $C \in \Gamma_{\text{Qual}}$ can recover SI whereas any set $C \in \Gamma_{\text{Forb}}$ is not able to recover SI. Let $\Gamma_0 = \{C \in \Gamma_{\text{Qual}}: C' \notin \Gamma_{\text{Qual}}$ for all $C' \subseteq C, C' \neq C\}$ be the collection of minimal qualified subset of PA. Let B_1 and B_0 are two collections which consist of $n \times m$ Boolean matrices used for constructing n shares of SI. The row of each matrix in both B_1 and B_0 corresponds to m sub pixels. For sharing a 1 (resp. 0) pixel in SI, randomly choose a matrix T from B_1 (resp. B_0) and assign row i of T to the corresponding positions of share S_i, $1 \leq i \leq n$. The matrices in the two collections B_1 (resp. B_0) consist of all column permutations of T_1 (resp. T_0). The vector obtained by bitwise OR operation to the rows of T corresponding to the elements in PA is represented as T_{PA}. Let $w(T_{PA})$ denotes the Hamming weight of the vector T_{PA}. The stacking corresponds to the bitwise operation between sub pixels in the shares (S_i).

Definition 1 [2]: Let $\Gamma = (PA, \Gamma_{\text{Qual}}, \Gamma_{\text{Forb}})$ be an access structure. Two collections B_1 and B_0 constitute a (Γ, m)—VCS if there exists a value $\alpha(m) > 0$ and a set $\{(PA, t_{PA})\}_{PA \in \Gamma_{Qual}}$ which satisfies the following conditions.

1. Any set $\{pai_1, pai_2, pai_3, ..., pai_q\} \in \Gamma_{Qual}$ can recover SI by stacking their shares. Formally, for any $T \in B_0$, $w(T_{PA}) \leq t_{PA} - \alpha.m$, whereas for any $T \in B_1$, $w(T_{PA}) \geq t_{PA}$.
2. Any set $\{pai_1, pai_2, pai_3, ..., pai_q\} \in \Gamma_{Forb}$ has no information on SI. Formally, the collections B_t, $t \in \{0, 1\}$ of $q \times m$ matrices obtained by restricting each

$n \times m$ matrix in $T \in B_t$ to rows $i_1, i_2, i_3, \ldots, i_q$ are indistinguishable in the sense that they contain the same matrices with the same frequencies.

The property 1 (resp. 2) ensures contrast (resp. security) of the scheme. The following is an example of (2, 3)—VCS with contrast $\alpha = \frac{1}{3}$ and $m = 3$, for $PA = \{pa_1, pa_2, pa_3\}$, where $\Gamma_{Qual} = \{\{pa_1, pa_2\}, \{pa_3, pa_2\}, \{pa_1, pa_3\}, \{pa_1, pa_2, pa_3\}\}$ and $\Gamma_{Forb} = \{\{pa_1\}, \{pa_2\}, \{pa_3\}\}$.

Example 1 Let the basis matrices be $T_0 = \begin{bmatrix} 1 & 0 & 0 \\ 1 & 0 & 0 \\ 1 & 0 & 0 \end{bmatrix}$ and $T_1 = \begin{bmatrix} 1 & 0 & 0 \\ 0 & 1 & 0 \\ 0 & 0 & 1 \end{bmatrix}$.

Then, when stacking the value of $w(T_{PA}) = 2$ is obtained for black pixel, and the value of $w(T_{PA}) = 1$ is obtained for white pixel.

3 A Review of Collusive Cheating Attacks and CIVCS's

3.1 DWtB and DBtW Attack—Horng et al. (2006) [4]

Let S_1, S_2, and S_3 are three distinct shares of SI, and they are distributed to each participant pa_1, pa_2, and pa_3, respectively. Let pa_1 and pa_2 are the adversaries, and pa_3 is the honest participant. During attack, pa_1 and pa_2 will create fake block F by predicting the share structure of pa_3. Let us explain a DWtB and DBtW attack using the basis matrices given in the example of Sect. 2. For 0 pixel, the block in the shares of pa_1, pa_2, and pa_3 is [1 0 0]. If the cheaters pa_1 and pa_2 collusively identify the structure of pa_3, they can create F using the blocks [0 1 0] and [0 0 1], respectively. The stacking of F and corresponding block in S_3 will result in a 1 pixel. This is DWtB attack. For the secret pixel 1, the blocks in the shares of pa_1, pa_2, and pa_3 are [1 0 0], [0 1 0], [0 0 1], respectively. If pa_1 and pa_2 collusively identify the structure of pa_3 to generate F as [0 0 1], the stacking of F with corresponding block in S_3 will results in a white pixel. This is DBtW attack.

3.2 CIVCS—Horng et al. (HCT) (2006) [4]

In this method instead of creating n shares and distribute it to each participants, $(n + l)$ shares are generated from SI, and randomly picked n shares are distributed to each participant. This scheme will protect pixel 1 but not pixel 0. The probability that adversaries can correctly guess the 1 pixel's in the honest participants share is $\frac{1}{l+1}$. This scheme can protect only DBtW attack, but not DWtB attack. Section 2.3 of paper [4] explains how complementary images can be used to make this scheme immune against DWtB attack.

3.3   Simple CIVCS—de Prisco et al. (DD1) (2010) [7]

Let T_1 (resp. T_0) be the basis matrices shown in example of Sect. 2, then the basis matrices for CIVCS are given by A_1 (resp. A_0) for 1 (resp. 0) pixel are

$$A_0 = \left[\begin{array}{c|c} \begin{array}{c} 0 \\ . \\ . \\ 0 \end{array} & T_0 \end{array} \right] \text{ and } A_1 = \left[\begin{array}{c|c} \begin{array}{c} 0 \\ . \\ . \\ 0 \end{array} & T_1 \end{array} \right]$$

respectively. This scheme also can protect only DBtW attack, but not DWtB attack.

3.4   Better CIVCS—de Prisco et al. (DD2), 2010

Let A_1 (resp. A_0) be the basis matrices of $DD1$ scheme given above. Let D be a matrix which contain all possible 2^n column vectors, then the basis matrices B_1 (resp. B_0) for 1 (resp. 0) pixel for better scheme are $[D \quad A_0]$ (resp. $[D \quad A_1]$). But it is shown in Sect. 4 of paper [11] that, this scheme is also vulnerable to collusive cheating.

4   Proposed (2, N)—CIVCS's

4.1   MS CIVCS—Modified (Sreekumar and Babusundar 2008) [15]

A uniform code of length t consists of precisely $\lceil \frac{t}{2} \rceil$ 1's and $\lfloor \frac{t}{2} \rfloor$ 0's. Let N_t denote the number of uniform codes of length t, where $N_t = \binom{t}{\lfloor \frac{t}{2} \rfloor}$. In the construction of (2, n)—VCS based on OR operation by Sreekumar and Babusundar [15], the $n \leq N_t$ shares are generated using uniform codes. The scheme [15] is vulnerable to DWtB in the case of more than two collusive adversaries for any value of n and vulnerable to DBtW attack in the case of $n - 1$ collusive adversaries only when $n == N_t$. But, in the case of 2 to $n - 2$ collusive adversaries, the scheme [15] is immune to DBtW attack when $n == N_t$. The following shows an example of vulnerability of the scheme [15].

Example 2 For constructing a (2, 6)-VCS using uniform codes [15], let us select the value of $t = 4$, which satisfies the condition $n \leq N_t$. The ($N_t = 6$) distinct uniform codes for the value of $t = 4$ are given in $UC = \{1100, 0110, 0011, 1001, 1010,$

0101}. Let $M = \begin{bmatrix} 1 & 1 & 0 & 0 \\ 0 & 1 & 1 & 0 \\ 0 & 0 & 1 & 1 \\ 1 & 0 & 0 & 1 \\ 1 & 0 & 1 & 0 \\ 0 & 1 & 0 & 1 \end{bmatrix}$ be a matrix constructed using UC. For sharing a

pixel 1, a randomly selected row from M is distributed to all the six participants. For sharing a pixel 0, each row of M is distributed to six corresponding participants. Here, DWtB attack is possible because when any $n-1$ cheaters (here 5) collude, they can predict the block of the victim as [1 0 0 1], if the matrix used for sharing a pixel 0 is

$\begin{bmatrix} 1 & 0 & 0 & 1 \\ 1 & 0 & 0 & 1 \\ 1 & 0 & 0 & 1 \\ 1 & 0 & 0 & 1 \\ 1 & 0 & 0 & 1 \end{bmatrix}$ .The five cheaters can conduct a DWtB attack by generating the

fake matrix $\begin{bmatrix} 1 & 1 & 0 & 0 \\ 0 & 1 & 1 & 0 \\ 0 & 0 & 1 & 1 \\ 1 & 0 & 1 & 0 \\ 0 & 1 & 0 & 1 \end{bmatrix}$. Here, DBtW attack is possible because when any

$n-1$ cheaters (here 5) collude, they can predict the block of the victim as [1 0 0 1],

if the matrix used for sharing a pixel 1 is $\begin{bmatrix} 1 & 1 & 0 & 0 \\ 0 & 1 & 1 & 0 \\ 0 & 0 & 1 & 1 \\ 1 & 0 & 1 & 0 \\ 0 & 1 & 0 & 1 \end{bmatrix}$ .The five cheaters can

conduct a DBtW attack by generating the fake matrix $\begin{bmatrix} 1 & 0 & 0 & 1 \\ 1 & 0 & 0 & 1 \\ 1 & 0 & 0 & 1 \\ 1 & 0 & 0 & 1 \\ 1 & 0 & 0 & 1 \end{bmatrix}$. Here, in

this example, the value of $n = 6$ and the value of $N_t = 6$. So when $n == N_t$, DBtW attack is possible by $n-1$ (here 5) cheaters.

Proposed construction: The following are the steps done by the dealer.

Step1: **If** the value of $n == N_t$, then M be the basis matrix of order $N_{t+1} \times (t+1)$ and each element in the matrices is different vectors of length $(t+1)$.
Else M be the basis matrix of order $N_t \times t$, and each row in the matrices is different vectors of length t.

Step2: Construct a set G which is a collection of different row permuted matrix of M.

Step3: This step is applicable to all pixels in SI. The dealer randomly selects a matrix from G and constructs another matrix K of order $n \times (t+1)$ or $n \times t$ based on the value of n. For sharing 1, the dealer uses any row permuted matrix K, and for sharing 0 the dealer selects any one row of matrix K and distribute to all n participants.

Theorem 1: The MS—(2, n) CIVCS is vulnerable to DWtB attack but immune to DBtW attack.

Proof Let TS be the share of honest participant, and let B be the block corresponding to 1 of SI, and W is a block corresponding to 0 of SI. The dealer selects a matrix M of order $N_m \times m$, where $m = \begin{cases} t+1 & if \quad n == N_t \\ t & Otherwise \end{cases}$, $n \geq 2$ and $t \geq 2$.

Here, when any d collusive participants (cheaters) combine, the probability for correctly guessing B in TS is $\frac{1}{N_m - d}$, and the probability for correctly guessing W in TS is 1.

This scheme is used as a BA-based CIVCS which can resist both DWtB and DBtW attack, if the shares of complementary secret image are also distributed to participants analogous to (2, $n + l$)—CIVCS constructed by Horng et al. [4]. Assume that SI = [1 0] be the secret image to be shared using (2, 3)—CIVCS. For constructing a CIVCS, we need to generate a complementary matrix of SI, say $SI' = [0\ 1]$. Both SI and SI' are shared among the three participants.

Example 3 For constructing a (2, 6)—CIVCS, let us select a matrix

$$K = \begin{bmatrix} 1 & 1 & 1 & 0 & 0 \\ 1 & 1 & 0 & 1 & 0 \\ 1 & 0 & 1 & 1 & 0 \\ 0 & 0 & 1 & 1 & 1 \\ 1 & 0 & 0 & 1 & 1 \\ 0 & 1 & 1 & 1 & 0 \end{bmatrix}$$ which satisfies the following criteria $n \leq N_t$. Here, if any

five collusive adversaries combine, they cannot predict the block corresponding to pixel 1. So the scheme is immune to DBtW attack. If complementary secret is also shared with the original secret, the scheme resists DWtB attack also.

4.2 Comparison of CIVCS's

Below Tables 1 and 2 show the comparison of proposed MS CIVCS with related works which are reported as secure in the paper [14] (CTH).

Table 1 Comparison of secure $(2, n)$—CIVCS's

Scheme	HCT	DD1	CTH	MS
Reconstruction operation	OR	OR	OR	OR
Type of CIVCS	BA	BA	SA	BA
Complimentary secret image	YES	YES	NO	YES
Complexity	$O(n)$	$O(n)$	$O(n)$	$O(\log n)$

Table 2 Pixel expansion of secure $(2, n)$—CIVCS's

N	HCT($l = 1$) $2 \times (n+l)$	DD1 $2 \times (n+1)$	CTH $(2 \times n) + 1$	MS $2 \times m$
2	6	6	5	6 ($t = 3$)
4	10	10	9	8 ($t = 4$)
6	14	14	13	10 ($t = 5$)
8	18	18	17	10 ($t = 5$)
10	22	22	21	12 ($t = 6$)
15	32	32	31	12 ($t = 6$)
18	38	38	37	12 ($t = 6$)

5   Conclusion

The order of pixel expansion for the existing secure blind authentication based $(2, n)$—CIVCS's is $O(n)$. Here, we proposed a $(2, n)$—CIVCS which can prevent $n - 1$ cheaters by modifying Sreekumar and Babusundar [15] uniform code scheme in which the order of pixel expansion is $O(\log n)$.

References

1. Naor, M., Shamir, A.: Visual cryptography. In: EUROCRYPT. Lecture Notes in Computer Science, vol. 950, pp. 1–12. Springer, Berlin Heidelberg (1994)
2. Ateniese, G., Blundo, C., DeSantis, A., Stinson, D.R.: Visual cryptography for general access structures. Inf. Comput. **129**(2), 86–106 (1996)
3. Yang, C.N., Laih, C.S.: Some new type of visual secret sharing schemes. In: National computer symposium, vol. 3, pp. 260–268 (1999)
4. Horng, G., Chen, T.H., Tsai, D.S.: Cheating in visual cryptography. Des. Codes Crypt. **38**(2), 219–236 (2006)
5. Tsai, D.S., Chen, T.H., Horng, G.: A cheating prevention scheme for binary visual cryptography with homogeneous secret images. Pattern Recogn **40**(8), 2356–2366 (2007)
6. Hu, C.M., Tzeng, W.G.: Cheating prevention in visual cryptography. IEEE Trans. Image Process. **16**(1), 36–45 (2007)
7. DePrisco, R., DeSantis, A.: Cheating immune threshold visual secret sharing. Comput J **53**, 1485–1496 (2010)

8. Tsai, D.S., Huang, C.C.: A new deterministic algorithm based cheating prevention scheme for visual cryptography. Hsiuping J. **20** (2010)
9. Liu, F., Wu, C., Lin, X.: Cheating immune visual cryptographic scheme. IET Inf. Secur. **5**(1), 51–59 (2011)
10. Wang, R.Z., Hsu, S.F.: Tagged visual cryptography. IEEE Sig. Process. Lett. **18**(11), 627–630 (2011)
11. Chen, Y.C., Horng, G., Tsai, D.S.: Cheating human vision in visual secret sharing. https://eprint.iacr.org/2011/631.pdf (2011)
12. Chen, Y.C., Horng, G., Tsai, D.S.: Comment on cheating prevention in visual cryptography. IEEE Trans. Image Process. **21**(7), 3319–3323 (2012)
13. Chen, Y.C., Tsai, D.S., Horng, G.: A new authentication based cheating prevention scheme in Naor-Shamir's visual cryptography. J. Vis. Commun. Image Representation **23**(8), 1225–1233 (2012)
14. Chen, Y.C., Tsai, D.S., Horng, G.: Visual secret sharing with cheating prevention revisited. Digit. Sig. Process. **23**(5), 1496–1504 (2013)
15. Sreekumar, A., Babusundar, S.: Uniform secret sharing scheme for (2, n) threshold using visual cryptography. Int. J. Inf. Process. **2**(4) (2008)

Securing Sensitive Data Information Through Multiplicative Perturbation Approach Based on UML Modeling

Anurag, Deepak Arora and Upendra Kumar

Abstract Mining mainly aims toward extracting hidden useful patterns from the large amount of data which is exploited by various organizations for their business planning. The organizations generally used to store sensitive data, during their mining processes, which they do not want to disclose either due to legal constraints or competition among themselves. Privacy preserving data mining is focused on hiding this sensitive information while still getting accurate mining results. This paper deals with UML modeling of multiplicative perturbation-based approach applied in privacy preserving data mining system. Multiplicative perturbation is the most popular technique nowadays, which mainly aimed to preserve multidimensional information while performing different mining operations. Authors have modeled the entire approach by using IBM Rational Software Architect tool. This standard model will facilitate researchers and organizations to design more specific and aligned business solutions in order to secure sensitive data by proposed UML models based on multiplicative perturbation approaches.

Keywords Multiplicative perturbation · Privacy preserving data mining UML modeling · Object-oriented modeling · Distribution reconstruction algorithm

Anurag (✉)
Amity Institute of Information Technology, Amity University, Lucknow Campus,
Lucknow, Uttar Pradesh, India
e-mail: anurag.smit@gmail.com

D. Arora
Department of Computer Science & Engineering, Amity University, Lucknow Campus,
Lucknow, Uttar Pradesh, India
e-mail: deepakarorainbox@gmail.com

U. Kumar
Department of Computer Science & Engineering, Birla Institute of Technology, Patna
Campus, Patna, India
e-mail: upendrakumarphdp@gmail.com

© Springer Nature Singapore Pte Ltd. 2018
K. Saeed et al. (eds.), *Progress in Advanced Computing and Intelligent Engineering*,
Advances in Intelligent Systems and Computing 563,
https://doi.org/10.1007/978-981-10-6872-0_25

1 Introduction

Model is a representation of something in a certain medium either in the same or different medium. A model depicts important aspects of the thing needed during modeling from certain point of view and simplifies or omits the rest. Many creative fields, Engineering, Architecture etc., use models. The software system is modeled in a modeling language such as UML. The models include semantics and notation that can take various forms and include both picture and texts [1]. Object-oriented technology is being used for modeling the software system. It organizes the software components into collections of discrete objects, which incorporates both data and possible actions over it [2]. It has gained popularity in recent years as it closely represents the problem domain which makes its program easier to create, design, and understand it. The objects in the system are less immune to change with the changing requirements and thus incorporate changes more easily. Inheritance and close association with objects in design facilitates the newer applications to reuse existing module and hence reduces development costs and cycle time [3]. UML is a standard graphical language for modeling the object-oriented software. It creates the visual model of the entire system [4]. As the organizations nowadays concern with protection of personal sensitive data from disclosure and still wish to obtain accurate results while performing mining operations, privacy preserving data mining is an evolving research topic. Compared to the other multidimensional approach, multiplicative perturbation mainly preserves the inner products or distance between data dimensions and thus most suitable for classification and clustering models. Hence, classification and clustering models show the similar accuracy than the original data models [5]. Since this approach preserves multidimensional information, so data mining algorithm could be directly applied on it without modifying any existing algorithm [6]. In this research work, authors have designed generalized UML model of multiplicative perturbation technique for protection of sensitive data so that the overall functionality of the entire system could be well analyzed and any error could be priorly detected and corrected before its actual implementation so that successful operations of the project could be achieved. The modeling allows the entire work to be organized into collection of independent discrete objects so that each of its functionality could be well analyzed and easily managed.

2 UML Modeling of Multiplicative Perturbation System

Multiplicative data perturbation approach involves different stakeholders. Use case diagram has been drawn to emphasize all the important scenario and cases of the system used by these actors. Authors have identified four major stakeholders in the system—Data Provider, Data Warehouse Server, Data Mining Server, and Decision Makers. Figure 1 shows the interaction of different stakeholders with the system for performing different use case operations. Data Provider performs the tasks of input data to be perturbed by the system. Perturbed data use case also includes specific multiplicative perturbation technique chosen by Data Provider for perturbing the raw data. The data is then transferred to DataWarehouseServer. DataWarehouseServer manager has the job of managing warehouse data, such as cleaning and integrating data from multiple data providers, converting data into the form suitable for mining, processing the query received from DataMiningServer, converting the data into aggregate from, and then sending it to Data Mining Server. Data Mining Server performs the tasks of sending the query to Data Warehouse Server, receiving the aggregated perturbed data from Data Mining Server, applying distribution reconstruction algorithm on the multiplicative perturbed data, comparing the multiplicative perturbed data with original data distribution, and finally performing mining operations. Data mining results than evaluated by the Decision Makers and then infers useful knowledge on the basis of mined result.

Figure 2 represents the generalized class diagram of the Privacy Preserving System. Authors have identified different classes and relationship, each having different roles and responsibilities, and all these classes collectively function together for achieving multiplicative perturbation operations. Class *DataProvider* saves its data in the *DataDepositorandPerturbizer* class. Different perturbation technique could be selected by the *DataDepositorandPerturbizer* class one at a time

Fig. 1 Use case diagram of multiplicative perturbation system

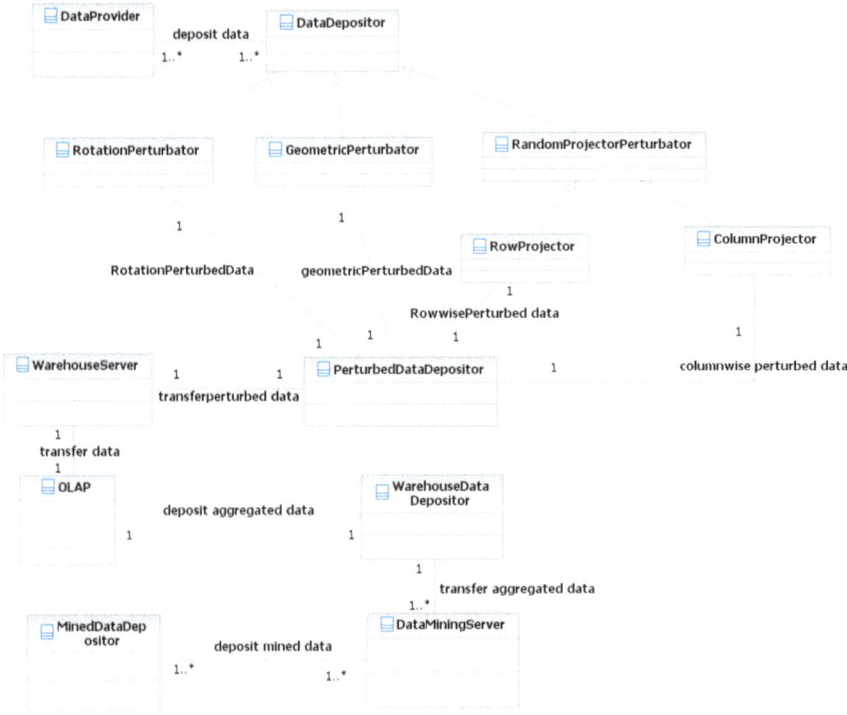

Fig. 2 Class diagram of multiplicative perturbation system

depending on types of data, data dimension corelation needed to be preserve during mining. Class *RandomRotationPerturbator* has the responsibility of applying Random Rotation Perturbation. Class *GeometricDataPerturbator* geometrically perturbs data. Class *RandomProjectorPerturbator* has the task of applying suitable Random Projection Perturbation on the raw data. Class *RowProjector* performs the job of row-wise projection of raw data. Class *ColumnProjector* has the tasks of column-wise projection of raw data. The perturbed data of various *DataProvider* is stored in *PerturbedDataDepositor* class. It then transfers to class *WarehouseServer*. Class *WarehouseServer* converts the data into the form suitable for mining. The data then transferred to the class *OLAP* for converting it into aggregate form. The aggregated data is then deposited in the *WarehouseDataDepositor* class. Class *DataMiningServer* sends the queries to the *WarehouseServer*. Class *WarehouseServer* fetches the appropriate data from *WarehouseDataDepositor* class and sends it to the *DataMiningServer*. *DataMiningServer* applies distribution recon-struction algorithm, compares it with original data distribution for checking its accuracy, and then applies data mining operations on the perturbed data. Table 1 illustrates the functionality of each of these classes.

Table 1 Attributes and functions of each class

	Class name	Attributes	Methods
1.	DataProvider	Name:string, Source Id:int	Set_connection(), putdata(),depositRawDataX()
2.	DataDepositorandPerturbizer	data: void()	getdata(),applySuitableApproach(),transferControl()
3.	RotationPerturbator	data: void()	generate_RandomOrthogonalMatrixR_{m*m}(), multiplyRX(), putdata(), getdata()
4.	RandomProjector	data: void()	applySuitableProjectionApproach()
5.	RowProjector	data: void()	check_k,generate_ProjectionMatrixP_{k*m},computeU, Getdata(),Putdata()
6.	ColumnProjector	data: void()	check_k,generate_ProjectionMatrixP_{n*k}ComputeV, Getdata(),Putdata()
7.	GeometricPerturbator	data: void()	generate_RandomOrthogonalMatrixR_{m*m}(), generate_translationMatrixT(), generate_NoisydataD, compute_sumR + T+D()
8.	DataWarehouseServer	data: void()	datacleaning(),dataintegration(),physicalSchematicDesign(), receivequery().fetchAggregateData(),sendAggregateData()
9.	WarehouseDataDepositor	data: void()	getdata(),putdata()
10.	OLAP	data: void()	getdata(),performDataAggregation(),putdata()
11.	DataMiningServer	data: void()	getdata(),sendquery(), receive_aggregated_data(), applyDistribution_Reconstruction_Algorithm(),comparewith_originol_distribution, apply_DataMiningAlgorithm()
12.	MinedDataDepositor	data: void()	Getdata()

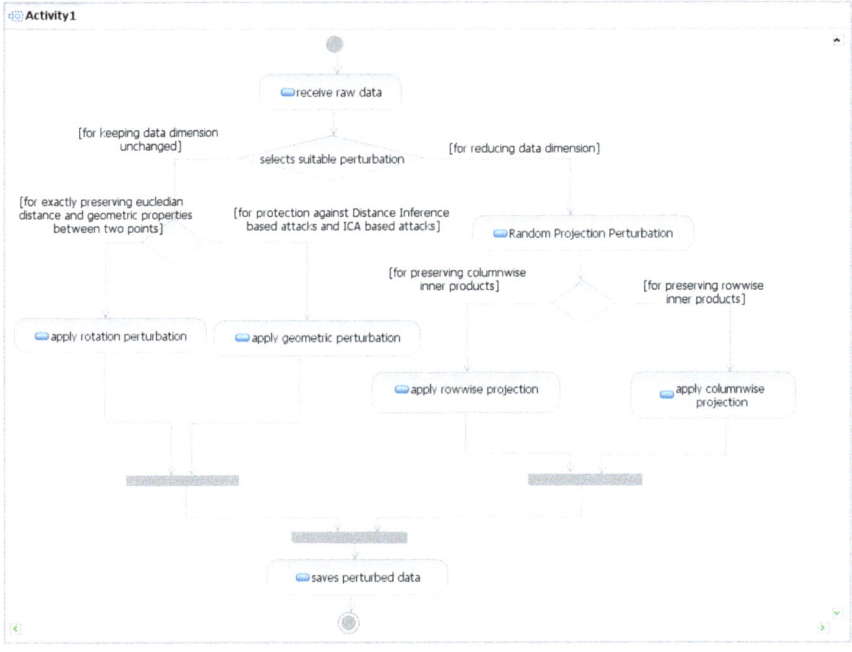

Fig. 3 Activity diagram for selection of suitable perturbation approach

Figure 3 illustrates the activity diagram for selection of suitable multiplicative perturbation approach. The data provided by the Data Provider is being checked for applying suitable multiplicative perturbation technique. Rotation Perturbation and Geometric Perturbation is suitable for situation in which user wishes to keep data dimension unchanged while preserving statistical properties [6]. Random Projection Perturbation is suitable for situation in which user wishes to keep both dimension and exact value of data elements confidential and still preserving statistical properties [7]. Random Rotation Perturbation technique is suitable for preserving the geometric properties of datasets such as Euclidian distance, inner product, and angles between the two vectors [7, 8]. Thus, most data mining models including classifier and clustering techniques on the transformed data get the similar accuracy than original datasets [8]. Random Rotation deals with exact distance preservation, and Geometric Perturbation deals with approximate distance preservation. Geometric Perturbation preserves important geometric properties of multidimensional data. It has been introduced to overcome the weakness of Rotation Perturbation and Projection Perturbation while preserving data quality. Random translation matrix has been added to address the attack on rotation center and hence add additional difficulty to ICA-based attack. Noise additive components address the weakness of distance inference attack [9]. It provides better preserve privacy in terms of data classifications [10]. It approximately preserves distance as distance between the pairs of points is disturbed slightly [11]. Projection Perturbation mainly deals with

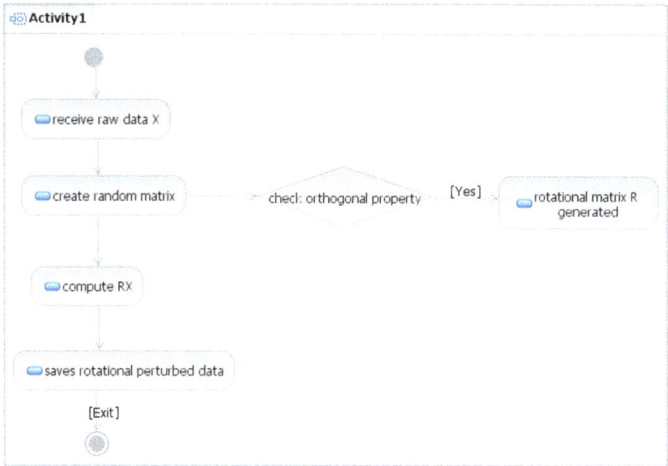

Fig. 4 Activity diagram for Rotation Perturbation approach

approximate distance preservation while reducing the data dimension, which is suitable for mining models. It mainly deals with preserving correlation coefficient, inner products, and Euclidian distance in the new data. Thus, by projecting it in the random space, its original form could be changed while preserving much of its underlying characteristics. Pair-wise distances between the two points could be maintained with arbitrarily small factor. Row-wise Projection Perturbation is suitable for preserving column-wise inner product properties, and Column-wise Projection Perturbation is suitable for preserving row-wise inner product properties [7].

Figure 4 shows the various activities performed during Random Rotation Perturbation. Rotation matrix generated in such a way that it should satisfy orthonormal (or orthogonal) properties. The resulting matrix is multiplied with the raw data [9], and the result is saved in *PerturbedDataDepositor* class for further mining operations.

Figure 5 shows the various activities performed during Geometric Perturbation. Orthonormal matrix R_{m*m} is generated and multiplied to the raw data X_{m*n} to generate perturbed data RX. Then, it creates the translation matrix T_{m*n}, adds it with RX, and saves it in perturbed data $G(X) = RX + T$. The resulted perturbed data is stored in the *PerturbedDataDepositor* class. Suitable noise could be added, if Data Provider knows that attacker has information about his original data. The noise to be added must have independent and identically distributed variables with mean 0 and small variance σ. The noise D_{m*n} is added with the perturbed data G (X), and the resultant data $G(X) = RX + T + D_{m*n}$ is saved in the *PerturbedDataDepositor* class.

Figure 6 illustrates the series of activities take place during Row-wise Random Projection Perturbation approach. It checks for the input value k, and if k < m, then further perturbation takes place. Data Provider generates the projection matrix P_{k*m}

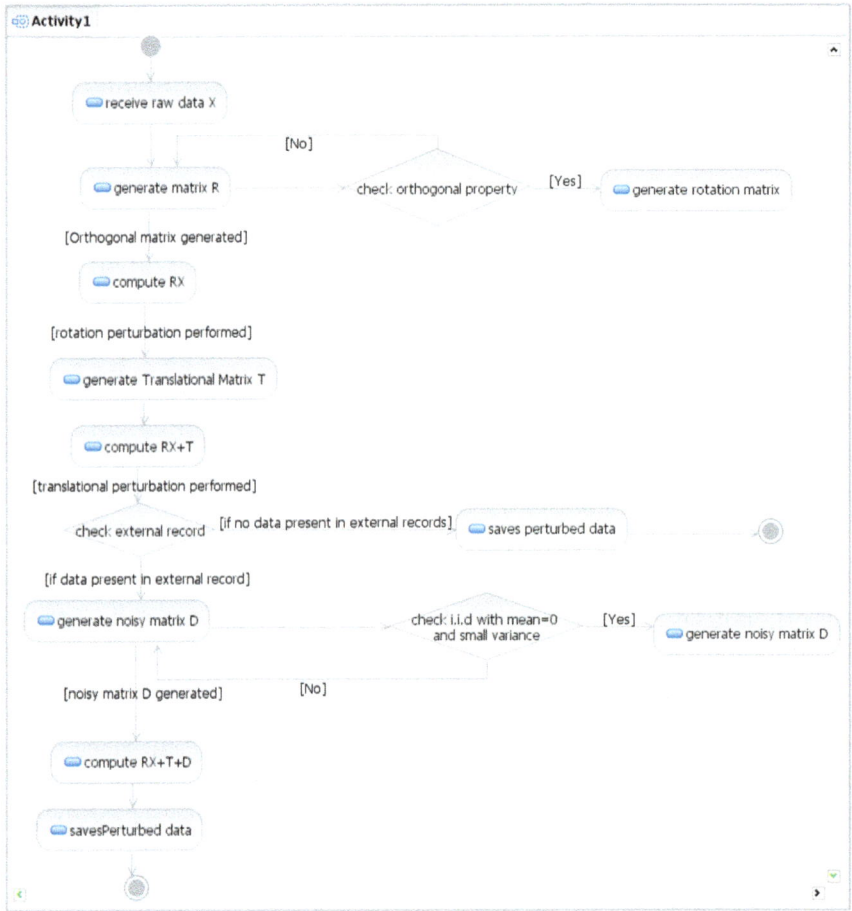

Fig. 5 Activity diagram for Geometric Perturbation approach

such that each entry p_{ij} of P must be independent and identically distribution chosen from a random unknown distribution with mean 0 and some variance σ_p^2. It then computes $U = \frac{1}{\sqrt{k\sigma}}PX$, [8] where X_{m*n} is the raw data. The resulting data U_{k*m} (where k < m) is stored in the *PerturbedDataDepositor* class.

Figure 7 demonstrates the column-wise perturbation takes place during Projection Perturbation. It generates the projection matrix P_{n*k}, where k < n. It computes $V = \frac{1}{\sqrt{k\sigma}}XP$, [8] where X is multidimensional raw data X_{m*n}. The resultant perturbed data V_{m*k} (where k < n) is stored in *PerturbedDataDepositor* class.

Figure 8 shows the sequence diagram of Random Rotation Perturbation. *DataProvider* saves the data in the *DataDepositorandPerturbizer* class. *DataDepositorandPerturbizer* class applies the Random Rotation technique for the raw data perturbation. Class *RandomRotatorPerturbator* generates orthonormal

Fig. 6 Activity diagram for Row-wise Projection Perturbation

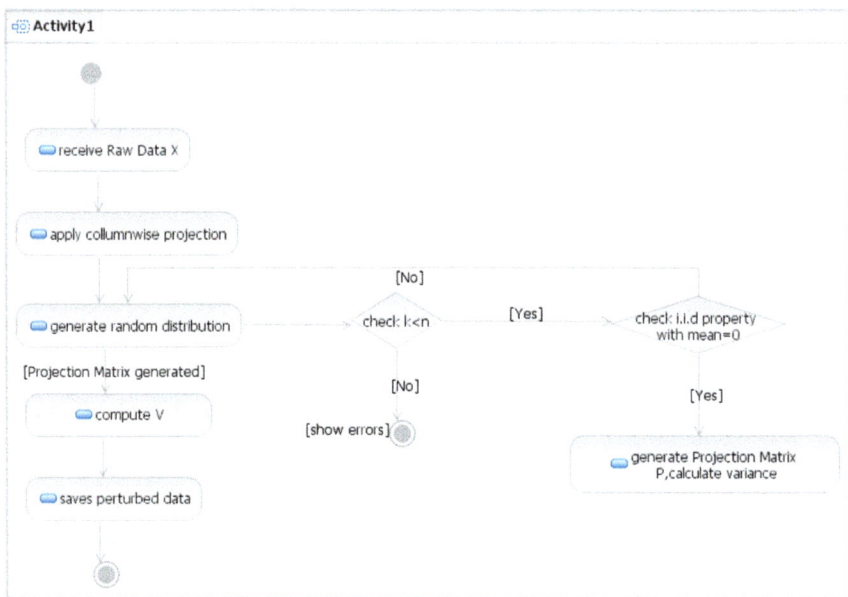

Fig. 7 Activity diagram of Column-wise Projection Perturbation

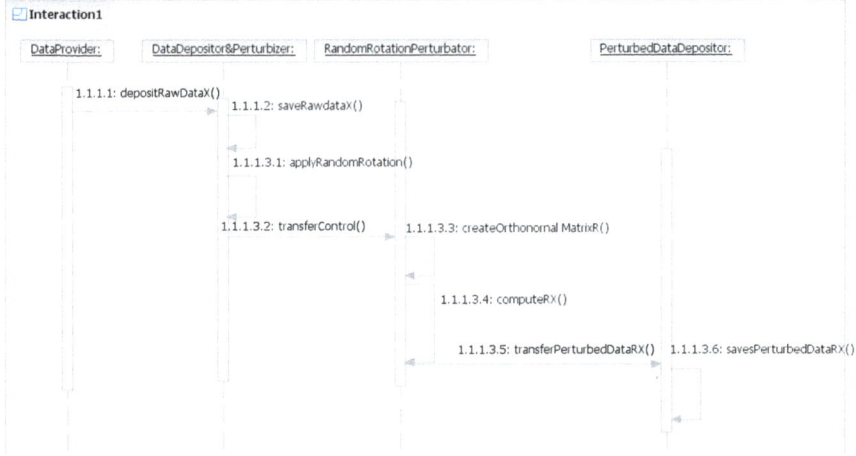

Fig. 8 Sequence diagram of Random Rotation Perturbation

matrix R. It then multiplies the matrix R with the raw data X. The resultant per-turbed matrix RX is saved in *PerturbedDataDepositor* class.

Figure 9 illustrates the sequence diagram of Geometric Perturbation approach. Data Provider saves its raw data in the *DataDepositorandPerturbizer* class. This class applies the Geometric Perturbation approach for the raw data Perturbation. Class *GeometricDataPerturber* generates the orthogonal matrix R and multiplies it with raw data X to generate the perturbed data RX. It then generates the translation matrix T_{m*n} and adds it to the above perturbed data. The resultant perturbed data G

Fig. 9 Sequence diagram of Geometric Perturbation approach

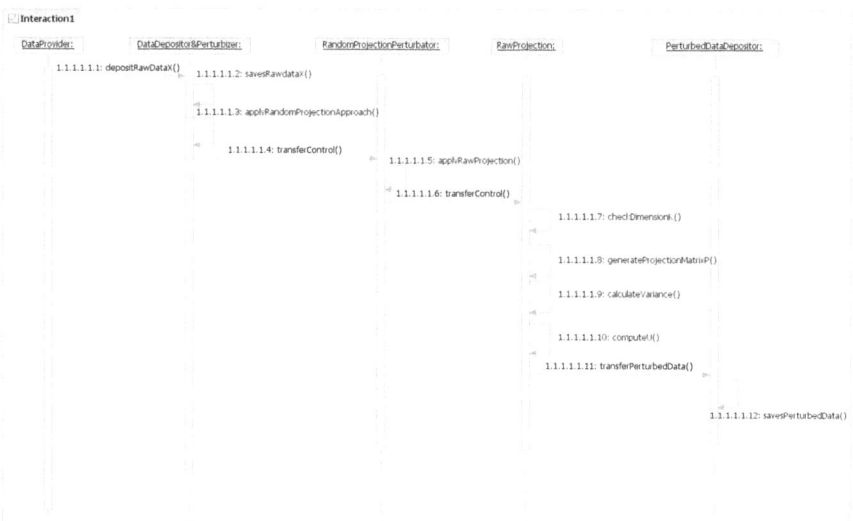

Fig. 10 Sequence diagram of Row-wise Projection Perturbation

(X) = RX + T is saved in the *PerturbedDataDepositor* class. Suitable noisy data D_{m*n} may be generated optionally and adds it to the above perturbed data G (X) = RX + T + D_{m*n}. The resulting geometrically perturbed data G(X) is saved in the *PerturbedDataDepositor* class.

Figure 10 depicts the sequence diagram of Row-wise Projection Perturbation. *DataDepositorandPerturbizer* class applies Row-wise Projection Perturbation. Class *RowwiseProjector* generates the suitable projection matrix R_{k*m}. It then computes σ_R and U, and the resulting data is saved in the *PerturbedDataDepositor* class.

Figure 11 represents the sequence diagram of Column-wise Projection Perturbation. *DataDepositorandPerturbizer* class applies Column-wise Projection Perturbation. Class *ColumnWiseProjector* class generates suitable projection matrix R_{n*k}. It computes σ_R and V. The resultant perturbed data is saved *PerturbedDataDepositor* class.

Figure 12 represents the sequence of operations between *DataWarehouseServer*, *DataMiningServer,* and *WarehouseDataDepositor*. Class *DataMiningServer* sends queries to class *DataWarehouseServer*. Class *DataWarehouseServer* accesses aggregated data from *WarehouseDataDepositor* class and transfers it to class *DataMiningServer*. Class *DataMiningServer* applies distribution reconstruction algorithm and data mining algorithm on the multiplicative perturbed data.

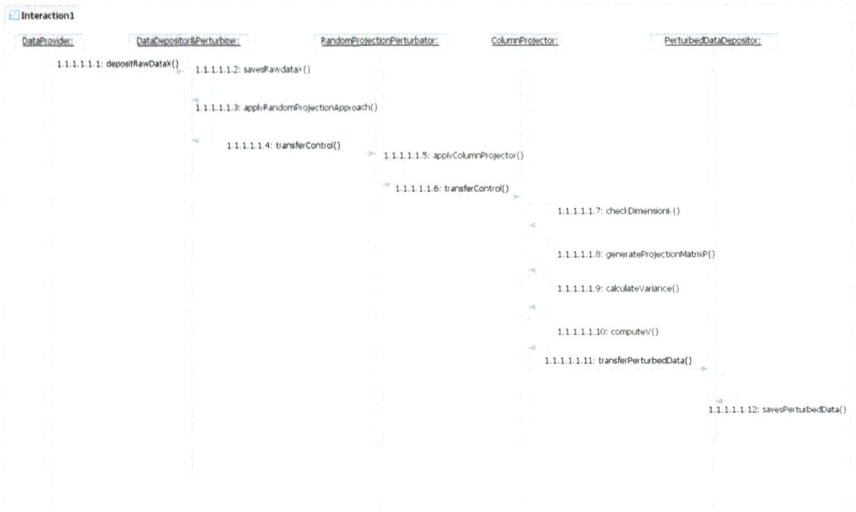

Fig. 11 Sequence diagram of Column-wise Projection Perturbation

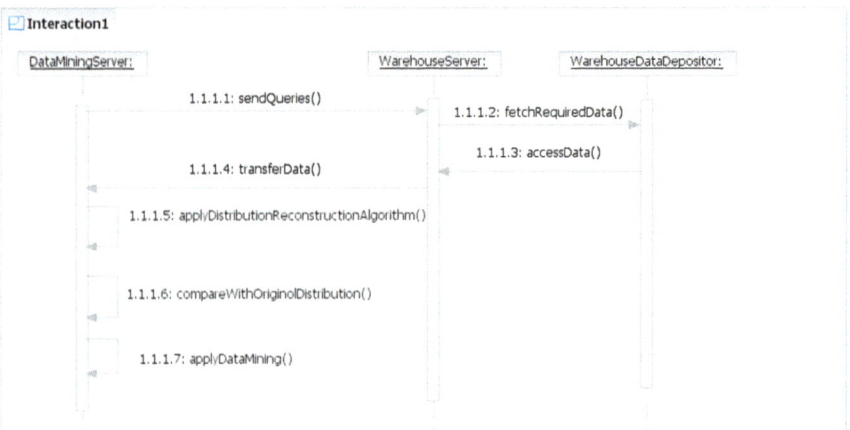

Fig. 12 Seqence diagram between classes *DataMiningServer, WarehouseServer,* and *WarehouseDataDepositor*

3   Concluding Remarks

In this paper, the author here has proposed the generalized model of multiplicative perturbation approach applied in Privacy Preserving System. The author here has analyzed both static and dynamic aspects of the entire system. Its graphical model will help the designer and developers to have deep insight into the whole system and discover and rectify the flaws, if any presents in the system prior to implementation or earlier phases of its development. This increases the probability of success of the entire project and reduces its development effort in terms of cost and overall development time.

Acknowledgements The authors are very thankful to respected Maj. Gen. K. K. Ohri AVSM (Retd.), Pro-Vice Chancellor, Amity University, Lucknow Campus, for providing excellent computation facilities in the university campus. Authors are also grateful to Brig. U.K. Chopra, Director, Amity Institute of Information Technology, Amity University, Lucknow Campus, for giving all best possible support to carry out the research work.

References

1. Rumbaugh, J., Jacobson, I., Booch, G.: The Unified Modeling language Referential Manual, 2nd edn. Addison Wesley, Pearson Education (2005)
2. Yadav, H.N.: Object Oriented C++ Programming. University Science Press, Laxmi Publications Pvt. Ltd. (2008)
3. Jalote, P.: An Integrated Approach to Software Engineering, 3rd edn. Narosa Publications House (2005)
4. Lathbridge, T.C., Laganiere, R.: Object Oriented Software Engineering, Tata Mc-Graw Hill Edition (2009)
5. Chen, K., Liu, L.: A Survey of Multiplicative Perturbation for Privacy Preserving Data Mining (2006)
6. Keyvanpour, M.R., Moradi, S.S.: Classification and evaluation the privacy preserving data mining techniques by using a data modification–based framework. Int. J. Comput. Sci. Eng. (IJCSE) **3**(2), 862–870, Feb 2011
7. Liu, K., Kargupta, H., Ryan, J.: Random projection-based multiplicative data perturbation for privacy preserving distributed data mining. IEEE Trans. Knowl. Data Eng. **18**(1), 92–106 (2006)
8. Lin, Z., Wang, J., Liu, L., Zhang, J.: Generalized Random Rotation Perturbation for Vertically Partitioned Data Sets, pp. 159–162. IEEE (2009)
9. Patel, D.J., Patel, S.: A survey on data perturbation techniques for privacy preserving in data mining. Int. J. Sci. Res. Dev **3**(01), 52–54 (2015)
10. Kavitha, S., Raja Vadhana, P.: Data privacy preservation using various perturbation techniques. Int. J. Innov. Res. Comput. Commun. Eng. **3**(2), 1039–1042, Feb 2015
11. Chen, K., Liu, L.: Geometric Data Perturbation for Privacy Preserving Outsourced Data Mining: Knowledge and Information Systems (2010)

Secret Communication Combining Cryptography and Steganography

Ayan Chatterjee and Asit Kumar Das

Abstract Recently, many applications are Internet-based where confidentiality is maintained by secret communication. Cryptography and steganography are two popular methods used for the purpose. The proposed method develops a new data hiding scheme combining the concepts of cryptography and steganography to enhance the security of communication. In cryptography, basic concept of differential and integral calculus is used during encryption and decryption, respectively, and in steganography, data is embedded in an image file and extracting the original message using discrete cosine transformation. This hybrid technique is sound as the number of secret keys used for encryption varies with the size of message, and at the same time generated keys are independent of each other which prevent hacking of all keys together. As a result, more security is enforced in the communication channel. The method is evaluated by measuring the distortion of originality of image file computing peak signal-to-noise ratio.

Keywords Cryptography · Steganography · Network security
Discrete cosine transform · Signal-to-noise ratio · Mean square error

1 Introduction

The word *security* signifies the quality of protection, i.e., to be free from any danger. Network security is protection of secret data among sender and intended receivers [1, 2] during wireless communication to maintain confidentiality, integrity, and availability of the data. Therefore, communication technique should be developed in such a way that any unwanted person cannot hack the secret data

A. Chatterjee (✉)
D.El.ED Section, Sarboday Public Academy, East Midnapore, India
e-mail: ayanchatterje2012@gmail.com

A. K. Das
Department of Computer Science & Technology, IIEST, Shibpur, India
e-mail: akdas@cs.iiests.ac.in

© Springer Nature Singapore Pte Ltd. 2018   281
K. Saeed et al. (eds.), *Progress in Advanced Computing and Intelligent Engineering*,
Advances in Intelligent Systems and Computing 563,
https://doi.org/10.1007/978-981-10-6872-0_26

[3–5]. Cryptography is the art of secret writing [1], and steganography is the art of hidden communication. In other words, cryptography scrambles a message so that it cannot be understood and steganography hides the message so that it cannot be visualized [2]. The word steganography is obtained from Greek dictionary, in which *stego* means 'keep secret' and *Graphy* means 'making words or writing' [3]. The general architecture of encryption and decryption of cryptography is shown in Fig. 1.

From Fig. 1, it is realized that the original message or plaintext which is required to be made secure will be encrypted through some encryption algorithm and generated secret key(s). Encrypted message will be sent through some unsecured medium to the receiver. Receiver will obtain the encrypted message called cipher text and apply proper decryption algorithm with secret key(s), to decrypt the cipher text for finding original message. This is the general architecture of cryptography. Various encryption and decryption techniques of cryptography [1, 3, 4, 6, 7] have developed previously. Among them, Advanced Encryption Standard (AES), Data Encryption Standard (DES), and Rivest-Shamir-Adleman (RSA) are the popular approaches due to their security concern. But the major disadvantage of basic cryptography model is that if anyone wants to hack the data from the unsecured channel, though he cannot realize the meaning of data, there is a possibility of existence of secret communication.

Steganography is another popular concept used to hide the secret communication [8, 9]. In this approach, the secret data can be inserted into an image, audio, or video with a minimum change or distortion of the original one. The distortion of the cover image is so less that the difference between it and stego image cannot be distinguished visually [7, 10, 11]. Generally, data is inserted into the least significant bit (LSB) positions of the image pixels or in the LSB positions of the pixels of video frame. The quality of a data hiding scheme depends on the rate of distortion between cover and stego image. The general architecture of steganography is shown in Fig. 2.

Various steganography techniques [3, 5, 12] were developed previously. Among them, Direct Least Bit Substitution (DLSB), Optimal Pixel Adjustment Procedure (OPAP), Pixel Indicator Technique (PIT), Pixel-Value Differencing (PVD), Selected Least Significant Bit substitution (SLSB), etc., were developed depending

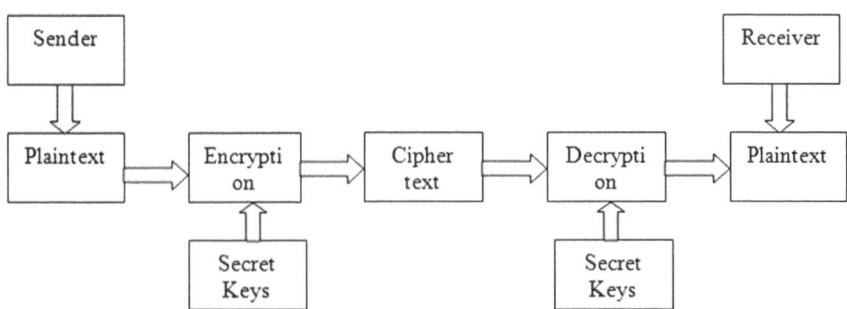

Fig. 1 Model of cryptography

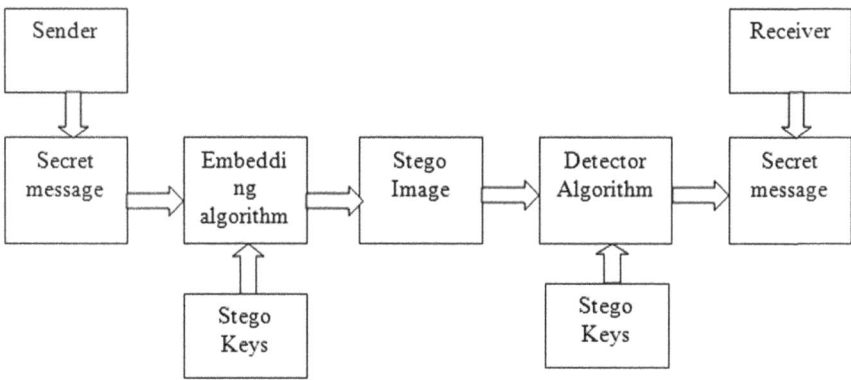

Fig. 2 Model of steganography

on spatial domain steganography and JSteg, Outguess, F3, F4, F5 algorithms [3] were developed depending on Transform domain steganography [3]. Transform domain is more secured than spatial domain. Some hacking techniques [3, 5] have also developed gradually. Among them, chi-square attack, visual attack, histogram attack, and blockiness are very common [3–5]. Generally, PSNR, SNR, etc., are parameters to check the quality of data hiding algorithms. Now a day, various hybrid techniques [3, 4, 13, 14] are developed combining cryptography and steganography, which are more secured than only a steganography approach. The general architecture of hybrid technique of cryptography and steganography is shown in Fig. 3.

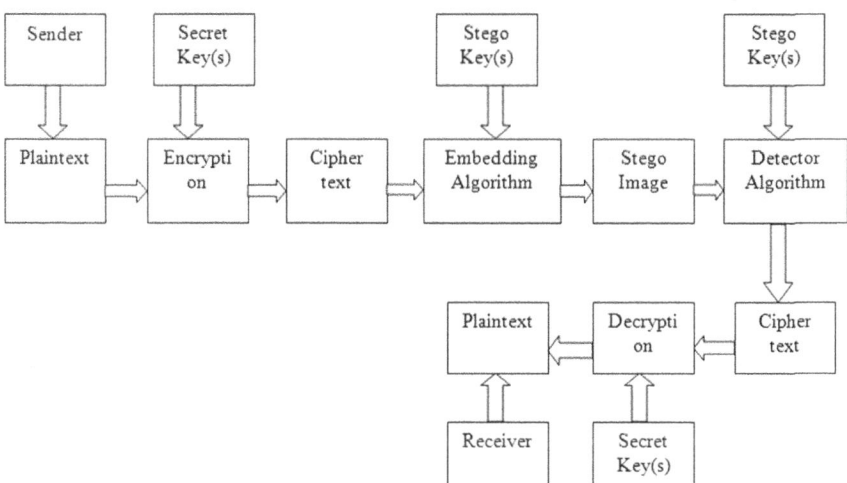

Fig. 3 Model of hybrid technique of cryptography and steganography

From Fig. 3, it can be realized that the general hybridized technique secrets the data at first using some encryption algorithm and secret key(s). Then the encrypted message is sent to the receiver hiding in an image or the frames of a video/audio.

In 2012, Deepak Singla and Rupali Syal proposed a hybrid scheme [6] to send a message safely and securely. The scheme combines RSA algorithm for cryptography DCT-based algorithm for steganography. This technique is used only for a fixed length message in a block to extract the data from image in the receiver end. In the same year, Rahul Jain and Naresh Kumar proposed another new scheme [15] with same demerits. In this scheme in an 8×8 block of an image, a 16-bit data is sent [15]. These techniques are not suitable when the length of messages varies in an image block. In the paper, we propose a novel approach where number of bits inserted in an $N \times N$ block of image is not always fixed.

In Sect. 2, we discuss our proposed data hiding scheme. In Sect. 3, some experiments with results are analyzed comparing with other approaches. Finally, the conclusion is made on Sect. 4.

2   Data Hiding Scheme

In this scheme, we have used basic idea of differential and integral calculus for encryption and decryption part of cryptography. The uniqueness of encryption is that the secret keys will be generated at the time of encryption depending on data and number of keys depending on length of data. In steganography part, the encrypted message, the secret keys, and number of secret keys obtained after encryption are embedded in image blocks using DCT transformation. This stego image is sent through the communication channel and receiver decrypts the encrypted message and secret keys using decryption algorithm. So, maintaining the basic architecture of hybridization of cryptography and steganography, secret communication is done between sender and receiver.

2.1   Encryption Method

Suppose, a message $a_1a_2, \ldots, a_n$ of length n is to be sent to the receiver. At first, a matrix is formed with unique secret values of the letters and is kept in sender and receiver site before communication. A temporary cipher text $A_1A_2, \ldots, A_n$ is generated corresponding to the message $a_1a_2, \ldots, a_n$ with the help of the secret matrix, where A_i is the position value of the letter a_i defined in the secret matrix. We define a polynomial function f(x) of degree $(n-1)$, given in Eq. (1).

$$f(x) = A_1 x^{n-1} + A_2 x^{n-2} + \cdots + A_{n-1} x + A_n \tag{1}$$

Now differentiating the function f(x), (n-1) times with respect to x the only constant term is obtained which is used as the cipher text (encrypted message) corresponding to the actual message. During the derivation process, the constant term (i.e., x free terms) in each step is stored as secret keys. In other words, the constant terms in are named as k_{n-1}, k_{n-2}, k_{n-3}, ... , k_1th key, respectively. So, after encryption part, the cipher text (encrypted message) and secret keys (required for decryption) are generated.

2.2 Embedding Method

In this part, the encrypted message (no. of secret keys) and secret keys are embedded in a cover image. At first, image blocks are taken from an image for inserting a secret data. Starting image block (K_3) is a secret key block in embedding method and is predefined in both sender and receiver site. In this block, the cipher text (encrypted message) is embedded. After K_2 no. of image blocks, the number of secret keys (required for decryption) is inserted and after next K_2 number of blocks, first key is inserted. This process is continued until all the keys (for decryption) are embedded. The value of K_2 is also a secret key among sender and receiver which is predefined among them. To embed data in each block, we use modified DCT and LSB substitution in images. At first, 2D DCT is applied to the RGB values of an N × N block of an image and quantization is performed using a standard quantization matrix. Then by zigzag scanning [6], convert N × N block into a one-dimensional array. But in this case, receiver cannot realize the length of the data. So, here a secret key K_1 is used to realize the length of the data. Convert the actual data and K_1 into BCD with equal number of bits (inserting 0 in the MSB positions, if required). Replace the LSB of DCT coefficients in odd positions of 1D array with data bits sequentially and even positions with the bits of K_1. Then convert 1D zigzag array back to N × N block and perform inverse DCT on that block and stego image block is ready. The same operation is performed in each selected image block with same key K_1, and all the blocks are combined to form stego image.

2.3 Extraction Procedure

This is the first part of receiver side work extracting the encrypted message with proper secret keys (made after encryption procedure) from the stego image. After getting the stego image, receiver takes all the image blocks sequentially from beginning. At first, K_3th block is selected for extracting the encrypted message. The values of secret keys K_3, K_2, K_1 (mentioned in Sect. 2.2) are predefined among sender and receiver. Extraction procedure also follows the same model, which is used for embedding purpose. To extract data from each block, DCT and LSB substitution is used. Initially, 2D DCT is applied to the RGB values of the $N \times N$ blocks of the image files. Then quantization using standard quantization matrix is performed, which is also used for encryption purpose. Then by zigzag scanning [6], convert $N \times N$ block into one-dimensional array. The DCT coefficients are checked, and LSB of these values are collected. The LSB values of the even positions of the array are selected until getting the secret key K_1 and position (p) of the last pixel value of K_1 is obtained, which helps to get the encrypted data completely. The receiver collects the LSBs of the pixel values of the positions 1, 3, 5, ..., p-1 of the image block in order to receive the encrypted data. The same process is followed for the block $K_3 + K_2$ to know the number of secret keys and for remaining blocks (i.e., block number $K_3 + 2K_2$, ..., $K_3 + nK_2$) for obtaining the secret keys which are required for decryption.

2.4 Decryption Method

After the extraction part, receiver gets the cipher text (C) or encrypted message, number of secret keys (ns), and secret keys $(K_1, K_2, ..., K_{ns})$ for decryption. Considering the cipher text C as a function (x), integration is done *ns* times with respect to *x*. The original function f(x) is obtained from integral constants replacing it by k_i after i-th integration. Finally, $A_1 A_2, ..., A_n$ is obtained and replaced with the corresponding letters, defined in the secret value matrix. Receiver gets the actual secret message collecting these letters sequentially.

Procedure

Sender side

Input: Secret message $a_1 a_2, ..., a_n$

   A cover image

Output: Stego image

Secret keys: A matrix with secret values of the letters

   Length of secret message indicating key to the receiver side (K_1)

   Block interval (K_2)

   First selected image block (K_3)

Algorithm

   Step 1: Replace the letters (a_i) of the secret message sequentially with secret values (A_i) sequentially. Symbolically, it can be written as

$$A_i \leftarrow \text{secret value } (a_i)$$

Step 2: Construct a polynomial of x with the secret values A_i as in the following:

$$f(x) = \sum_{i=0}^{n-1} A_{n-i} x^i \dots \tag{2}$$

Step 3: *for r = 1 to (n-1)*

Calculate $f^r(x)$

Where $f^r(x)$ is the rth derivative of f(x).

Step 4: Take the value of $f^{n-1}(x)$ as cipher text or encrypted message C. Symbolically,

$$C \leftarrow f^{n-1}(x)$$

Step 5: Generate (n−1) secret keys in the following manner

$k_{n-1} \leftarrow A_{n-1}$
for i = 1 to (n − 2)
$k_i \leftarrow (n - i - 1)! * A_i$

Where k_i is the ith secret key for decryption part.

Step 6: Take the image blocks of the cover image and select K_3 th image block at first for inserting cipher text C.

Step 7: Perform 2D DCT on that block.

Step 8: Perform quantization on that block.

Step 9: Perform zigzag scan to convert NXN block into one-dimensional array d.

Step 10: Replace the LSB of the DCT coefficients with data bits of secret data and key K_1 in the following manner:

(a) Convert the secret data and key K_1 into BCD with equal no. of bits each (i.e., insert '0' in the higher significant bit positions, if required).

(b) Insert the secret data bits in the LSB of odd position DCT coefficients of d sequentially.

(c) Insert the bits of K_1 in the LSB of even position DCT coefficients of d sequentially.

Step 11: Convert 1D array d back to NXN block.

Step 12: Perform inverse DCT on that block and stego image is ready.

Step 13: Leave K_2 no. of blocks.

Step 14: Insert no. of secret keys formed at the time of encryption and perform the operations step 7 to step 13.

Step 15: Insert the secret keys formed at the time of encryption one by one and perform the operations from step 7 to step 13 every time.

Step 16: Combine the image blocks and stego image is ready.

Receiver side

Input: Stego image

Output: Secret message $a_1a_2, \ldots, a_n$

Secret keys: A matrix with secret values of the letters

Length of secret message indicating key to the receiver side (K_1)

Block interval (K_2)

First selected image block (K_3)

Algorithm

Step 1: Take the image blocks of the stego image and select K_3 th block at first for extracting cipher text C.

Step 2: Take the particular NXN block from that image for inserting data.

Step 3: Perform 2D DCT on that block.

Step 4: Perform quantization on that block.

Step 5: Perform zigzag scan to convert NXN block into one-dimensional array.

Step 6: Check the DCT coefficients and take LSB of these values sequentially.

Step 7: POP the even positions LSB values sequentially until getting the secret key and check position (p) of the last DCT coefficients, at which the key is obtained fully.

Step 8: POP the corresponding odd positions LSB values sequentially and concatenate LSB values of the positions 1, 3, 5, ..., p−1 sequentially and take cipher text C.

Step 9: Leave K_2 no. of image blocks.

Step 10: Perform the operations from step 2 to step 9 and obtain the no. of secret keys (ns) for decryption.

Step 11: Do the operations from step 2 to step 9 until obtaining ns (no. of secret keys for decryption).

Step 12: Evaluate length of the message n as n← ns + 1.

Step 13: Calculate position values of i-th letter of the actual text P_i as

$$P_1 \leftarrow C/(n-1)!$$
$$for\ i = 2\ to\ n$$
$$P_i \leftarrow k_{i-1}/(n-i)!$$

Step 14: Replace the secret values (P$_i$) sequentially with letters (a$_i$) defined in the secret value matrix of the letters and print the letters sequentially. In other words,

for i = 1 to n
a$_i$ ← letter (P$_i$)
print a$_i$

3 Experiments and Results

In this section, we will analyze the security of the scheme comparing with some previous schemes with some standard JPEG images and a message 1500 bits. In steganography, there are two types of security analysis. In other words, the secret data can be hacked from secured technique in two different ways. First one is that hacker cannot trace the procedure of the particular scheme at a particular moment of communication. Only from stego image, he wants to hack the secret data. To check this type of security, the only fact is quality of stego image. So, we use PSNR and MSE as parameters to check the security from first type attack. The working formulas of PSNR and MSE are in the following:

$$PSNR = 20 \, \log_{10} \left(\frac{MAX_i}{\sqrt{MSE}} \right) \dots \tag{3}$$

$$MSE = \frac{1}{mn} \sum_{i=0}^{m-1} \sum_{j=0}^{n-1} \|I(i,j) - K(i,j)\|^2 \dots \tag{4}$$

Here, the symbols hold their usual meanings. Here, we observe that PSNR value is dependent on MSE. So, in this analysis, we only show PSNR to show efficiency of the scheme from first type unintentional attack (Table 1).

In the PSNR table, we observe that our proposed scheme is good enough from first type attack comparing with other schemes. Here, the PSNR values of the images are near about same to the LSB-DCT scheme, which is best among others.

The second type attack is that unintended receiver traced the particular procedure at a particular moment of hidden communication. We should also check that a

Table 1 Comparison of PSNR values of some images in different schemes

Images	PSNR values		
	Modulus	LSB 4th bit	LSB-DCT/ proposed
Mountain. jpg	46.32	52.73	55.43
Dog. jpg	48.54	53.43	57.48
Flower. jpg	48.64	52.57	56.51
Tree. jpg	47.33	52.41	57.44

particular scheme is how much protected from that type attack. The security of a scheme is parameterized with weights of secret keys for this type of security analysis. In our proposed scheme, the number of secret keys is not fixed in all the cases and that is dependent on length of message. These keys are not dependent among each other. So, our technique is very much protected from both of first and second type attack.

4   Conclusion

In this paper, a hybridized approach of information security is developed during wireless communication by mixture of cryptography and steganography. The concept of differential and integral calculus has improved the encryption and decryption of information as well as DCT increases the quality of stego image (measured through PSNR) of transformed domain image steganography. This approach is attractive for the dynamic use of independent secret keys and becomes secured from different types of statistical attacks. Also, this approach is equally and efficiently applicable for long data maintaining the basic requirements of cryptography and steganography. Instead of using image, the data can be sent through the audio and video frames and a comparative study may be considered as a future scope of this work.

References

1. Pachghare, V.K.: Cryptography and Information Security. PHI Learning Private Limited, New Delhi (2011)
2. Joseph Raphael, A., Sundaram, V.: Cryptography and steganography—a Survey. Int. J. Comp. Tech. Appl. 2(3), 626–630. ISSN: 2229-6093
3. Rawat, P., Pandey, A.K., Singh Kushwala, S.: Advanced image steganographic algorithms and breaking strategies. National Seminar on Recent Advances in Wireless Networks and Communication, NWNC-2014, IJCA (2014)
4. Rajyaguru, M.H.: Cryptography—combination of cryptography and steganography with rapidly changing keys. Int. J. Technol. Adv. Eng. 2(10) (2012). ISSN: 2250-2459
5. Bloisi, D.D., Iocchi, L.: Image Based Steganography and Cryptography, Dipartimento di Informatica e Sistemistica, Sapienza University of Rome, Italy
6. Singla, D., Syal, R.: Data security using LSB & DCT steganography in images. IJCER, 2(2), 359–364 (2012). ISSN: 2250-3005
7. Dutt, D., Hegde, V., Bose, P.T.: AAKRITI^{ed}: an image and data encryption—decryption tool. Int. J. Comput. Sci. Inf. Technol. Res. 3(2), 264–268 (2015). ISSN: 2348-1196 (print), ISSN: 2348-120X (online)
8. Jain, R., Kumar, N.: Efficient data hiding scheme using lossless data compression and image steganography, IJEST 4(08) (2012). ISSN: 0975–5462
9. Soleimanpour-Moghadam, M., Talebi, S.I.A.M.: A novel technique for steganography method based on improved algorithm optimization in spatial domain. Iraninan J. Electr. Electron. Eng. 9(2) (2013)

10. Mehndiratta, A.: Data hiding system using cryptography & steganography: a comprehensive modern investigation, IRJET **2**(1)(2015)
11. Chakrapani, G., Reddy, V.L.: Optimized videotape steganography using genetic algorithm. IJCS **15** (2014)
12. Walia, Dr. E., Jain, P.: An analysis of LSB & DCT based steganography. Glob. J. Comput. Sci. Technol. **10**(1) (Ver 1.0) (2010)
13. Gokul, M., Umeshbabu, R., Vasudevan, S.K., Karthik, D.: Hybrid steganography using visual cryptography and LSB encryption method. Int. J. Comput. Appl. **59**(14) (0975–8887), (2012)
14. Roy, S., Venkateswaran, P.: Online payment system using steganography and visual cryptography. In: IEEE Students Conference on Electrical, Electronics and Computer Science, 978-1-4799-2526-1 (2014)
15. Deb, K.: A fast and elitist multiobjective genetic algorithm: NSGA-II, IEEE Trans. Evol Comput. **6**(2) (2002)

Biometric-Based System for Intrusion Detection and Prevention Mechanism for Energy Efficiency in Wireless Sensor Networks

Gauri Kalnoor and Jayashree Agarkhed

Abstract Wireless sensor networks (WSNs) are a new and most recent technology which is used in many critical applications. WSN is a network which is deployable anywhere in the environment. Due to this nature of WSN, an unattended network is more probable for vulnerable and unavoidable attacks. WSN is comprised of huge different types of nodes which are distributed within the network. Some of the nodes have sensing technique for detecting unusual behaviour in the network. Detection and prevention mechanisms are used for securing the network, with the consideration of energy efficiency. Biometric-based system is most efficient way of detecting an intruder which is one of the major techniques of pattern matching. Since WSN is more probable vulnerable to various harmful security attacks as it has the nature of broadcasting the transmission data and has very low capacity of computation, an Intrusion detection system (IDS) is used to detect such mobile attackers or intruders. Intrusion prevention system (IPS) uses a mechanism which enables the reduction of consumption of energy in sensor nodes of WSN.

Keywords Fingerprint · Biometric system · IDS · IPS · Pattern matching
Energy · Efficiency

1 Introduction

In a deployment field, also known as a target field of WSN, large set of tiny, low-powered battery devices called sensor nodes are distributed. An infrastructure-less wireless network is formed once all the nodes are deployed in the network. The most critical applications of WSNs include military-based, health care, smart homes and other real-time applications. In WSN, a wireless-based

G. Kalnoor (✉) · J. Agarkhed
P.D.A College of Engineering, Gulbarga 585102, Karnataka, India
e-mail: kalnoor.gauri@gmail.com

J. Agarkhed
e-mail: jayashreeptl@yahoo.com

© Springer Nature Singapore Pte Ltd. 2018
K. Saeed et al. (eds.), *Progress in Advanced Computing and Intelligent Engineering*,
Advances in Intelligent Systems and Computing 563,
https://doi.org/10.1007/978-981-10-6872-0_27

communication is used between every pair of nodes within their ranges and the routing of data is done towards the nearest base station. The path chosen for communication by the nodes is multi-hop routing. Since the nodes can be distributed randomly in the required field, WSN is easy to be deployed in the given area. These large set of nodes are dropped into the field, and once the nodes start coordinating with the neighbouring nodes, a network is formed linking to a particular nearest base station which exists within the range of the wireless network. The security and privacy of the information gathered in WSN play an important role for all critical applications. The network is deployed in unattended and unsecured field with important data collected based on the application it provides. This is one of the challenging issues of WSN. A hierarchy of such nodes is formed in the heterogeneous WSN based on their capabilities: sensor nodes, cluster heads (CHs) and the base stations (BSs). Sensor nodes are low powered with very limited capability and also limited memory size. These nodes communicate with each other in a group called as a cluster, and finally, they communicate with CH, which has high memory space and battery life. CHs also have powerful processing capabilities of data and antenna.

The sensor nodes consist of most critical or sensitive data and operate in an unattended field. This makes necessary for providing security mechanisms. Some preventive measures which need minimum energy requirement are considered to provide security for WSN [1]. Some of the important characteristics WSN comprises are lack of bandwidth, minimum data storage and less energy consumption [2–4]. Intrusion prevention mechanisms are provided to overcome some severe constraints of resources. For example, if an intruder is a moving object, then IDS is set in such a way that an alarm packet is generated and sent to all the nodes in WSN. These alarm packets may flood throughout the network with high-energy consumption.

The best security in WSN can be provided by giving access to any node based on authentication and digital certificates in which signature verification is done by using public key due to their large size. Using hash functions, the unique way of security is the technique of biometric. In this technique, SHA1 hash function [5] is used and the algorithm known as DES can be used. In password-based technique, most of the passwords are simple and easy to guess. This makes the attacker break the password easily, and thus, a system with biometric technique is considered as a system for recognition of patterns. Here, the data for biometric is acquired from each node. In this system, set of features are extracted from the data gathered and is compared with the set of templates stored in the databases [6].

In WSN, the field of target for an attack to take place, physically maximum number of nodes can be compromised and later captured by the attacker, and both the CHs and the sensor nodes compromise to the attacker. Then the attacker captures the sensitive data that is stored. But the BS is assumed not to be compromised by the attacker. Thus, to prevent such situation or to detect the intruder, a biometric-based authentication scheme is used in WSN.

2 Related Work

Most of the researchers have proposed some of the mechanisms to eliminate the security attacks that are harmful to the sensor network. In this section, the relevant literature discussed.

In [7], the authors have proposed the puzzle-based technique for detecting intruders. In this technique, the client-based puzzle protocol is used, where a server accepts request for connection only in the normal operation and if it does not come under any type of attack. But once the server is trapped by an attacker, a unique puzzle is forwarded to each client of the network.

The authors in [8] have discussed authentication protocol using client-based security and mainly uses digital signature. In this technique, the authors have used one-way hash function as a partial collision with puzzle as CPU-bound. It is claimed that the proposed technique uses minimum CPU lifecycle and is a light-weight puzzle.

In [9–11], the authors have proposed many algorithms, where the performance and accuracy of the system for fingerprint recognition is discussed. The most common issue faced is the compensations for deformations of non-linear patterns. This may lead to the image of fingerprint having real-world distortions. Memory and computations of processor intensively are other important issues faced by the proposed algorithms.

Most reliable and discriminating approach proposed by authors in [12] explains that the minutiae matching of patterns is the most important challenge nowadays. In this approach, large distortions are caused considering fingerprint matching with different set of rotations. The authors have also proposed different unique features for matching the fingerprint patterns. The known fingerprint patterns can be easily identified by the IDS. In the discussed approach, the patterns which are known are stored in the databases for future reference.

In [13], the most popular industry-related systems are proposed by the authors, which are referred as automatic fingerprint identification systems (AFISs). This system deals with the sensors that have capability of capturing the fingerprint images. The ability to compare and match the biometric data that originates from different sensor nodes is the most possible assumptions made by the authors in biometric systems. This system has the great challenge for increasing the matching process speed without compromising with the accuracy of the detected unauthorized patterns. The authors in [14] have explained about identifying mobile intruders with less overhead by completely isolating the intruders and preventing the network from possible attacks. The system for prevention of intruders is called as green firewall, and it compares the broadcast messages that are flooding in the sensor network. Here, the approach discussed shows how efficiently the alarm packet transmissions reduce the attack with prevention mechanisms and thus minimizing the amount of energy consumed in the sensor nodes within the network.

The technique proposed by the authors in [15] is the intrusion detection based on semantics and patterns which are mainly formulated with different types of attacks. In this approach, the sink detects in the coverage area regarding the events occurred and the state transition of each node in the specific coverage area of WSN. The tool known as another tool for language recognition (ANTLR) was proposed for detecting semantic-based intrusions and analysis of events occurred.

In [16], the proposed scheme uses misuse-based and anomaly-based IDS. The profiles are stored based on pre-defined attacks of known patterns. The expected behaviour is defined based on the profiles, but the unusual behaviour or deviations are detected only by using anomaly-based IDS. The energy efficiency is the major challenge in this proposed scheme.

In the recent technologies, the system used for intrusion detection is based on fuzzy rules where the binary bits are encoded using own data set that is wireless and the data mining techniques with knowledge discovery. The authors in [17] proposed the technique using different genetic algorithms where the network audit data sets are performed based on the rule's fuzziness.

A string matcher technique is proposed by the authors in [18] where every bit is compared with the nth bit of the features, which is of one-character size. The state of the patterns identified is stored in a vector known as pattern-match vector (PMV), and then the ith bit is compared with ith pattern. This matched pattern is computed in the vector using logical AND operation.

In [19], the authors explain pattern identification techniques that identify and store patterns of known signatures and already previously detected intrusions based on the matched activities in the system of information and databases. The proposed techniques are more efficient and accurate compared to other techniques for detection known patterns and intrusions.

In [20], the authors have discussed the ordered property based on Markov chain and results in high performance compared to other techniques. The detection of an intruder is assisted by ordering of events consisting of audit data. Based on the properties such as frequency, ordering and duration, the audit data are classified. In this method, probabilistic-based techniques are used for detection of an intruder.

3   Biometric-Based IDS

WSN is used in most critical applications and is more vulnerable to harmful and malicious attacks. Due to its deployment in unattended environment, the nodes in the field of probable attack can be easily captured by physically being attacked by harmful intruders or enemies. Some assumptions are made in the proposed biometric-based IDS.

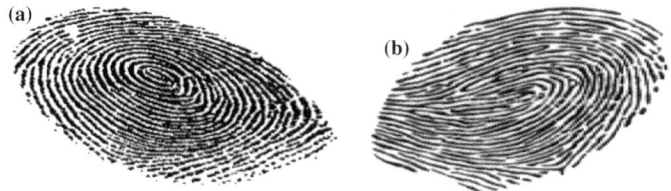

Fig. 1 Example fingerprint authentication: **a** Left-aligned fingerprint. **b** Right-aligned fingerprint

- Both the sensor nodes and the CHs or group heads (GHs) can be captured or compromised to an attacker.
- Due to criteria of cost, the hardware which is tamper resistant is not equipped in any of the nodes of the sensor network. Thus all related stored, sensitive data of the nodes and cryptographic information can be known to the attacker if any of the nodes are captured.
- No BS becomes compromised in any case of WSN.

The four phases of the scheme are discussed based on user-based authentication. In this scheme, verification of a user is done through biometric-based system which uses a significant template of biometric for verifying the matching of patterns. In this system, the pattern of each user is stored in the specified template.

This template is given as input by the user in the system of biometric. The input will then be matched with the specific template that is saved in the user's database. Fingerprint pattern template is one of the best example for biometric-based systems. In this example, the user needs to place his/her finger on the scanning device which may be at different angles. Thus, the alignment can be towards left or right side [21] as shown in the Fig. 1.

If a match of template occurs, then the user authentication is said to be valid and thus user can access some of the allocated resources in WSN. Symmetric key-based algorithm and one-way hash function algorithm that is considered as secured is represented as h(.). This function is mainly used for encrypting and decrypting the key. In the algorithm using secure hashing function, the 128-bit long message is considered as the secured key for encrypting or decrypting in the algorithm using DES as shown in Fig. 2.

The phase I of biometric IDS system is called the **Pre-Deployment phase**. In this phase, the BS is assigned with a unique identifier ID_{ij} to each of the sensor node S_i that belongs to a cluster and also registered to a particular cluster head C_J. The BS also randomly generates unique master key ID MK_i for each sensor node and MK_j for each of its assigned CH. Once the network is deployed in the field of target, the system is loaded with set of unique IDs and the master keys.

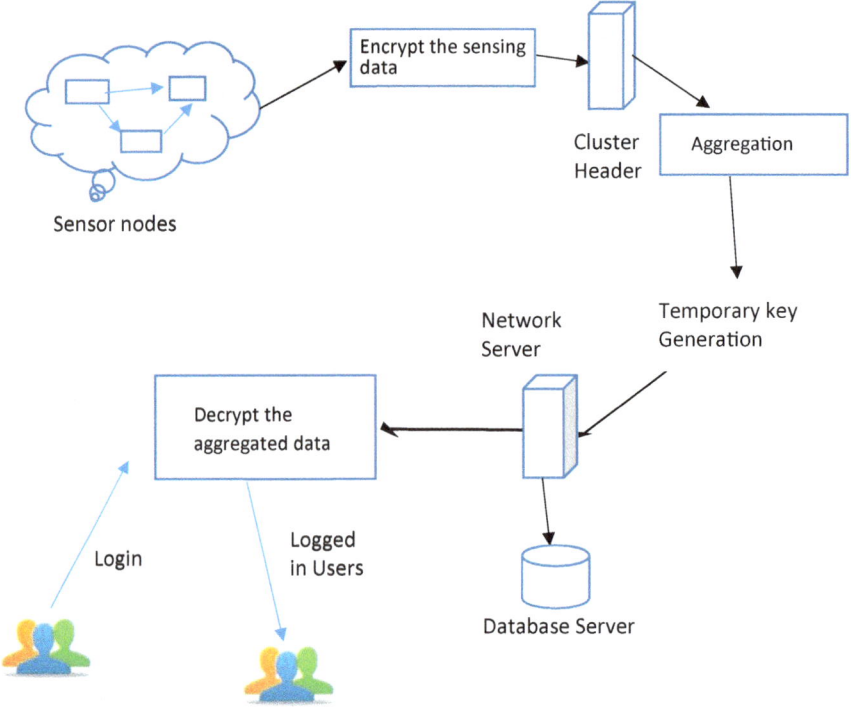

Fig. 2 Architecture of biometric system

The **Phase for Registration** is the second phase of the biometric system in WSN. The user U_i first needs to register to its BS, before any kind of data can be accessed from the particular node of sensors. The registration is done by the input from the user which can be his/her personal biometric information B_i through a device. This information is stored as a password PW_i and identity ID_i of the user U_i. The next step in this phase is the features of biometrics are converted using one-way hash function by the assigned BS for user.

The **Phase for Login** is a phase whenever any user needs to access the data and requires to input its password and identity information in the form of the biometrics. In this phase, the verification of user's information is done by comparing the input with the stored information of the biometric template. If the test is passed, the system allows user to access the data from the sensor node. Otherwise, the user is not allowed to login the system as the pattern is not matched.

Algorithm 1 explains different phases of the biometric system for intrusion detection.

Algorithm 1: Secure Encryption and hash function based algorithm

Step 1: Start
Step 2: Input from the User U_i, biometric password B_i.
Step 3: Verification of the User comparing with biometric template.
Step 4: The user's biometric features is transformed into Biometric template using one-way hash function. $f_i = F(B_i)$
Step 5: BS computes and encrypts the information as

$$e_i = h(PW_i \| f_i)$$
And $r_i = $ h $(ID\| \| PW_i \| \| f_i) \, XOR \, h \, (\, ID_i\| \|X)$

Step 6: BS selects m set of sensor nodes randomly $S_1, S_2, ... \, S_m$ by number of different clusters and then Authentication path is assigned by calculating root value of a Particular node as $(ID_i$, route) pair of a network.
Step 7: BS stores the parameters $ID_i, r_i, f_i,$ h(.), e_i and finally Route pair of m sensor nodes (route, ID_i).
Step 8: if input pattern matched
		Access allowed
	Else
		Deny access and verify again.
Step 9: End

---

4 Energy-Efficient Intrusion Prevention

In this section, the entire procedure of prevention technique of an intruder is discussed; mainly whenever the intruder/attacker escapes from the IDS and moves on to attack the target node.

Before the intruder reaches the target node, the IDS detects the intruder, but once it enters the sensor network, the intruder would have moved through its path for an attack. Thus, a second line of defence is considered to be our prevention system. The node which detects an intruder is considered to be a node of decision. This node broadcasts the pre-alarm packet to all its neighbouring nodes. Once the pre-alarm packet is reached to each node, they start recording the intruder in their greylist and starts monitoring the network. As the intruder passes on through the path in the network, more pre-alarm packets would be sent to the decision node till its threshold value is exceeded. Then, the node of decision assumes that an intruder has entered and able to access the cluster. It starts sending packet of alarm on one-node hop. This helps cluster nodes for defending themselves against an intruder by simultaneously sending the alarm packet to all other neighbouring nodes in the cluster.

The alarm packets are broadcasted only to the nodes which are close to the intruder. There is minimal overhead by using this technique of intrusion prevention.

Thus, it effectively minimizes the consumption of energy of nodes and sensor nodes in WSN.

Algorithm 2 explains the mechanism of intrusion prevention which reduces the energy consumption, once an intruder is detected.

Algorithm 2: Electing node decision Algorithm
Step 1: Start
Step 2: Each node waits for packet consisting of "Node for Decision: node ID" message.
Step 3: If no node recieves the message within time T
        Each node checks its resources allocated
        Checks the battery availabilty
        If resources are enough then
        Braodcast " Node for Decision: Node ID"
      Else
     Request for resource allocation
      Endif
    Endif
Step 4: Broadcasted node is elected as a decision node
Step 5: Decision node monitors the network and prevents from Intruder if found.
Step 6: End

The energy consumption is minimized using the above algorithm where the intrusion is prevented before any attempt of attack is possibly made by an intruder.

5 Analysis of Security of Biometric System

In this section, we analyse the security against different types of attacks in biometric system for intrusion prevention in WSN.

5.1 Replay Attack

Suppose an attacker tries to interpret the system using a request valid message to login with the input (ID_i, PW_i, C_J) provided in login phase and then tries to make an attempt for logging into the BS with a replay of same input message as $\left(ID_i, PW_i, C'_j\right) = (ID_i, PW_i, C_J)$. Once the BS receives the request of login message, firstly, it will start computing the key symmetric by using hash function of ID and the secret value 'X'. This value of X will be known to the BS only. The message C_J is decrypted by the BS and verifies if the random number matches the random number of the user U_i that is stored in its database. After verification, the BS

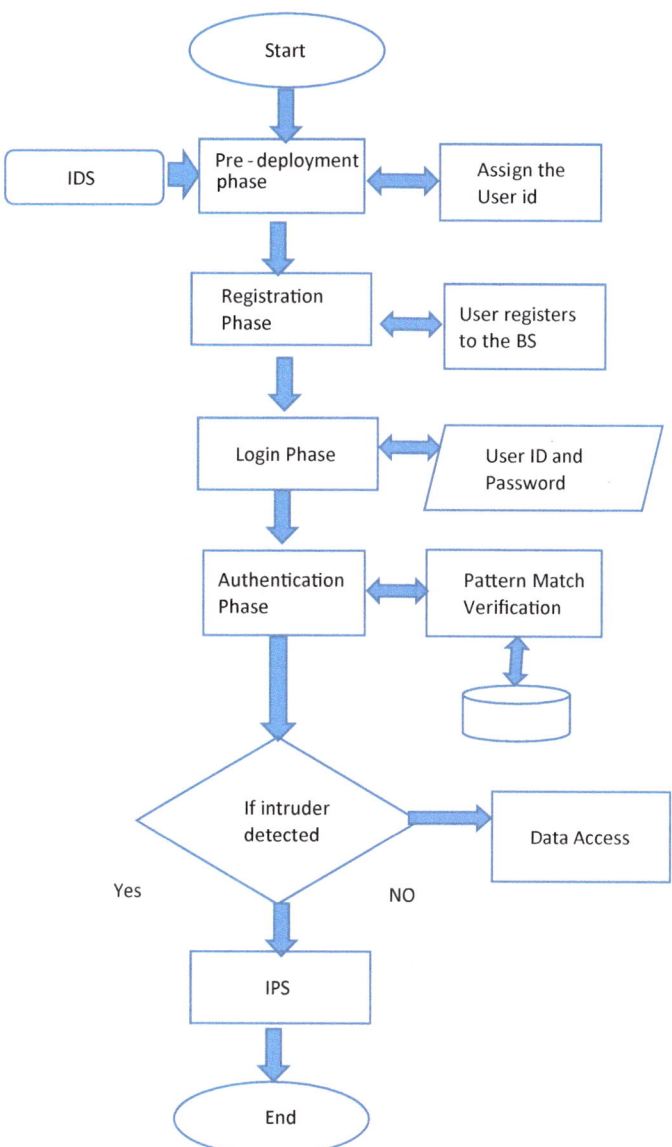

Fig. 3 Flowchart for biometric IDS in WSN

considers the login input as a replay message and thus discards the message without giving any alarm to the nodes of the network.

In our proposed system, we use random nonce which is unique to each user and is generated by each sensor node and known only to the BS. This prevents the replay attack that can occur in future since random nonce is stored/saved in the database of the network's BS.

5.2 Change Password Attack

It is very difficult to guess the password and change it, as the attacker needs to go through the biometric system for verification. The attacker cannot change the password as it has to match with the biometric template of user collected and saved in the database. Suppose an attacker attempts to change the password to login to the network for access, the mismatch of attacker's template of biometric with original user's biometric template will occur. In our proposed scheme, this attack is very rare to occur; an attacker is required to pass the verification of old password.

6 Design of Energy-Efficient Biometric System for Intrusion Detection and Prevention

The flowchart in Fig. 3 explains different phases of the biometric IDS. The user ID and password is assigned to the user only when a user registers to a particular BS of the CH. The random nonce number is also assigned to each user which is unique and prevents the intruder from harming the system in WSN. The two types of security attacks in WSN considered using this technique are replay attacks and change password attack. After login phase, authentication of the user is verified.

In this phase, if a user fails authentication, then intruder is detected by biometric IDS and data is not accessible for that user. And the pre-alarm packet is sent by the decision node to all its neighbouring nodes including sensor nodes.

7 Performance Results and Discussions

The performance of the sensor network with obtained results is discussed and explained.

7.1 Performance Analysis

This section gives the description of detection rate using biometric IDS with analysis, and prevention of intruder is applied with minimum overhead and energy consumption.

Table 1 Performance results

Biometric methods	Detection rate (%)	Energy efficiency (%)
Fingerprint verification	95	84.7
Face recognition	96	90.15
Signature	80	79
Username and password login verification	87	94

Suppose an intruder enters any one of the cluster, the broadcasting is done by every node in the cluster by sending the pre-alarm packet and adds the detected intruder node into its blacklist.

The overhead of a particular node is minimized in each cluster and the results are discussed in the Table 1.

8 Conclusion

A mechanism for intrusion prevention provides energy efficiency in WSNs is proposed. This is called as a green firewall which mainly isolates the intruder from the surrounding nodes. Two types of lists are considered which are known as blacklist and greylist. Two types of packets are pre-alarm packet and alarm packet. To reduce the consumption of energy, a local propagating mechanism is designed. A theoretical analysis is considered and is compared with the traditional flooding broadcast blacklist. The future work may be done to improve the performance of a green firewall, by its extensive simulation performed with five representative scenarios: static, crossing movement, short-distance movement, long-distance movement and multiple intruders. The results show that the green firewall can provide low control overhead by reducing the number of alarm packet transmissions.

References

1. Wu, J., Hu, H., Uysal, M.: High-rate distributed space-time-frequency coding for wireless cooperative networks. IEEE Trans. Wirel. Commun. **10**(2), 614–625 (2011); Rassam, M.A., Maarof, M.A., Zainal, A.: A survey of intrusion detection schemes in wireless sensor networks. Am. J. Appl. Sci. **9**(10), 1636 (2012)
2. Conti, M., Di Pietro, R., Mancini, L., Mei, A.: Distributed detection of clone attacks in wireless sensor networks. IEEE Trans. Dependable Secure Comput. **8**(5), 685–698 (2011)
3. Keung, Y., Li, B., Zhang, Q.: The intrusion detection in mobile sensor network. In: Proceedings of the Eleventh ACM International Symposium on Mobile Ad Hoc Networking and Computing, pp. 11–20. ACM, Sept 2010
4. Zhu, T., Mohaisen, A., Ping, Y., Towsley, D.: DEOS: dynamic energy-oriented scheduling for sustainable wireless sensor networks. In: INFOCOM, 2012 Proceedings IEEE, pp. 2363–2371. IEEE, Mar 2012

5. Zhu, T., Towsley, D.: E 2 R: energy efficient routing for multi-hop green wireless networks. In: 2011 IEEE Conference on Computer Communications Workshops (INFOCOM WKSHPS), pp. 265–270. IEEE, Apr 2011

6. Zhu, T., Zhong, Z., He, T., Zhang, Z.L.: Energy-synchronized computing for sustainable sensor networks. Ad Hoc Netw. **11**(4), 1392–1404 (2013)

7. Chen, T.H., Shih, W.K.: A robust mutual authentication protocol for wireless sensor networks. ETRI J. **32**(5), 704–712 (2010)

8. Alizadeh, M., Hassan, W.H., Zamani, M., Karamizadeh, S., Ghazizadeh, E.: Implementation and evaluation of lightweight encryption algorithms suitable for RFID. J. Next Gener. Inf. Technol. **4**(1), 65 (2013)

9. Alizadeh, M., Shayan, J., Zamani, M., Khodadadi, T.: Code analysis of lightweight encryption algorithms using in RFID systems to improve cipher performance. In: 2012 IEEE Conference on Open Systems (ICOS), pp. 1–6. IEEE, Oct 2012

10. Amiri, E., Keshavarz, H., Alizadeh, M., Zamani, M., Khodadadi, T.: Energy efficient routing in wireless sensor networks based on fuzzy ant colony optimization. Int. J. Distrib. Sens. Netw. (2014)

11. Gharooni, M., Zamani, M., Mansourizadeh, M., Abdullah, S.A.: Confidential RFID model to prevent unauthorized access. In: 2011 5th International Conference on Application of Information and Communication Technologies (AICT), pp. 1–5. IEEE, Sept 2011

12. Ghazizadeh, E., Zamani, M., Ab Manan, J.L., Alizadeh, M.: Trusted computing strengthens cloud authentication. Sci. World J. (2014)

13. Ghazizadeh, E., Zamani, M., Pashang, A.: A survey on security issues of federated identity in the cloud computing. In: 2012 IEEE 4th International Conference on Cloud Computing Technology and Science (CloudCom), pp. 532–565. IEEE, Dec 2012

14. Shohreh, H., Mazdak, Z., Roza, H.: Dynamic monitoring in ad hoc network. In: Applied Mechanics and Materials, vol. 229, pp. 1481–1486. Trans Tech Publications (2012)

15. Janbeglou, M., Zamani, M., Ibrahim, S.: Improving the security of protected wireless internet access from insider attacks. Adv. Inf. Sci. Serv. Sci. **4**(12)

16. Kiani, F., Amiri, E., Zamani, M., Khodadadi, T., Manaf, A.A.: Efficient intelligent energy routing protocol in wireless sensor networks. Int. J. Distrib. Sens. Netw. **2015**, 15 (2015)

17. Araghi, T.K., Zamani, M., Mnaf, A.B.A.: Performance analysis in reactive routing protocols in wireless mobile Ad Hoc networks using DSR, AODV and AOMDV. In: 2013 International Conference on Informatics and Creative Multimedia (ICICM), pp. 81–84. IEEE, Sept 2013

18. Araghi, T.K., Zamani, M., Manaf, A.A., Araghi, S.K.: An access control framework in an Ad Hoc network infrastructure. In: Advanced Computer and Communication Engineering Technology, pp. 747–754. Springer International Publishing (2015)

19. Mohebbi, K., Ibrahim, S., Zamani, M., Khezrian, M.: UltiMatch-NL: a web service matchmaker based on multiple semantic filters. PLoS one **9**(8), e104735 (2014)

20. Sadeghian, A., Zamani, M.: Detecting and preventing DDoS attacks in botnets by the help of self triggered black holes. In: 2014 Asia-Pacific Conference on Computer Aided System Engineering (APCASE), pp. 38–42. IEEE, Feb 2014

21. Sadeghian, A., Zamani, M., Manaf, A.A.: A taxonomy of SQL injection detection and prevention techniques. In: 2013 International Conference on Informatics and Creative Multimedia (ICICM), pp. 53–56. IEEE, Sept 2013

A Distinct Cyberattack: Phishing

G. Prisilla Jayanthi

Abstract The cyberattack happens to be a nation concern—as the target can hold the confidential matter of military or a commercial business with the intellectual assets and viable secrets that pave the way to challenge. The majorities of the professionals are aware of the cyberattacks on the data center space but are helpless and have a severe impact on the two domains. The prime domain, information technology which refers to the information processing—servers, storage arrays. The other one is about the physical infrastructure that controls the first domain. The cyberattacks on the network can be mitigated through the educative programs for the employees and training the staffs with the certain basic security policies. Introducing the preventive techniques can defend the sensitive data from the hackers.

Keywords Cyberattacks · Cyber criminals · Phishing · Sensitive data

1 Introduction

In the world of Internet of things, the leak of sensitive data increases. As the technology improves, one can discover number of hackers. The researchers have controlled with the several approaches and opinions that cybercrime hold; instructions for the customers are given to defend themselves, their companies, and vendors who rely on well-organized, reliable, and secure control structure and network setup [1]. In social engineering, phishing is a tool to deceive a person to get the personal identity, credit card details, sensitive information regarding any company, or national information. The study reveals that it does not matter how many encryption techniques, firewalls, or authentication methods an organization has, but a person behind the system is prey of phish as in Fig. 1.

G. Prisilla Jayanthi (✉)
Department of Computer Science Engineering, Geethanjali College
of Engineering and Technology, Cheeryala, Hyderabad, India
e-mail: prisillaj28@gmail.com

© Springer Nature Singapore Pte Ltd. 2018
K. Saeed et al. (eds.), *Progress in Advanced Computing and Intelligent Engineering*,
Advances in Intelligent Systems and Computing 563,
https://doi.org/10.1007/978-981-10-6872-0_28

Fig. 1 Avoid being a prey for phish

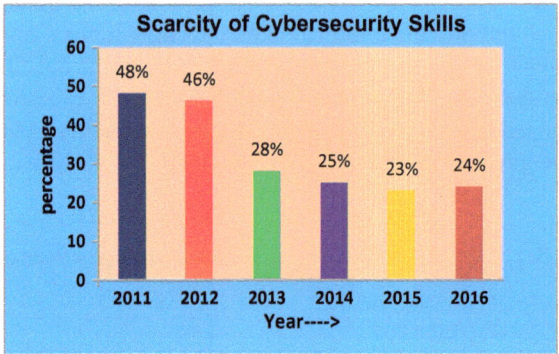

Fig. 2 Scarcity of cybersecurity skills

The phishing attack involves three phases; firstly, the victims receive the phish mail. Secondly, the victim opens the web site involved in the phish mail thus installs the malware software and the sensitive information is given, and thirdly the criminal monetizing the stolen information [2]. Figure 2 shows the scarcity of IT cybersecurity skills from the year 2012–2016 in any organization or in present scenario.

2 Suspicious Activity

The spoof e-mails aim to obtain the sensitive and confidential information, passwords, or account numbers. The e-mail misleads and tricks to click on a false link or ask to check the attachment given or make call to a phone number. When a scammer

Table 1 Difference between a counterfeit and legitimate e-mail

Counterfeit e-mail	Legitimate eBay e-mail
False e-mails are generally sent to a person to get the confidential information	These e-mails do not request to give a trustworthy evidence
They provide a crucial tone for account suspension and threatens if one does not update their information	All the messages bidding info will be present in the Messages tab in the eBay. Block the mails which appears to be uncertain
Attachments are included in this mail. On opening attachments, the malicious code infects the system(s)	The e-mails never include attachments
The greetings are given with the statement as "Attention eBay member"	These mails greet a person by the initial and latter name they recorded in eBay account that exists with the unique login name

sends an SMS message to our phone number with a bogus or fake phone number or URL is known as smishing. This type of phishing comes through phone call or SMS. Yet another type of fraudsters uses an automated system to make voice calls say for account problem and ask for account information. This is called vishing.

False e-mails and Web site seem to appear that they were sent from genuine firms. Lets take for illustration the four steps to know if it is legitimate e-mail from eBay as shown in Table 1.

3 Impact of Phishing

Phishing has great concern over the financial and commercial sectors [3]:

1. **Direct Financial Loss**: In this type of the fraud site, a consumer or business can have a financial loss with the theft of financial data.
2. **Loss of community conviction in the cybernet**: The public's trust on online financial sites is damaged through phishing.
3. **Complications in Law Implementation**: The situation where a phisher from one country hosts his phishing Web sites to the computer of other countries.
4. **Incentives for Cross-Border Tasks by Illegal Groups**: Few clues that illicit assemblies are engaging hackers to create phishing e-mails and fake sites and to develop malicious code for practicing phishing attacks.

4 Anti-phishing Training

The most unique and powerful embedded training system is Phish-Guru which communicates customers from being a prey for phishing e-mails. As the customers become a prey for the attack by the clicking on fake URL, then training message is delivered by the following e-mail. The training material presents the customer with the steps to follow and avoid phishing attack and explains in what way it would be easy for lawbreakers to execute the attacks. The study on training reveals that Phish-Guru improves user's ability to recognize phishing mails and sites.

One of the best approaches is giving training and educating customers regarding phishing. Posting of the various articles related to phishing on Web sites either with the help of non-profit and business partnership. Web-based test allowing the users to understand their own knowledge of phishing conducted by an interactive technique in Fig. 3 [4]. Figures 4 and 5 illustrates anti-phishing Phil game which helps the user to understand the fake Web sites or URL as they make an attempt to click the links provided [4].

Many such interactive videos can be developed and advertised for the user educative purpose.

Secondly, users are provided by a number of tools for warning and to identify the Web sites which are fraud and the interfaces that help people notice that they may be on a phishing Web site as shown in Table 2. The three weaknesses are (a) the people require to install special software, (b) customers either do not realize or respond according to the prompts given by toolbars, and (c) the study reveals maximum of the anti-phishing toolbars may miss only 20% of the phishing sites and found to be accurate [4].

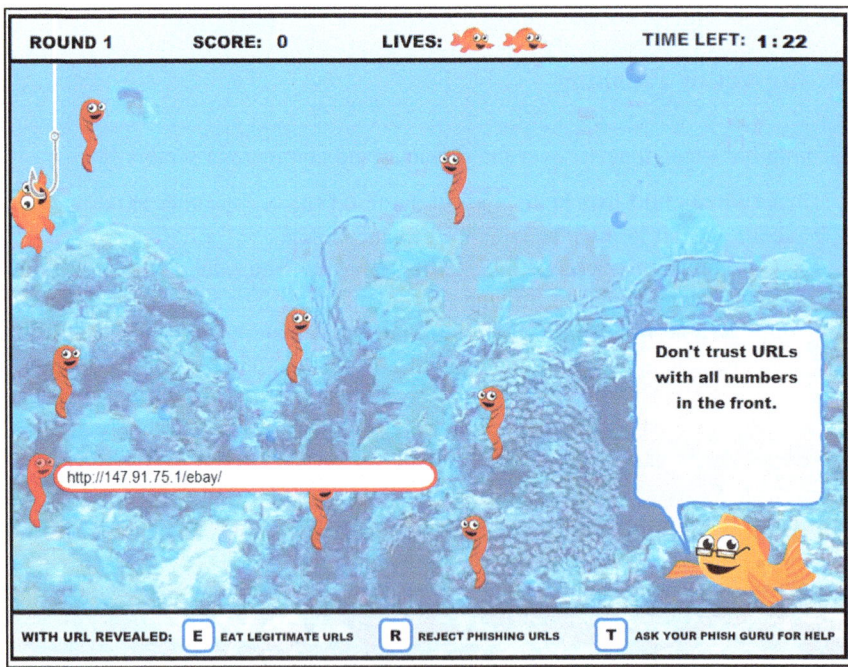

Fig. 3 Anti-phishing phil. The URL being displayed in screenshot and the lower-right corner features a tip from the Phish-Guru fish

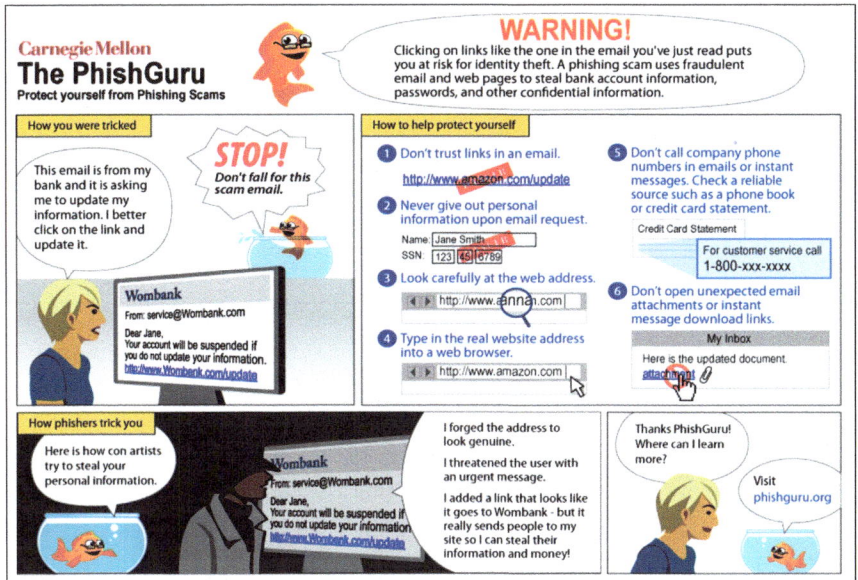

Fig. 4 One of the Phish-Guru techniques to educate the users and have proven to be effective in communicating the instructions to people

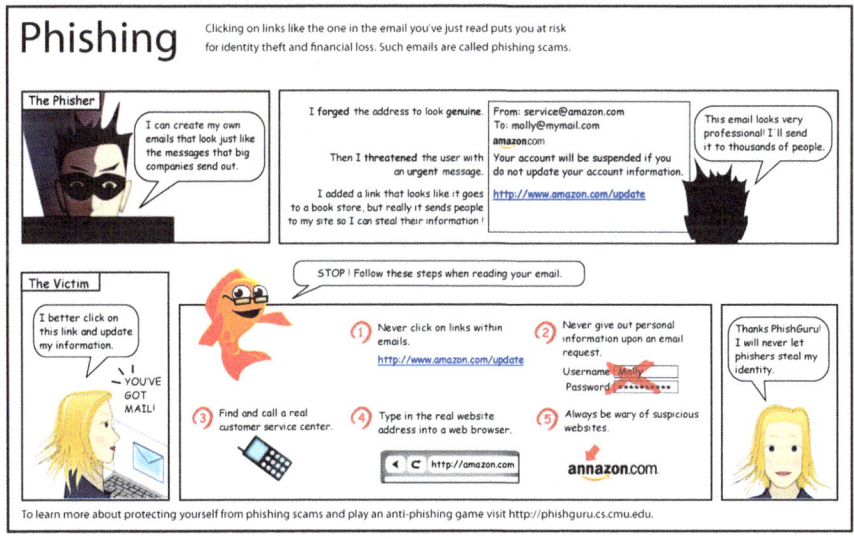

Fig. 5 Phish-Guru techniques to educate customers

Table 2 Focus on domain and URLs

Examples	Message
http://169.34.211.10/ PayPal/login.html	URLs containing numbers in the front are not to be trusted
http://future-signing.ebay. com.ttps.uk/	Looking at the word ebay.com in the website one should not be fooled, this site belongs to ttps.uk
http://www.msn-clarify. com/ http://www.ebay-debit. com/login.php	If a hyphen is found in the enterprise name, then it is a scam site Domains should not contain any security associated keywords

Finally, the risk provides protection without requiring any awareness on the part of users. The action includes detecting phishing sites, deleting phishing e-mails automatically, and shutting down of the Web sites. The existing tools are able to detect e-mails and phishing Web sites stay online to trap unsuspecting victims [4].

5 The Phil Game

The Phil game's key objective is to impart the customers three things: (A) to recognize phishing URLs, (B) where to search the trusted or untrusted sites in browsers, and (C) how to identify legitimate sites [5]. The search engine is supposed to be a dynamic tool in recognizing fake Web sites. In interactive environments, the training methods are established by the approach of games which are extremely encouraging and inspiring for the user's education.

Within any organization, humans are the weakest link that presents an opportunity for cybercriminal to penetrate the company. But the first and top line of defense is the employees; the company can protect its sensitive information using resilient security education program. The cyberattacks can be mitigated by simple methods, but people fail to consider what appears to be an elementary precaution—as applying patches, strong passwords and use antivirus, antispyware, and firewall software. The employees are treasured as an asset to the organization and valued as the instrument of the company that builds associations with its customers.

Cybercriminal view the employees to be the path of the least resistance. Cybercriminals always are sure of the fact that employees pave the way for 42% of confidential data loss measured the largest single data loss cause [6]. In case, if one wanted to access to the clients, records of the employee, social engineering tactics targeting employees are often the easiest way to penetrate an organization. The training of the cyber skills plays an essential role by educating the employees through interactive videos or classes.

In order for the organization to reach the best possible defense, it requires a cohesive solution that has the following components:

- E-mail filter—Configured to block phishing attacks, spoofed senders, malicious file types, and known bad files.
- Web filter—Updated in near real time for phishing e-mails, malicious URL's, and known bad files that have been identified in recent e-mails.
- Containerization—Configured to intercept unknown files and to place a containment wrapper around them before they are delivered to end points.
- Malware sandbox—Configured to scan and analyze unknown files for malicious activity.
- SIEM—Configured and tuned to provide alerts on suspicious activity (Fig. 6).

The few innovative technologies that give multi-layered defenses against the cyberattacks on any networks in an organization.

- Automatic vulnerability scanning and patch management
- Automatic Exploit Prevention
- Zeta-Shield
- Whitelisting Lab
- Kaspersky Security Network
- Glass-wire firewall

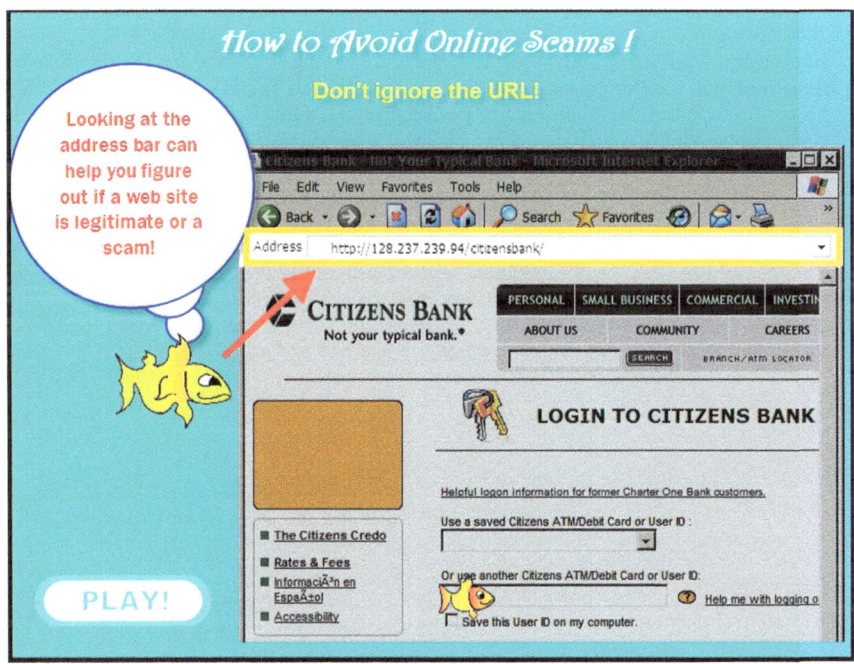

Fig. 6 Interactive game to phishing Web site

6 Conclusion

The cybercriminals can infiltrate the organization by various tricky techniques either to access the sensitive and confidential data or destroy the information. Through vulnerabilities in business software that have been used/running with the click on the malware-infected e-mails or Web site by the employees and many of the business are attacked via insecurities in the self-company-owned Web sites. Hence, all of the organizations are likely to invest in robust cybersecurity measures, and the key requirement is to educate the employees about the risk which they would encounter on cyberattack. This helps to gain wider visibility of the network, and see that everything in the network applies the security measures appropriately.

Acknowledgements My sincere thanks to Praful Chandra, HOAT Cadet Training Wing, MCEME, for his valuable support.

References

1. Hesse, D., Anderson, T.: INL Cyber Security Research-Defending the Network Against Hackers
2. Hong, J.: The state of phishing attacks. Commun. ACM. **55**(1). https://doi.org/10.1145/2063176.2063197
3. Report on Phishing: Binational Working Group on Cross-Border Mass Marketing Fraud, Oct 2006
4. Kumaraguru, P., Rhee, Y., Acquisti, A., Cranor, L.F., Hong, J., Nunge, E.: Protecting People from Phishing: The Design and Evaluation of an Embedded Training Email System (2007)
5. Sheng, S., Magnien, B., Kumaraguru, P., Acquisti, A., Cranor, L.F., Hong, j., Nunge, E.: Anti-Phishing Phil: The Design and Evaluation of a Game that Teaches People Not to Fall for Phish. http://repository.cmu.edu/isr/22 (2007)
6. Beware the invisible cybercriminals: cyberespionage and your business. http://usa.kaspersky.com
7. Kumaraguru, P., Cranshaw, J., Acquisti, A., Cranor, L., Hong, J., Blair, M.A., Pham, T.: School of phish: a real-world evaluation of anti-phishing training. In: Symposium on Usable Privacy and Security (SOUPS), 15–17 July 2009
8. Kumaraguru, P., Sheng, S., Acquisti, A., Cranor, L.F., Hong, J.: Lessons From a Real World Evaluation of Anti-Phishing Training (2006)

Part III
Emerging Techniques in Computing

Designing and Simulation Tools of Renewable Energy Systems: Review Literature

Prashant Kumar and Sanjay Deokar

Abstract Renewable energy hybrid system is one of the most promising, economical, and reliable options for electrifying rural areas. Hybrid energy systems are being utilized for minimizing usage of fossil fuels to reduce environmental effects. But analysis of hybrid system is quite complex; therefore, it requires software tools for design, analyze, and optimization of software. Motive of the paper is to introduce various major software tools required for design, analyze, and optimization of hybrid system. The software discussed are HOMER, RET Screen, PVsyst, iHOGA.

Keywords Hybrid energy · Software tools · Renewable energy

1 Introduction

In this advanced era, population is increased day by day, which leads to power crisis. Consistent and secure power supply is necessary for economic development. In most of rural areas, electricity is not there due to the extension of grid connection is economical; here we can install stand-alone hybrid system to electrifying rural areas. Fossil fuel reservation is a worldwide problem, and they pollute environment by emission. On the other hand, renewable energy sources are available at free of cost in nature and can produce electricity from them. But only one renewable energy-based system is economical and not reliable because they are seasonal and not with us throughout the year continuously; hence, we are approaching toward hybrid system.

Hybrid system solution is complex, and thus, it requires software tools which can be used for design, analysis, simulation, economical, and optimization planning [1–5]. Nowadays simulation has got worldwide application. Therefore, software become cheap and easy to learn. A number of software have been developed to overcome problems regarding to optimum designing or sizing which leads to

P. Kumar (✉) · S. Deokar
ZCOER, Electrical, Pune, India
e-mail: Prashant2685@gmail.com

© Springer Nature Singapore Pte Ltd. 2018
K. Saeed et al. (eds.), *Progress in Advanced Computing and Intelligent Engineering*,
Advances in Intelligent Systems and Computing 563,
https://doi.org/10.1007/978-981-10-6872-0_29

315

Table 1 Free version software tools [2]

Software name	Manufacturer/developing institute	Cost
HOMER	Mistaya Engineering, Canada	Free
RETScreen	Natural Resources Canada	Free
PVsyst	Institute of Environmental Sciences (ISE), University of Geneva, Switzerland	Free
iHOGA	University of Zaragoza, Spain	EDU version is free

Table 2 Priced version software tools [2]

Software name	Manufacturer/developing institute
INSEL	German University of Oldenburg
TRNSYS	University of Wisconsin and University of Colorado
iGRHYSO	University of Zaragoza, Spain

installation cost high. In this paper, we are going to discussed HOMER, RETScreen, PVsyst, iHOGA [1–3].

1.1 Available Software for Hybrid System Simulation

Some major hybrid system simulation software tools are tabulated in Table 1, with their manufacturer or developing institutions and availability (Table 2).

2 Software Tools

2.1 HOMER

HOMER is the acronym for Hybrid Optimization Model for Electric Renewable. It is most widely used software and developed in 1993 [6]. HOMER does both optimization and sensitivity analysis. With the help of HOMER, system performs the energy balance calculation considering several numbers and sizes of component. A sorted list of configuration result based on the Total Net Present Cost (TNPC) has been displayed on the software. Sensitivity analysis determines varying factors such as wind speed, fuel cost. HOMER displays simulation results in a tabular and graphical form on the basis of possible configurations. These results assist in comparison of different configurations and evaluation based on their economic merits. HOMER Pro 3.1 is released on January 20, 2015; it has full

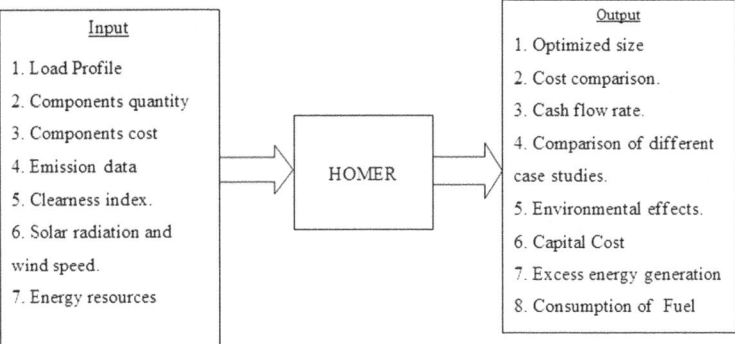

Fig. 1 Schematic representation of HOMER [1, 2]

HOMER 2 capability and much more to HOMER Pro. HOMER Pro 3.1 is a free upgrade, and it includes two new modules—advanced grid and hydrogen. HOMER is in use throughout the world, and over 110,000. Version 3.1 of HOMER Pro includes major performance like adds graphing capability and user can his own battery and wind turbine and many other new features and capabilities. HOMER 3.3 is released on July 2015 [2, 7] (Fig. 1).

2.1.1 Analysis Capabilities

1. Economic Analysis
2. Technical Analysis
3. PV System
4. Wind System
5. Generator
6. Storage device
7. Bio-energy
8. Hydro energy.

2.1.2 Advantages

1. It has hourly data handling capacity.
2. It provides results in graphical and tabular form.
3. HOMER is easy to understand, freely accessible.
4. Graphs can import in users documents.
5. It can consider sales or purchase of electricity from grid.
6. Computational time is very less.

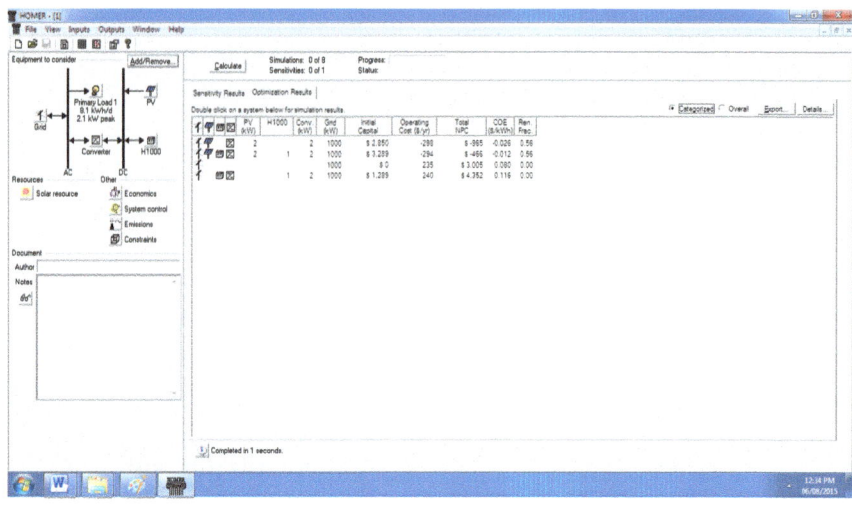

Fig. 2 Simulation result of HOMER [2]

2.1.3 Disadvantages

1. It allows using only imperial units.
2. Time series data in a form of daily average cannot be imported.
3. It does not have analysis capabilities in thermal system.
4. Fluctuation in bus voltage is not taken into account (Fig. 2).

2.2 RETScreen

Renewable Energy Technologies Screen (RETScreen) is developed in Canada by Ministry of Natural Resources for evaluating economical and environmental cost. It is freely available tool [2, 5]. It has database of a global climate more than 5000 ground stations energy resource maps, hydrology data, product data. Two versions of RETScreen are available, RETScreen Plus, and RETScreen 4. Energy performance study is done by RETScreen Plus. RETScreen Plus is a Windows-based energy management software tool. RETScreen 4 is based on Microsoft Excel and works on energy project analysis. It provides energy analysis, cost analysis, sensitivity or risk analysis, emission analysis, financial analysis for user [2, 7] (Fig. 3).

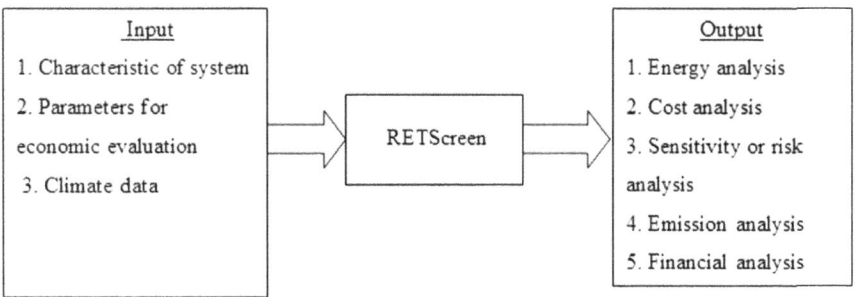

Fig. 3 Schematic representation of RETScreen [1, 2]

2.2.1 Analysis Capabilities

1. Economic analysis
2. Technical analysis
3. PV system
4. Wind system
5. Storage device.

2.2.2 Advantages

1. Main strength is financial analysis
2. Easy to use due to EXCEL-based software.

2.2.3 Disadvantages

1. Data or files cannot be imported
2. Lots of limitations for search
3. Less data input
4. Generator, hydro energy, bio-energy, thermal system analysis cannot be performed (Fig. 4).

2.3 PVsyst

PVsyst is used for grid-connected, stand-alone, pumping, and DC-grid systems. It performs sizing, simulation, and data analysis. Working platform of this software is Windows 7, Windows 8, Vista [2]. PVsyst gives multiple things to clarify shading

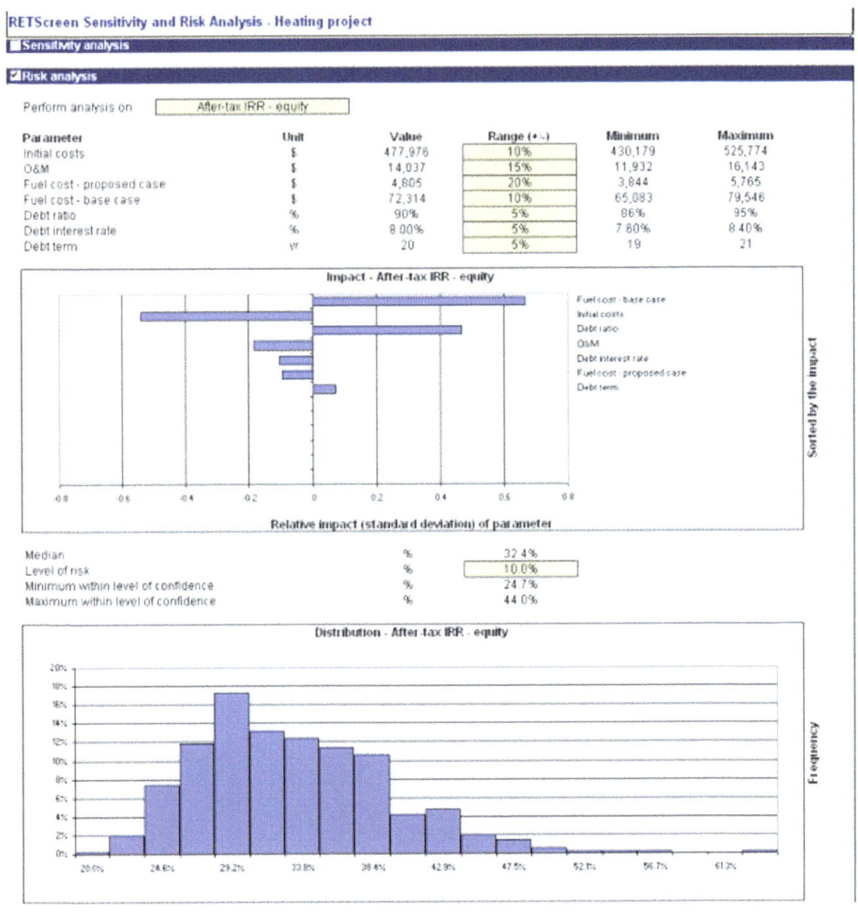

Fig. 4 Sensitivity and risk analysis result of RETScreen [2]

parameters and capability to define multiple PV fields and simulate PV systems with assorted directions. This gives output as graphs of the behavior of components, behavior of electrical PV array under partial shading, on graph comparison with clear day model, generation of meteo hourly synthetic files from monthly values, quick meteo calculations on hourly meteo plots, and irradiation calculation. Its latest Version 6.35 is released on 03.24.2015 [2, 8] (Fig. 5).

2.3.1 Analysis Capabilities

- Economic analysis
- Technical analysis
- PV system.

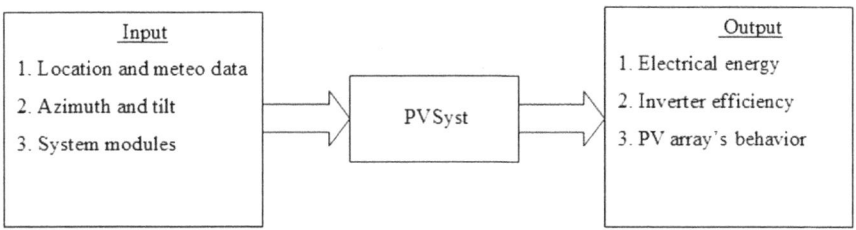

Fig. 5 Schematic representation of PVsyst [2]

2.3.2 Advantages

- It draws loss diagram to identify the weaknesses of the system design.
- This system analyzes the real running parameters of the system, identify very little misrunning.
- All result graphs, tables can be exported to other software.
- Rough estimation of cost available.

2.3.3 Disadvantages

- Hidden parameters were not modifiable (bug).
- Error in the simulation in V 6.33 and 6.34.
- Sometimes errors in the display of number of strings (Fig. 6).

2.4 iHOGA

It is modified Hybrid Optimization by Genetic Algorithm. iHOGA is developed by University of Zaragoza, Spain. It is formerly called as HOGA. It is the modified version of the iGRHYSO. There are two versions of iHOGA, EBU, and PRO [2, 5]. EDU version of iHOGA is freely available to have some limitations, but PRO is a priced version without any limitations. iHOGA is available only in Spanish language. It only runs in Windows XP, Vista, 7 or 8. This software is used for temperature effect on photovoltaic generation and production by wind turbines. It calculates life cycle emissions. This tool also considers different types of sales/purchase of electricity from grid. Data exportation is available in this software. iHOGA gives highly accurate models. It uses multi- or mono-objective optimization using genetic algorithm. Version 2.2 is released on November 2013 [2] (Fig. 7).

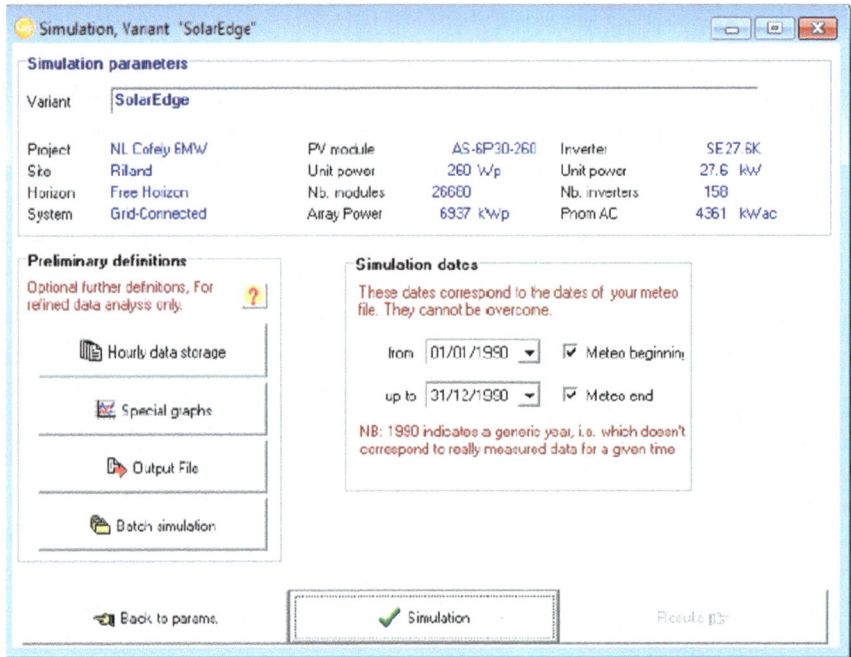

Fig. 6 Simulation result of PVsyst [2]

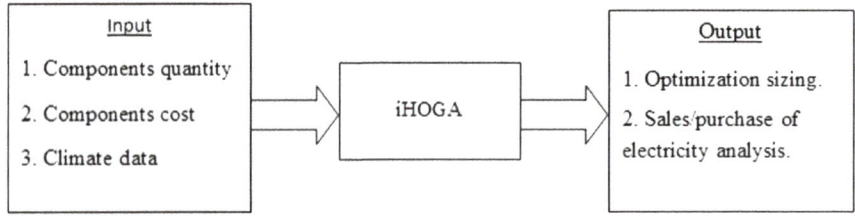

Fig. 7 Schematic representation of iHOGA [2]

2.4.1 Advantages

- It uses genetic algorithms to optimize hybrid systems in lesser time
- It does probability analysis
- Very accurate models for resources, for components and for economical calculations
- It optimizes the slope of the PV panels
- Any case of net-metering can be simulated by iHOGA.

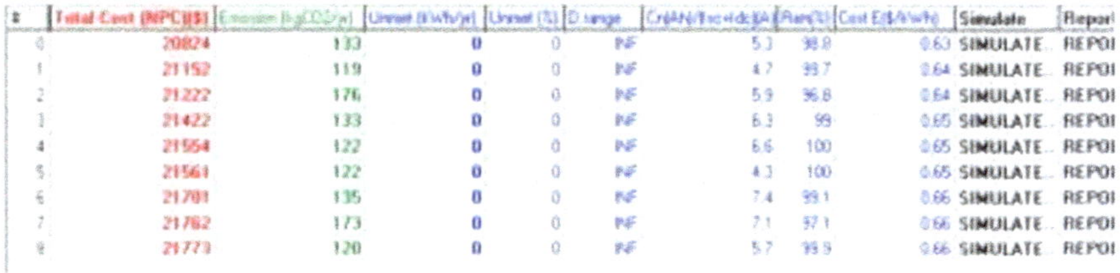

#	Total Cost (NPC)[$]	Emission [kgCO2/y]	Unmet [kWh/y]	Unmet (%)	D.range	Cr(AN/Tech4dc)[A]	Ren[%]	Cost E[$/kWh]	Simulate	Report
0	20824	133	0	0	INF	5.3	98.9	0.63	SIMULATE	REPOI
1	21152	119	0	0	INF	4.7	99.7	0.64	SIMULATE	REPOI
2	21222	176	0	0	INF	5.9	96.8	0.64	SIMULATE	REPOI
3	21422	133	0	0	INF	6.3	99	0.65	SIMULATE	REPOI
4	21554	122	0	0	INF	6.6	100	0.65	SIMULATE	REPOI
5	21561	122	0	0	INF	4.3	100	0.65	SIMULATE	REPOI
6	21701	135	0	0	INF	7.4	99.1	0.66	SIMULATE	REPOI
7	21762	173	0	0	INF	7.1	97.1	0.66	SIMULATE	REPOI
8	21773	120	0	0	INF	5.7	98.9	0.66	SIMULATE	REPOI

Fig. 8 Simulation result of iHOGA [2]

2.4.2 Disadvantage

- EDU version has limitations of total average daily load.
- For activation of license, Internet connection is required (Fig. 8).

3 Conclusion

In this era, there are lots of advancements in the field of software programming that is for analyzing hybrid system at free of cost. All the above-mentioned software tools in this paper are user-friendly. We can analyze the hybrid system very fast with the help of these software tools. So this paper will be helpful in understanding the handling of available software.

References

1. Kumar, P., Bade, A., Patil, D., Jadhav, A., Patil, S.: Optimal design solution for distributed power generation using HOME. In: 3rd International Conference on Electronics and Communication Systems, pp. 1048–1051 (ICECS 2016)
2. Kumar, P.: Analysis of hybrid systems: software tools. In: International Conference on Advances in Electrical, Electronics, Information, Communication and Bio-Informatics (AEEICB16), pp. 329–332
3. Ellabban, O., Abu-Rub, H., Blaabjerg, F.: Renewable energy resources: current status, future prospects and their enabling technology. Renew. Sustain. Energy Rev. **39**, 748–764 (2014). Elsevier
4. Elhadidy, M.A., Shaahid, S.M.: Parametric study of hybrid (wind + solar + diesel) power generating systems. J. Renew. Energy **21**, 129–139
5. Kumar, P.: Optimal design configuration using HOMER. Proc. Technol. **24**, 499–504 (2016)
6. Zhao, D., Li, Y., Liu, Y.: Optimal design and sensitive analysis of distributed generation system with renewable energy sources. In: China International Conference on Electricity Distribution, pp. 456–460, (CICED 2014). ISSN: 2161-7481

7. Zoulias, E.I., Lymberopoulos, N.: Techno-economic analysis of the integration of hydrogen energy technologies in renewable energy -based stand-alone power systems. Renew. Energy **32** (4), 680–696 (2007)
8. Markvart, T.: Sizing of hybrid PV-wind energy system. Solar Energy **57**(4), 227–281

Design of a Medical Expert System (MES) Based on Rough Set Theory for Detection of Cardiovascular Diseases

Sangeeta Bhanja Chaudhuri and Mirzanur Rahman

Abstract A MES is developed which simulates the methodology of a medical practitioner's way of detecting Cardiovascular diseases. The challenge of missing data
has been solved by most frequent value imputation method. The issue of continuous
attributed data is solved by entropy-based discretization method. The model seems
to predict the presence or absence of heart disease from minimal attribute set thus
minimizing redundancy. Reducts has been extracted from the features by considering quality of approximation. Learning by Example Module, Version 2 (LEM2)
algorithm based on Rough Set has been used for rule induction and generation. Two
Rough Set classifiers are designed, namely RSC-1 and RSC-2. Each classifier is fed
with all features and then with reducts separately as inputs. A comparative study is
performed between the classifiers, and result shows that RSC-1 whose input is locally discretized reduct performed better with an accuracy of 84.46 ± 5.24 percentage.

Keywords Data mining · Rough set · Heart disease · Classification · Machine
learning · Artificial intelligence · LEM

1 Introduction

Cardiovascular disease (CVD) defines conditions that affect our cardiac health. Cardiac disease involves narrowing of blood channels that might cause cardiac arrest or
cardiac stroke [1]. A study has been conducted over the past 17 years and recently
reported in New England Journal of Medicine that maximum mortality cases related
to CVD are found in second and third world countries (e.g. India) because of igno-
rance and inadequate treatment [2]. Due to this alarming rate of increase in CVD,

S. B. Chaudhuri (✉) · M. Rahman
Department of Information Technology, Gauhati University, Guwahati, Assam, India
e-mail: bhanjachaudhurisangeeta@gmail.com
URL: http://www.gauhati.ac.in

M. Rahman
e-mail: mirzanurrahman@gmail.com

© Springer Nature Singapore Pte Ltd. 2018
K. Saeed et al. (eds.), *Progress in Advanced Computing and Intelligent Engineering*,
Advances in Intelligent Systems and Computing 563,
https://doi.org/10.1007/978-981-10-6872-0_30

325

there is a need for interdisciplinary research so that patients at an early stage can be detected by the experts without any confusion.

The analysis of innumerable symptoms and risk factors makes the diagnosis of heart diseases challenging. There is a high chance of biasness in the manual diagnosis process. The advancement of Artificial Intelligence, Data Mining and Machine Learning has contributed interdisciplinary exploration in the medical field and led to the emergence of expert systems. Motivated by the need of an automated system, a Rough Set-based model has been proposed here, which is able to diagnose heart disease with maximum efficiency. Imprecise information cannot be used to distinguish or classify objects. There lies the significance of Rough set Theory (RSTh) which handles imprecise dataset. This work has been done using RSTh and ROSE2 software [3, 4].

The flow of work is systematically ordered as follows: Sect. 2 elaborates the intended approach and implementation, result analysis is detailed in Sect. 3. At last, the manuscript is wind up and future research has been proposed in Sect. 4.

2 The Proposed Methodology and Implementation

The objective of the proposed technique is to (i) handle missing data using most frequent value imputation method [3, 4], (ii) discretize it using entropy-based discretization [5, 6], (iii) extract significant features (reducts), (iv) develop two Rough Set classifiers such as RSC-1 and RSC-2, (v) evaluate and compare their efficiency. For RSC-1, the inputs are the reducts and all the features which has been discretized by local discretization method (LD). Reducts and all the features are fed separately into the RSC-1 to test their performance. The same method is applied for RSC-2 where the inputs are the reducts and all the features which have been discretized using global discretization method (GD). The model for this technique is shown in Fig. 1. The results are discussed in Sect. 3.

(i) *Clinical Data Acquisition:* Real-time heart dataset (Cleveland) is considered here for study. It has been taken from UCI, [7, 8]. The experiments refer to 14 attributes. The 14th attribute refers to the class level, which represent the absence or presence of cardiac disease.

(ii) *Algorithm for Missing data handling:* For each dataset of size b × c where b indicates number of rows and c indicates number of columns, the following steps need to be performed:

   (1) For a given attribute, histogram of values is calculated.
   (2) Then, the values are sorted according to their decreasing frequencies.
   (3) The first value is selected to complete the blank spaces.

(iii) *Algorithm for Discretization*:

   (1) Entropy of data is calculated using Eq. (1).

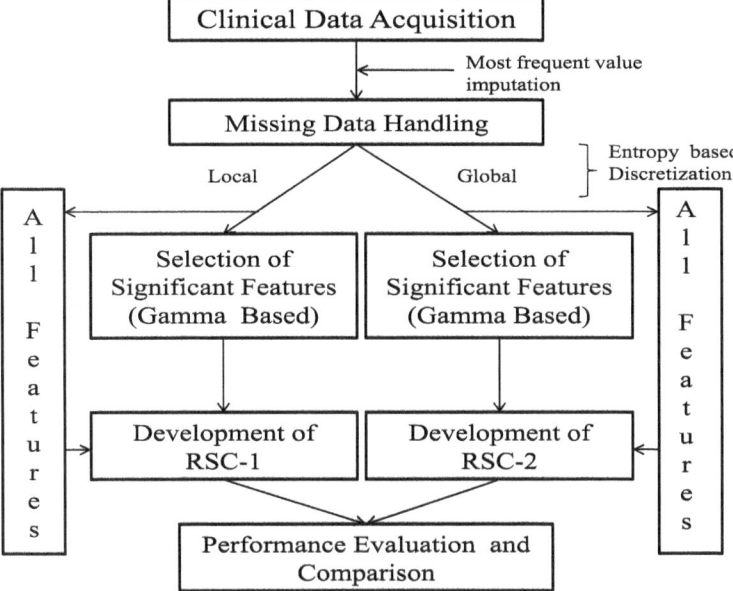

Fig. 1 Schematic representation of the intended approach

(2) For each potential split in the data, the following steps are performed:
    (2.1) Entropy is calculated for each potential bin as shown in Eq. (1).
    (2.2) Net entropy for the split is computed using Eq. (2).
    (2.3) Entropy gain is calculated as shown in Eq. (3).
(3) The split with the highest entropy gain is selected.
(4) Recursive or iterative partition is performed on each split until termination criteria is met.

The termination criteria might be: (i) reaching a specified number of bins. (ii) entropy gain falls below a range. (iii) both (i) and (ii) can be used.

The formula for entropy calculation is:

$$Entropy(D) = - \sum_{j=1}^{r} p_j log_2 p_j \qquad (1)$$

where j spans between 1 and r. r is the number of classes, p is the likelihood of getting specific class for the given bin of interest. The lower the entropy, the better is the bin.

The net entropy can be calculated by:

$$NetEntropy(D) = \frac{|D1|}{D} Entropy(D1) + \frac{|D2|}{D} Entropy(D2) \qquad (2)$$

where $\frac{|D1|}{D}$ and $\frac{|D2|}{D}$ are the proportions of the bins size and *Entropy*(D1) and *Entropy*(D2) are the bin's entropy, respectively.

The entropy gain is calculated by the formula:

$$Gain(D) = Entropy(D) - NetEntropy(D) \qquad (3)$$

The discretization process takes place based on the information of the local norm file [9, 10]. As an illustration, let us consider attribute A4 whose intervals are (–inf, 107, <107, 143 and <143, +inf). The first interval captures all values less than 107. The second interval captures values greater than or equal to 107 and less than 143. The third interval captures values 143 or above. Hence, the original dataset is replaced by 0, 1 or 2 depending on the intervals they fall into. This dataset from a Rough Set perspective is easier to handle than the continuous one.

2.1 Rough Set

In RSTh [9, 10], the data are represented in a tabular form known as information system table (IST). IST is a 2D Table or matrix consisting of rows and columns. The rows represent the objects (cases or observations), and the columns represent the attributes (condition attribute). There is an extra attribute which shows the class for each observation. The matrix has been depicted as $IST = (U, F \cup G)$ where U depicts non-infinite set of objects, F depicts the set of predictors and G depicts the class label.

T-Indiscernibility relation, $(Ind_y(T))$: If $T \subseteq F$, there lies corresponding equivalence relation $Ind_y(T)$. Let $b \in F$, $T \subseteq F$, then indiscernibility of T may be stated as:

$$Ind_y(T) = u, v \in U \times U : \forall b \in T, b(u) = b(v) . \qquad (4)$$

Lower approximation (LApp): The T—$LApp$ or positive area of set B, $\underline{T}(B)$, signifies union of all equivalence classes in $[u]_T$ which exactly matches target set X. The equation may be stated as:

$$\underline{T}(B) = u \in U : T(X) \subseteq u . \qquad (5)$$

Upper approximation (UApp): The T—$UApp$ or negative area of set B, $\overline{T}(B)$, signifies union of all equivalence classes in $[u]_T$ which might not exactly match target set X [9].

$$\overline{T}(B) = u \in U : T(X) \cap u \neq \emptyset . \qquad (6)$$

Reduct, RED(T): It might be defined as the minimum subset of attributes that provides similar quality of classification (γ) as provided by the whole set of attributes in T. The collection of attributes recurrent within all reducts is known as Core. To calculate reduct, determining γ of target Set X is essential. This can be computed by applying the formula as given below:

$$\gamma_T(X) = \frac{\sum_{i=1}^{n} cardinality(\underline{T}(B_i))}{cardinality(U)} \tag{7}$$

where i spans between 1 and n [9].

Algorithm to find Reducts:

(i) Indiscernibility relation of the whole attribute is calculated using Eq. (4). The coefficient gamma (γ) is calculated using Eq. (7).
(ii) The target sets are fixed.
(iii) The lower approximations are computed based on the target set using Eq. (5).
(iv) The indiscernibility and gamma are calculated for each subset of attributes. The (γ) of the whole attributes is compared with the (γ) of the subsets of attributes. The similar values give desired reducts.

Rule Generation and Validation using LEM2: LEM2 is a *minimal covering rule* that induces minimal set of decision rules [5, 9–12]. Let Z be the collection of decision rules induced for the target class C. Accepting rule $z \in Z$ as discriminant, set of rules Z are said to be minimum if it defines the concept C as given below:

(1) The condition part of each rule is minimal
(2) $U_{z \in Z}[Z] = C$
(3) Absence of any rule $\acute{z} \in Z$ so that $Z - \acute{z}$ covers condition (1) and (2).

The complete algorithm can be found in [12–14]. LEM2 can handle only pre-discretized data. For *rule validation*, k-fold cross validation has been employed here to generate and evaluate multiple sets of rules. In the present scenario, ten fold cross validation (stratified) has been employed to evaluate training accuracy. In stratified cross validation, the distribution of decision classes in each fold is the same as in the entire set. The parameters of the algorithms [3–5] used are as follows: (i) Majority Threshold: 21% (ii) NDTM Method: Strength of rules (iii) Minimum Similarity: 50% (iv) partially matched rule: all. For each classification, a confusion matrix is generated from which the evaluation metrics such as accuracy and error has been calculated.

3   Result Discussion

The data has been preprocessed, and two classifiers has been developed namely RSC-1 and RSC-2. This is basically a 2-class problem, hence two tentative grades has been assigned, 0 for *no disease* and 1 for *diseased*. Two cases for each classifier has been studied here as follows: (i) Classifier using all features (ii) Classifier using reducts.

3.1 Results of RSC-1

Classifier using all features: In this case, the input for RSC-1 consists of all the thirteen condition attributes and one decision attribute from LD Dataset. During rule induction, RSC-1 generated fifty-one certain rules and one approximate rule. Figure 2a top left shows the confusion matrix generated during validation. It pictorially displays the classification and the misclassification rates. The matrix can be interpreted as: out of 164 cases from *class* 0, 137 cases are correctly classified. Hence, for *class* 0, the accuracy is 83.64 ± 9.34%. Again, out of 139 cases from *class* 1, 117 cases are correctly classified with an accuracy of 84.23 ± 13.06%. The average accuracy is found to be 83.89 ± 7.76%.

Classifier using reducts: Three reducts has been generated. The attributes sets are AT1, AT2, AT3, AT5, AT7, AT8, AT9, AT10, AT12, AT13 for Reduct1; AT1, AT2, AT3, AT4, AT5, AT7, AT8, AT9, AT10, AT12 for Reduct2; and AT1, AT3, AT5, AT7, AT8, AT9, AT10, AT11, AT12, AT13 for Reduct3, respectively. All the reducts consisted of ten condition attribute and one decision attribute. Reduct1 generated sixty-one certain rules and one approximate rule during validation. It outperformed other results with an accuracy of 84.46 ± 5.24%. The classifier seems to correctly interpret 141 patients as *not diseased* and 115 patients as *diseased* as shown in

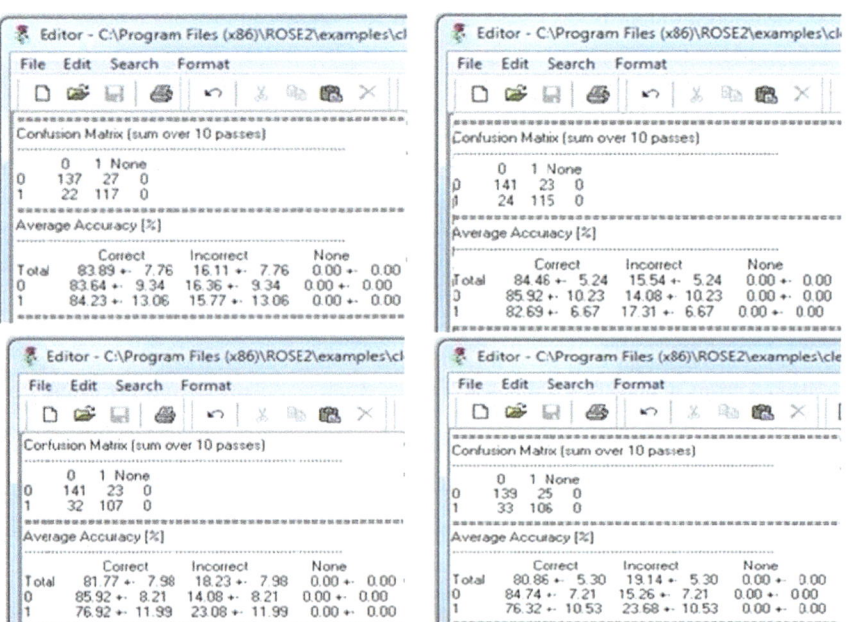

Fig. 2 Classification Results: **a**. using all features—LD, **b**. using Reduct1—LD, **c**. using all features—GD, **d**. using Reduct2 - GD.

Fig. 2b top right. Reduct2 and Reduct3 shows an accuracy of about $82.84 \pm 3.85\%$ and $81.23 \pm 7.30\%$, respectively.

3.2 Results of RSC-2

Classifier using all features: During rule induction, RSC-2 generated fifty-seven certain rule. Classification result shows that for *class* 0, the accuracy is $85.92 \pm 8.21\%$, and for *class* 1, it is $76.92 \pm 11.99\%$. The average accuracy is found to be $81.77 \pm 7.98\%$ (Fig. 2c bottom left).

Classifier using reducts: Fifty-seven reducts has been generated and three has been considered here for study. The attributes sets are AT1, AT4, AT5, AT6, AT7, AT8, AT9, AT11, AT12 for Reduct1; AT1, AT2, AT5, AT7, AT8, AT9, AT10, AT12, AT13 for Reduct2; and AT1, AT2, AT3, AT4, AT8, AT12, AT13 for Reduct3, respectively. Reduct1 gave an accuracy of $76.88 \pm 6.97\%$. Reduct2 gave an accuracy $80.86 \pm 5.30\%$ (Fig. 2d bottom right). Reduct3 gave an accuracy 79.19 ± 9.17.

An in-depth study exhibited that the LD reducts detected the disease more accurately than the GD reducts or the GD data.

4 Conclusion and Future Research

Automation of medical diagnosis process is being approached here. A model has been proposed which will be able to handle missing values, deal with continuous attributed data, and classifying the data which will act as a diagnostic system and simulate the work of a doctor. Rough Set has been found to be an efficient method in preprocessing and classify the data. It is noteworthy here that the records with missing values are preserved here; it is imputed and used for classification. None of the records are deleted. The study reveals that the features such as *age, Chest pain, Serum cholesterol, rest ECG, Thalach, Exang, Old − peak and Ca* are core attributes, and their removal might disturb the quality of classification. These symptoms seem to contribute maximum towards cardiac diseases. They might be used for prediction of cardiac diseases. In this study, LEM2 algorithm (mathematical principles of RSTh) is applied for rule induction and generation for the intelligent analysis of medical data.

It can be seen from the experiments that RSC-1 with Reduct 1 is being able to detect heart disease with an accuracy of $84.46 \pm 5.24\%$ and found to be better than RSC-2 with a maximum accuracy of 81.77 ± 7.98. Hence, the reducts are found to be an important part of this experiment which contributes towards decision-making process. Discretization method acts significant as a wrong method of discretization might lead to loss of important information. The result can be interpreted as the LD data gave better result than GD data. This study is related to heart disease of the human beings, and if a patient goes underdiagnosed, it will be a serious health threat to him/her than being overdiagnosed. Experiments conducted using entropy-based discretization shows that the underdiognosed case in cases of RSC-1 has been reduced much and there lies the significance of the study. The more encouraging fact is that the misclassified patients can be detected from the set-up and can be studied further. The study finally concludes that the expert system developed can be used to detect heart patients in a hospital in alliance with medical practitioners so that it can be improved further. The accuracy of the system can be improved using modified and advanced version of the algorithm used. The principal aim is to develop full-fledged expert system for the real-life use.

Acknowledgements We would like to thank Dr. Szymon Wilk [Asst. Professor, Poznan University of Technology] who provided insight as well as expertise, that greatly assisted the Study.

References

1. Mayo Clinic. http://www.mayoclinic.org/diseases-conditions/heart-disease/basics/definition/con-20034056
2. World Health Organization, Global status report on noncommunicable diseases 2010. http://www.who.int/nmh/publications/ncd_report_full_en.pdf
3. Predki, B., Slowinski, R., Stefanowski, J., Susmaga, R., Wilk, S.: ROSE—software implementation of the rough set theory. In: Polkowski, L., Skowron, A. (eds.) Rough Sets and Current Trends in Computing, LNAI, vol. 1424, pp. 605–608. Springer, Berlin (1998)
4. Predki, B., Wilk, S.: Rough set based data exploration using ROSE system. In: Ras, Z.W., Skowron, A. (eds.) Foundations of Intelligent Systems, LNAI, vol. 1609. pp. 172–180. Springer, Berlin (1999)
5. Grzymala-Busse, J.W., Stefanowski, J.: Three discretization methods for rule induction. Int. J. Intell. Sys. **16**, 29–38 (2001)
6. Fayyad, U.M., Irani, K.B.: Multi-interval discretization of continuous-valued attributes for classification learning. In: Proceedings of the 13th International Conference on Machine Learning, pp. 1022–1027. Morgan Kaufmann, San Francisco (1993)
7. UCI Machine Learning Repository Irvine, CA: University of California, School of Information and Computer Science. http://archive.ics.uci.edu/ml
8. Aha, D., Kibler, D.: Instance-based prediction of heart-disease presence with the Cleveland database. Technical Report, University of California, Irvine, Department of Information and Computer Science (1988)
9. Pawlak, Z.: Rough sets. Int. J. Comput. Info Sci. **11**, 341356 (1982)
10. Pawlak, Z.: Rough Sets: Theoretical Aspects of Reasoning about Data. Kluwer Academic Publishers, Netherlands (1991)
11. Grzymala-Busse, J.W.: LERS—a system for learning from examples based on rough sets. In: Slowinski, R. (ed.) Intelligent Decision Support Handbook of Applications and Advances of the Rough Sets Theory, pp. 3–18. Kluwer Academic Publishers, Dordrecht (1992)
12. Stefanowski, J.: On rough set based approaches to induction of decision rules. In: Polkowski, L., Skowron, A. (eds.) Rough Sets in Data Mining and Knowledge Discovery, pp 500–529. Physica-Verlag (1998)
13. Chan, C.C., Grzymala-Busse, J.W.: On the two local inductive algorithms: PRISM and LEM2. Found. Comput. Decis. Sci. **19**, 185–204 (1994)
14. Grzymala-Busse, J.W., Lakshmanan, A.: LEM2 with interval extension: an induction algorithm for numerical attributes. In: Tsumoto, S. (ed.) Proceedings of the Fourth International Workshop on Rough Sets, Fuzzy Sets and Machine Discovery, pp. 67–73. Tokyo (1996)

Relevant Slicing of Feature-Oriented Programs

Madhusmita Sahu and Durga Prasad Mohapatra

Abstract We propose an approach for computing relevant slices for feature-oriented programs. Our approach is based on an intermediate representation of the program called *dynamic feature dependence graph* (DFDG). We have named our proposed approach *feature relevant slicing* (FRS) algorithm. Our approach first computes the dynamic slice for a test case corresponding to the desired program location. Then, it determines the potential dependency for the desired program location, and eventually, it computes relevant slice using FRS algorithm.

Keywords Relevant slice · Potential dependency · Feature-oriented programming (FOP) · Feature relevant slicing (FRS) algorithm

1 Introduction

Program slicing technique is used for the extraction of specific computation-related program statements. It aids in several software engineering activities like debugging, program verification, and testing. A slicing criterion, consisting of a statement and a variable of concern, is used for performing program slicing. *Relevant slicing* is one technique to prioritize test cases in a test suite in order to achieve maximum coverage of software. If a statement in the relevant slice is changed, then it is necessary to rerun the modified software just by using the test cases whose relevant slices consist of a changed statement. The *feature-oriented programming* (FOP) paradigm identifies functionalities of software in the form of *features*. The basic idea of FOP is to develop the software by the composition of features.

M. Sahu (✉) · D. P. Mohapatra
Department of CSE, National Institute of Technology,
Rourkela 769008, Odisha, India
e-mail: madhu_sahu@yahoo.com

D. P. Mohapatra
e-mail: durga@nitrkl.ac.in

© Springer Nature Singapore Pte Ltd. 2018
K. Saeed et al. (eds.), *Progress in Advanced Computing and Intelligent Engineering*,
Advances in Intelligent Systems and Computing 563,
https://doi.org/10.1007/978-981-10-6872-0_31

The organization of the remainder of the paper has been done as follows. A brief introduction to feature-oriented programming (FOP) and few definitions regarding understanding of our approach are provided in Sect. 2. Section 3 presents our proposed approach on computation of relevant slices for feature-oriented programs. Section 4 discusses few related works. Section 5 furnishes conclusion of the paper along with future work.

2 Basic Concepts

Christian Prehofer [10] was the pioneer to coin the term feature-oriented programming (FOP). FOP is an extension of the object-oriented programming (OOP) paradigm. Subclassing technique is used to support incremental development in OOP. But FOP enables compositional programming to resolve feature interactions. User requirements on the software system are satisfied by *features*, which are the basic building blocks. The details of FOP can be found in [2–4, 10, 11].

Example 1. Consider the program for calculating the gross salary of an employee in an organization. If his basic salary is above or equal to Rs 20000, then DA = 25%, HRA = 20%, and Increment = 8% of basic salary. If his basic salary is above or equal to Rs 10000 and less than Rs 20000, then DA = 20%, HRA = 15%, and Increment = 5% of basic salary. Otherwise, DA = 10%, HRA = 5%, and Increment = 2% of basic salary. Calculate his gross salary.

The various features supported by Example 1 are given in Fig. 1. The Base feature provides the *basic salary*, the DA feature calculates the *DA*, the HRA feature calculates the *HRA*, the Inc feature calculates the *Increment*, and the Gross feature calculates the *gross salary* of an employee. When an employee is provided with any of the DA, HRA, and Increment features, then he must be provided with other two features. Also, the employee may not be provided with any of the DA, HRA, and Increment features. Features are composed from left to right in order. The leftmost feature is top mixin layer, the next feature is the next layer, and so on. The order in which the layers are composed, the execution is performed in reverse order. The Super () method causes the transfer of control to its parent layer. The Jak program for Example 1 is given Fig. 5. The configuration file for composing the features given in Fig. 1 is shown in Fig. 2. The files generated after composition of the features in the order given in Fig. 2 for the program given in Fig. 5 are depicted in Fig. 6.

Fig. 1 Features supported by Example 1

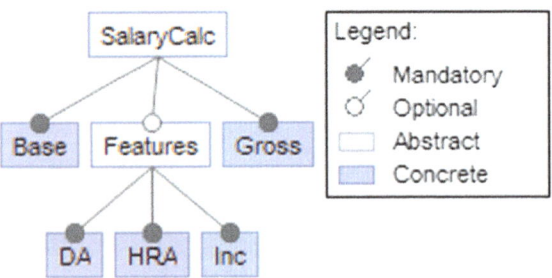

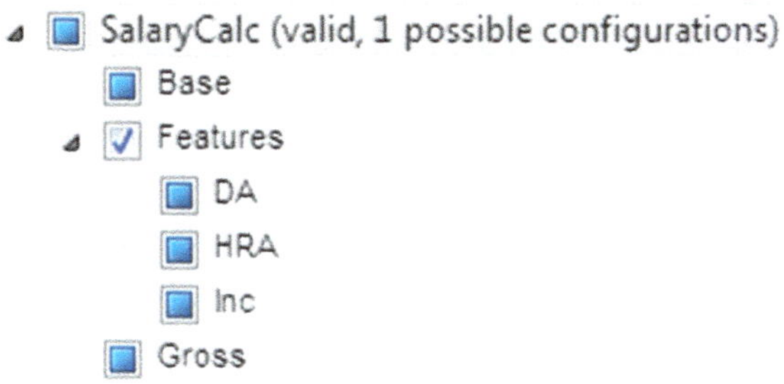

Fig. 2 Features used for composition

Fig. 3 Example program

```
1:   read(a, b, c);
2:   int x=0, y=0, z=0;
3:   x := a + 1;
4:   y := b + 1;
5:   z := c + 1;
6:   int w := 0;
B1:  if x > 3 then
B2:    if z > 4 then
7:       w := w + 1;
     endif
     endif
B3:  if y > 5 then
8:     w := w + 1;
     endif
9:   write(w);
```

Below, we discuss some definitions for understanding of our work. The terms *node*, *vertex,* and *statement* are used interchangeably in this paper. Also, we use the terms *input* and *test case* interchangeably.

Fig. 4 Relevant slice of Fig. 3 with respect to slicing criterion < {a = 1,b = 5, c = 4}, 9, w>

```
1:   read(a, b, c);
2:   int x=0, y=0, z=0;
3:   x := a + 1;
4:   y := b + 1;
5:   z := c + 1;
6:   int w := 0;
B1:  if x > 3 then
B2:    if z > 4 then
7:       w := w + 1;
     endif
     endif
B3:  if y > 5 then
8:     w := w + 1;
     endif
9:   write(w);
```

```
import java.util.Scanner;
public class Sal {
    double gs,bs,sda,shra,sinc;
    void input() {
    Scanner sc=new Scanner(System.in);
    System.out.print("Enter Basic Salary: ");
    bs=sc.nextDouble();
    sc.close();
    }
    void comp() {
    sda=0;
    shra=0;
    sinc=0;
    }
    void output() {
    System.out.println("Basic Salary is: "+bs);
    }
}
```

(a) Base/Sal.jak

```
public refines class Sal {
    void comp(){
    Super().comp();
    if(bs>=20000)
    sda=0.25*bs;
    else if(bs>=10000 && bs<20000)
    sda=0.2*bs;
    else
    sda=0.1*bs;
    }
    void output(){
    Super().output();
    System.out.println("DA is: "+sda);
    }
}
```

(b) DA/Sal.jak

```
public refines class Sal {
    void comp(){
    Super().comp();
    if(bs>=20000)
    shra=0.2*bs;
    else if(bs>=10000 && bs<20000)
    shra=0.15*bs;
    else
    shra=0.05*bs;
    }
    void output(){
    Super().output();
    System.out.println("HRA is: "+shra);
    }
}
```

(c) HRA/Sal.jak

```
public refines class Sal {
    void comp(){
    Super().comp();
    if(bs>=20000)
    sinc=0.08*bs;
    else if(bs>=10000 && bs<20000)
    sinc=0.05*bs;
    else
    sinc=0.02*bs;
    }
    void output(){
    Super().output();
    System.out.println("Increment is: "+sinc);
    }
}
```

(d) Inc/Sal.jak

```
public refines class Sal {
    void comp(){
    Super().comp();
    gs=bs+sda+shra+sinc;
    }
    void output() {
    Super().output();
    System.out.println("Gross Salary is: "+gs);
    }
    public static void main(String[] args) {
        Sal s=new Sal();
        s.input();
        s.comp();
        s.output();
    }
}
```

(e) Gross/Sal.jak

Fig. 5 Jak program to calculate gross salary of an employee for Example 1

Relevant Slice: A *relevant slice* of a program comprises all statements that get executed under a test case affecting the desired program point and all statements that may affect the desired program point, if modified. Thus, a relevant slice contains the dynamic slice and the statements that have the potential to influence the slicing criterion [5]. The relevant slice of Fig. 3 with respect to slicing criterion $< \{a = 1, b = 5, c = 4\}, 9, w >$ is depicted in Fig. 4 as bold text.

Fig. 6 Files generated after composition of features in the order given in Fig. 2 (Sal.java)

```java
import java.util.Scanner;
c1  abstract class Sal$$Basic {
        double gs,bs,sda,shra,sinc;
m2      void input() {
s3      Scanner sc=new Scanner(System.in);
s4      System.out.print("Enter Basic Salary: ");
s5      bs=sc.nextDouble();
s6      sc.close();
        }
m7      void comp() {
s8      sda=0;
s9      shra=0;
s10     sinc=0;
        }
m11     void output() {
s12     System.out.println("Basic Salary is: "+bs);
        }
    }
c13 abstract class Sal$$DA extends  Sal$$Basic {
m14     void comp(){
s15         super.comp();
s16         if(bs>=20000)
s17         sda=0.25*bs;
s18         else if(bs>=10000 && bs<20000)
s19         sda=0.2*bs;
s20         else
s21         sda=0.1*bs;
        }
m22         void output(){
s23         super.output();
s24         System.out.println("DA is: "+sda);
        }
    }
c25 abstract class Sal$$HRA extends  Sal$$DA {
m26     void comp(){
s27     super.comp();
s28     if(bs>=20000)
s29     shra=0.2*bs;
s30     else if(bs>=10000 && bs<20000)
s31     shra=0.15*bs;
s32     else
s33     shra=0.05*bs;
        }
m34     void output(){
s35     super.output();
s36     System.out.println("HRA is: "+shra);
        }
    }
c37 abstract class Sal$$Inc extends  Sal$$HRA {
m38     void comp(){
s39         super.comp();
s40         if(bs>=20000)
s41         sinc=0.08*bs;
s42         else if(bs>=10000 && bs<20000)
s43         sinc=0.05*bs;
s44         else
s45         sinc=0.02*bs;
        }
m46         void output(){
s47         super.output();
s48         System.out.println("Increment is: "+sinc);
        }
    }
c49 public class Sal extends  Sal$$Inc {
m50     void comp(){
s51     super.comp();
s52     gs=bs+sda+shra+sinc;
        }
m53     void output() {
s54     super.output();
s55     System.out.println("Gross Salary is: "+gs);
        }
m56     public static void main(String[] args) {
s57         Sal s=new Sal();
s58         s.input();
s59         s.comp();
s60         s.output();
        }
    }
```

Potential Dependency: Potential dependency between a predicate statement, p, and the desired program point, u, exists if the dynamic slice of u does not contain p, and p has the potential to affect the variable at u upon modification at p. This dependency is represented by a *potential dependence edge* in the dependence graph from the node representing p to the node representing u.

Def(v): A node u in DFDG is said to be a *Def(v)* node if u defines a variable v. For example, in DFDG depicted in Fig. 8, nodes *s17, s19, and s21* are *Def(sda)* nodes.

Use(v): A node u in DFDG is said to be a *Use(v)* node if u utilizes a variable v. For example, in DFDG depicted in Fig. 8, node *s24* is *Use(sda)* vertex.

3   Proposed Approach

This section discusses our proposed approach. Our approach has been named *feature relevant slicing* (FRS) algorithm. Our FRS algorithm first calculates dynamic slice for a slicing criterion. Then, it determines potential dependencies among the predicate statements that are not included in dynamic slice and the desired statement. At last, the potential dependent predicate statements and the statements on which these predicate statements are data dependent are included to the computed dynamic slice to compute the relevant slice. We have adopted the technique of Sahu et al. [11] to construct the dynamic feature dependence graph (DFDG).

Let $dyn\_slice(u)$ denote the dynamic slice using slicing criterion $< i,\ s,\ v >$, where v is the variable at statement s representing vertex u, and i is the input. Let p_1, $p_2, \ldots, p_k$ be the predecessor vertices of u in DFDG. Then, the dynamic slice using slicing criterion $< i,\ s,\ v >$ is given by

$$dyn\_slice(u) = \{u, p_1, p_2, \ldots, p_k\} U\ dyn\_slice(p_1) U\ dyn\_slice(p_2) U \ldots U\ dyn\_slice(p_k).$$

Let $rel\_slice(u)$ denote the relevant slice of a variable v using a slicing criterion $< i,s,v >$ and dynamic slice $dyn\_slice(u)$. Let p be the predicate node on which u is potential dependent and d be the node on which p is data dependent. Then, the relevant slice for u is computed as follows:

$rel\_slice(u) = dyn\_slice(u) U \{\{p,d\}|u$ is potential dependent on p and p is data dependent on $d\}$.

3.1   Feature Relevant Slicing (FRS) Algorithm

We present our proposed FRS algorithm in pseudocode form in Algorithm 1.

Algorithm 1: Feature Relevant Slicing (FRS) Algorithm

Input: Feature-oriented program with selected features, slicing criterion
     consisting of input data, statement number, and variable

Output: Required relevant slice

1. Compose the feature-oriented program with selected features.
2. Execute the composed program for an input specified in the slicing criterion.
3. Store each statement in an execution trace file according to execution order after it gets executed.
4. If loops are present in the program, then

   (a) Store each statement under loop in an execution trace file after every time it gets executed.

5. If a method is invoked more than once, then

   (a) Store each statement of the method in an execution trace file after every time it gets executed.

6. For every statement in the execution trace file, do

   (a) Create a node in DFDG.

7. For every instance of a statement in the execution trace file, do

   (a) Create a separate node in DFDG.

8. For each vertex created in DFDG, do

   (a) If vertex m controls the execution of vertex n, then

      i. Add *control dependence edge* from m to n, $m \rightarrow n$.

   (b) If vertex n utilizes a variable defined at vertex m, then

      i. Add *data dependence edge* from m to n, $m \rightarrow n$.

   (c) If vertex m invokes a method defined at vertex n in same mixin layer, then

      i. Add *method call edge* from m to n, $m \rightarrow n$.

   (d) If vertex n in a mixin layer utilizes value of a variable defined at vertex m or returned from vertex m in another mixin layer, then

      i. Add *mixin data dependence edge* from m to n, $m \rightarrow n$.

   (e) If vertex m in a mixin layer invokes a method defined at vertex n in another mixin layer, then

      i. Add *mixin call edge* from m to n, $m \rightarrow n$.

   (f) If vertex m in child mixin layer invokes a method defined at vertex n in parent mixin layer in response to Super() method at m, then

      i. Add *refinement edge* from m to n, $m \rightarrow n$.

9. For all nodes u before computing relevant slices, do

    (a) Compute $Def(v)$ and $Use(v)$ for all variables v.
    (b) Set $dyn\_slice(u) = \varphi$.
    (c) Set $rel\_slice(u) = \varphi$.

10. Compute the dynamic slice for slicing criterion $< i,s,v >$ as

$$dyn\_slice(u) = \{u, p_1, p_2, \ldots, p_k\} U \ dyn\_slice(p_1) U dyn\_slice(p_2) U \ldots U dyn\_slice(p_k),$$

where vertex u symbolizes statement s, and p_1, p_2, ..., p_k are the predecessor vertices of u.

11. For node u, do

    (a) Find a node $p \ \epsilon \ C$ on which s is control dependent, and p evaluates to *false* during execution.

    //C is set of predicate nodes, and S is set of $Def(v)$ vertices where v is the variable in slicing criterion $< i,s,v >$. C also contains nodes pertaining to *else* statement.

    (b) If p is not executed, then

        i. Find another node $q \ \epsilon \ C$ that controls execution of p and q evaluates to *false* during execution.

    (c) Repeat Step 11b to get a node $r \ \epsilon \ C$ such that r evaluates to *false* during execution.

    (d) If p is an *else* statement and does not get executed, then

        i. Set $r = p$.

    (e) If r does not exist in DFDG, then

        i. Create a node for r in DFDG.

    (f) If r is data dependent on a node k and k does not exist in DFDG, then

        i. Create a node for k.

    (g) For newly created nodes, do

        i. Add required dependence edges using Step 8.

    (h) If node u is data dependent on a node $z \ \epsilon \ C$ and z is not included in $dyn\_slice(u)$, then

        i. Set $t = u$.
        ii. Add *potential edge* from r to t, $r \rightarrow t$.

12. For a slicing criterion $< i,s,v >$, do

(a) Compute $rel\_slice(u) = dyn\_slice(u)$ U $\{\{p,d\}|u$ is potential dependent on predicate node p and p is data dependent on node $d\}$, where u symbolizes statement s.

3.2 Working of FRS Algorithm

Now, we explain the working of FRS algorithm by employing an example. Table 1 depicts test cases for the program provided in Fig. 6.

Suppose, the program given Fig. 6 is executed test case T2. The statements executed under T2 are $m56, s57, s58, m2, s3, s4, s5, s6, s59, m50, s51, m38, s39,$ $m26, s27, m14, s15, m7, s8, s9, s10, s16, s18, s19, s28, s30, s40, s42, s43, s52, s60,$ $m53, s54, m46, s47, m34, s35, m22, s23, m11, s12, s24, s36, s48,$ and $s55$. These executed statements are kept in an execution trace file depicted in Fig. 7.

The DFDG of program given in Fig. 6 using execution trace shown in Fig. 7 is shown in Fig. 8.

Now, for our slicing criterion $< \{bs = 15000\}, s24, sda >$, the dynamic slice is computed as

$$dyn\_slice(s22) = \{s22, s19, m22\} U\ dyn\_slice(s19) U\ dyn\_slice(m22).$$

The above expression is evaluated recursively to yield the dynamic slice at statement $s24$. The dynamic slice contains the statements corresponding to the following set of nodes:

$\{m2, s3, s5, m14, s16, s18, s19, m22, s24, m26, s27, m34, s35, m38, s39, m46,$ $s47, m50, s51, m53, s54, m56, s57, s58, s59, s60\}$.

This is also depicted as bold vertices in Fig. 9.

Using FRS algorithm, we can see that the set C contains the statements $s16, s18,$ $s20, s28, s30, s32, s40, s42,$ and $s44$.

$$Thus\ C = \{s16, s18, s20, s28, s30, s32, s40, s42, s44\}.$$
$$Def\,(sda) = \{s17, s19, s21\}\ and\ \ Use(sda) = \{s24\}.$$
$$Thus, S = \{s17,\ s19,\ s21\}.$$

Table 1 Test cases for the program given in Fig. 6	Test Case	Input	Output			
		bs	da	hra	inc	gs
	T1	22000	5500	4400	1760	33660
	T2	15000	3000	2250	750	21000
	T3	8000	800	400	160	9360

```
m56(1): public static void main(String[] args) {
s57(2): Sal s=new Sal();
s58(3): s.input();
m2(4): void input() {
s3(5): Scanner sc=new Scanner(System.in);
s4(6): System.out.print("Enter Basic Salary: ");
s5(7): bs=sc.nextDouble();
s6(8): sc.close();
s59(9): s.comp();
m50(10): void comp(){
s51(11): super.comp();
m38(12): void comp(){
s39(13): super.comp();
m26(14): void comp(){
s27(15): super.comp();
m14(16): void comp(){
s15(17): super.comp();
m7(18): void comp() {
s8(19): sda=0;
s9(20): shra=0;
s10(21): sinc=0;
s16(22): if(bs>=20000)
s18(23): else if(bs>=10000 && bs<20000)
s19(24): sda=0.2*bs;
s28(25): if(bs>=20000)
s30(26): else if(bs>=10000 && bs<20000)
s31(27): shra=0.15*bs;
s40(28): if(bs>=20000)
s42(29): else if(bs>=10000 && bs<20000)
s43(30): sinc=0.05*bs;
s52(31): gs=bs+sda+shra+sinc;
s60(32): s.output();
m53(33): void output() {
s54(34): super.output();
m46(35): void output(){
s47(36): super.output();
m34(37): void output(){
s35(38): super.output();
m22(39): void output(){
s23(40): super.output();
m11(41): void output() {
s12(42): System.out.println("Basic Salary is: "+bs);
s24(43): System.out.println("DA is: "+sda);
s36(44): System.out.println("HRA is: "+shra);
s48(45): System.out.println("Increment is: "+sinc);
s52(46): System.out.println("Gross Salary is: "+gs);
```

Fig. 7 Execution trace for test case T2 in Table 1

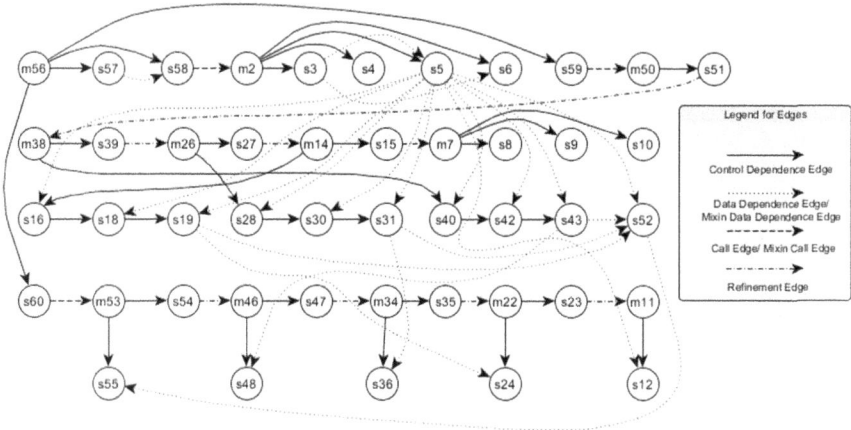

Fig. 8 Dynamic feature dependence graph (DFDG) of the example program given in Fig. 6

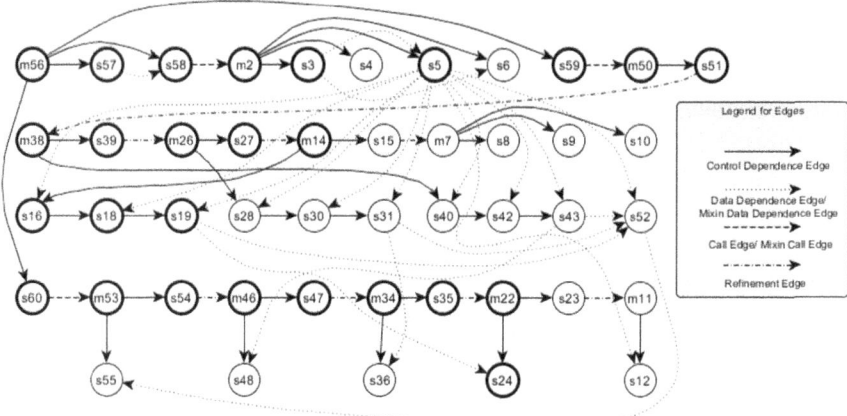

Fig. 9 Dynamic slice using slicing criterion < *{bs = 15000}, s24, sda*>

Node *s24* is data dependent on nodes *s17*, *s19*, and *s21* as these nodes are in *Def (sda)*. Nodes *s17* and *s21* are not included in *dyn_slice(s24)*. Nodes *s17* and *s21* are control dependent on nodes *s16* and *s20*, respectively. Node *s16* is evaluated to *false*, and node *s20* is not executed during the execution of test case T2.

Thus, $p = \{s16, s20\}$ and in turn, $r = \{s16, s20\}$.

Nodes *s16* and *s20* are data dependent on node *s5*, and node *s5* is already in *dyn_slice(s24)*.

Since node *s24* is data dependent on nodes *s17* and *s21* and these nodes are in Set *S* and are not included in *dyn_slice(s24)*, we get $t = s24$.

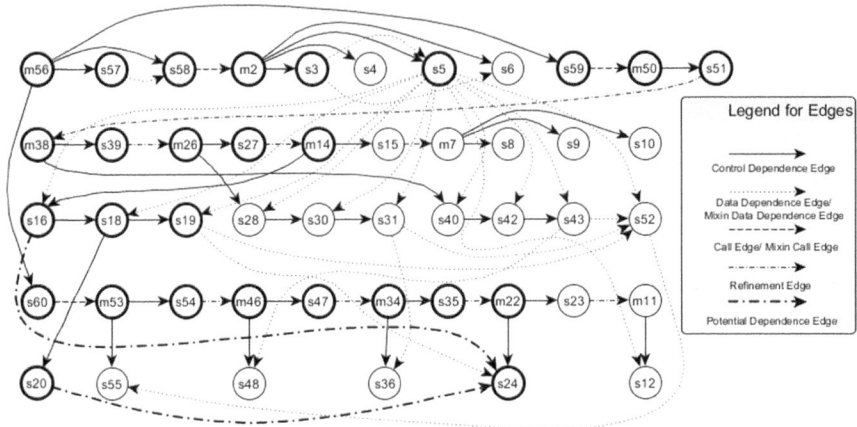

Fig. 10 Relevant slice using slicing criterion < *{bs = 15000}, s24, sda*>

Nodes *s16* and *s20* are to be created in DFDG. Since DFDG contains node *s16*, it is not required to create node *s16*. Only node *s20* is created in DFDG. Potential edges are added from node *s16* to node *s24* and from node *s20* to node *s24*.

The relevant slice using slicing criterion < *{bs = 15000}, s24, sda* > is computed as follows:

$$rel\_slice(s24) = dyn\_slice(s24)U\{s16, s20\}.$$

Evaluation of this expression results in the following set of nodes:
{m2, s3, s5, m14, s16, s18, s19, s20, m22, s24, m26, s27, m34, s35, m38, s39, m46, s47, m50, s51, m53, s54, m56, s57, s58, s59, s60}.

The statements corresponding to these vertices form the desired relevant slice. The relevant slice is depicted as bold vertices in Fig. 10.

4 Comparison with Related Work

Korel et al. [9] first proposed the notion of relevant slice. They used the rationale that a faulty program component produced an incorrect result for a correct input to localize faults. A forward global method for computing relevant slices was developed by Gyimothy et al. [6] using a *program dependence graph*. Jeffrey et al. [7] suggested an approach for the prioritization of test cases using coverage requirements.

Apel et al. [2] introduced a language for FOP in C++ namely FeatureC++ and discussed few problems of FOP languages in implementing program families. Apel et al. [3] discussed FeatureC++ along with its adaptation to aspect-oriented programming (AOP) concepts. Apel et al. [1] discussed drawbacks of crosscutting modularity and the inability of C++ to support it. They also focused on solutions ease evolvability of software. Batory [4] presented basic concepts of FOP and a subset of the tools of the *Algebraic Hierarchical Equations for Application Design* (AHEAD) tool suite. Sahu et al. [11] suggested a technique to compute dynamic slices for feature-oriented programs. Their technique first composed the selected features of feature-oriented programs. Then, they used an execution trace file and a dependence-based program representation, namely *dynamic feature-oriented dependence graph* (DFDG). The dynamic slice was computed by traversing DFDG in breadth-first and depth-first manner and mapping the traversed vertices to the program statements.

All these works [6–9] are based on computing relevant slices for intraprocedural programs. They cannot handle feature-oriented programs. We have computed relevant slice for feature-oriented programs.

5   Conclusion and Future Work

We have presented an approach to compute relevant slices for feature-oriented programs. Initially, dynamic slice for a test case corresponding to the desired program location is computed. Then, potential dependency for that desired program location is determined, and eventually, relevant slice is computed using FRS algorithm. The implementation of our approach is in process. We shall also apply our approach to some other feature-oriented languages like Fuji, FeatureC++, and FeatureHouse. In future, we shall try to extend our work to calculate relevant slices for object-oriented programs, aspect-oriented programs, etc.

References

1. Apel, S., Leich, T., Rosenmuller, M., Saake, G.: Combining feature-oriented and aspect-oriented programming to support software evolution. In: Proceedings of the 2nd ECOOP Workshop on Reflection, AOP and MetaData for Software Evolution (RAM-SE), pp. 3–16. School of Computer Science, University of Magdeburg (July 2005)
2. Apel, S., Leich, T., Rosenmuller, M., Saake, G.: FeatureC++: feature-oriented and aspect-oriented programming in C++. Technical report (2005)
3. Apel, S., Leich, T., Rosenmuller, M., Saake, G.: FeatureC++: on the symbiosis of feature-oriented and aspect-oriented programming. In: Proceedings of the International Conference on Generative Programming and Component Engineering (GPCE'05), pp. 125–140. Springer (2005)
4. Batory, D.: A tutorial on feature oriented programming and the AHEAD Tool Suite. In: Proceedings of the 2005 International Conference on Generative and Transformational Techniques in Software Engineering (GTTSE'05), pp. 3–35. Berlin, Heidelberg (2006)
5. Chauhan, N.: Software Testing: Principles and Practices, 3rd edn. Oxford University Press (2012)

6. Gyimothy, T., Beszedes, A., Forgacs, I.: An efficient relevant slicing method for debugging. ACM SIGSOFT Softw. Eng. Notes **24**(6), 303–321 (1999)
7. Jeffrey, D., Gupta, N.: Test case prioritization using relevant slices. In: Proceedings of 30th Annual International Computer Software and Applications Conference (COMPSAC'06), pp. 411–420. IEEE, 17–21 September 2006
8. Jeffrey, D., Gupta, N.: Experiments with test case prioritization using relevant slices. J. Syst. Softw. **81**, 196–221 (2008)
9. Korel, B., Laski, J.: Algorithmic software fault localization. In: Proceedings of the Twenty-Fourth Annual Hawaii International Conference on System Sciences, pp. 246–252 (1991)
10. Prehofer, C.: Feature-oriented programming: a fresh look at objects. In: Proceedings of 11th European Conference on Object-Oriented Programming (ECOOP), pp. 419–443 (1997)
11. Sahu, M., Mohapatra, DP.: Dynamic slicing of feature-oriented programs. In: Proceedings of 3rd International Conference on Advanced Computing, Networking and Informatics (ICACNI 2015), pp. 381–388. Bhubaneswar (2015)

Efficient Tuning of COCOMO Model Cost Drivers Through Generalized Reduced Gradient (GRG) Nonlinear Optimization with Best-Fit Analysis

Surjeet Dalal, Neeraj Dahiya and Vivek Jaglan

Abstract The software effort estimation phase is especially critical in the software development phase. This phase is principally oriented on manipulation of the values of the cost drivers and scale factors. Also, most of the models depend on the size of the project, and a diminutive alteration in the size directs to the in proportion alterations in the effort. Miscalculations of the cost drivers have even additional ear-splitting data as a result too. In this paper, the approach of generalized reduced gradient (GRG) nonlinear optimization with best-fit analysis has been applied to tune the COCOMO model cost drivers so that level of accuracy can be achieved. This proposed methodology has been observed more efficiently in providing the software effort estimation through the help of minimizing MRE value. We have applied this methodology on NASA 63 data sets. We have shown the comparison between the estimated MRE and actual MRE of the data sets. We have also exposed the evaluation between the estimated MMRE and actual MMRE.

Keywords Nonlinear optimization · Best-fit analysis · Software cost estimation MRE

1 Introduction

In the year of 1981, Dr. Berry Boehm had proposed the constructive cost estimation model (COCOMO) which is being used in various aspects of the software effort estimation. It is a process for estimating the expenditure of the whole software development process. In accordance with Dr. Berry Boehm, the COCOMO model can be applied in the software cost estimation process through the following phases as given below:

S. Dalal (✉) · N. Dahiya
Department of Computer Science & Engineering, SRM University, Sonipat, Haryana, India
e-mail: profsurjeetdalal@gmail.com

V. Jaglan
Department of Computer Science & Engineering, Amity University, Manesar, Haryana, India

© Springer Nature Singapore Pte Ltd. 2018
K. Saeed et al. (eds.), *Progress in Advanced Computing and Intelligent Engineering*,
Advances in Intelligent Systems and Computing 563,
https://doi.org/10.1007/978-981-10-6872-0_32

- Basic COCOMO model for small-size projects
- Intermediate COCOMO model for medium-size projects
- Detailed COCOMO model for large-size projects

This model estimates the effort cost with the help of two main equations. The first equation for calculating the development effort has been given below:

$$\text{Development Effort: ManOfMonth} = a * \text{KDSI} * b \tag{1}$$

It is measured in the terms of ManOfMonth unit which indicates the total work or effort done by one individual person. In this model, ManOfMonth consists of 152 h per person-month. The value of ManOfMonth may diverge from the typical by 10–20 %. Subsequently, TDEV will be calculated with the help of the following equation as given below:

$$\text{Efforts and Development Time (TDEV): TDEV} = 2.5 * \text{ManOfMonth} * c \tag{2}$$

The value a, b and c define levels of the development [1]. The basic COCOMO estimation model is given by the subsequent terminologies:

$$\text{Effort} = a_1 * (\text{KLOC}) * a_2 * \text{PM} \tag{3}$$

$$\text{Tdev} = b1 * (\text{Effort}) * b2 * \text{Months} \tag{4}$$

The effort evaluation is articulated in units of person-months (PM). The COCOMO model has the following advantages as given below [2–6]:

- This model seems translucent that means that individual can check the whole working that is not found in other models like SLIM
- In this model, the cost drivers are particularly accommodating to the estimator to realize the impact of different factors that involve project costs.

Although this model has a lot of advantages, but it has the following disadvantages as given below:

- This model is applied to guesstimate the cost and timetable of the project, opening from the design point and till the ending of assimilation segment. For the enduring segments, a split evaluation model should be used.
- This model is not an ideal pragmatic model. The suppositions made at the establishment may differ as time progress in the process of building the project.
- For the purpose of calculating the cost with revised parameters, fresh calculations of estimating effort are being required. Past calculation cannot be used in the case of revising the cost drivers and attributes. It is a time-consuming process.
- This model works on the principle of fixed and constant user requirements during the software development life cycle; any alteration in the requirements is not accommodating for computation of cost of the project.

- There is meaning differences between the various categories of this model, for example, basic and intermediate mode [7].
- This model is not appropriate for non-sequential, speedy growth, engineering, reuse case models [8].

Hence, we can say that the concert of the COCOMO model can be superior with competent tuning of its cost drivers.

2 Related Work

Helm Jim [9] correlated actual data with COCOMO estimated values and determined if the COCOMO method accurately reflected documented program expenditures. Because space-borne micro processing was a relatively new arena, the primary constraint associated with developing a model was the limited available database. It supported a statistical analysis which is presented along with a discussion on calculated COCOMO results. In the analyses, the use of nonparametric statistics for small samples was addressed. Wilcox on signed-rank and Kendall-rank statistics supported distribution-free analyses of the data.

Leung and Fan [10] provided a wide-ranging summary of software cost estimation methods as well as the current advances in the field. As a quantity of these models relied on a software size approximation as input, they first offered an outline of common size metrics. They then highlighted the cost estimation models that had been projected and used effectively.

Ismae [11] discussed the use of constructive cost model (COCOMO II) to estimate the cost of software engineering. This model allowed us to calculate approximately the cost, endeavour and scheduling and planning innovative software development.

Rollo [12] proposed a substitute use of the COCOMO model to help in the task of evaluation. The commonly established technique of valuation using a functional sizing technique was to be found the approximation on preceding project data, where those projects for a homogeneous set with the project under learning. The principal complicatedness was to uncover an adequately standardized set of projects.

Živadinović et al. [13] presented the most relevant methods and models for effort estimation used by software engineers in the past four decades.

Al Qmase and Qureshi [14] focused on the lead of the COCOMO model. It supplementary consisted of its two subordinate models called COCOMO I and II. The findings to date showed that COCOMO II was supplementary precise in decisive time and cost for the triumphant termination of a software project than other models for an analogous appliance for case in point task manager.

3 COCOMO Cost Drivers

The early estimation made in the COCOMO II model is attained by means of a set of attributes (cost drivers) that replicate following facts as given below [15–18]:

- Product distinctiveness such as the indispensable system consistency and product involvedness.
- Computer distinctiveness such as implementation time or memory constrictions.
- Personnel uniqueness such as programming language dexterity, those take the familiarities and the capabilities of the people functioning of the project into account.
- Project uniqueness of the software development, such as the IDE that is obtainable and the growth agenda.

These drivers have critical impacts in the effort estimation phase and define the whole software development process. Each scale driver has a variety of evaluation levels at extra high, very high, high, nominal, low and very low. Each ranking level has a weight, W, and the precise significance of the weight is called a scale factor.

4 Generalized Reduced Gradient (GRG) Nonlinear Optimization

Nonlinear Optimization is being identified as nonlinear programming. It has confirmed itself as a constructive method to diminish costs and to maintain other intentions, particularly in the refinery diligence. Nonlinear programming knocks on the companies' doors. Mixed integer nonlinear optimization, an area under persistent enlargement, is now establishing itself in many branches, e.g. the progression diligence or economic services, and it certainly has much to offer for the future.

GRG algorithm is being used for solving general optimization problems with nonlinear constrictions and/or a nonlinear intention. It is the most widely used program of its type with a long record of successful applications in numerous miscellaneous areas. GRG is used by universities and conglomerate worldwide and is the essential statistical system for numerous dedicated optimization products.

The most important initiative in GRG method is that beneath various circumstances, it is potential to integrate impartiality constrictions unswervingly into the intended meaning, by this means tumbling the dimensionality of the sculpt and converting it to an unrestrained form. Hence the predicament has abridged dimensioned and can be solved by the gradient method. It is easiest to integrate egalitarianism constraints into the intended purpose if they are linear. The ornamented items are enduring constants at the prearranged position [19]. The accurateness of the rough calculation declines as the user stirs beyond absent from where the ascent of the utility is estimated. If the curvature is exceedingly nonlinear, then the accurateness of the rough calculation disgraces still earlier.

4.1 Steps in GRG Algorithm

The main purpose of GRG algorithm is to solve the optimization problem. This algorithm is capable of optimizing the problem by solving its sub-problems consisting of nonlinear constraints. The basic steps of this algorithm are given below:

- Initially select a preliminary point A, value of tolerances and other parameters.
- Apply the gradient method for finding the updated value of A1 based on linearly estimated value of the nonlinear constraints around the initial value of A.
- Calculate the difference of these values of A and A1. If the difference is diminutive sufficient, then finalize the updated value as the solution.
- Now the value of preliminary point change, i.e. A1 ← A.
- Go to Step 2.

The primary component is to renovate all discrepancy constraints into disinterest constraints by the adding of apposite slack and spare variables. The second part is a diminutive harder. As the algorithm proceeds, it needs to take a decision in which inequalities are bouncy. There exists the heuristic vigorous set strategy for making this decision, based on, e.g. how close a constriction is to be fulfilled at the present point.

5 Proposed Cost Drivers Tuning Methodology

The proposed methodology is oriented with generalized reduced gradient (GRG2) nonlinear optimization with best-fit analysis approach. This proposed methodology advocates the integration of generalized reduced gradient (GRG) nonlinear optimization with best-fit analysis for efficient tuning of the cost drivers of COCOMO model. The proposed methodology is oriented to achieve the following goal given below:

Minimize the function of calculating MRE of given software effort estimation parameters:

$$f(15 \, Cost \, Drivers, LOC, a, b, Actual \, Effort) = (Estimated \, Effort \sim ActualEffort)/Actual \, Effort$$

$$(5)$$

where

$$EAF = C1 * C2 * C3 * \, ... * C15$$
$$\text{Estimated effort} = a * LOC * b * EAF$$

And constraints C1 (RELY) >=0.75 & C1 (RELY) <=1.4 as so on..............
For all other remaining 14 cost drivers, the constraints are being defined. The proposed GRG2 nonlinear optimization generates the minimized value of the above

function of calculating the magnitude of relative error (MRE) value by calculating the optimized values of 15 cost drivers, and best-fit analysis helps to find these optimized values to their nearby standard values.

6   Results

The measurement of accuracy of software effort estimation defines how close the estimated result is with its actual value. Software cost estimates play significant role in delivering software projects. Hence, this proposed methodology has been oriented on the concept of the mean magnitude of relative error (MMRE), to evaluate the opulence of calculation systems. Comparisons can be made across data sets and prediction model types. The proposed model is validated by NASA 63 project data sets. These are one of the most analysed data sets. The GRG algorithm is being applied with aiming minimization of the MRE function. These data sets include 63 historical projects with 15 cost drivers of the software development effort. The proposed methodology has been implemented with the help of MATLAB.

6.1   Data Analysis

The significance of 15 cost drivers can be shown by their impact on MMRE of efforts on original 63 NASA data sets (Fig. 1; Table 1).

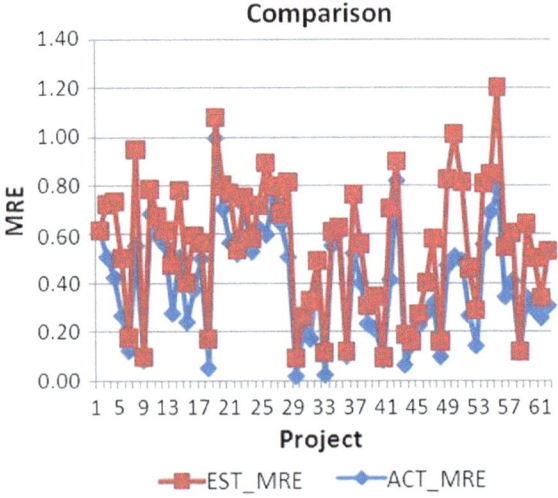

Fig. 1 Estimated MRE versus actual MRE of NASA 63 data sets

Table 1 MMRE of NASA 63 projects for various project modes

Number of projects	MMRE for proposed method	MMRE for COCOMO
63	0.15	0.40

7 Conclusion

The work carried out in the paper is to minimize the errors during the software estimation process using the COCOMO model. The proposed algorithm is oriented on the generalized reduced gradient (GRG) nonlinear optimization with best-fit analysis capability. The above-mentioned results express that applying the proposed technique in the software effort estimation process minimizes the value of MMRE value from 0.40 to 0.15. It has improved the accuracy level of the software estimation process using the COCOMO model. Tuning of COCOMO cost drivers has been done with the help of modified GRG nonlinear optimization. The order of occurrence of various cost drivers has a significant impact on the overall efforts in project estimation. The employment of the proposed algorithm for other appliance in various engineering fields can also be investigated in the future.

References

1. Furulund, K.M., Moløkken-Østvold, K.: Increasing software effort estimation accuracy—using experience data, estimation models and checklists. In: 7th International Conference on Quality Software (QSIC'07), 342–347 (2007)
2. Molokken, K., Jorgensen M.: A review of software surveys on software effort estimation. In: International Symposium on Empirical Software Engineering (ISESE'03), 220–230 (2003)
3. Ferrucci, F., Gravino C., Oliveto R., and Sarro F.: Genetic programming for effort estimation: an analysis of the impact of different fitness functions. In: 2nd International Symposium on Search Based Software Engineering (SSBSE'10), 89–98 (2010)
4. Sheta, A.F.: Estimation of the COCOMO model parameters using genetic algorithms for NASA software projects. J. Comput. Sci. 2(2), 118–123 (2006)
5. Boehm, B.W.: Software Engineering Economics. Prentice Hall, IEEE (1984)
6. Magne, J., Shepperd, M.: A systematic review of software development cost estimation studies. IEEE Trans. Softw. Eng. 33(1), 33–53 (2012)
7. Braga P.L., Oliveira A.L.I., Meira S.R.L.: A GA-based feature selection and parameters optimization for support vector regression applied to software effort estimation, In: 23rd Annual ACM Symposium on Applied Computing (SAC'08), 1788–1792 (2008)
8. Harman, M., Jones, B.F.: Search-based software engineering. Inf. Softw. Technol. 43(14), 833–839 (2012)
9. Helm Jim, E.: The viability of using COCOMO in the special application software bidding and estimating. IEEE Trans. Eng. Manage. 39(1), 42–58 (1992)
10. Leung, H., Fan, Z.: Software cost estimation. In: Handbook of Software Engineering And Knowledge Engineering. World Scientific Pub. Co, River Edge, NJ (2004)
11. Ismaeel, H.R.: Software engineering cost estimation using COCOMO II model. Al-Mansour J. (10), 86–111 (2007)

12. Rollo, A.L.: Functional size measurement and COCOMO – A synergistic approach. In: Proceedings of the IEEE International Conference on Industrial Engineering and Engineering Management (IEEM '07), pp. 1256–1260. Singapore, December 2009

13. Živadinović J., et al.: Methods of effort estimation in software engineering. In: International Symposium Engineering Management and Competitiveness 2011 (EMC2011), June 24–25, (2011)

14. Al Qmase, M.M., Qureshi, M.R.J.: Evaluation of the cost estimation models: case study of task manager application. Int. J. Mod. Educ. Comput. Sci. 5(8),1–7 (2013)

15. Clarke, J., Dolado, J.J., Harman, M.: Reformulating software engineering as a search problem. IEE Proc. Softw. **150**(3), 161–175 (2003)

16. Jørgensen, M., Grimstad, S.: Avoiding irrelevant and misleading information when estimating development effort. IEEE Softw. **25**(3), 78–83 (2008)

17. Lederer, A.L., Prasad, J.: A causal model for software cost estimating error. IEEE Trans. Softw. Eng. **24**(2), 137–148 (1998)

18. Basha, S., Dhavachelvan, P.: Analysis of empirical software effort estimation models. Int. J. Comput. Sci. Inf. Secur. **7**(3), 68–77 (2010)

19. Huang, X., Capretz, L.F., Ren, J.: A neuro-fuzzy model for software cost estimation. In: Third International Conference On Quality Software (QSIC'03), 12–19 (2003)

A Smarter Approach for Better Lifestyle in Indian Societies

Agam Agarwal, Divyansh Goel, Ankit Tyagi, Anshuli Aggarwal
and Rohit Rastogi

Abstract This research paper deals in digitally solving all the general problems of residents living in Indian apartments. It ensures the high standard of facilities which they get in their apartments. The application proposed in this paper will make the society more luxurious and standardized. The modules like REAP for praising the service providers, voting system for the selection process of members of the society, and community engagement module for spreading information/notice among all the residents will make the lifestyle of the people smarter and easy going. It ensures high comfort for the residents and promotes digitalization in the society.

Keywords REAP · Society curriculum · Service provider · Park alert

1 Introduction

Apartments have become the first priority for the majority of new home buyers in metropolitan cities. Services such as plumber and electrician are available on call which will help to reduce the hassle of maintaining the home. Moreover, in Indian apartments, there is availability of convenience stores with apartment complex which helps you to shop for basic essential/groceries easily.

A. Agarwal (✉) · D. Goel · A. Tyagi · A. Aggarwal · R. Rastogi
Department of Computer Science and Engineering,
ABES Engineering College Ghaziabad, Ghaziabad, India
e-mail: agamaggarwal11@gmail.com

D. Goel
e-mail: divugoel@gmail.com

A. Tyagi
e-mail: ankittyaagi@gmail.com

A. Aggarwal
e-mail: anshuliaggarwal@yahoo.in

R. Rastogi
e-mail: rohit.rastogi@abes.ac.in

© Springer Nature Singapore Pte Ltd. 2018
K. Saeed et al. (eds.), *Progress in Advanced Computing and Intelligent Engineering*,
Advances in Intelligent Systems and Computing 563,
https://doi.org/10.1007/978-981-10-6872-0_33

In this paper, we present an application through which you can reward or praise any service provider or vendor which will help in the acknowledgment of his work. This application sorts and handles all types of problems and issues that came in front of the residents of the society like purchasing of goods from the market and being an active member in society curriculum. With this application, now the solution of all the problems lies under a single roof and the resident need not to go elsewhere to rectify his/her problem.

2 Literature Survey

Some of the existing approaches related to this work have been there which focuses on reserving your parking area likewise

- [1] Divyansh Goel, Agam Agarwal, and Rohit Rastogi presented in their paper "A Novel approach for Residential," a smarter approach to sort out parking-related problems and to give better lifestyle and smarter solution to avail the society maintenance services.
- [2] Lalitha Ayer; Manali Tare; Renu Yadav; Hetal Amrutia in their paper "Android Application for Vehicle Parking System: "Park Me" presented a mobile application to examine the number of free slots for parking in an area. In their application, driver can pre-book a parking slot and administrator allocates the slots to the users in the queue.
- [3] Soumya Banerjee; Hameed Al-Qaheri proposed a paper "An Intelligent Hybrid scheme for optimizing parking space: A tabu metaphor and rough set based approach." In this, they solve the parking area problem using a search technique, i.e., tabu search assisted by rough set.
- [4] Akmarina Izza; Mokeri stated a paper "Smart Parking via RFID Tag" in which they used the RFID tag to alert the driver from wrong parking and vacant parking space along with the indication of the vehicle location.
- [5] Ramneet Kaur and Balwinder Singh presented "Design and Implementation of Car parking system on FPGA" which includes two basic modules modeled in HDL, one to identify the visitor and another for checking the slot status. These modules are implemented on FPGA.
- [6] Kun-Chan Lan; Wen-Yuah Shih in their paper "An intelligent driver location system for smart parking" presented a phone-based system which will track the driver's path to know when the parking slot can get free.

3 Our Approach

Overcoming the limitations and drawbacks of above papers, we come up with the smart and digital solution to all the problems under a single roof. This application includes these modules:

- ParkAlert with Calling
- REAP
- Voting System
- Community Engagement

3.1 ParkAlert with Calling

We have added the calling feature in the ParkAlert module which is given in [1]. Previously, only message is sent to the problem-creating user's mobile number but now if the user fails to come within 15 min of sending the message, then an automated call button will be generated and the affected user can directly call the user whose vehicle is creating the problem. If the user fails to come within 30 min even after calling, then a ticket will be generated against the problem-creating user and he/she has to pay some amount of penalty for the same.

3.2 REAP (Reward, Earn, Acknowledge, and Praise)

In this module, residents of the society who are availing any kind of service can acknowledge or praise the service providers and vendors from whom they have purchased the goods by giving stars (gold, silver, and bronze), which will decide the overall ranking of all of them, and according to that, their pay scale will be decided. If there are a large number of negative reviews for a particular service provider or vendor, then they will be blacklisted from the society and further action will be taken accordingly.

3.3 Voting System

This module deals with the voting regarding any hot topic or issue raised in the society. For this, admin will post the issue which will be shown in all the profiles of the residents and residents can give their points in favor or against the issue or topic.

This module also deals with the selection procedure of the various members like president, society maintenance in-charge.

3.4 Community Engagement

This module deals with spreading information regarding any notice or upcoming activity in the society so that residents can participate in the events and can know about the notices and information. This module also promotes digitalization as notices and information are now sent directly to the profiles of the residents, and there is no paperwork involved in the process.

3.5 Extra Services

This consists of all the other services that are required for the better lifestyle in the society like society maintenance bill payment, chatting with the other residents, shopping from the nearby stores. With this, the resident need not to explore anything elsewhere. All the services and solution to their problems lie in the application itself.

4 Implementation

4.1 Reap

For this, the resident has to search the name of the service provider or vendor in the search field, and the corresponding results will be shown from the database and he/she can simply choose among different stars and can give the suitable karma and reason for that. Every star has some points according to which the ranking will be decided and the pay scale of the service provider will decide accordingly. Admin of the society can revoke the acknowledgment and karma if he finds something mischief or wrong (Fig. 1).

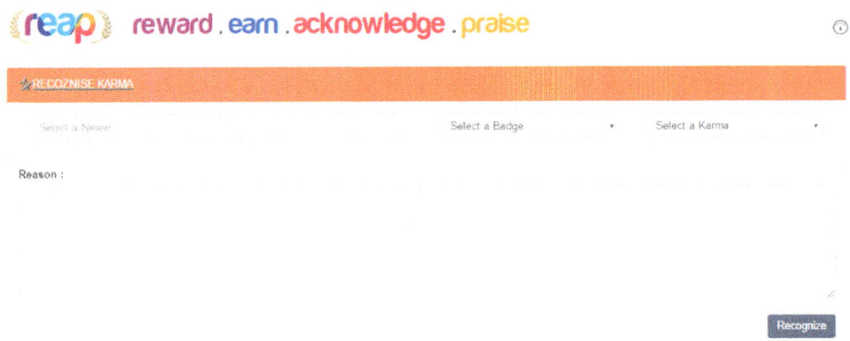

Fig. 1 Reward page in resident's profile

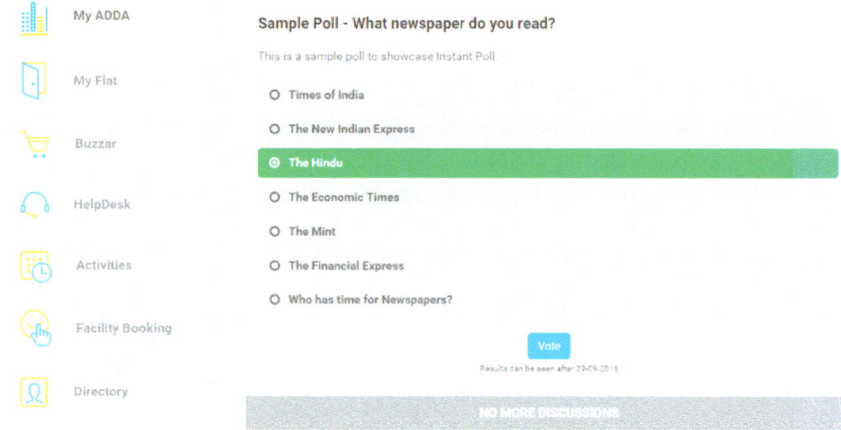

Fig. 2 Voting system for the selection procedure for any issue/problem

4.2 Voting System

This module works as a social networking page, and the latest hot topic will be shown as a newsfeed in the resident's profile from where he/she can give their suggestions and their viewpoint and their opinion whether they are in favor or against the issue.

This module also helps in the selection procedure of the president and other important members of the society. Admin of the society will post the list of the eligible candidates in the residents' profile, and from there, residents can vote for the desired candidate (Fig. 2).

4.3 Community Engagement

This module is only accessible by the admin of the society. If admin wants to share any information, then he can simply broadcast that message and it will be shown in the profile of all the residents from which they will get updated.

This is divided into two parts

- **Transactional Message:**

If admin wants to send information or notice to a particular group of people, then he can limit the message to those only so that others will not get that information.

- **Promotional Message (Broadcast):**

If any notice or information is to be shared among all the residents of the society, then with the help of this module it can be done in a single click and that information will be shown in profiles of residents (Fig. 3).

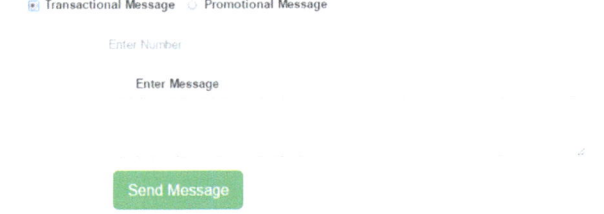

Fig. 3 Messaging module for spreading information/notice among the residents

5 Result and Analysis

After analyzing the previous results of the application and considering the new results after adding new functionalities, we come up with the graph showing the boom and increasing the popularity of our application in Pratap Vihar Society, Ghaziabad.

The application shows the usefulness and correct implementation of all the jobs that are done in the society which increases standard of living of the residents (Fig. 4).

5.1 Benefits of Our Approach

- It ensures high quality of the service provided by the service providers and vendors in the society.

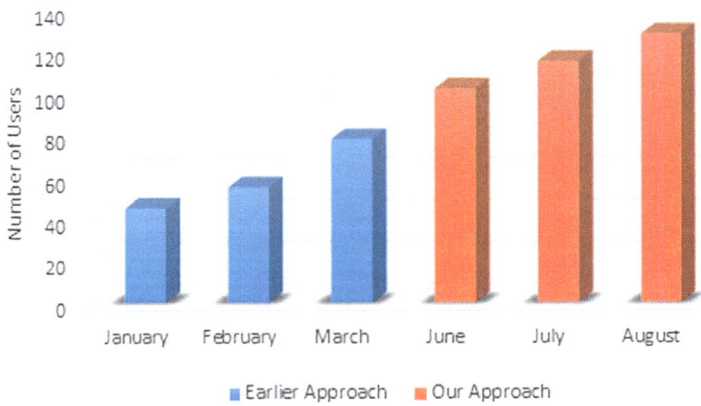

Fig. 4 Analysis and result in Pratap Vihar Society, Ghaziabad

- Residents will have easy access to all the notices and information digitally.
- A solution of all the generic problems of the residents now solved under one roof.
- The addition of calling feature in ParkAlert module made it more convenient for the affected user to reach to the problem-creating user.
- Fast and user-friendly interaction between resident and service provider and vendor.

5.2 Limitations

- In case if the owner of the car is not carrying the mobile registered in the database, then the alert will not be delivered.
- Wrong and inappropriate acknowledgment can also be sent which will affect the profile of the service provider or vendor.
- Voting can be biased if any cyber-crime activity is involved during the digital voting process.

6 Future Scope

- If the application spreads across a large number of societies, then parking problem arising outside the society can also be rectified.
- More authentications can be done to get a secured application and unbiased voting can be done.
- After acknowledgment by the residents, it will be thoroughly verified by the admin so that it will be removed if wrong praise is done.

7 Conclusion

We are trying to make the facilities provided in Indian apartments more comfortable for the residents to use them in more efficient manner. An application that sorts out all the basic problem in using these facilities of the apartment in a digital and friendly for residents. REAP module, voting system, and shopping app from nearby stores ensure that the resident need not to go anywhere. This also provides high quality of services and standard way of living. So, this paper deals in providing more luxurious and hassle-free lifestyle of increasing residents in Indian cooperative society.

Acknowledgements We are very thankful to all the members involved in this paper. We are very grateful to the Head of the Department, Prof. (Dr.) Shailesh Tiwari, HOD-CSE for his kind support and guidance at every single step and also to Educational ABESEC Infrastructure.

References

1. Goel, D., Agarwal, A., Rastogi, R.: A novel approach for residential. In: International Conference on Advances in Computer Science and Application (2016)
2. Iyer, L., Tare, M., Yadav, R., Amrutia, H.: Android application for vehicle parking system: park me. Int. J. Innov. Adv. Comput. Sci. 3(3). ISSN 2347-8616
3. Banerjee, S., Al-Qaheri, H.: An intelligent hybrid scheme for optimizing parking space: a tabu metaphor and rough set based approach. Egypt. Inform. J. **12**, 9–17
4. Akmarina Izza, M.: Smart parking via RFID tag. Project Report. UTeM, Melaka, Malaysia
5. Kaur, R., Singh, B.: Design and implementation of car parking system on FPGA. Int. J. VLSI Des. Commun. Syst. **4**(3) (2013)
6. Lan, K.-C., Shih, W.-Y.: An intelligent driver location system for smart parking. Expert Syst. Appl. **41**(5), pp. 2443–2456 (2014)

Characterization of Human Knowledge for Intelligent Tutoring

Neelu Jyothi Ahuja

Abstract Seismic data interpretation and subsurface mapping are key skills to analyze subsurface geology. They form the basis for the decision concerning hydrocarbons exploration and extraction. Interpreting a seismic graph, with perfection, needs expert knowledge. The knowledge of seismic data interpretation used in exploration industry is largely individualistic, with each human expert using his/her own set of mental database of interpretation rules developed over years of experience. For the lack of appropriate structure and formalization, this essential body of knowledge is unable to smoothly percolate to the next generation of seismologists, who are expected to deliver reasonable accuracy in their interpretations, almost immediate to their induction. Characterization of human knowledge is the process of structuring, formalizing and transforming the nature of the knowledge from tacit form to explicit form. Current work presents design and development of intelligent system to characterize this knowledge and deliver it using tutoring strategy exclusively devised as per the learner adjudged learning preference. This prototype additionally also measures learner's performance and facilitates learning gain. The system has been tested with 16 participants, and the resultant performance is recorded.

Keywords Intelligent tutoring system · Seismic data interpretation
Intelligent tutoring · Tutoring strategy · Pedagogy

1 Introduction

Seismic images are extensively used by seismologists to delineate subsurface geology [1]. This interpretation is manual and is often reliant on the interpreter's own skill and proficiency. Thus, as a result, complex and similar seismic images are

N. Jyothi Ahuja (✉)
Department of Computer Science and Engineering, Centre for Information Technology,
University of Petroleum and Energy Studies, Dehradun, India
e-mail: neelu@ddn.upes.ac.in

© Springer Nature Singapore Pte Ltd. 2018   363
K. Saeed et al. (eds.), *Progress in Advanced Computing and Intelligent Engineering*,
Advances in Intelligent Systems and Computing 563,
https://doi.org/10.1007/978-981-10-6872-0_34

interpreted in a different way by different seismologists. This uncertainty is partly due to complex geological structure and partly because of the absence of thumb rules of interpretation, further increasing dependency on human expert. As an effort in this area, Ahuja and Diwan (2010) have developed a rule-based expert system Seis Expert to assist in the process of seismic data interpretation under two steps: manual interpretation involving visual examination and analytical interpretation involving computations [2].

New and novice seismologists take a number of years to come to the efficiency level of being able to interpret a seismic section to a considerable degree of accuracy. This is a constant challenge, throughout risking the interpretation process and the decisions (primarily drilling decisions) based on it. Significant developments in computer industry and digital technologies, over the years, have been limited to seismic image quality enhancement and extended digitization, leaving exploration industry starving for interpretation knowledge in an appropriate and easy-to-learn form. The inexperienced seismologists are expected to deliver near-accurate interpretation, almost immediately after their induction, resulting in not so satisfactory performance. The effort is in terms of overload of information, supplied in an inappropriate sequence, leading to incomplete understanding, thereby resulting in incompetent interpretation. On the other hand, the expertise available with established human experts of the field goes largely under or non-utilized.

Intelligent tutoring systems (ITS) as instructional systems incorporate AI (Artificial Intelligence) techniques providing tutors with "what", "whom" and "how" of teaching. The design and development of such tutors lie at the intersection of discipline of computer science, cognitive psychology and educational research [3]. This intersection denoted to as cognitive science has gained immense focus in the current years, for the very reason that the challenge of tutor is to make learning happen, given the complexity of learning process. Complexity is also due to the reason that each learner learns differently "*One shoe cannot fit everybody*".

The novelty of the present work is structuring and formalization of knowledge of seismic data interpretation, which was lacking to this day, and its subsequent delivery. The work demonstrates characterization of human knowledge of seismic data interpretation and its delivery through an ITS developed for this purpose with an objective to facilitate smooth percolation of this rare expensive knowledge to novice seismologists as per their learning preferences improvising on the learning gain.

2   Related Work

The section below provides a comprehensive coverage of evolution of ITS ranging from initial computer-assisted instruction systems to much advanced systems of current times, incorporating more and more intelligence mechanisms in terms of delivering the learning material as per learner's learning profile.

The first intelligent tutoring system (ITS) was modernized in the year 1950 in the form of (CAI) computer-aided instruction [4]. In this ITS system, the pre-decided course material or frames is presented in linear fashion, i.e. they sequentially organize the frames in linear fashion, so that the learners can follow the step-by-step procedure to achieve the desired goal.

An advanced intelligent computer-aided instruction (ICAI) system developed later [5] aided control over the content shown and helped to effectively interact with the learner. In the 1980s era, Carbonell's introduced AI techniques with CAI (Computer-Assisted Instruction) to overcome the limitations of generative CAI. In 1982, Sleeman and Brown presented ITS as coaches, consultants, problem-solving monitors and laboratory instructors. They presented a survey of progressive development of ITS [6].

Andes Intelligent Tutoring System developed to teach physics [7] uses Bayesian network for decision-making. InterMediActor utilizes fuzzy inference technique [8] and navigation graph data structure to sequence the course material. The fuzzy inference mechanisms and fuzzy sets have been used for making use of learner's ability and knowledge of subjects for deciding the course content for the learner. SQL Tutor is used to teach SQL through an agent developed using artificial neural network [9, 10]. C++ tutor is a rule-based ITS to teach the concept of computer programming language 'C++' [11]. CIRCSIM tutor offers a dialogue-based mechanism to teach physiology [12], where the communication with the learner is in the mode of dialogues. [13] have presented a record of ITS research over past years and have reported its primary focus on modular and fine-grained curriculum delivery, customization for different student populations, customized presentation and assessment and development of rich knowledge repository for instructors to tutor and remediate the learners. ViSMod is another ITS for visually representing Distributed Bayesian Student Model [14].

3 Development of Tutoring System

In order to discover under earth, oil and gas deposits, a variety of exploration methods have been developed over the past years. Seismic reflection methods yield the most precise and accurate results [1] and play a prominent role in search for suitable geological structures, where commercially viable amounts of oil and gas deposits may have been lodged. Interpretation of seismic images enable seismic expert's to infer subsurface geology and information regarding accumulation at depths of hundreds or thousands of feet below the ground surface. Geophysicists utilize their prior knowledge (gained with training and experience) to apply a concept and interpret. At times, they even formulate a new concept. Lack of formal knowledge-base of interpretation rules causes dependency on human experts, additionally, hindering much-needed training/imparting of knowledge to the forthcoming generation. This warrants the need to formalize the interpretation knowledge and facilitate it to reach to younger generation of seismologists, in the form, that best matches their learning profiles.

A knowledge-based tutoring system for seismic data interpretation integrated with intelligent tutoring has been designed and developed, soliciting interpretation knowledge and facts of the field from seismic experts, with a purpose to deliver customized learning experience to each learner. The system can be used by the experienced seismologists as well for additional assistance in the field during interpretation process.

3.1 Architecture

The design is modular with five distinct components interlinked together to deliver the desired objective of the system. The components are—Domain (or content) module, learner module, tutoring module (tutoring engine), pedagogical module, knowledge assessment module and a user-friendly user interface. The learner logs into the system and is subjected to a pre-tutoring phase, which includes a profiling test. The performance in the test sets the learner's profile to any of the three stereotype profiles ('beginner', 'proficient' and 'expert') priori marked with the tutor. Subsequent to this, a custom-tailored tutoring strategy is devised, specific to the learner's profile, according to which the learning progresses. The sections below present each component in detail. Figure 1 presents the schematic representation of the components.

3.2 Components

3.2.1 Tutoring Module (Engine)

Tutoring engine, implements the intelligent tutoring aspect into the system. It interconnects with the other modules, viz, Pedagogical module, Learner module, Domain/Content module and Knowledge Assessment module. The working of tutoring engine involves tutoring of a learner according to a particular initially assigned learning profile ('beginner', 'proficient' and 'expert'), which may undergo a change, subject to re-assignments as and when deemed necessary. For these decisions, it depends on the input received from other modules. This part of the component holds decision rules written using one of the development tools JESS (Java Expert System Shell) [15].

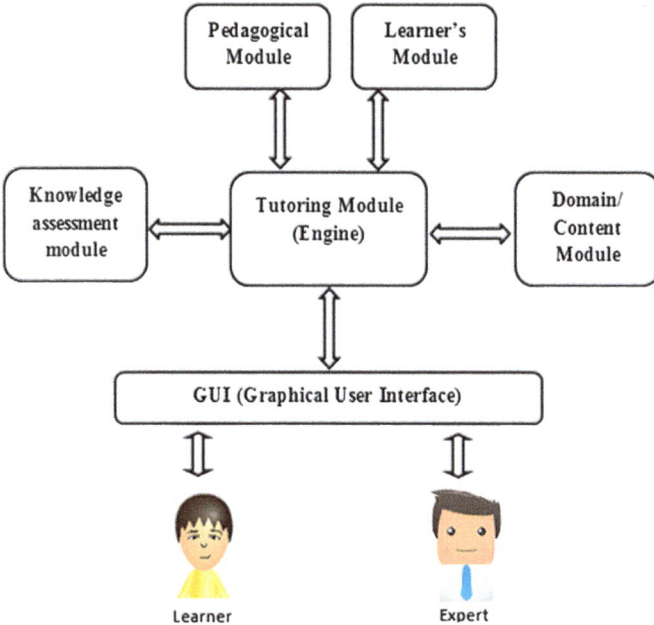

Fig. 1 Components

3.2.2   Learner Module

The learner's module accumulates and updates learner data and his/her performance throughout the tutoring process. The learner's information is stored in structured representation form (frame-based manner), presently, under two categories:

1. Personal data of learner (Id, Name and e-mail).
2. Progress Record: These are the records related to the learner's activities and interactions during learning sessions. This data is made available to the learner module by the knowledge assessment module.

3.2.3   Knowledge Assessment Module

It determines the learner's progress by identifying what the learner has learned during the process. It helps the pedagogical module to devise a tutoring strategy accordingly and helps the engine module to re-align/re-organize the learning material served by the content module. The learner's performance, progress management and resultant decision-making have been modelled in the present system through fuzzy inference mechanism. The learning progress is calculated using parameters and compared with a progress marker. If it reaches to a set threshold mark, accordingly the pedagogical module will trigger changes of tutoring strategy, update of the learner profile and the pedagogy (as applicable).

The net progress of a learner (NP) and progress impact factor (PIF) have been taken as measures and calculated using Eqs. (1) and (2) presented below:

$$NP_j = h_j - l_j - pp_j \tag{1}$$

$$PIF = NP_j / t_{th} \tag{2}$$

where h_j = highest score obtained by the jth learner, l_j = Lowest score obtained by the jth learner, pp_j =Previous Progress of jth learner, NP_j = Net Progress of jth learner, m = Progress Marker, t_{th} = Threshold Value.

This performance detail is further provided to the pedagogical module for design of tutoring strategy which is handed over to tutoring engine for execution. The section below presents fuzzy variables, fuzzy rules and fuzzy membership function. Figure 2 represents the fuzzy membership function.

Fuzzy Variables:

PL = Positive large, PS = Positive Small, NL = Negative Large, NS = Negative Small

Z_0 = Constant.

Fuzzy Rules:

(1) If NP is PL, then m is decreased by 1
(2) If NP is PS, then m is decreased by PIF
(3) If NP is NL, then m is increased by 1
(4) If NP is NS, then m is increased by PIF
(5) If NP is Z_0, then m is increased by some constant.

The combined value is fuzzified using a set of fuzzy variables {PL, PS, NL, NS, Z_0}. The fuzzified value represents the performance of the learner. Fuzzy inference system monitors learner performance in real time and classifies progress rate as negligible (no progress), little, average or excellent progress. It computes the

Fig. 2 Input membership function

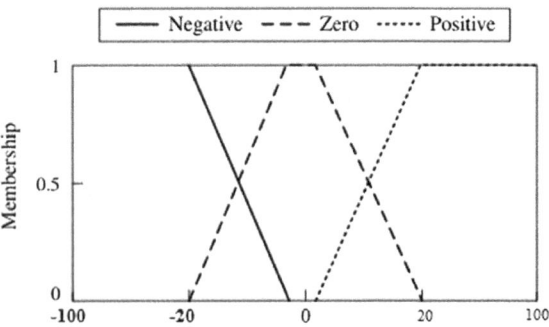

net progress of learner using values of the highest score, lowest score and previous progress from the learner's previous state of performance.

3.2.4 Content Module

This module maintains the subject matter, the learning material structured into modules and sub-modules. Further, each of the sub-modules has units, sub-units and chapters/topics. Each chapter presents the objective at the beginning and summaries the expected learning outcomes at the end. The organization of chapters is based on increasing complexity with basic (straightforward), tutoring material lined up in graphical, guided-tour style, videos explaining concepts, key-points coverage 'How to interpret' simplified seismic section exercises, etc. being marked for delivery under 'beginner' stereotype profile, (keeping tough technical concept, away, as of now), and slightly complex tutoring material lined up in textual, graphical, combination of textual and graphical styles, descriptive notes, problem-solving exercise patterns, guided interpretation for seismic sections, etc. being marked for delivery under 'proficient' stereotype profile. Further, advanced level of learning material is lined up in descriptive, elaborative, textbook form, power points, complex problem-solving exercises, real-scenarios, seismic section interpretation case studies, 'to be' interpreted seismic sections, being marked for delivery under 'expert' stereotype profile. The content module presents itself in the form of a dashboard.

The major contribution of this module is the characterization of interpretation knowledge of seismic sections. This knowledge is tacit, and hence, its acquiring, structuring, formalization and representation in explicit form is the challenge that has been addressed, along with delivery in a manner that best suits the learner's learning profile so as to ensure smooth percolation and maximum imbibe. The well-organized and structured content made available qualifying to be repository in this rare field of study is another quantifiable derivable of the present work.

3.2.5 Pedagogical Module

Learning efficiency varies significantly from individual to individual (i.e. every learner has his/her own distinct pattern/style of learning). Providing all the tutoring material in one style is not called for, especially, due to the fact that though proficiency expected of each learner is reasonably similar, the learning style/pattern is largely individualistic and almost personalized to each learner. Also, it is not very straightforward to ascertain a particular learning style for a learner. It is next to impossible to do it spontaneously. It involves a greater degree of probing into learners' cognitive abilities, personal learning patterns, appealing instructional styles, acceptable level of complexity, etc. in order to devise a tutoring strategy and execute it with a suitable delivery mechanism and tool.

On the other hand, it is also a challenge to decide a target delivery style for a given learning concept. It may involve careful study of the concept to be able to convey its understanding to different profiles of target audience of learners in a custom-tailored manner. This module plans a pedagogical flow of instruction for tutoring material. This involves aligning/realigning the different learning concepts in a particular sequence to be displayed on the dashboard. This process is termed as devising a tutoring strategy and hands over to the tutoring engine to execute it for a given learner [16]. This is done in the present system by using data structures queues and de-queues and pointer to link various tutoring concepts and different delivery modes of a given concept.

3.2.6 User Interface/Tutoring Dashboard

It acts as an interface between the learner and tutor. The learner can log on to the tutor and get started with the learning process. As the tutoring sessions proceed, the tutor increasingly adapts to the learner's style and accordingly modifies the material delivery mechanism, with an objective to enhance learning experience and improve overall learning of the given topic. The learner is constantly updated about his/her performance, obtained scores, profile change, the profile change history, module coverage statistics, etc. The complete interaction is brought by this interface presented as a tutoring dashboard in a very user-friendly and interactive manner. The interface has been designed with JDK 1.7 using Eclipse interactive development environment (IDE) console.

4 Implementation Notes

This section presents the functionality of prototype tutoring system through an example tutoring session. This tutoring system was tested over a period of four months, with 16 participants, who logged on to the system and underwent tutoring on 'seismic data interpretation'. In the pre-tutoring phase, an initial test was conducted. Participants scoring less than 50% were allotted a 'beginner' profile, if the score obtained was between 50 and 80% 'Proficient' profile was allotted and if the score was above 80%, the learner was allotted 'expert' profile.

It was observed that eleven participants were allotted 'beginner' profile, three of them were allotted 'proficient' and the remaining two were adjudged 'expert' profile. Each of them registered as 'new users', logged in and started their tutoring sessions. As the tutoring progressed, if at all, the participant desired to pause his sessions to be resumed at a later point of time, the system offers to bookmark the particular point. Later, when the participant wishes to resume back, the system facilitates navigation to the bookmarked point within the dashboard.

The performance and learner progress were recorded by the system and used for profile/pedagogy change as applicable. Three of the participants, earlier allotted a

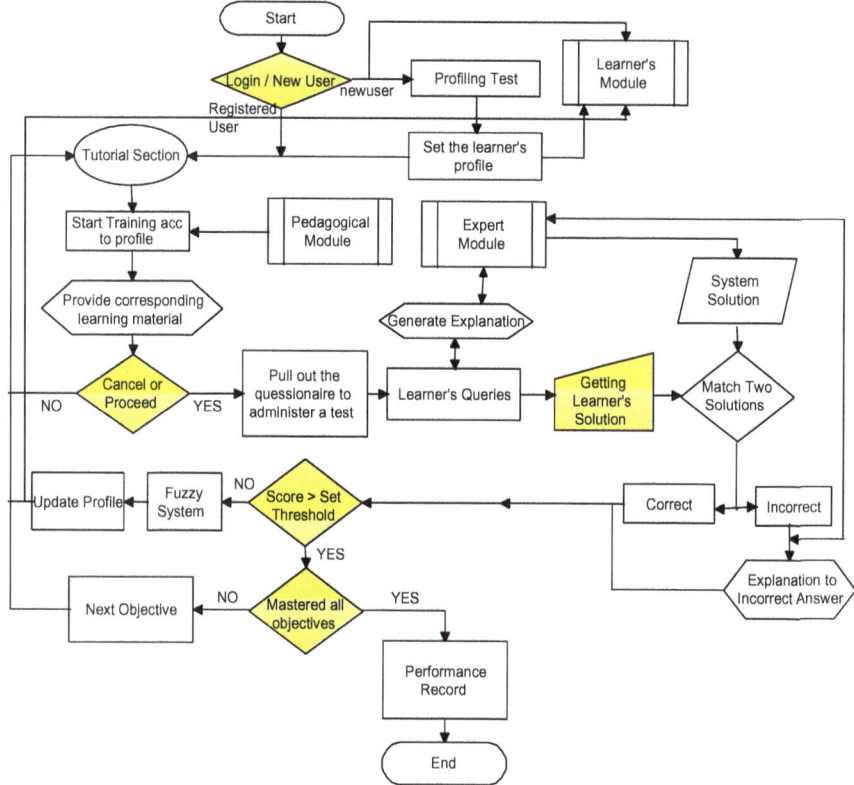

Fig. 3 Flow of work in SEIS-Tutor

'beginner' profile, over the third, fourth and seventh module (out of a total study material of ten modules), respectively, obtained an upgrade to 'proficient' profile. There was no change of profile for rest of the thirteen participants. The full functionality is being shown in Fig. 3.

5 Conclusion

The developed intelligent tutoring system characterizes human knowledge and presents it in explicit form. It intelligently identifies most suitable learning style, monitors learner's activities and performance, switching between profiles and offering personalized instruction, thereby significantly increasing the learning gain in imparting 'seismic data interpretation and subsurface mapping' skills to an individual. It offers rich collection of learning material, problem-solving exercises (acting as a knowledge repository) and presents them interactively by providing response on the learner's actions, detecting learner's fallacies, strong areas and weak areas, helping the learner in gaining mastery in domain by self-assessment and by guiding the learner with "what concepts to apply", "when to apply", "why to apply" and "how to apply". Learning was assessed by testing the system with participants.

6 Future Work

Being a potential area of research, it offers future extension directions which can be explored. Further proposed enhancement includes the 'Natural Language dialogue' (NL) module which will facilitate an enriched delivery and effective adaptation of the learning material decreasing the communication gap between humans and computer tutors. An essential aspect of enhancement is additional content development with focus on clarity of the domain and added ease of access. Additional to the different learning levels, there is an impulsive need to incorporate innovative learning styles, as well. Further, machine learning techniques can be used to generate production rules from seismic images and domain background knowledge. This will impart an enriched functionality to the system by generating rules automatically within a defined time frame and provide non-programmers and domain experts with an important tool for development of intelligent tutoring systems and authoring tools.

Acknowledgements The author thankfully acknowledges the management of University of Petroleum and Energy Studies (UPES), Dehradun, India, for supporting and granting permission to publish this work.

References

1. Gadallah, M.R., Fisher, R.L.: Applied Seismology: A Comprehensive Guide to Seismic Theory and Application, pp. 2–10 (2005)
2. Ahuja, N.J., Diwan, P.: An expert system for Seismic data interpretation using visual and analytical tools. Int. J. Sci. Eng. Res. **3**(4), 1–13 (2012)
3. Nwana, S.N.: Intelligent tutoring systems: an overview. Artif. Intell. Rev. **4**, 251–277 (1990)
4. Skinner, B.F.: Teaching Machines. Science 128
5. Woolf, B.P., McDonald, D.D.: Building a computer tutor: design issues. IEEE Computer
6. Sleeman, D., Brown, J.S.: Introduction: intelligent tutoring systems. In: Intelligent Tutoring Systems. Academic Press, pp. 1–11 (1982)
7. Conati, C., Gertner, A., VanLehn, K., Druzdzel, M.: On-line student modeling for coached problem solving using Bayesian networks. In: Proceedings of the Sixth International Conference on User Modeling, Vienna, pp. 231–242 (2002)
8. Kavcic, A., Pedraza-Jimenez, R., Molina-Bulla, H., Valverde-Albacete, F.J., Cid-Sueiro, J., Navia-Vazquez, A.: Student modeling based on fuzzy inference mechanisms. In: The IEEE Region 8 EUROCON 2003: Computer as a Tool, pp. 379–383 (2003)

9. Wang, T., Mitrovic, A.: Using neural networks to predict student's performance. In: Proceedings of International Conference on Computers in Education, pp. 969–973 (2002)
10. Mitrovic, A.: An intelligent SQL tutor on the web. Int. J. Artif. Intell. Educ. **13**(2–4), 171–195 (2003)
11. Baffes, P., Mooney, R.: Refinement-based student modeling and automated bug library construction. J. Artif. Intell. Educ. **7**(1), 75–116 (1996)
12. Evens, M.W., Brandle, S., Chang, R., Freedman, R., Glass, M., Lee, Y.H., Shim, L.S., Woo, C.W., Zhang, Y., Zhou, Y., Michael, J.A., Rovick, A.A.: CIRCSIM-Tutor: an intelligent tutoring system using natural language dialogue. In: Proceedings of the Twelfth Midwest AI and Cognitive Science Conference, Oxford, pp. 16–23 (2001)
13. Chakraborty, S., Roy, D., Basu, A.: Development of knowledge based intelligent tutoring system. In: Sajja, Akerkar (eds.) TMRF e-Book, Advanced Knowledge Based Systems: Model Applications and Research, vol. 1, pp. 74–100 (2010)
14. Zapata-Rivera, D., Greer, J.: Interacting with inspectable Bayesian student models. Int. J. Artif. Intell. Educ. **14**, 127–168 (2004)
15. Friedman-Hill. E.: Jess in Action: Java Rule-Based Systems. Manning Publications Co. (2003)
16. Jeremic, Z., Devedzic, V.: Design pattern ITS: student model implementation. In: Proceedings of the IEEE ICALT 2004 Conference, CA: IEEE Computer Society, pp. 864–865 (2004)

A Conceptual Software Model of E-Business to Analyze the Growth Rate of a Product

Sumanta Chatterjee, Sneha Roy Chowdhury and Kaveri Roy

Abstract Electronic commerce is the trading or facilitation of trading of products and services using computers and latest technology. Here, focus is on developing software for e-marketing to provide a new platform which will give benefit to the already existing way of e-marketing. In this, ways are given out to analyze the sales progress of a particular product in the market. Based on different customer's demand studied, certain parameters are found which allows us to calculate the growth of a product. A model will improve the process of selling a product by managing customer's demand along with the important parameters by integrating online and offline marketing skills and also improve e-marketing plan. The main objective is to anticipate the demand based on some factors. After implementing e-marketing plan and strategies, the sale is supposed to be enhanced.

Keywords E-marketing · Demand analysis · Growth rate · E-marketing campaign · Population density

1 Introduction

Marketing has always been a part and participle of our lives from earlier times. E-marketing is the technique of marketing a product or service using the Internet. It includes both indirect marketing and direct response marketing elements and uses different technologies to help connect businesses to their customers.

E-marketing envelopes all the activities a business conducts with the help of the Internet with the objective of developing its brand identity and attracting new business.

S. Chatterjee (✉) · S. R. Chowdhury · K. Roy
Department of Computer Science and Engineering,
JIS College of Engineering, Kalyani, West Bengal, India
e-mail: itsmrs8@gmail.com

© Springer Nature Singapore Pte Ltd. 2018
K. Saeed et al. (eds.), *Progress in Advanced Computing and Intelligent Engineering*,
Advances in Intelligent Systems and Computing 563,
https://doi.org/10.1007/978-981-10-6872-0_35

If e-marketing is implemented correctly, the return on investment (ROI) can exceed that of traditional marketing strategies. E-marketing also aids in enhancing the demand of the product in the market.

Demand is the volume of goods that consumers or buyers are willing and capable to buy at a given price in a given time period while everything else remains the same.

The demand analysis and the demand theory are important to the business enterprises. The success or failure of business firms depends mainly on its ability to generate resources by satisfying the demand of consumers. The importance of demand analysis in business decision can be explained under following headings:

- Sales forecasting—Whenever demand is high, sales will be high, and whenever demand is low, sales will also be low.
- Pricing decisions—When the demand for the product is high, the company can charge high price. On the contrary, when the demand is low, the firm cannot increase price.
- Marketing decisions—It measures and analyzes the forces that determine demand.
- Production decisions—The quantity a company can produce depends on its capacity, but how much it should produce depends on demand.
- Financial decisions—When the demand for a product is high and growing, the needs for additional finance will be greater.

Marketing is another facet which must be taken care of at the time of launching a new product in the market. To develop products and services in one's business which meet the needs of one's target, market planning is needed. Good marketing helps one's customers to understand why a product/service is better than or different from the competition. A good marketing plan can help in reaching the target audience, boosting customer base, and increasing bottom line. It is always required when seeking funding and helps in setting realistic, clear, and measurable objectives for the business.

2 Proposed Model

The project is based on implementation of the following proposed model which is based on demand analysis, customer's feedback, and field survey done over about 100 customers to determine the way in which they judge a product available in the market.

From the proposed model, factors affecting a product's growth in the market are available. The factors are analyzed and are used to calculate the threshold value (K). Using this growth, rate of the product can be determined.

The following factors are taken into consideration while designing the below-shown model to calculate the growth rate of the product:

1. If it is an old product which is already present in the market, the following factors are to be considered:

   a. E-marketing campaign (EMC)
   b. Customer's intension (CI)
   c. Population density (PD)
   d. Historical records (HR)
   e. Per capita income (PCI)
   f. Competition with similar product (CSP)

2. If it is a new product and not introduced in the market till now, then the following factors are to be considered-

   a. Objective of the product.
   b. Application of appropriate strategy for proper advertisement of the product.
   c. E-marketing campaign (EMC)
   d. Customer's intension (CI)
   e. Population density (PD)
   f. Historical records (HR)
   g. Per capita income (PCI)
   h. Competition with similar product (CSP)

Here, it can be seen that if the product is already available in the market, evaluation of the product can be started from customers' feedback itself; else if it is a new product, at first the product's objective is to be determined and proper advertisement is to be done before it is launched in the market. After launching it, similar steps are to be followed like in the case of an old product.

The factors considered are as follows:

2.1 E-marketing Campaign (EMC)

E-marketing campaign is a method of reaching the consumers directly via Internet technology, e.g., email marketing, online advertisements, and discounts. It is an important factor in finding out the growth of a product.

2.2 Population Density (PD)

Population density is an important factor for determining the growth rate of a product. Determining the population density of a particular area determines the type of product that may be sold in the particular area. Depending on it, the amount of the products required, the price of the product, etc. are determined.

2.3 Analysis of Historical Records (HR)

Another factor to be considered in determining success of a product depends on the historical records of such products in market. Here, two factors are considered:

2.3.1 Brand Value

The reputation of the brand in the market or on the customers plays key role in the sale of the product. If it is an old and reliable brand, more people are likely to buy its product other than buying products of a new novice brand.

2.3.2 Product Value

The acceptance history of a particular type of product in the market determines the chances of acceptance of similar type of new product to be launched in the market in the near future.

2.4 Analysis of Per Capita Income (PCI)

Per capita income is the average income earned per person in a given area. If a product is launched in an area where the price of the product is not affordable as per the per capita income of people living there, then the product is not going to be sold there. So per capita income is a deciding factor in the sale of a product launched in the market.

2.5 Customer's Intention (CI)

Customers' intension can be determined by observing the activity of a customer on a particular Web site. This remains as one of the main criteria for maintaining positive growth rate of any product in the market.

Now and then every company does a customer satisfaction survey, by different means. It also helps companies to show the customer a customized list of products in which he might be interested. Customers' intension can be measured in terms of the following:

- Price
- Features offered
- Quality of the product

2.6 Competition with similar products (CSP)

Competition with other similar products is a key factor to be considered. Comparison can be done either on the basis of price of the product or features of the product or on its quality or reputation of its brand in the market or the discount available on the product. Customers do compare products on the basis of the above criteria, so being prepared for such comparisons would be a pretty much intelligent decision (Fig. 1).

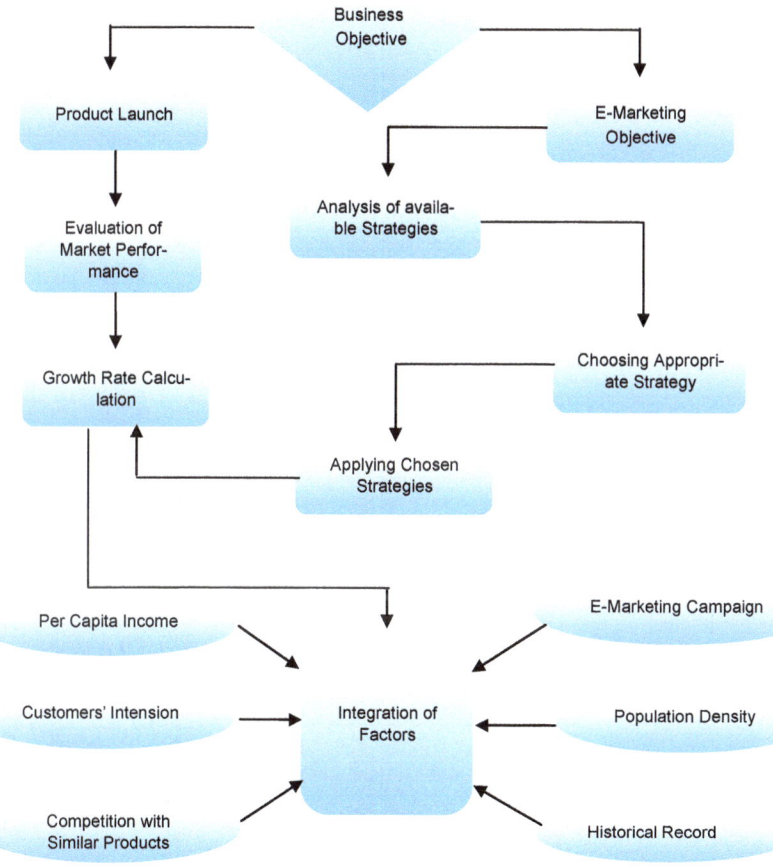

Fig. 1 Proposed model to analyze the growth rate of a product

3   Analysis of Proposed Model

3.1   *Training Data*

(Figs. **2**, **3**, Table **1**)

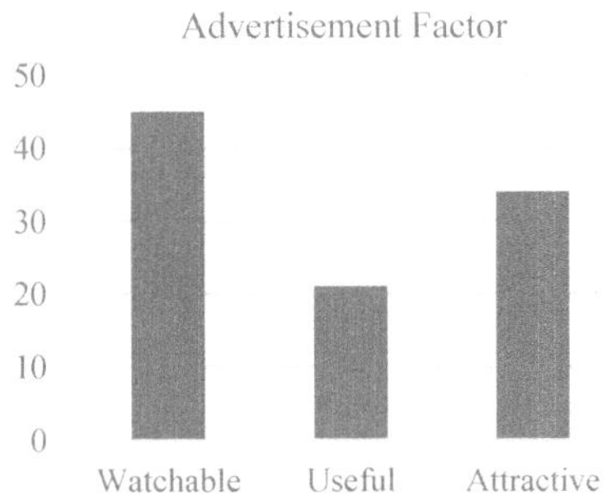

Fig. 2 Graph showing how many people bought a particular product by seeing its advertisement on Internet

Fig. 3 Graph showing how many people bought a particular product on the basis of different offers and discounts offered

Table 1 Training data showing different customer's opinion on different factors and the corresponding action taken

Serial No.	EMC	PD	HR	PCI	CI	CSP	Decision Made
1	Y	Y	N	Y	Y	Y	X
2	N	Y	Y	N	Y	Y	X
3	N	Y	Y	Y	N	Y	X
4	N	Y	Y	N	N	Y	F
5	Y	N	N	Y	Y	Y	X
6	Y	N	Y	Y	Y	Y	X
7	N	Y	N	Y	N	Y	F
8	Y	N	Y	N	Y	Y	X
9	Y	Y	N	Y	N	Y	X
10	Y	N	N	N	N	Y	F

Signs used: Y—yes or positive value, N—no or negative value, X—product is bought, F—product is not bought

3.2 Graphs Made on the Basis of Training Data

According to Fig. 4, calculated overall impact of e-marketing campaign on product sale = 82.40%

Thus, analyzing these data, it can be understood that e-marketing campaign affects the sale of a product significantly.

According to Fig. 5, calculated overall impact of population density on product sale = 17.03%

This means out of 100 approximately 17 people buy a product depending on population density factor (Figs. 6, 7).

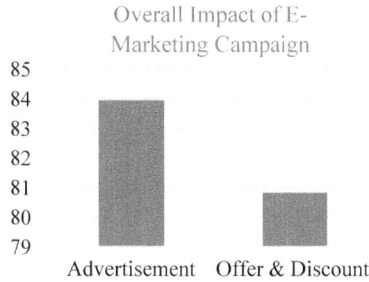

Fig. 4 Graph for overall impact of EMC on the sale of the product

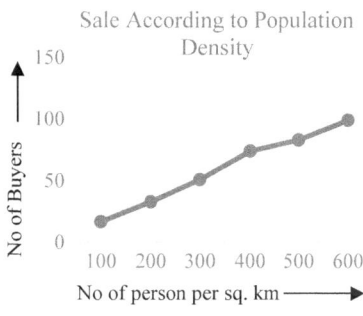

Fig. 5 Graph shows increase in sale according to population density

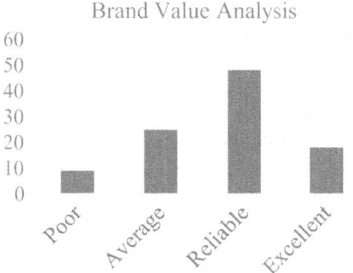

Fig. 6 Graph for brand value analysis

Fig. 7 Graph for product
value analysis

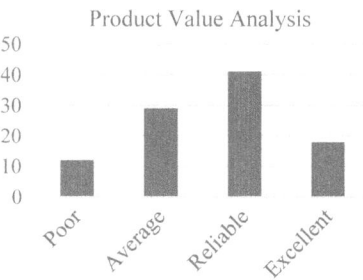

Fig. 8 Graph for historical
record factor

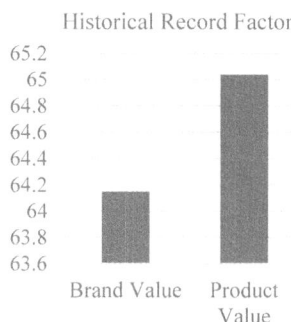

According to Fig. 8, calculated overall impact of historical record on product
sale = 64.59%

This means out of 100 approximately 64 people buy a product depending on the
products' historical record.

According to Fig. 9, calculated overall impact of per capita income on product
sale = 24.6%

This means out of 100 approximately 24 people buy a product depending on per
capita income of the region (Figs. 10, 11, 12).

According to Fig. 13, calculated overall impact of customers' intension on
product sale = 60.55%

Fig. 9 Graph showing how
many people buy a particular
product depending on their
per capita income

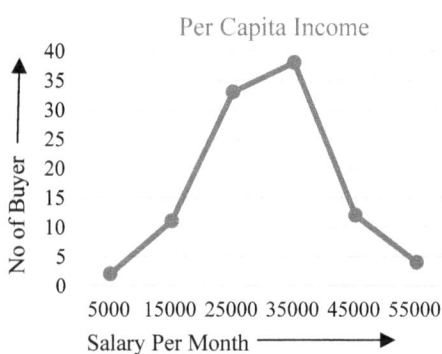

Fig. 10 Graph based on price factor

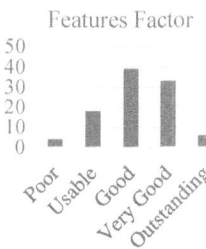

Fig. 11 Graph based on features factor

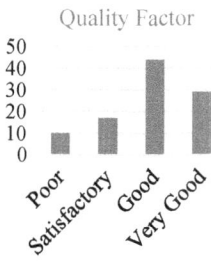

Fig. 12 Graph for quality factor

Fig. 13 Graph for overall impact of customer's intension on product sale

This means out of 100 approximately 60 people buy a product depending on customer intension factor.

According to Fig. 14, calculated overall impact of competition with other similar product on product sale = 43.03%

This means out of 100 approximately 43 people buy a product after comparing it with other similar products (Fig. 15).

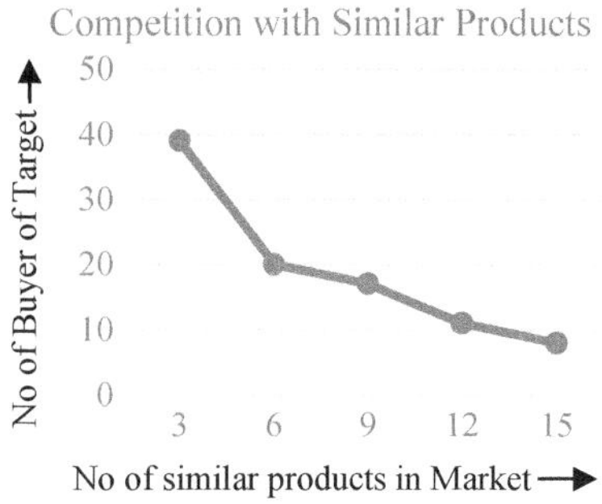

Fig. 14 Graph for overall impact of competition with similar product on sale

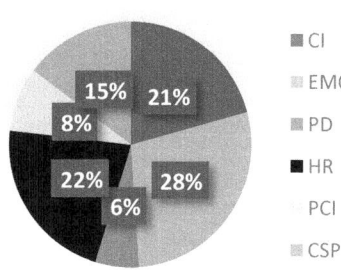

Fig. 15 Pie chart showing contribution of the factors in the growth rate

3.3 Calculation and Formula

PROPOSED FORMULA FOR GROWTH RATE CALCULATION:

On the basis of the survey done and the factors considered, the below formula is proposed for growth rate calculation of a product.

$$\text{Growth} = K * \sqrt{\left(\left(CI^2 + EMC^2 + HR^2 + PD^2 + PCI^2 - CSP^2 \right) / 6 \right)} \qquad (1)$$

The value of K is found to be 1.35(approx.) whose calculation is shown below:

$$
\begin{aligned}
K &= \text{Growth} / \left(\sqrt{\left(\left(CI^2 + EMC^2 + HR^2 + PD^2 + PCI^2 - CSP^2 \right) / 6 \right)} \right. \\
&= \left((3*60.55) + (2*82.4) + (3*17.03) + (2*64.59) + (3*24.6) \right. \\
&\quad \left. - (5*43.03) \right) / \left(\left(\sqrt{3666.30 + 6789.76 + 290.0 + 4171.87 + 605.16 - 1851.5} \right) / 6 \right) \\
&= 64.22 / 47.43 \\
&= 1.35
\end{aligned}
$$

$$\mathbf{K = 1.35}$$

Now, this can be treated as the threshold value. Products with growth rate factor equal to this or above this will be considered to have a good growth in the market.

A value slightly below this is also acceptable, but a much smaller value shows that the product is not in a good condition.

4 Challenges

Implementing this model might have a lot of challenges, but they can be sorted out with careful observation and hard work. This model is almost fully dependent on "Customer's Feedback" and "Growth rate of a Product," and until and unless correct inputs are provided, this model will remain unimplemented and growth rate calculation is also not possible. If this model is successfully implemented, this model will play an important role in supporting the middlemen like the wholesalers and dealers in the business-to-business (B2B) model of e-commerce in checking the status of the product in the market and decision making about whether to take dealership of a particular product or not.

Secondly, proper training is required to carry out the process and proper implementation of the model especially for those who do not have basic idea about how the e-commerce models work and what exactly e-commerce is.

Moreover customer's feedback should be analyzed and handled carefully because it is the basis of the whole model.

5 Conclusion

With the evolution of new technologies every day, this model will surely aid as a new and trustworthy technology to the day-by-day changing electronic field of commerce. It is a highly economical way of marketing because the cost spent on launching a product and then checking its progress is far higher than simply using this model to determine the demand of the product before even launching it. Previously, there was no supporting system to support the middlemen and the dealers, but with the introduction of this model, the middlemen will be highly benefitted.

6 Future Scope of Development

The model is a new concept in the field of e-commerce because never before the middlemen were given this much importance that a model solely dedicating to them will be proposed. Now, since this is a proposed work, its proper implementation using software can be considered as a future development. Along with this, more factors can be included which may determine the growth more accurately.

References

1. Gan-geshwer, D.K.: E-commerce or internet marketing: a business review from indian context. Int. J. u e–Serv. Sci. Technol. **6**(6), 187–194 (2013). Department of Mathematics, Bhilai Institute of Technology (BIT), Durg, (C.G), India. dgangeshwar@yahoo.co.in. https://dx.doi.org/10.14257/ijunesst.2013.6.6.17, https://www.sersc.org/journals/IJUNESST/vol6_no6/17.pdf
2. Sudame, P., Sivathanu, B.: Challenges affecting the organized retail sector. APJEM Arth Prabhand: J. Econ. Manag. **2**(8) (2013). H.O.D., C.P. & Berar College, Tulsibag, Nag-pur, India. Assistant Professor, Indira Institute of Management, Wakad, Pune, India. ISSN 2278-0629. https://prj.co.in/setup/business/paper114.pdf
3. Bansal, M.P., Maan, M.V.K., Rajora, M.M.: Rural retailing in india—a changing paradigm. Int. J. Adv. Res. Comput. Sci. Softw. Eng. **3**(11) (2013). GZSPTU Campus, India ASSTT. Prof. GZSPTU Campus, India, November 2013 ISSN: 2277 128X. Research Paper Available online at: https://www.ijarcsse.com, https://www.ijarcsse.com/docs/papers/Volume_3/11_November2013/V3I11-0276.pdf
4. Erp in wholesale and distribution: a logical solution to logistical challenges; research report 20 Sep 2013. https://aberdeen.com/research/8632/si-wholesale-distribution-logistics/content.aspx
5. Uma, K.K., Sankarasubramanian, R.: Business Intelligence System–a survey. Int. J. Adv. Res. Comput. Sci. Softw. Eng. Res. **4**(9) (2014). Scholar, Erode Arts and Science College, Erode, India 2 M.Sc., PBDCSA., M.Phil., Associate Professor, Erode Arts and Science College, Erode, India ISSN: 2277 128X. Research Paper Available online at: https://www.ijarcsse.com, https://www.ijarcsse.com/docs/papers/Volume_4/9_September2014/V4I9-0372.pdf
6. https://northcarolinadeportal.com/eMarketing/emarketing-plan/

MINION: A Following Robot Using Ultrasonic Wave

Intisar Reza Abir, Iffat Sharmim Shanim and Nova Ahmed

Abstract Our aim is to use a machine-driven helping figure which can solve the problems of carrying heavy goods by following us. Keeping this conception in our mind, we built 'MINION' which could be the better viable answer to these problems of carrying goods when there is no one around. Our robot is built on chassis rover tank with four sonar sensors as receivers which are faced in four directions for forward, right, backward and left sides. It follows a transmitter which gives signal continuously to the receivers on the robot. A person carries the transmitter and makes the robot follow him/her. It moves itself with the help of transmitter–receiver using ultrasonic wave so that it has no need of manual input. An infrared sensor is set up on the robot to maintain a specific distance between them. Both MINION and the transmitter module use DC battery, and it is easy to operate and thus is useful in numerous applications.

Keywords Sonar sensors · Arduino · Motor shield · DC motors
Infrared sensor

I. R. Abir (✉)
Albattross, Flat A4, House 4, Road 20, Sector 7, Uttara,
Dhaka 1230, Bangladesh
e-mail: bijoy71@rocketmail.com

I. S. Shanim
Golden Dream, Flat A5, House 87, Road 2, Block B,
Bashundhara R/A, Dhaka 1229, Bangladesh
e-mail: isanim93@gmail.com

N. Ahmed
Golden Dream, Plot 15, Block B, 1229, Bashundhara R/A,
Dhaka 1229, Bangladesh
e-mail: nova.ahmed@northsouth.edu

© Springer Nature Singapore Pte Ltd. 2018   387
K. Saeed et al. (eds.), *Progress in Advanced Computing and Intelligent Engineering*,
Advances in Intelligent Systems and Computing 563,
https://doi.org/10.1007/978-981-10-6872-0_36

1   Introduction

As we know, currently the whole world is running based on technology. We use machines and technology to be independent in our day-to-day works. Everyone is engaged with their own works so that no one has time for others. In our daily life, sometimes we face many problems that require help from others, specifically when no one is around us. One problem is carrying heavy goods. Carrying heavy goods can be very hard for a person trying to solve multiple tasks single-handedly. In that case, we can use this 'MINION' robot by following us as well as carrying goods up to 8 kg. People easily use this robot for a reasonable cost of making and operate it by taking the transmitter of this robot without giving any manual input.

This robot 'MINION' can be used in those places where the problem of carrying heavy goods creates. This problem occurs mainly during carrying disaster relief from one place to another, climbing mountains with heavy goods, carrying patients when adequate number of nurses is not available in hospital. This robot can also be used in the shopping mall.

2   Equipments

The equipments for making 'MINION' are so available that it can be made easily. The list of equipments that we have used to make this robot and its transmitter is shown in Table 1.

Table 1 Table of essential equipment quantity for MINION

Equipments	Quantity
Arduino Mega	1
Arduino UNO	1
DC motor	4
Motor shield	1
Battery	2 (1 lipo 11v 2200 mAh, 1 lion 9 v 800 mAh)
Ultrasonic sensor	5 (1 transmitter 4 receiver)
IR sensor	1
Wire	28
Partex sheet	1 (11" × 22")

3 Working Procedure

MINION starts working by getting the signal at the range of 200 cm (6.5 feet) from the transmitter which gives ultrasonic signals constantly with the help of Arduino UNO connected with it. The ultrasonic signals are received by the receivers which are set up on the robot. After getting the signals, the receivers send the information to the Arduino Mega which is connected with the motor shield. Then, it receives the data and determines which direction the signal comes from or the less distance of signal comes from, and it sends instructions to the motor shield.

The motor shield receives signal from it and drives each motor according to the signal. The robot maintains a specific distance between the obstacle and the robot itself by the infrared sensor. The robot has several receivers facing different directions (Fig. 1).

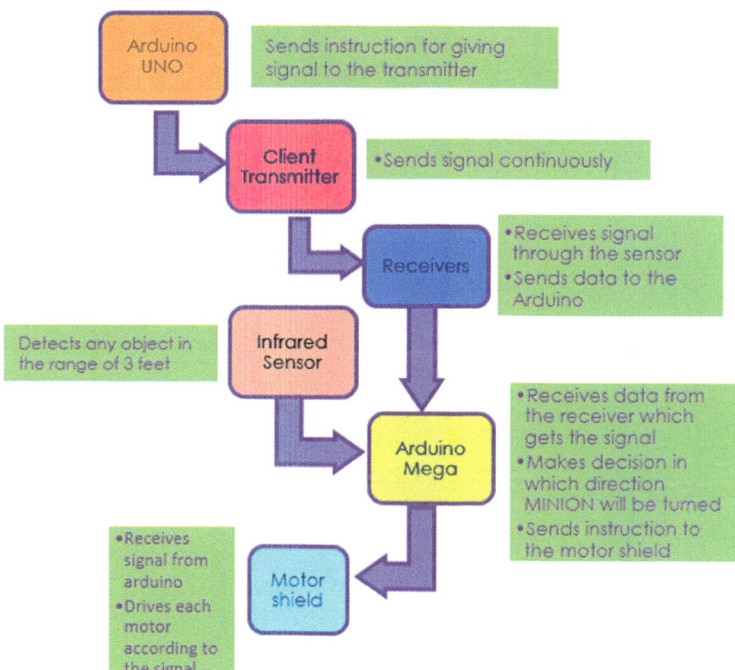

Fig. 1 MINION's working step

4 Robot Description

The major parts of building MINION are hardware part and programming part.

4.1 Hardware Part

At first, MINION is built by attaching Arduino Mega and motor shield on the chassis rover tank and then sonar sensors as four receivers, IR sensor connected to

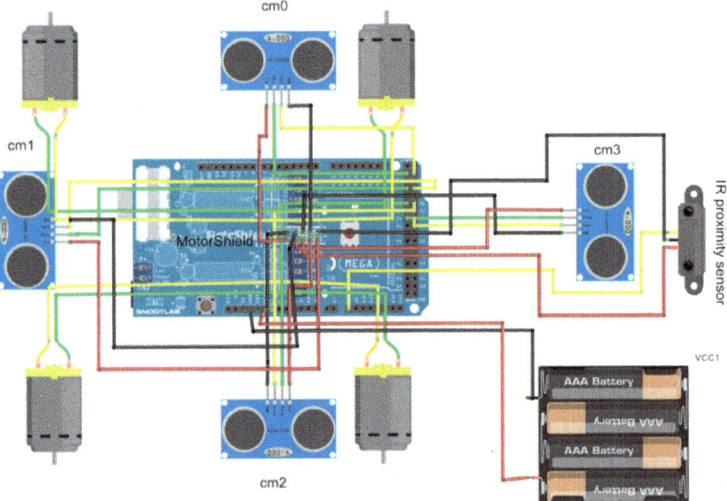

Fig. 2 Overall MINION connection diagram

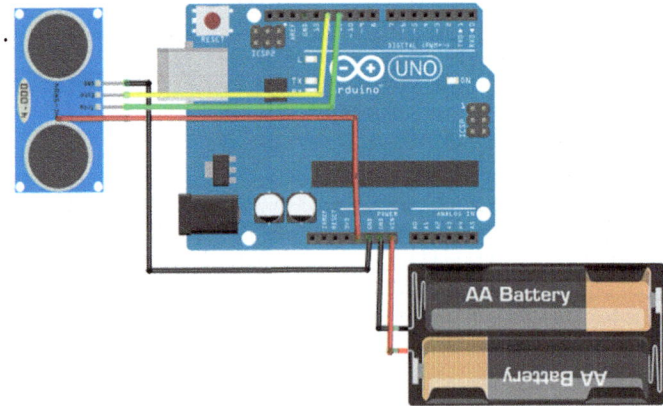

Fig. 3 Transmitter module connection diagram

the Arduino Mega. The transmitter is built with the sonar sensor joined to Arduino UNO. The connections of the robot and the transmitter are shown in Figs. 2, 3

Wheel set-up: The robot moves around using 4 DC motors as wheels. The wheels on both sides are connected with an elastic belt which gives MINION a better menu variability on rough terrain. The motors are powered by 7.4 V power supplied by the motor shield. The right side motors are connected to the M2 slot of the motor shield, and the left side motors are connected to the M4 slot. The motor shield is connected to the Arduino Mega 2560. It takes the analog 0 to 13 pin of Arduino as input. We are using motor shield L293D to move the robot. The motor shield gets its power from Arduino.

Receiver set-up: Receivers are connected to the digital input–output port of the Arduino. The receivers are named as 0, 1, 2 and 3 cm. The trigger pin of these receivers are connected to 22, 26, 30 and 34 number pins, respectively, and the echo pins are connected to 23, 27, 31 and 35 pins, respectively. We are using 1 cm at the back, 3cm at the front, 0 cm on the right side of 3 cm and 2 cm on the left side of 3cm. The four receivers are placed in such a manner that the angle between 3 to 1 cm is 180°, angle between 2 to 0 cm is 180°, the angles between 1 to 2 cm and 1 to 0 are 90°, and the angles between 3 to 0 cm and 3 to 2 cm are also 90°. These pins power up the receivers. We built four compartments to place the ultrasonic sensors, Arduino Mega and the motor shield and also the DC motors. The DC motors are placed in the first (bottom) compartment. The second compartment has Arduino Mega, motor shield and wires to connected elements. The third compartment is a 31 cm by 39 cm platform that resides the ultrasonic sensors, and this platform has four compartments to isolate each ultrasonic sensor, so that there can be no interference when receiving signals. The forth (top) compartment which also has the same dimensions as the third compartment is used to keep goods for the robot to carry. In the design, the fourth compartment in the third platform is acting as waveguides for the transmitted signals. We are also using an IR sensor to detect obstacle so that the robot can stop the collision with it. It is placed in front of the robot outside the second compartment, just under the 3 cm ultrasonic receiver sensor.

IR sensor set-up: We are also using an IR sensor to detect obstacle so that the robot can stop the collision with it. It is placed in front of the robot outside the second compartment just under the 3 cm ultrasonic receiver sensor. The IR sensor power-up wires are also connected to the 5 V pin of the motor shield.

Overall set-up: The overall wiring of the robot is given in the diagram below. The whole circuit is powered by one 11 V 2200 mAh battery.

Transmitter set-up. The transmitter, however, is a very easy case. All we need here is the HCSR04, and we are using as transmitter to transmit signal. For this, we are using Arduino UNO and a 9 V battery. The transmitting HCSR04 has its trig and echo pins connected to pins 12 and 11, respectively. The VCC and GND pins are connected to 5 V and GND pin of the Arduino, respectively.

4.2 Programming Part

For the programming part, first the algorithm is set which is shown in Fig. 4. The programme codes for Arduino IDE 1.3.5 and used newping library for ultrasonic receiver and transmitter and AFmotor library to run the motor.

Logic: At first, we decided to take digital input of the receivers to determine which direction the robot should go, but this started creating problems when we were in a closed room where the ultrasonic sound transmitted from the transmitter echoed to other receivers, thus making the robot confused, so we came up with another solution to determine the distance. The logic was to decide the shortest distance of the transmitter from the receiver when the receiver is getting signal. As we all know, the HCSR04 uses the sound travelling time in the air to determine the distance; the longer it takes for the reflected signal to arrive, the longer the distance will be. Now, if the transmitter is placed in front of a receiver, it will get direct signal, but other receivers will get echoed signal. As a result, the receiver in front of the transmitter will get signal faster, while other receivers will get signal later. So while calculating the distance, the receiver in front of the transmitter will always give lower output to the other receivers. Hence, the robot will be able to confirm that the transmitter is in front of it, and it can move or turn to that transmitter. Now, the robot needs to learn when to stop the following transmitter. For that, we are

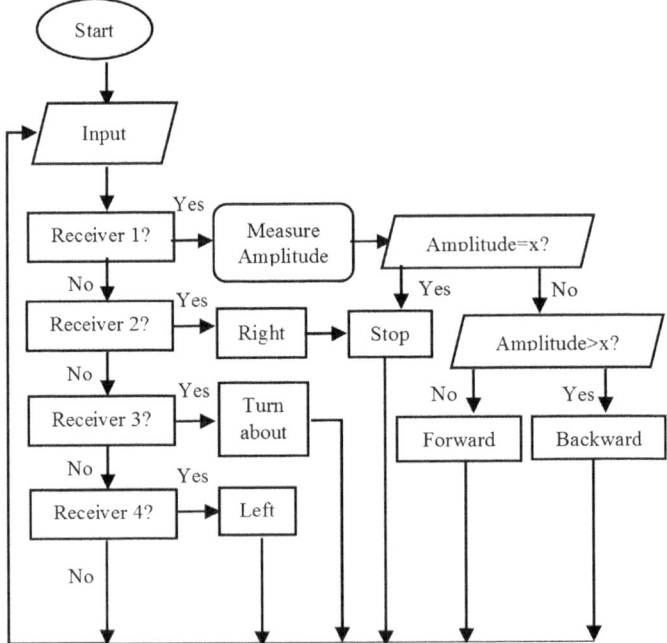

Fig. 4 Chart of programme algorithm

using IR proximity sensor, and when the analog reading of the IR sensor crosses a certain limit, the robot determines that it is close enough to the transmitter or an obstacle, so the robot will stop moving.

Transmitter programming: The code used to generate signal from the transmitter is given below. It only generates signal, so no logical understanding is needed. It is a simple distance measuring code from the Arduino library.

```
#include<NewPing.h>#define TRIGGER_PIN  12  // Arduino
pin tied to trigger pin on the ultrasonic sensor.
#define ECHO_PIN    11 // Arduino pin tied to echo pin
on the ultrasonic sensor.
#define MAX_DISTANCE 200 // Maximum distance we want to
ping for (in centimeters). Maximum sensor distance is
rated at 400-500cm.
NewPingsonar(TRIGGER_PIN, ECHO_PIN, MAX_DISTANCE); //
NewPing setup of pins and maximum distance.
voidsetup() {
  Serial.begin(115200); // Open serial monitor at 115200
baud to see ping results.
}
voidloop() {
  delay(50);  // Wait 50ms between pings (about 20
pings/sec). 29ms should be the shortest delay between
pings.
  Serial.print("Ping: ");
  Serial.print(sonar.ping_cm()); // Send ping, get
distance in cm and print result (0 = outside set distance
range)
  Serial.println("cm");
}
```

5 Results

MINION was tested for two different working environments, indoor and outdoor. It gives different results when it starts receiving signal from the transmitter in mentioned spaces.

5.1 Result Analysis for Indoor Signal

The reason behind this result for indoor is that the transmitted ultrasonic signal echoes and all the receivers are receiving the signal. In that case, one of four

Fig. 5 Receivers signal result graph (indoor)

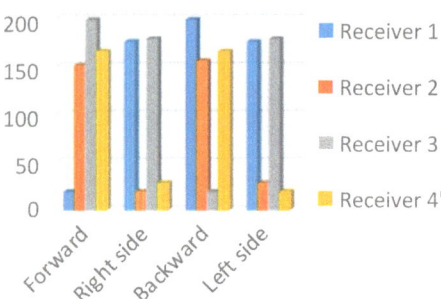

Fig. 6 Receivers signal result graph (outdoor)

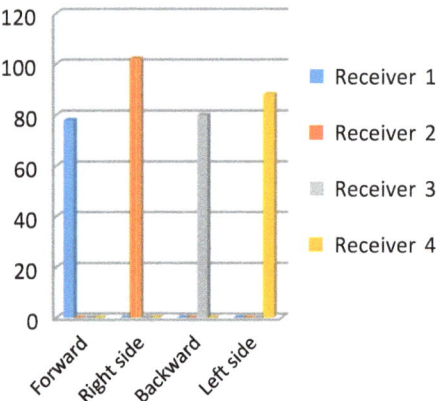

receivers gives the instruction to the Arduino Mega according to less distance of receiving signal. In Fig. 5, the graph of receiving signals of the receivers is shown for the indoor case.

5.2 Result Analysis for Outdoor Signal

On the other hand, as there is no echo from transmitting signal for outdoor place, only one receiver gets the transmitting signal and the values of this signal distance are more than zero. Other three receivers get no signal that means the signalling distance values are zero. The result for outdoor is shown in Fig. 6.

6 Future Work

For further work with this robot, RFID will be added to the robot and the transmitter for identifying that specification so that the robot can detect the received signal from its own transmitter. Like RFID, we can attach LED to the robot because of the dark place, and 3D camera with flashlight can be linked for surveillance purpose and getting a clear video recording.

7 Conclusion

In this digital world, this MINION can become more useful part and parcel in our life. People also feel more relaxed in working and travelling with heavy goods without any tension because they have MINION. We hope building MINION will get success in future and turn into the most important robot in the whole world.

References

1. Available: https://en.wikipedia.org/wiki/Infrared
2. Labourbehindthelabel.org: Bangladesh building collapse kills over 1100 workers: primark and mango labels found. http://www.Labourbehindthelabel.org/news/item/1140-bangladesh-building-collapse-kills-over-80-workers-primark-and-mango-labels-found
3. India-forums.com: One killed, several trapped in Bangladesh coal mine accident (2010). https://www.india-forums.com/news/bangladesh/247744-one-killed-several-trapped-in-bangladesh-coal-mine-accident.htm
4. Petersen, J.K.: Handbook of Surveillance Technologies, 3rd ed. pp. 495. USA (2012)
5. Petersen, J.K.: Handbook of Surveillance Technologies, 3rd ed. pp. 494. USA (2012)
6. Subramanian, Ramkumar, Rao, G.P., Sudarsan, Jayabharathi, R.: Semi-autonomous underwater surveillance robot. Int. J. of Electron and Electr. Eng., 7(1), 13–18, (2014)
7. Divya, V., Dharanya, S., Shaheen, S., Umamakeswari, A.: Amphibious surveillance robot with smart sensor nodes. Indian J. Sci. Technol. 6(5) (2013)

Genetic Algorithm-Based Approach for RNA Secondary Structure Prediction

Pradnya S. Borkar and A. R. Mahajan

Abstract In recent years, bioinformatics has become an essential subject for molecular biological study. The various available algorithms are used for analyzing and integrating biological data. Among many biological statistics RNA (ribonucleic acid) is one of the most important as it is used in protein synthesis. In computational molecular biology, the optimal secondary structure prediction of large RNA is a problem being faced today. RNA sequences of some virus are very large in number which requires a large amount of time for secondary structure prediction. Consequently, parallelization of algorithm is one of the solutions to diminish time consumption. This paper proposes the algorithm GAfold for predicting secondary structure of RNA on shared memory multicore architecture. The various RNA sequences as an input have been taken from Gutell database. For calculating minimum free energy, thermodynamic model is used and the outcomes are compared with existing algorithms.

Keywords RNA secondary structure · Genetic algorithm · Multicore architecture

1 Introduction

In recent times, the size of data in various fields such as medical sciences, computational biology, and bioinformatics has increased to such proportions that a lot more time is needed to process this agglomerated data. To manage this voluminous

P. S. Borkar (✉)
Computer Science and Engineering Department, Priyadarshini J.L. College
of Engineering, Nagpur, Maharashtra, India
e-mail: pmgajbhiye@gmail.com

A. R. Mahajan
Information Technology Department, Government Polytechnic,
Nagpur, Maharashtra, India
e-mail: armahajan@rediffmail.com

© Springer Nature Singapore Pte Ltd. 2018   397
K. Saeed et al. (eds.), *Progress in Advanced Computing and Intelligent Engineering*,
Advances in Intelligent Systems and Computing 563,
https://doi.org/10.1007/978-981-10-6872-0_37

data in a short span of time, concurrent processing is needed where the application runs in parallel, and the result can be achieved in short time. Due to the advancement in computer architecture, concurrent processing is now possible, and multicore architecture is one of the findings of new technological era which provides this facility. It is the platform to run the application in parallel. Bioinformatics is an essential part of molecular biological study in which the various available algorithms are used for analyzing and integrating biological data. Among many biological statistics, RNA (ribonucleic acid) is one of the most important as it uses in the protein synthesis. In computational molecular biology, the optimal secondary structure prediction of large RNA is a problem being faced today. RNA is an important factor in protein synthesis. The various types of RNA are ribosomal, messenger RNA mRNA), and the transfer RNA (tRNA). The RNA nucleotides comprise of nitrogenous base, the sugar, and the phosphate group. Adenine (A), cytosine (C), guanine (G), and uracil (U) are the four different bases of nitrogenous base in RNA. At the backbone of RNA, ribose five atom carbon-sugar are present which is counted from 1' through 5' and it is attached by two phosphate groups in 3' and 5', respectively. These bases are attached to the five-carbon sugar in 1' position [1]. The RNA molecule becomes charged molecule because of negative charged phosphate group in the backbone [2]. Due to this negative charge, the RNA molecule inside the living cells is not stable. Therefore, to obtain more steadiness, double-stranded RNA gets formed by folding back the single-stranded RNA on itself. The RNA primary structure is a string of four bases {A, C, G, U}, and the secondary structure of RNA is derived by pairing these four bases according to Watson–Crick and Wobble rules as {(A,U), (U,A), (G,C), (C,G), (G,U), (U,G)}. The 2D structure also called as secondary structure of RNA is formed by pairing these bases. The medical researcher uses the RNA molecule by analyzing its secondary structure for discovering the antiviral drugs on diseases like aids, ebola, cancer. While folding back, the RNA molecule releases some energy and this released energy is called free energy, and the RNA secondary structure is said to be optimal structure if it has minimum free energy (MFE). The pairing among four bases creates various types of loops such as stack loop, hairpin loop, interior loop, bulge loop, multiloop, external loop. The minimum free energy (MFE) of complete secondary structure is the summation of free energies of all loops. The Mfold [3], UNAfold, and RNAfold [4] are algorithms used by many biological researchers for analyzing RNA secondary structure. Simfold [5] also have been developed for the same problem solving. These algorithms predict secondary structure for RNA molecules having less than 1000 nucleotides accurately, but for longer RNA sequences, the accuracy is not up to the mark. According to thermodynamic hypothesis, the optimal RNA secondary structure is a structure with minimum free energy. We have implemented a parallel approach called GAfold for predicting secondary structure which runs on shared memory multicore architecture. This program runs on shared memory architecture for 40000 length of nucleotides on 04, 08, 28 core symmetric multiprocessor system. The GAfold is based on dynamic programming approach, and the optimization is carried out using genetic algorithm. The GAfold algorithm is parallelized at coarse-grain level parallelism. In this

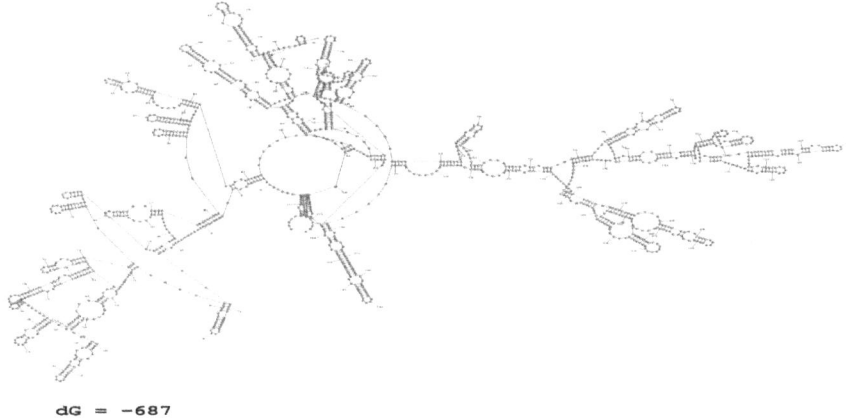

dG = -687

Fig. 1 RNA secondary structure of 1474 nucleotides

approach, the optimal energy of subsequences gets calculated in parallel which in turn calculates the optimal energy of full sequence, i.e., the optimal energy of the complete sequence is the summation of energy of each loop. RNA folding can be achieved by various methods. Here we have taken evolutionary method. Evolutionary method is an optimization technique, and its basic goal is to find out the best optimum result in any given problem. In the bioinformatics field, genetic algorithm plays an important role when it comes to the enlargement and improvement in the prediction of RNA secondary structure. It is an adaptive method for solving optimization problems based on its fitness value function. Figure 1 shows the RNA secondary structure of 1474 length of nucleotides.

2 Related Work

The various sequential and parallel algorithms have been developed for the prediction of RNA secondary structure. Waterman and Smith [6] used the mathematical analysis to predict RNA secondary structure as a function of free energy and they also included nearest neighbor parameter. They have used iterative method for a search over the entire configuration space of the RNA molecule. They focused on the two main problems first, the search to be performed on the entire RNA configuration space is large and assigning free energies to substructure is the second problem. Tinoco's approach [7] has been modified by Waterman and Smith [6] by defining the base pairing matrix. Tinoco et al. [7] noted the strategy for finding minimum free energy structure that the pairings for all structures present in base pairing matrix, and the optimal structure is represented by some set of non-overlapping antidiagonal strings of 1's. It occurred that direct observation or some counting schemes [8, 9] are required to infer maximal pairing configuration for short sequences. Number of

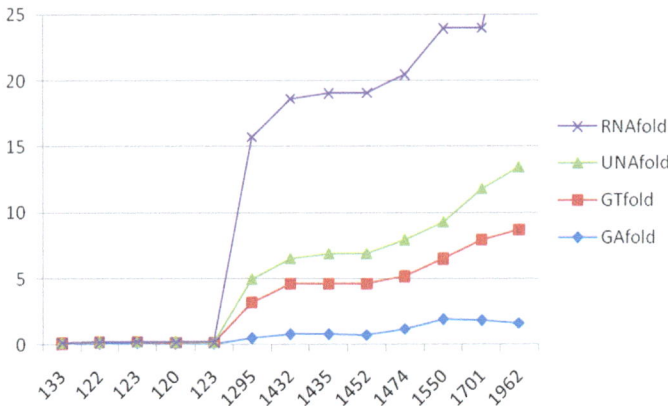

Fig. 2 Sequential time comparison (in sec) of GAfold, GTfold, UNAfold, and RNAfold

attempts is carried out to construct an algorithm to search for the optimal combination of such antidiagonal strings [8–10] but none of it searches the entire configuration space. Co-fold was found to be an optimal alignment algorithm based on Sankoff's algorithm which was developed for prediction of RNA secondary structure has the time complexity as $O(n^4\zeta\ (n))$ and $\zeta(n)$ can converge to $O(n)$. The RNAalifold algorithm computed the RNA secondary structures from multiple alignments with modified energy models. The time complexity of RNAalifold is $O\ (m \times n^4 + n^3)$, and space complexity is $O(n^2)$ where n is the RNA sequence length and m is the number of sequence alignments. RNAfold algorithm is an implementation of Zuker's prediction algorithm [11] based on MFE model and the time and space complexity of this algorithm is $O(n^3)$ and $O(n^2)$, respectively. Mfold [3] and RNAfold [4] are the sequential methods for the corrected RNA secondary structure prediction with $O(n^4)$ execution time and $O(n^2)$ spatial storage complexities (Fig. 2).

Many existing sequential algorithms have high computational complexities and also have limitation on length of RNA primary sequences. Therefore, various parallel approaches have been implemented by researchers to meet computational complexities like algorithms using multicore architecture, algorithms to run on GPU, parallel methods using MPI, and the implementation on FPGA. GTfold [12] is the most scalable design, and it is implemented on multicore CPU. The parallel implementation of GTfold algorithm is carried on 16-core dual symmetric multiprocessor system. Out of various types of loops as mentioned in Sect. 1, internalloop and multiloop are time-consuming loops. GTfold improved the internal loop speedup algorithm (ILSA) that helped to reduce the time complexity from $O(n^4)$ to $O(n^3)$. Due to the highly powerful and parallel structure, the GPU becomes potential approach in the field of computational biology and the high computational complex algorithms get speedup on GPU. Many parallel algorithms have been developed on GPU to speedup the RNA secondary structure prediction. In this paper, we have explained GAfold algorithm which is implemented on shared memory multicore architecture.

3   GAfold Algorithm

Secondary structure prediction with minimum free energy is an optimization problem. The GAfold algorithm implements parallelism for shared memory multicore architecture. The GAfold is implemented using genetic algorithm, and the fitness function is to calculate the free energy as it is specified that the secondary structure which has minimum free energy is the optimal structure. The algorithm is implemented on OpenSUSE Linux 11.3 operating system, also run on Red hat linux, and Ubuntu 14.4 using OpenMP 3.1 and g ++ compiler. The input RNA sequences are taken from Gutell database. The GAfold algorithm works as follows. In below algorithm, pragma block is written for parallel execution. The genetic algorithm is one of the best ways for solving an optimization problem. In the era of bioinformatics, genetic algorithm provides the way of natural genetics and creates the population of individuals. Generally, genetic algorithm works on the principle of selection and evolution to produce some optimized solution to a given problem. In GA, there are several chromosomes (individuals) and set of chromosomes forms population. Some pairs of chromosomes with good fitness value produce new individuals using some operations like crossover, mutation.

GAFold Algorithm

---

1. read four parameters, sequence file, population size, no. of iteration & no. of cores
2. Begin
   #pragma omp parallel shared(sequence, initialRNA population)
     create initial population()
     end
3. call generate Looppoint()
4. find the base pair and in turn loop formed by this base pairs
5. free energy of secondary structure = $\sum$ free energy of all loop
6. find bestindividual(optimal secondary structure)
      begin
   #pragma omp parallel shared (population, newRNApopulation)
      do crossover
      end
   return bestindividual(optimal secondary structure)

After applying these operations, new individual will be considered to produce next generation according to fitness value. Creation of initial population method finds the different RNA secondary structure with its minimum free energy. To find different RNA secondary structures from the same input sequence by n threads, the same input sequence is shared on shared memory architecture. The threads first generate the loop points, i.e., randomly selects the position of i and j in such a way that $1 < i < j < N$. N indicates the sequence length. The threshold value between i and j is set as 30 to have some elements between i and j so that some loop can be formed. If the base pair is found between i and j then store these loop point into loop point vector. If base pair does not get form then new values of i and j get are selected. After getting the loop points, the formation of various loops can be checked between these loop points. At the same time, the free energy associated with these different loops gets calculated. The initial population is then sorted according to the ascending order of MFE. The crossover method selects the first two individuals from this sorted list of initial population and finds the new individual. In crossover method, to avoid the dependency, the access pattern is selected as shown in Fig. 3. In GAfold algorithm, individuals are RNA secondary structures and its array forms the population, and from this array of population, the RNA secondary structures with lowest minimum free energy get selected and after

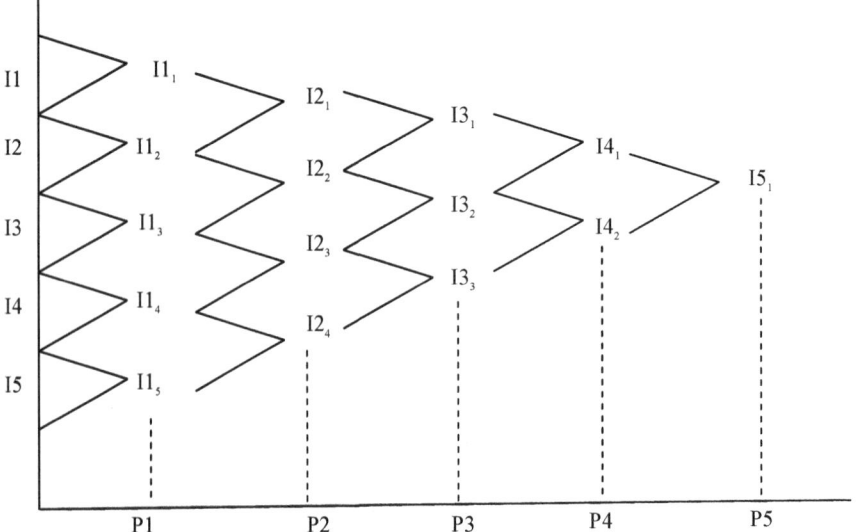

Fig. 3 Access pattern in crossover method

applying the crossover operation, new RNA structures are created which are stored in another array and the same procedure keeps getting repeated till the secondary structure with lowest minimum free energy is obtained. The dynamic scheduling is used for both the parallelization modules. The API OpenMp 3.1 is used as it is specially meant for shared memory architecture.

3.1 Minimum Free Energy

While folding back, the RNA molecule releases some energy and this released energy is called free energy. The RNA secondary structure which has minimum free energy is the optimal structure. The pairing among four bases creates various types of loops. These various types of loops are stack loop, hairpin loop, interior loop, bulge loop, multiloop, etc. The minimum free energy (MFE) of complete secondary structure is the summation of free energies of all loops.

$$MFE \ of \ RNA \ Secondary \ Structure = \sum free \ energy \ of \ all \ loops$$

4   Experimental Analysis

The empirical studies are carried out on Dell Precision Workstation Tower 79, Intel (R) Xeon(R) CPU E5-2695 v3 @ 2.30 GHz, 28 Cores with operating system Red Hat Enterprise Workstation release 7, Dell Precision Workstation Tower T5500, 08

Table 1 Free energy (kcal/mol) comparison of GAfold, GTfold, RNAfold, and UNAfold for 16S rRNA sequences

Accession number	Length	GAfold	GTfold	UNAfold	RNAfold
X00794	1962	−857.2	−741.9	−722.7	−746.6
X54253	701	−204.6	−149	−141.3	−149.03
X54252	697	−199.9	−142.5	−137.5	−142.52
Z17224	1550	−607.4	−564.8	−549.1	−565.12
X65063	1432	−628.3	−582	−570.8	−581.94
Z17210	1435	−749.2	−761.9	−626.6	−762.7
X52949	1452	−754	−802.7	−794.5	−804.4
X98467	1295	−549.8	−487	−460	−489.31
X59604	1701	−634.3	−573	−491.4	−574.7
K00421	1474	−702	−687	−682.1	−687.01

Table 2 Sequential time (in sec) comparison of GAfold, GTfold, RNAfold, UNAfold for 16S rRNA sequences

Length	GAfold	GTfold	UNAfold	RNAfold
133	0.064	0.065	0.014	0.015
122	0.076	0.107	0.012	0.013
123	0.109	0.109	0.012	0.012
120	0.064	0.107	0.013	0.015
123	0.08	0.08	0.013	0.012
1295	0.5	2.68	1.75	10.8
1432	0.8	3.83	1.89	12.13
1435	0.803	3.82	2.3	12.14
1452	0.766	3.85	2.29	12.2
1474	1.2	3.96	2.75	12.54
1550	1.89	4.62	2.78	14.67
1701	1.84	6.11	3.85	12.18
1962	1.65	7.06	4.75	23.94

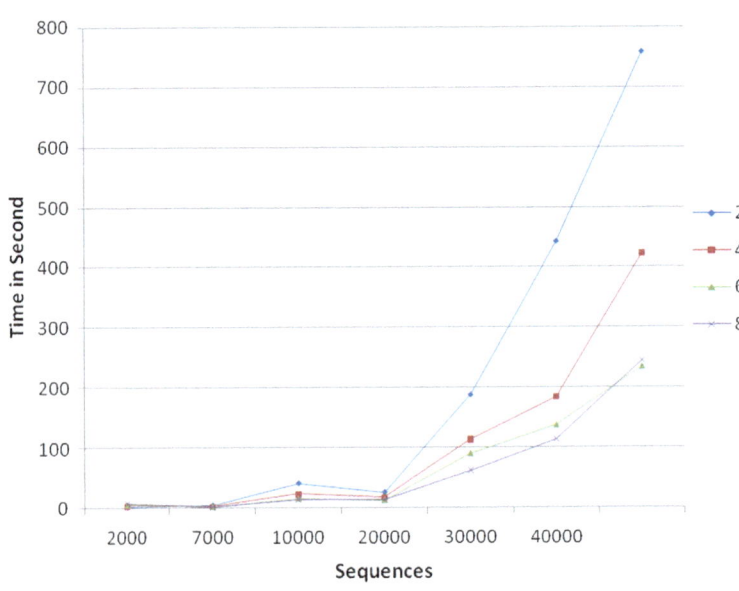

Fig. 4 Cores performance comparison of GAFold algorithm on 8 core machine

Cores with operating system Ubuntu14.04, and Intel(R) 04 Core(TM) i5-4200U CPU@ 1.6 GHz 2.3GHz with operating system OpenSUSE Linux 11.3. The RNA sequences as an input are taken from Gutell database. The following experimental results show the minimum free energy and execution time comparison.

Table 3 Speedup factor on various multicore architecture

Multicore architecture	Speedup factor
28 Core	≈ 3.9
8 Core	≈ 3.09
4 Core	≈ 2.5

Table 4 Time (in sec) comparison of GAfold on 08 core machine

Accession number	Sequence Length	No. of core			
		2	4	6	8
X007941	1962	5	2.78	2.095	1.79
ENST00000608442	6795	41.517	23.87	15.631	13.795
ENST00000454051	9793	26.36	17.28	13.012	13.869
ENST454051_2	19744	188.01	113.99	90.13	62.24
ENST454051_3	29537	443.3	184.64	137.307	113.445
ENST454051_4	39171	758.55	423.52	234.545	245.489

Table 1 shows the minimum free energy of GAfold, GTfold, RNAfold, and UNAfold where the minimum free energy of GAfold is miminum for maximum number of RNA sequences as compared to GTfold, RNAfold, and UNAfold, and Table 2 shows the sequential running time (in sec) of GAfold Vs GTfold, RNAfold, and UNAfold on single core.

The execution of GAfold, GTfold, RNAfold, and UNAfold is carried on 04, 08, and 28 core machine. The speedup factor on various multicore architectures is achieved as follows (Fig. 4; Table 3).

Table 4 shows the parallel running time on 08 core machine, the execution time required for 39171 nucleotides on 06 core machine is 234.5 s, whereas on 08 core machine it is 245.4 s, this may happen due to the communication overhead between processors. The following Table 5 shows the parallel execution time of GAfold on 28 core machine (Fig. 5).

Table 5 Time (in sec) comparison of GAfold on 28 core machine

Accession number	Length	1	2	4	6	8	10	12	14	16	18	20	22	24	26	28
X98467	122	0.11	0.08	0.06	0.06	0.05	0.05	0.06	0.05	0.04	0.06	0.05	0.04	0.05	0.08	0.08
K00421	1474	1.58	1.11	0.74	0.52	0.72	0.67	0.59	0.62	0.60	0.6	0.57	0.52	0.5	0.47	0.45
X007941	1962	7.13	4.9	2.78	2.09	1.79	1.83	1.90	1.85	1.76	1.73	1.56	1.46	1.25	1.09	1.02
ENST00000452079	6358	30.9	10.9	8.59	5.57	7.19	6.03	6.67	5.63	4.03	3.08	2.32	2.05	2	1.8	1.6
ENST00000458001	7169	29.3	15.0	9.71	7.56	7.24	7.48	9.64	8.5	8.33	8.2	8.07	7.9	7.68	7.2	7.01
ENST00000454051	9793	35.4	26.3	17.2	13.0	13.86	12.3	12.3	14.9	18.0	14.2	13.6	12.5	11.5	10.5	9.70

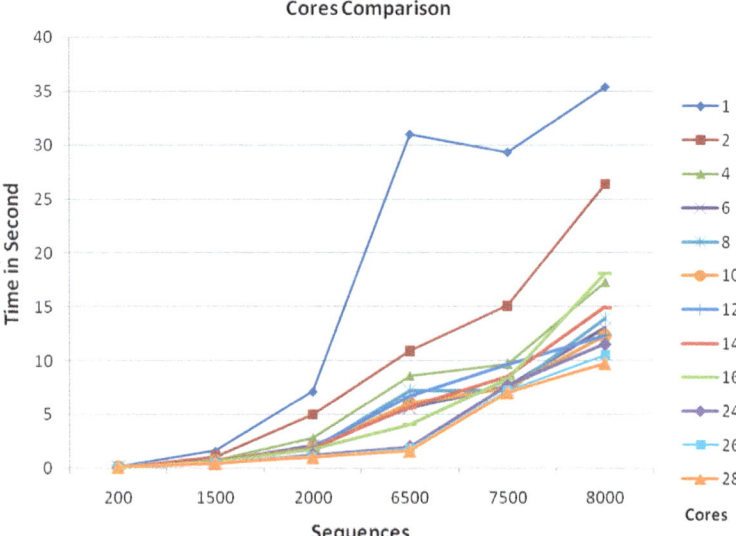

Fig. 5 Cores performance comparison of GAFold algorithm on 28 cores machine

5 Conclusion

In this paper, the overview of RNA secondary structure prediction using genetic algorithm is given. The GAfold algorithm is implemented on multicore architecture for shared memory environment. The experimental results show the time comparison in seconds for sequential and parallel execution. The algorithm is run on 04 core machine, 08 core machine, as well as on 28 core machine with approximate speedup factor as 2.5, 3.09, and 3.9, respectively.

References

1. Ra'e, M., Al-Khatib, R.A., Rashid, N.A.A.: A comparative taxonomy of parallel algorithms for RNA secondary structure prediction. Evol. Bioinform. 6, 27–45 (2010)
2. Draper, D.E.: A guide to ions and RNA structure. RNA **10**(3), 335–343 (2004)
3. Zuker, M.: Mfold web server for nucleic acid folding and hybridization prediction. Nucl Acids Res. **31**(13), 3406–3415 (2003)
4. Hofacker, I., Fontana, W., Stadler, P., Bonhoeffer, L., Tacker, M., Schuster, P.: Fast folding and comparison of RNA secondary structures. Monatsh. Chem. **125**, 167–188 (1994)
5. Andronescu, M., Aguirre-Hernandez, R., Condon, A., Hoos, H.: RNAsoft: a suite of RNA secondary structure rediction and design software tools. Nucleic Acids Res. **31**(13), 3416–3422 (2003)
6. Waterman, M., Smith, T.: RNA secondary structure: a complete mathematical analysis. Math. Biosci. **42**(3–4), 257–266 (1978)
7. Tinoco Jr., I., Ahlenbeck, O.C., Levhe, M.D.: Nature **230**, 362 (1971)

8. Klambt, D., Richter, O.: J. Theor. Biol. **58**, 319 (1976)
9. Lapidus, R., Rosen, B., Heppcrle, R.: J. Theor. Biol. **64**, 587 (1977)
10. Pipas, J.M., McMahon, J.E.: Proc. Nut. Acad. Sei. U.S.A. **72**, 2017 (1975)
11. Zuker, M., Stiegler, P.: Optimal computer folding of large RNA sequences using thermodynamics and auxiliary information. Nucleic Acids Res. **9**(1), 133–148 (1981)
12. Mathuriya, A., Bader, D.A., Heitsch, C.E, Harvey, S.C.: GTfold: a scalable multicore code for RNA secondary structure prediction. In: Proceedings of the 2009 ACM synposium on Applied computing, Honolulu, Hawaii (2009)

E-Commerce—Consumer Perceptive Model for Online Purchases

Ahmad Khalid Nazari and Archana Singh

Abstract Online shoppers get ample opportunity to buyers to shop online. The Web gives online shoppers extra channel for looking data of items and administrations, despite everything. But still, it has various issues to be addressed, basically the risk and fraud affecting the attitude and number of sales online. The paper explores the consumer perceptive model toward online shopping. It gives the insight of the factors affecting using structural equation modeling. The trust relationship is hugely affected between the customer and the online merchandiser if the risk to fraud vendor or product increases.

Keywords Customer perceptions · Trust relationship · Online shopping
Risk factor

1 Introduction

The progression of innovation, more quick and helpful shopping diverts are created lately. For instance, customers could purchase books from online book shops, they could arrange motion pictures tickets through the Web request frameworks, and they could purchase anything from the online stores regardless of where they are presently. Numerous better approaches for Internet shopping are acquainted with Taiwan as of late. In the shopping Web sites, the customer can access items in various categories with various options under one roof; this reduces the hunt stress of clients. In the electronic trade, Internet shopping is the new path utilized as a part of shopping channel [1]. The business technology business-to-client (B2C), the Web has significantly changed the perception and functional processes. For instance, online stores, which actualize digital marketing, sort out and keep up a

A. K. Nazari (✉) · A. Singh
Amity University Uttar Pradesh, Noida, India
e-mail: Khalidnazari0002@gmail.com

A. Singh
e-mail: archanaelina@gmail.com

© Springer Nature Singapore Pte Ltd. 2018
K. Saeed et al. (eds.), *Progress in Advanced Computing and Intelligent Engineering*,
Advances in Intelligent Systems and Computing 563,
https://doi.org/10.1007/978-981-10-6872-0_38

business system to convey their items and administrations to advertise on the contrary the customary business models enhance the capacity to make items and convey administrations [2]. What's more, the capacity of Web gives propelled look abilities, and Internet shopping model can give stock to shoppers crosswise over locales and national wilderness [2].

In spite of the fact that Web gives online shoppers extra channel for looking at the data of items and administrations, despite everything it has, a few issues to be explained. In case, when the customers shop on the Web, they cannot feel and touch the truth of the items or administrations what they require. They might stress over the security of transmitting charge card data by means of the Web [3]. It is an impediment that individuals see absence of security and protection on the Web in the reception of electronic business [4]. The trust relationship between clients and suppliers on Web that the suppliers of items and administrations on Web could offer more secure instruments to diminish saw danger of the online customers.

E-trade gives a stage to the trading of things online. This can keep running from asking for on the Web, through online movement of paid substance, to spending plan ary trades, for instance, improvement of money between records. Elizabeth Goldsmith and others (2000) reported that the general grouping of e-business can be isolated into two segments:

1. E-stock: offering products and administrations electronically and moving things through dissemination channels, for instance through Internet looking for basic needs, tickets, music, garments, equipment, travel, books, blooms, or blessings.
2. E-money: managing an account, platinum cards, shrewd cards, saving money machines, phone and Internet keeping money, protection, monetary adminis-trations, and home loans online (Elizabeth Goldsmith and others, 2000) [5].

As another promoting channel of trade, Web shopping incorporates a bigger number of hazards and risks than standard ones. Two basic reasons why customers don't purchase things or organizations on the Web are security of Web shopping and insurance of individual information. Before clients purchase the things or organizations on the Web, buyers cannot by any stretch of the imagination check the way of them. Buyers also feel slightness when they pass on the cash related in addition, singular information (e.g., recognizing confirmation numbers and charge card numbers) by method for the Web. Online trade consolidates differently inherent vulnerabilities. For example, shameful acts of estimating, infringement of security, transmissions of off base data, unapproved tracks of exchanges, and unapproved employments of charge cards [27–31] showed that apparent danger or misfortune would adversely impact saw value toward Internet shopping. [32] portrays that apparent hazard contrarily impacted both saw handiness toward Internet shopping and saw convenience to shop on the Web.

2 Related Work Done

This study endeavors to explore perils in Web shopping is understood as the shopping ways of customers or locales used for Web getting trades [6]. Web shopping media is one kind of the virtual retail stores have the taking after two characteristics: (1) stock cost correlations and utilization of data among choice things then again organizations and (2) brisk access of destinations of Web stores [7]. The Web shopping could diminish the time which the purchasers spend on shopping [2]. What's more, the sites of virtual stores give progressed looking abilities, and they are much of the time used to pursuit data of items or administrations some time recently obtaining [8]. By and large, Web shopping is a proficient channel for shopping. Web shopping is another channel to buy items or administrations on the Web. The curiosity to shoppers may come about in a few issues. Past examination has pointed out a relationship between the evident risk of another shopping channel and the choice of securing using that channel [2]. While customers see peril in most obtaining decisions, non-store securing decisions tend to have a bigger measure of seen risk associated with them [10]. Web shopping is one of non-store purchasing channels on the Web. Pires et al. [11] reported that Web shopping is late information advancement related sort of direct showcasing and is correspondingly seen as higher peril or hardship by clients and observed that risk contradicted customers are less disposed to shop on the Web. Since Liebermann and Stashevsky [4] proposed saw threat of clients, various specialists have analyzed the issue and have presented different created definitions [12–16]. Seen danger is a conceivably past measure of shopper seen convenience and saw usability toward obtaining on the Web. Seen danger is a development of seen circumstance [17] that has been characterized in different ways. Investigating earlier open deliberation on the meanings of saw danger of shoppers, Cunningham [14] restated that the researchers have favored the two noteworthy segments as a proper meaning of saw hazard: the likelihood of a misfortune and the subjective sentiment unfavorable outcomes. Seen hazard relates basically to looking and picking data of items or administrations before acquiring choices [18]. On the off chance that the genuine obtaining encounters of online clients contrast from their acquiring objectives, they will see higher danger [13]. [19] delineated that apparent danger relied on the subjective instability of the results. For every obtaining choice, the purchasers will have a few purchasing objectives or expected results of buying items or administrations. A few sorts of saw danger have been broadly utilized as a part of past exploration [12, 20, 21]. Case in point, monetary danger is the potential fiscal misfortune that customers may experience after obtaining specific items or administrations. Execution danger is seen as the probability that an item executes not surprisingly. Physical danger is identified with safe issues emerging from utilizing the item, particularly those straightforwardly identified with well-being and security. Mental danger is the likelihood that the chosen item will be steady with the shopper's mental self-portrait. Social danger is thought to be the impression of critical others toward the items or administrations. Accommodation hazard remains for the added substance hazardous hindrances which the customer will experience when they buy the items or administrations.

3 Research Methodology—Customer Perception Model

We conducted an experimental analysis of consumer perspective on online purchasing. We tried to deduct the perception of the customer shopping online. The focus of this model is the association between intention, fear, and strengths of a customer purchasing online. We conducted a survey using questionnaire asked many questions to the consumers who were frequent on online purchase. After the survey, we found correlation between all attributes then found that the factors shown in figure were highly correlated. The demographic details of the customer are shown in the table below. We explored the market of fashion apparels and footwear Web sites. We collected data in two forms: primary and secondary. The primary data were collected by the survey, and secondary data were collected from the fashion Web sites myntra.com and fashionandyou.com, and we picked the customer reviews by looking at the customers' attitude toward online purchase. The data were unstructured review or star rating, and we converted those unstructured data (like star rating of a product, positive, and negative adjectives used) into proper format. The respondents returned the responses handwritten and as well as via email. Eventually, 240 respondents were interviewed. In Table 1, it detailed the customers' frequency of online shopping. We used two experiences, risk associated with online shopping and perceived ease of use of online shopping (Fig. 1).

Table 1 Profile of number of respondents

Profile of number of respondents (240)	
Question	Count (%)
Age (years)	
19–21	21
22–24	33
25–30	38
More than 30	8
Gender	
Male	23
Female	67
Frequency of online shopping	
Three times a week	34
Five times fortnightly	42
More than five times in a month	24

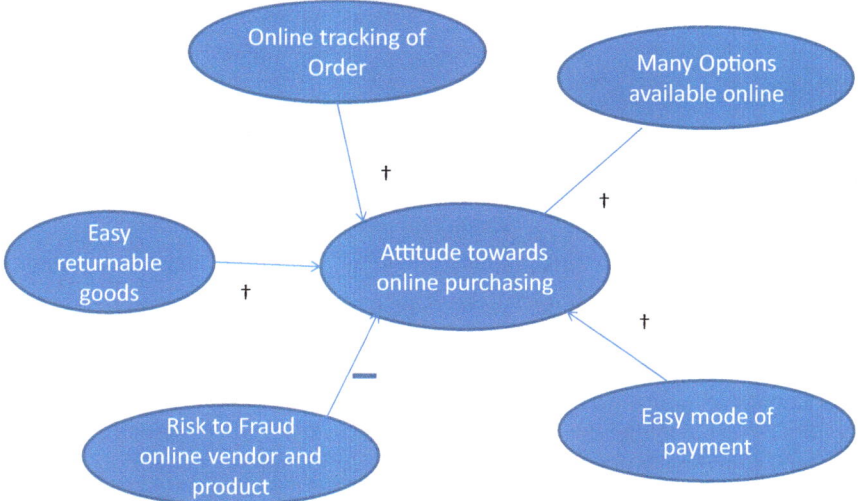

Fig. 1 Customer perceptive research model

4 Results and Discussion

We calculated the reliability and validity by using Cronbach's alpha and explora-tory factor analysis listed below Fig. 2 (Table 2).

All result scales are unidimensional and reliable. We estimated the model's parameters using the scales and structural equation modeling (SEM). The goodness of fit calculates the good fit with the data, and consequently we analyzed all the parameters.

The result shows the strong positive relation between attitude toward online purchase and many option available online, i.e., 0.91. The fashion footwear options available online also show the value 0.88. Another parameter which is also showing significant predictor in both the cases is risk to fraud online vendor and product.

Reliability coefficients for each construct		
	Fashion apparel (n=223)	Fashion Footwear's (n=212)
Many options available online	0.91	0.87
Easy returnable goods	0.65	0.62
Easy mode of payment	0.89	0.81
Risk to fraud online vendor and product	0.87	0.85
Online tracking of order	0.81	0.83

Fig. 2 Reliability coefficients and goodness of fit values

Table 2 Goodness of fit measure

Goodness of fit measure	Acceptable value	Fashion apparel	Fashion footwear
Chi-square	NS	$P < 0.001$	$P < 0.001$
RMSEA	<0.08	0.05	0.05
Goodness Fit index	Threshold not specified	0.90	0.88
Normed Fit index	0.90	0.91	0.92
Tucker-Lewis index	0.90	0.96	0.96

The research suggested that predictors of the attitude toward online purchasing are both perceived ease of use and risk associated with it. The factor, easy returnable goods in fashionable apparels and footwear both returned the negative impact on the attitude of online purchase. Today's customer feels comfortable with many options available to make the payment like cash on delivery, and also, the customer is happy to the track order online.

5 Conclusion

The study propelled customer attitude and apprehensions toward online shopping. To increase the percentage of customer acquisition in online shopping channel, the extant research shows that the supplier and distributors should improvise the online selling practices. The behavior study of customer shows that the intention and attitude toward online shopping are negatively associated with risk or fraud associated and various options available enhancing the purchasers. The merchandisers must reduce the risk or fraud to keep the brand name and value in the market of e-commerce.

Declaration

Authors have obtained all ethical approvals from appropriate ethical committee and consent from the individuals for publication who participated in this study.

References

1. Li, Y.-H., Huang, J.-W.: Applying theory of perceived risk and technology acceptance model in the online shopping channel. World Acad. Sci. Eng. Technol. **53** (2009)
2. Park, J., Lee, D., Ahn, J.: Risk-focused e-commerce adoption model: a cross-country study. J. Glob. Inf. Technol. Manag. **7**(2), 6–30 (2004)
3. Bhatnagar, A., Misra, S., Rao, H.R.: On risk, convenience and internet shopping behavior, association for computing machinery. Commun. ACM **43**(11), 98–105 (2000)
4. Liebermann, Y., Stashevsky, S.: Perceived risks as barriers to internet and e-commerce usage. Qual. Mark. Res. **5**(4), 291–300 (2002)

5. Davis, F.D.: Perceived usefulness, perceived ease of use, and use acceptance of information technology. MIS Q. **13**(3), 319–340 (1989)
6. Monsuwe, T.P.Y., Dellaert, B.G.C., Ruyter, K.D.: What drives consumers to shop online? A literature review. Int. J. Serv. Ind. Manag. **15**(1), 102–121 (2004)
7. Chiravuri, A., Nazareth, D.: Consumer trust in electronic commerce: an alternative framework using technology acceptance. In: Seventh Americas Conference on Information Systems (2001)
8. Dollin, B., Dillon, S., Thompson, F., Corner, J.L.: Perceived risk, internet shopping experience and online purchasing behavior: a New Zealand perspective. J. Glob. Inf. Manag. **13**(2), 66–88 (2005)
9. Tan, S.J.: Strategies for reducing consumers' risk aversion in internet shopping. J. Consum. Mark. **16**(2), 163–180 (1999)
10. Huang, W., Schrank, H., Dubinsky, A.J.: Effect of brand names on consumers' risk perceptions of online shopping. J. Consum. Behav. **4**(1), 40–50 (2004)
11. Pires, G., Stanton, J., Eckford, A.: Influences on the perceived risk of purchasing online. J. Consum. Behav. **4**(2), 118–131 (2004)
12. Mitchell, V.W.: Consumer perceived risk: conceptualizations and models. Eur. J. Mark. **33**(1), 164–196 (1999)
13. Bettman, J.R.: Perceived risk and its components: a model and empirical test. J. Mark. Res. **10**(2), 184–190 (1973)
14. Cunningham, S.M.: The major dimensions of perceived risk. In: Cox D.F. (ed.) Risk Taking and Information Handing in Consumer Behavior, pp. 82–108. Harvard University Press, Boston (1967)
15. Dowling, G.R., Staelin, R.: A model of perceived risk and intended risk handling activity. J. Consum. Res. **21**(1), 119–134 (1994)
16. Dowling, G.R.: Perceived risk: the concept and its measurement. Psychol. Mark. **3**(3), 193–210 (1986)
17. Cox, D.F., Rich, S.U.: Perceived risk and consumer decision-making: the case of telephone shopping. J. Mark. Res. **1**(4), 32–39 (1964)
18. Rindfleisch, A., Crockett, D.: Cigarette smoking and perceived risk: a multidimensional investigation. J. Public Mark. **18**(2), 159–171 (1999)

Data Rate and Symbol Error Rate Analysis of Massive MIMO for 5G Applications

Shipra Singh, Sindhu Hak Gupta and Asmita Rajawat

Abstract In this paper performance of the Massive MIMO is analyzed in terms of SER and SNR values. SER and high data rate is the current requirement for the 5G technology based on the observed result it may be concluded that Massive MIMO has lower SER values and this result is compared with the current MIMO technology. It is observed that SER values for Massive MIMO is comparatively very less than the MIMO system. Therefore, it will be beneficial to opt Massive MIMO concept for the next generation technology.

Keywords Massive MIMO · Generations · Bit error rate · Symbol error rate

1 Introduction

Wireless communication face a major disadvantage due to multipath fading. At the time of modeling or developing any communication system multipath fading is a key parameter feature of radio communication that needs to be taken into account. Any of the object that provides a reflective surface is responsible for the multipath fading. Multipath fading proves disadvantageous as it results in phase distortion and ISI (Inter-Symbol Interference). This problem can be solved by using various diversity techniques, equalization methods and multiplexing. Out of the mentioned methods diversity to huge extent can resolve the problems created by the multipath fading. Diversity can be achieved with the help of MIMO as it has the caliber of providing spatial diversity.

S. Singh (✉) · S. H. Gupta · A. Rajawat
Amity University Uttar Pradesh, Noida, India
e-mail: shiprasingh755@gmail.com

S. H. Gupta
e-mail: shak@amity.edu

A. Rajawat
e-mail: arajawat@amity.edu

© Springer Nature Singapore Pte Ltd. 2018 417
K. Saeed et al. (eds.), *Progress in Advanced Computing and Intelligent Engineering*,
Advances in Intelligent Systems and Computing 563,
https://doi.org/10.1007/978-981-10-6872-0_39

With the increasing number of users demand for higher data rates also increases, therefore Multiple-Input Multiple-Output (MIMO) emerges as an elementary feature for wireless communication. MIMO has engrossed much attention since last ten years and is considered as one of the best technique which may act as a fundamental technique for latest standard techniques. These techniques form the base of today's modern wireless communication system including LTE (Long Term Evolution) and Wi-MAX [1]. Conventional system which is named as Single Input and Single Output (SISO) system uses only a single trans-receiver antenna for signal transmission and reception at the transmitter and receiver. This system is not much efficient to provide substantial spectral gains and energy efficiency because of the only one antenna at both transmitter and receiver. Therefore need of a new technique came into the picture and concept of MIMO has been introduced. In MIMO systems multiple antennas can be used at both the sender and receiver side. Therefore, compared to conventional system i.e. SISO, MIMO system can accomplish abundant assets in spectral gains, power and energy efficiency. MIMO arrangement uses Time, frequency, space diversity and spatial multiplexing schemes which enable MIMO system to become advantageous [2]. MIMO system uses many antennas for the transmission and reception purpose and its spatial multiplexing technique enables it to provide higher data capacity by carrying traffics through various paths. Like every techniques MIMO system also have some disadvantages. First, Multiplexing gain is limited. Second, its multiplexing gain may disappear altogether in case of transmission medium unfavorable condition and strong interference at the cell edges. Extensive research in MIMO field leads to development of MASSIVE MIMO concept. MASSIVE MIMO is proposed as a latest technology for the wireless communication and considered as advancement in MIMO technology. It is used to overcome the disadvantages of MIMO [3–5].

LSA (Large Scale Antenna), VLM (Very Large MIMO), Full Dimensional MIMO etc. are the other name for the MASSIVE MIMO [6]. In MASSIVE MIMO system, base station is composed with several hundreds of low power trans-receivers to provide high spatial multiplexing gain than the conventional MIMO system [7]. For example, In Massive MIMO the number of trans-receivers at the base station is much larger compare to users in the system, therefore by using simple pre-coding and detection (uplink) techniques the effect of noise and interference can be mitigated at the base station which provides low complexity in signal processing and increase the performance facilities towards the users as well as base station [8–10].

MASSIVE MIMO traffic handling capability is more because it uses both Time and Frequency Division Duplexing. Since, it uses inexpensive low power components it proves to be cost effective and considered as a benefit of the Massive MIMO, other benefits of the Massive MIMO include robustness to interference and intentional jamming. MASSIVE MIMO increases the rate of data flow and reliability of link performance, as more trans-receiver antennas for transmission and reception, there is more possible paths for signals, it is one of the key features of the 5G that is required. Therefore use of Massive MIMO in 5G can contribute an important role by providing higher data rate.

Main contribution of this research work is that it has been shown that Massive MIMO attains a high data rate and lower SER in comparison to MIMO. This feature makes it most suitable for 5G technology. Further paper has been divided as follows: Section 2 represents the model for the system. Section 3 contains the performance, simulation result and Sect. 4 i.e. the last section represents conclusion.

2 System Model

The model which has been considered is given by Chuang et al. [2]. By using extra antennas Massive-MIMO enables to transmit and receive the signals into ever a smaller region by focusing the signal energy. For the model, it has been considered that as by using maximum aggregation data rate condition Massive MIMO can be able to provide services to multiple users with the higher data rate. Data rate for any u-th user can also be easily determined by equation number (1). Therefore, after analyzing it according to the massive MIMO system, it is found that data rate increases for the user as compared to the MIMO system. Consider a multiuser system, where U represent user entity and S represent subcarriers and where $U = \{1, 2, 3, …, U\}$ and $S = \{1, 2, 3, …, S\}$ which represents the number of the users and subcarriers respectively and D represent the Data transmission rate of the users. Therefore, the rate of data transmission for the U-th user Du is as-

$$Du = \frac{W}{S} \sum_{s=1}^{S} c_{u,s} \log_2(1 + SNR_{u,s}).\tag{1}$$

where W is the total assign frequency band for the network, $c_{u,s}$ represents the subcarrier assignment index which indicates whether the s-th subcarrier is occupied by the u-th user. Thus, the resultant equation is shown as follows-

$$D_{total} = \frac{W}{S} \sum_{u=1}^{U} \sum_{s=1}^{S} c_{u,s} \log_2(1 + SNR_{u,s})$$

$$c_{u,s} \in \{0,1\}, \forall u, s \tag{2}$$

$$\sum_{u=1}^{U} c_{u,s} = 1, \forall s$$

Therefore, Massive MIMO technology uses arrays of low-gain resonant antennas and is able to replace the convention base stations system. According to the Massive MIMO principle, arrays of the antennas distributedly to cover the entire city with a multitude of distributed antennas that serves many users.

Figure 1 shows the conventional MIMO system consists with 2, 4, and 8 antennas as well as Massive MIMO system which consist with array of many low profile antennas.

Fig. 1 Conventional MIMO system and MASSIVE MIMO system

3 Performance and Simulation Results

3.1 BER Performance for MIMO

The rate of occurrence of errors in any transmission media is known as a Bit Error Rate. BER depends on the transmission medium and BER will be insignificant and not affect much to system performance if, the medium is good enough between the source and destination and presence of noise is low, and this implies higher SNR ratio.

Figure 2 depicts the BER performance of MIMO for $m = 4$ and $n = 2$ at SNR = 35 db. Here m represents the number of receive antenna and n as users respectively.

The graph shows that initially at lower value of SNR (*Signal to Noise ratio*), SER (*Symbol Error Rate*) has higher value but as the value of SNR increases SER starts reduces.

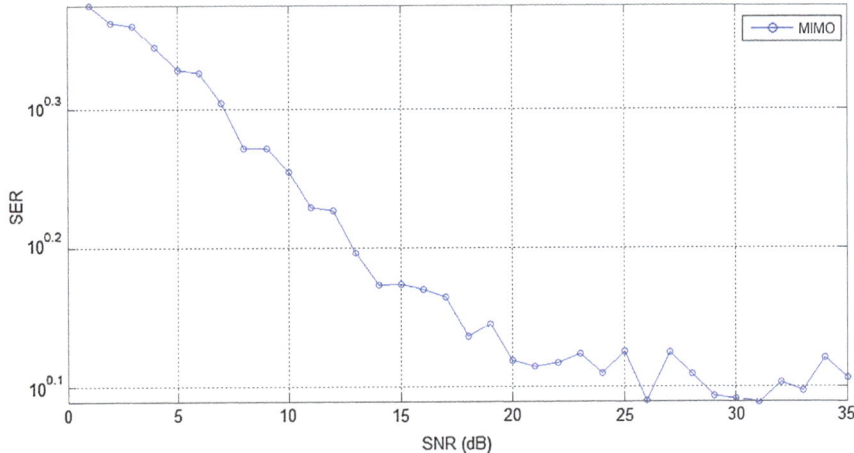

Fig. 2 BER performance of MIMO system for $m = 4$, $n = 2$

3.2 BER Performance for Massive MIMO

The above graph shows the relationship between SER and SNR for MASSIVE MIMO. The graph is drawn for the value of m = 130 and n = 16 at SNR of 35 db where m represents the number of receive antenna and n as users respectively. Figure 3 depicts that initially at lower value of SNR (1 db), SER has value of 1.947 but as SNR value start increasing SER values start decreasing and reaches to a value of 0.004 (at SNR value of 9 db).

3.3 BER Performance for MIMO and MASSIVE MIMO

Table 1 shows the BER performance for the MIMO and MASSIVE MIMO. Graph is plotted between SER and SNR respectively, SER stands for Symbol Error Rate and SNR stands for the Signal to Noise Ratio.

As it can observed from the graph as well as from the table that as the SNR value increases the SER value starts decreasing in both the cases i.e. for MIMO as well as for Massive MIMO but the decrement in the value of SER in Massive MIMO is more compared to MIMO. It may analyzed with table that at SNR = 9 db SER has value 0.004 for Massive MIMO and SER has value 1.848 for MIMO which is 99.78% more compared to Massive MIMO. Therefore, it can be concluded here that SER can be reduced to a great extent by using Massive MIMO instead of MIMO.

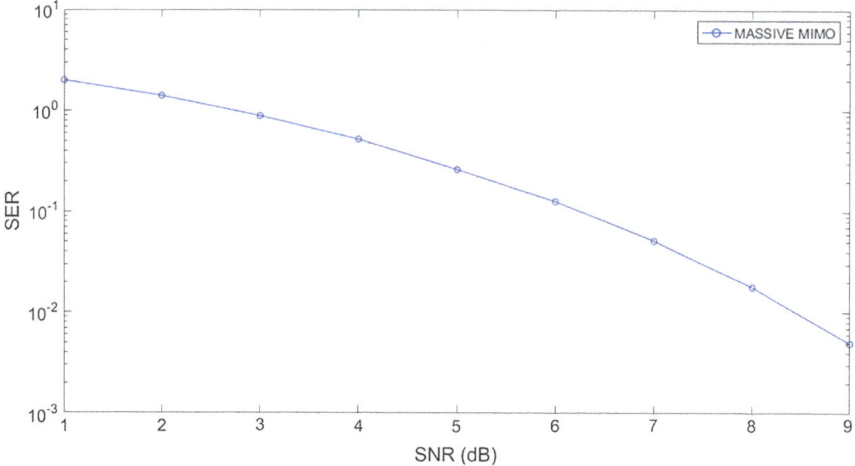

Fig. 3 BER performance for MASSIVE MIMO for m = 130, n = 16

Table 1 BER performance for MIMO and Massive MIMO

Sr. no.	SNR	SER for MIMO	SER for MASSIVE MIMO	% difference in SER
1	1	2.365	1.947	17.67
2	2	2.34	1.366	41.62
3	3	2.307	0.89	61.42
4	4	2.216	0.532	75.99
5	5	2.016	0.295	85.37
6	6	1.985	0.115	94.21
7	7	1.916	0.045	97.65
8	8	1.818	0.014	99.23
9	9	1.848	0.004	99.78

3.4 Data Rate Performance of Massive MIMO

The above graph shows the relationship between number of users, data-rates for the users and the sub-carriers which is drawn with the help of the maximum aggregation data-rate equation. The graph is drawn for the value of number of users U = 127, number of sub-carrier S = 10. In the graph, it is observed that the data-rates for users are high in this concept. Therefore, it may be concluded that using a Massive MIMO is an upcoming technology for 5G. It will be implementable for the 5G technology because achieving higher data-rate is a key requirement for the 5G technology which can be fulfilled with the concept of Data-rate for Massive MIMO (Fig. 4).

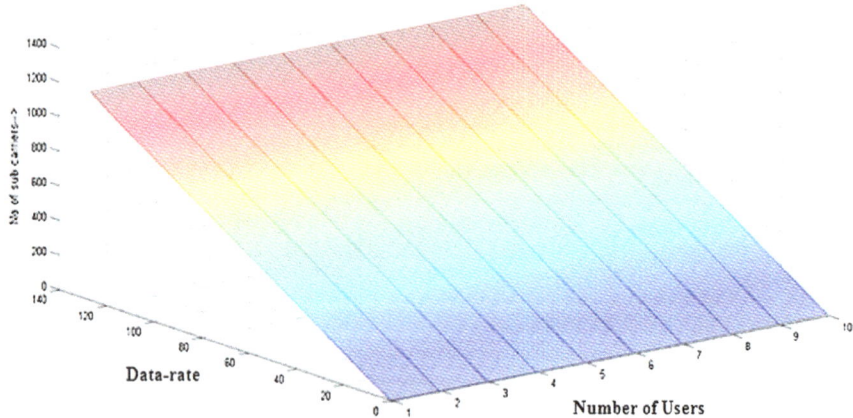

Fig. 4 Data-rate for MASSIVE MIMO

4 Conclusion

Although MIMO system has many advantages but it has certain limitation too of low multiplexing gain which makes it less efficient for the next generation. The drawbacks of the MIMO system can be overcome by Massive MIMO concept. Massive MIMO has the ability of the higher multiplexing gain and the higher data-rate. In the current work performance of Massive MIMO is analyzed in terms of SER and SNR. It is observed that the SER is very low for Massive MIMO at the higher values of the SNR. Also the observed SER value is very low compared to the MIMO SER values and makes Massive MIMO more efficient compared to the MIMO system. Low SER and higher data-rate is the key requirement for the 5G technology. Therefore, it can be concluded based on the observed results that use of the Massive MIMO will be more beneficial for 5G for getting higher data rate at low SER values.

References

1. Mietzner, J., Schober, R., Lampe, L., Gerstacker, W.H., Hoeher, P.A.: Multiple-antenna techniques for wireless communications—a comprehensive literature survey. IEEE Commun. Surv. Tutor. Sec. Quart. **11**(2), 87–105 (2009)
2. Chuang, M.-C., Chen, M.C., Yeali, S.: Resource management issues in 5G ultra dense small-cell networks. In: 2015 International Conference on Information Networking (ICOIN), Cambodia, pp. 159–164 (2015)
3. Caire, G., Shamai, S.: On the achievable throughput of a multi-antenna Gaussian broadcast channel. IEEE Trans. Inf. Theory **49**, 1691–1706 (2003)
4. Vishwanath, S., Jindal, N., Goldsmith, A.: Duality, achievable rates, and sum-rate capacity of Gaussian MIMO broadcast channels. IEEE Trans. Inf. Theory **49**, 2658–2668 (2003)
5. Gesbert, D., Kountouris, M., Heath, R., Chae, C., Salzer, T.: From single user to multiuser communications: shifting the MIMO paradigm. IEEE Signal Process. Mag. **24**, 36–46 (2007); Communications—a comprehensive literature survey. IEEE Commun. Surv. Tutor. **11**, 87–105 (2009)
6. Boccardi, F., Heath, R.W., Lozano, A., Marzetta, T.L., Popovski, P.: Five disruptive technology directions for 5G. IEEE Commun. Mag. **52**(2), 74–80 (2014)
7. Marzetta, T.L.: Noncooperative cellular wireless with unlimited numbers of base station antennas. IEEE Trans. Wirel. Commun. **9**(11) (2010)
8. Ngo, H.Q., Larsson, E.G., Marzetta, T.L.: Energy and spectral efficiency of very large multiuser MIMO systems. IEEE Trans. Commun. **61**(4), 1436–1449 (2013)
9. Wang, C.X, et al.: Cellular architecture and key technologies for 5G wireless communication networks. IEEE Commun. Mag. **52**(2), 122–130 (2014)
10. Larsson, E.G., et al.: Massive MIMO for next generation wireless systems. IEEE Commun. Mag. **52**(2), 186–95 (2014)

Prediction of Compressive Strength of Concrete Using M5' Model Tree Algorithm: A Parametric Study

Sarthak Jain and S. V. Barai

Abstract Concrete mix proportions proposed by empirical equations do not account for slight modifications. Mix design becomes primitive in case of introduction of new parameters as research progresses such as accounting for new kinds of admixtures, super-plasticizers or binders. In practice and theory, effect of age on compressive strength is correlated. Metric of concrete compressive strength with age is an important criterion in prediction problem. Prediction problem parameterisation as quantity or ratios is a controlling model choice decision. The existing codes of practice do not account for a standardised process of evaluating compressive strength. Considering the strength prediction problem here as a classification domain of input space, it is modelled using M5' model tree algorithm. The study conducted shows the performance promised by such a model to be accurate within statistical error.

Keywords Concrete compressive strength · M5' model trees · Prediction model

1 Introduction

Concrete is a mix of cement, water and aggregates by proportion. Booming construction industry requires investment of large amount of natural resources. Controlled choice of optimised mix design ensures economic use of natural resources and contributes to sustainable construction. Batching of concrete is a standard process followed according to the construction codes and advent into optimising concrete mix design calls for analysis of performance parameters such as slump, flow, compressive strength. In practice, it is usually not possible to test each and every sample proposed by theoretical analysis due to economic and time constraints, especially when

S. Jain (✉) · S. V. Barai
Indian Institute of Technology Kharagpur, Kharagpur 721302, West Bengal, India
e-mail: sarthak.jain@iitkgp.ac.in
URL: http://www.iitkgp.ac.in

S.V. Barai
e-mail: skbarai@civil.iitkgp.ernet.in

© Springer Nature Singapore Pte Ltd. 2018
K. Saeed et al. (eds.), *Progress in Advanced Computing and Intelligent Engineering*,
Advances in Intelligent Systems and Computing 563,
https://doi.org/10.1007/978-981-10-6872-0_40

the number of controlling independent variables is large. With observation, it is seen that usually most experimental variables can be split into dependent and independent variables hence forming a control set. An empirical relationship can be established between the independent variables and dependent variables through observation and analysis of test data. This gives rise to prediction and estimation modelling for parameters using test data which can be used to model various testable physical properties in terms of measurable physical parameters. Concrete is prepared as per mix design determined using standard codes of practice. The prediction models can easily determine the target strength of the concrete depending on the input parameters provided to the model. Prediction modelling strategy has to be clearly identified in terms of the inputs to be taken for the model. The works done aim at identifying how inputs have to be taken for a prediction problem in purview of concrete compressive strength. In general, concrete is prepared as per mix design from codes of practice. Such models aim at producing better estimation models for preparing concrete by controlling the materials used which shall be the model input parameters. The modelling aims at producing linear reproducible equations which can be easily used to estimate the compressive strength of concrete on the basis of the selected input parameters.

2 Literature Survey

Concrete compressive strength can be modelled by various methods which have been actively used in various applications. Conventional methods such as linear regression models have been used in modelling fatigue model fatigue strength of plain, ordinary and lightweight concrete subjected to compressive stress [9] and logistic regression models have been used in mapping of landslide hazard at Selangor, Malaysia [7]. These methods aim at producing linear predictors by way of error reduction in continuous domain or binary variables, respectively. Unconventional methods work by the principle of mapping data into nonlinear relationships by way of artificial neural networks like in a research conducted on predicting landslide for a susceptibility study. Bui et al. [3] where the target variable is represented in terms of a weighted sum of inputs expressed by a nonlinear mapping. Genetic expression programming searches the program space to find a function fitting the data given as shown in a study done to model prediction of velocity field [6]. Decision trees search the data space working as a classifier on the input variables to split the space into regions and fit them with linear models as demonstrated by the study conducted on predicting heart disease [4]. Traditional methods of regression analysis are not able to capture the effect of variation in various types and qualities of materials such as addition of fly ash, admixtures or aggregate replacement since they employ error minimisation. Non-traditional methods employ various other soft computing tools such as ANN which tend to give a complex non-mathematical, symbolic relation between input and target variables. Use of gene expression programming employs search through program space and since concrete is a nonlinear model, identifying such relations becomes a hefty task. Also, these methods become computationally demanding and hence time-consuming. Decision trees result in fewer factors of significance than traditional regression and neural networks, hence with a simpler structure produce accurate results as for predicting electrical energy consumption [10]. Hydrologic study of flood forecasting done in China with M5' model tree algorithm and neural networks shows that training of these models is faster than ANN, and they provide greater control over parameters' selection through tree-like linear models [8]. Prediction of wave height is a machine-learning problem, and comparison of instance-based learning methods ANN and M5' model trees shows that M5' model trees are able to produce piecewise linear models in form of simple, understandable rules in contrast with ANN where one has to set parameters in relation to number of layers, parameters, etc. [5]. Another example of application of this method is modelling water level–discharge relationship for an Indian River where again model trees seem to be a better choice in context of providing results at par with ANN in terms of accuracy but being more transparent and understandable [2]. A recent study conducted on predicting concrete's modulus of elasticity using this method showed reasonable performance over ANN [1]. Hence, the model tree method of predicting the properties is chosen for the current paper. The main objective of the paper is to carry out systematic parametric study considering various concrete related parameters and develop M5' models and based on the best performance, proposed the best model for the problem domain.

3   M5' Model Tree Algorithm: Background

M5 model works on building a tree with splitting at values of attributes to provide linear models through regression analysis at those split nodes. This method employs division of the data space into regions on the basis of input predictors which are then fit as linear models to produce the expected target values. The splitting criterion is based on maximising standard error reduction [11] of attribute and hence SDR, i.e. standard deviation reduction is computed for the attribute as given by Eq. (1)

$$SDR = sd(T) - \sum_i \frac{T_i}{|T|} \times sd(T_i) \tag{1}$$

T is the set of examples that reach the node and T1, T2 ... are the examples that reach the node. After the tree is built using linear regression models in attributes, the terms in sub-tree of an attribute are computed for least error and are dropped accordingly. This process of "pruning" the tree takes place according to the factor given by expression (2)

$$\frac{n+v}{n-v} \tag{2}$$

where n is number of training examples reaching that node and v is the number of parameters that represent class value at that node. The tree is pruned so long as the expected error reduces at the leaves. To account for any sharp discontinuities, the model is corrected by smoothing through applying a factor p' given by Eq. (3).

$$p = \frac{np + kq}{n + k} \tag{3}$$

Here k is a constant, q is value predicted at node, p is prediction passed from below node and n is number of training instances below the node. Thus, prediction accuracy is increased as the model accommodates for any sharp changes in prediction by smoothing out the values. Hence, the deviations are not very large from expected.

4   Model Development

The prediction model is being modelled considering concrete composed of natural aggregates, fly ash, silica fumes and different proportions of admixtures and water cement. Parameterisation of study in terms of normalising data by taking as ratios, considering age of concrete as an input and variation in number of splitting regions has been done in order to identify a well-suited model for predicting concrete compressive strength. Data has been collected from existing studies and works on natural aggregate concrete pertaining to variation of compressive strength of concrete with properties and quantities of constituting ingredient materials and age at which the strength has been tested. About 1030 data points are being taken by referring to published work on natural aggregate concrete [12–15]. For the current problem, following input parameters are being considered for the concrete: slag, cement, water, fly ash, coarse aggregate, fine aggregate, super plasticiser, age. The target output is the mean compressive strength of the concrete measured in MPa at given age in days. The model has been developed with the data set implementing the M5' model tree algorithm using data modelling procedure in R package as per input parameter choice designated in Table 1. The developed prediction model has been considered with taking 90% of data points as training sets and 10% of data points as testing sets. Mean error has been computed for the model thus developed as given by Eq. (4)

$$Error = \frac{|Target - Predicted|}{Target} \times 100\% \tag{4}$$

The process is repeated 10 times considering random splitting of the data into 9:1::train:test subsets of data, and the errors are computed. This ensures that the no bias error arises in the model during development. Since the model tree is controlled by number of splitting regions defined, the model has been defined for the number of divisions ranging from minimum of 1 to a maximum of 100. For ease of understanding, the models have been denoted as follows:

Table 1 Model development designation and parameters used

Model	Input parameters	Remarks	Output parameter
MLP1	S, C, F, W, CA, FA, SP, Age	C, S, &F as separate entities	CS
MLP2	B, W, CA, FA, SP, Age	C, S, &F as sum given by B	CS
MLP3	S, C, F, W, CA, FA, SP	C, S, &F as separate entities	28-Day CS
MLP4	B, W, CA, FA, SP	C, S, &F as sum given by B	28-Day CS
MLP5	S, C, F, W, CA, FA, SP, Age	Parameters taken as ratios; C, S, &F as separate entities	CS
MLP6	B, W, CA, FA, SP, Age	Parameters taken as ratios; C, S, &F as sum given by B	CS
MLP7	S, C, F, W, CA, FA, SP	Parameters taken as ratios; C, S, &F as separate entities	28-Day CS
MLP8	B, W, CA, FA, SP	Parameters taken as ratios; C, S, &F as sum given by B	28-Day CS

Acronyms

B	Binder
S	Slag
C	Cement
W	Water
F	Fly Ash
CA	Coarse aggregate
FA	Fine aggregate
SP	Super-plasticizer
CS	Compressive Strength (MPa)

5 Results and Discussion

The simulations for models have been carried out using R software, and studies have been performed for mean error and correlation between predicted values of compressive strength and actual values.

5.1 Study on Variation in Compressive Strength with the Age of Curing and with the Input Parameters Taken as Ratios

Figure 1 shows model performances with respect to increase in number of regions. The model performance has been reasonable within allowed statistical variation (Refer Table 2). For 95% confidence, the margin of error for a random sample population is given by Eq. (5)

Fig. 1 Mean error variation
with number of regions

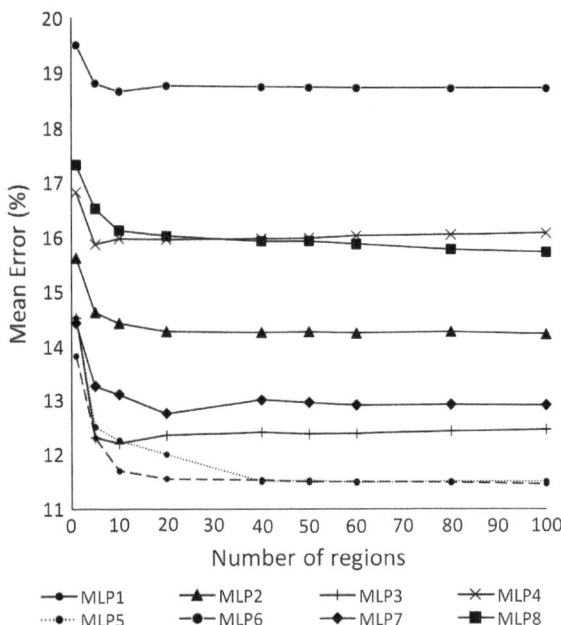

Number of regions

Table 2 Error, standard deviation and correlation for models

	Age of concrete included				Only 28 days concrete strength included			
	Non-ratio		Ratio		Non-ratio		Ratio	
Concrete model	MLP1	MLP2	MLP5	MLP6	MLP3	MLP4	MLP7	MLP8
CS mean	9.03	8.75	8.7	9.5	4.92	6.17	9.73	6.18
CS SD	10.51	13.49	11.66	15.52	8.20	10.67	7.66	9.72
CS R	0.964	0.953	0.963	0.951	0.926	0.859	0.858	0.886
CS RMSE	3.29	3.73	3.34	3.81	3.39	4.61	4.61	4.21

$$\text{Margin of error at 95\% confidence} = \frac{0.98}{\sqrt{n}} \qquad (5)$$

Hence for the present sample size of 1030, an error margin of 5% is obtained. Various observations can be derived out of the modelling experiment which are enlisted as follows:

- Error reduces and correlation between target and predicted values increases with increasing the number of splitting regions in the model. (Refer Fig. 3–Fig. 10)
- Reducing the number of attributes used in developing the model increases the error. (Refer Table 2: comparing MLP1 to MLP2, MLP3 to MLP4, MLP5 to MLP6 and MLP7 to MLP8)

Table 3 Variable usage for models

	Age of concrete included			
	Non-ratio		Ratio	
Variables	Conditions	Model	Conditions	Model
Age	97	93	95	62
Cement	51	97	30	62
Slag	44	91	24	47
Water	40	72	70	55
CA	7	68	17	25
FA	18	74	23	30
Ash	7	78	29	23
SP	26	60	7	25

- Age is an important variable for prediction problem posed which controls the compressive strength (Refer Table 3). This suggests that the curing age is a highly important factor in terms of the prediction of compressive strength of concrete.
- Taking the input parameters in terms of ratios rather than quantities of material does not contribute significantly to the performance of the model (Refer Fig. 3–Fig. 6). This suggests that the performance of the model is invariant of the input type taken for the models such as quantity or ratios to each other.

Considering the generated models, with correct choice of modelling strategy and a larger data set, the performance of prediction can be improved further to produce highly reliable system of prediction. On the basis of the parametric study, it has been identified that material quantity should be in form of quantities per metric cube of concrete defined as separate inputs; the splitting regions should be taken at a level of 20 and age of concrete should be considered to characterise a reliable prediction model.

6 Closing Remarks

M5' model tree algorithms are well suited for nonlinear prediction problems and perform within reasonable domain of error within the given data set. Statistical studies done with the M5' model tree algorithm on concrete composition gives a better insight into choice of modelling methodology in terms of the number of splitting regions to be taken as well as the normalisation for the data. Conducted parametric study proves useful in developing prediction models with reasonable accuracy for concrete. With a stochastic inference from the parametric variation study, key takeaways are in the form on invariance with data normalisation and high contributing effect of number of splitting regions to prediction performance for a decision tree-based modelling method. Concrete properties can be modelled in terms of its constituent materials and hence, prediction and mix design become fairly easy. With a reasonably large data set, a reliable prediction model can be produced for the concrete model.

References

1. Behnood, A., Olek, J., Glinicki, M.A.: Predicting modulus elasticity of recycled aggregate concrete using M5' model tree algorithm. Const. Build. Mat. **94**, 137–147 (2015)
2. Bhattacharya, B., Solomatine, D.P.: Neural networks and M5 model trees in modeling water level-discharge relationship for an Indian river. ESANN Proc. 407–412 (2003)
3. Bui, D.T., Tuan, T.A., Klempe, H., Pradhan, B., Revhaug, I.: Spatial prediction models for shallow landslide hazards: a comparative assessment of the efficacy of support vector machines, artificial neural networks, kernel logistic regression, and logistic model tree. Landslides 1–18 (2015)
4. Dangare, C.S., Apte, S.S.: Improved study of heart disease prediction system using data mining classification techniques. Int. J. Comput. Appl. **47**(10), 44–48 (2012)
5. Etemad-Shahidi, A., Mahjoobi, J.: Comparison between M5' model tree and neural networks for prediction of significant wave height in lake superior. Ocean Eng. **36**(15), 1175–1181 (2009)
6. Gholami, A., Bonakdari, H., Zaji, A.H., Akhtari, A.A., Khodashenas, S.R.: Predicting the velocity field in a 90 open channel bend using a gene expression programming model. Flow Measur. Instrument. **46**, 189–192 (2015)
7. Lee, S., Pradhan, B.: Landslide hazard mapping at Selangor, Malaysia using frequency ratio and logistic regression models. Landslides **4**(1), 33–41 (2007)
8. Solomatine, D.P., Xue, Y.: M5 model trees and neural networks: application to flood forecasting in the upper reach of the Huai River in China. J. Hydrol. Eng. **9**(6), 491–501 (2004)
9. Tepfers, R., Kutti, T.: Fatigue strength of plain, ordinary, and lightweight concrete. InACI J. Proc. **76**(5) ACI (1979)
10. Tso, G.K., Yau, K.K.: Predicting electricity energy consumption: a comparison of regression analysis, decision tree and neural networks. Energy **32**(9), 1761–1768 (2007)
11. Wang, Y., Witten, I.H.: Induction of Model Trees for Predicting Continuous Classes. Department of Computer Science, University of Waikato pp. 128–137 (1996)
12. Yeh, I.C.: Modeling of strength of high-performance concrete using artificial neural networks. Cement Concr. Res. **28**(12), 1797–1808 (1998)
13. Yeh, I.C.: Design of high-performance concrete mixture using neural networks and nonlinear programming. J. Comput. Civil Eng. **13**(1), 36–42 (1999)
14. Yeh, I.C.: A mix proportioning methodology for fly ash and slag concrete using artificial neural networks. Chung Hua J. Sci. Eng. **1**(1), 77–84 (2003)
15. Yeh, I.C.: Analysis of strength of concrete using design of experiments and neural networks. J. Mater. Civil Eng. **18**(4), 597–604 (2006)

Adsorption of CO_2 Using Modified ZSM-5 Zeolite in Petrol Engines

P. Baskara Sethupathi, M. Leenus Jesu Martin and J. Chandradass

Abstract Global warming effects on earth are caused by several factors. To understand the overall effects of global warming on earth, we have to understand the contributions and effects of each component of the planet. The exhaust gases from vehicles, power plants, and other sources are building up in the atmosphere, acting like an unimagined thick blanket over our planet. It has been clearly identified that additional effective technologies are needed to control CO_2 in the atmosphere. In the current study, absorption of CO_2 is achieved in a petrol engine using modified ZSM5. The reduction of CO_2 by absorption on Cu-ZSM5 and Li-ZSM5 is compared.

Keywords Li-ZSM5 · Cu-ZSM5 · Petrol engines · Exhausts emissions Adsorption of CO_2

1 Introduction

1.1 Exhausts Emissions Norms

The planet is getting warmer day by day. Most climatologists suggested that the final decade of the twentieth century as the hottest in the past millennium. Even slight alterations in worldwide temperature will cause a series of weather extremes and alter the climatic patterns of the planet [1]. In the last 10–15 years, the growing concern over global warming has led to the promulgation of legislation on fuel

P. Baskara Sethupathi (✉) · M. Leenus Jesu Martin
Department of Automobile Engineering, SRM University, Chennai, India
e-mail: sethupathi.b@ktr.srmuniv.ac.in

M. Leenus Jesu Martin
e-mail: hod.auto@ktr.srmuniv.ac.in

J. Chandradass
SRM University, Chennai, India
e-mail: chandradass.j@ktr.srmuniv.ac.in

© Springer Nature Singapore Pte Ltd. 2018
K. Saeed et al. (eds.), *Progress in Advanced Computing and Intelligent Engineering*,
Advances in Intelligent Systems and Computing 563,
https://doi.org/10.1007/978-981-10-6872-0_41

433

consumption for commercial vehicles. In the USA, emission standards (US Federal Register, Volume 76, No. 179, September 15th, 2011) have been introduced for green house gases (GHG), namely carbon dioxide (CO_2), nitrous oxide (N_2O), and methane (CH_4) emissions. Japan and China are introducing fuel consumption regulations. During 2012 and 2013, the European Commission (EC) gathered expert opinion from relevant stakeholders and is expected to publish proposals for control of carbon dioxide in vehicles. With importance being given to carbon dioxide reduction, the future norms in India are expected to have strict regulations on carbon dioxide emission. Hence, the carried out research will assist companies in making this technology commercial.

2 Design Concept and Engine Overview

2.1 *Engine Overview*

See (Tables 1 and 2).

2.2 *Design Concept*

In the first case, the engine is first attached to a three-way catalytic converter. The data is recorded. Then the after system carbon dioxide trap is attached. For comparison of Cu-ZSM5 and Li-ZSM5, two separate carbon dioxide traps are made

Table 1 Engine overview

Engine make	Briggs and Stratton
Engine model no	20S232-0036-F1
Displacement	305 cc
Bore	3.12 in.
Stroke	2.44 in.
Max power	10 hp
Exhaust specifications	Comes with stock muffler
Fuel type	Gasoline

Table 2 Engine dynamometer test data

Power	10 hp
Torque	19.8 Nm
Exhaust back pressure	12.6 mmHg (Idle rpm) 32.9 mmHg (Max rpm)
Exhaust temperature	586 °C (Idle rpm) 704 °C (Max rpm)

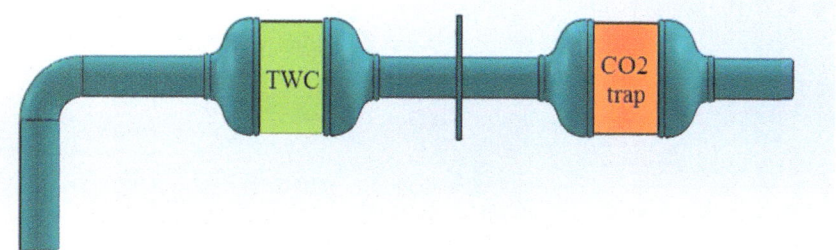

Fig. 1 Representation of after treatment TWC arrangement

which can be connected to the three-way catalytic converter with the help of a flange. Exhaust gases for each case are recorded and graphs are plotted (Fig. 1).

2.3 Adsorbent Selection Criteria

The key characteristic parameters used for the selection of absorbent are highlighted below:

- Disintegration temperature of the absorbent
- Desorption temperature;
- Limiting reaction (saturation pressure)
- Back pressure
- Number of active spots remaining post-process
- Commercial availability and feasibility

Based on these parameters, we evaluated activated carbon: BEA zeolite and ZSM-5 zeolite. Additionally, we zeroed down to ZSM-5 zeolite as the base material because of its higher affinity toward absorbing/adsorbing carbon materials. Its higher disintegration, desorption temperature, and availability show the practical usage (Fig. 2).

2.4 Properties of Modified ZSM-5 Zeolite (Cu-ZSM5/ Li-ZSM5)

Properties of modified ZSM-5 zeolite are as follows:

- Empirical formula Cu/LinAlnSi96–nO192 · 16 H₂O 19 < n < 96.

Fig. 2 ZSM-5 zeolite pellets

- The crystallographic unit cell of modified ZSM-5 has 96 T sites (Si or Al), 192 O sites, and a number of compensating cations depending on the Si/Al ratio that ranges from 12 to infinity [2].
- Modified ZSM-5 of any pore size can adsorb and absorb carbon dioxide and hydrocarbons to some extent. It has a pore size ranging from 4.6 to 5.4 angstroms.
- No deterioration upto temperature of 1100 °C.
- Modified ZSM-5 zeolite limiting factor is calculated as 1.75.
- Inexpensive and available in varied sizes, powder, extrusions.
- For any adsorbent/absorbent, limiting factor for the adsorption–desorption cycle must be in the range of 1.5–2.0. Modified ZSM-5 zeolite is in adequate limit of 1.75.

2.5   Design Specification

2.5.1   Three-Way Catalytic Converter

1. Volume of substrate: 540 cc
2. Material of substrate: cordierite (ceramic)
3. Substrate configuration:

   3.1. Cells per square inch: 400
   3.2. Wall thickness: 6.5 mil
   3.3. Geometric surface area: 2.74 $m^2\ lt^{-1}$

3.4. Open frontal area: 0.765%

4. Back pressure: 5 kPa (tested)
5. Converter casing material: SUS 409

2.5.2 Zeolite Adsorbent

1. Adsorbent material: ZSM 5
2. Amount of modified zeolite: 100 gm
3. Type of packing: loosely packed with wire mesh
4. Wire mesh configuration:

    4.1. Number of wire mesh layers: 3
    4.2. Weight of each layer: 33.8 gm
    4.3. Material of wire mesh: SUS 304

5. Material of casing: SUS 409
6. Total weight: 573 gm

2.5.3 Calculations for Adsorbent

Converter volume = Volume flow rate/Swept Volume [3]
  Volume flow rate = Swept volume × Number of intakes per hour = 0.000305
  × 1900 × 60 = 34.77 m^3/h
  Density = 1.43 kg/m^3
  Mass Flow Rate = Volume flow rate × density = 47.63 kg/h
  Space velocity = 34.77/0.00065 = 53492 h^{-1}

3 Manufacturing

3.1 Preparation of ZSM5 Pellets in Cuso$_4$ Solution

A solution of CuSO$_4$ and Li using water was made. The ZSM-5 pellets were soaked in this solution for 24 h. The immersed pellets were then subjected to a temperature of 120 ÂC. Once the pellets absorbed the CcSO$_4$ solution, then they were left to dry (Fig. 3).

Fig. 3 Zeolite treatment

Fig. 4 Packaging of zeolite

4 Fabrication

Wire mesh pouches were made, and these pellets were then filled in them. The purpose of the wire mesh pouch was to hold the pellets in place and place them efficiently inside the reactor chamber. The chamber was sealed using welding (Figs. 4, 5 and 6).

5 Exhausts Emissions Results

Graphs showing the reduction of gases in three setups.

Fig. 5 Converters outer shell

Fig. 6 Cu converter

5.1 Back Pressure Results

The experiment for back pressure was conducted at Sharda Motor Industry Ltd. (Figs. 7, 8, 9 and 10 and Tables 3, 4, 5, 6). The graph between mass flow rate and pressure obtained is under (Fig. 11, 12, 13 and 14 and Table 7).

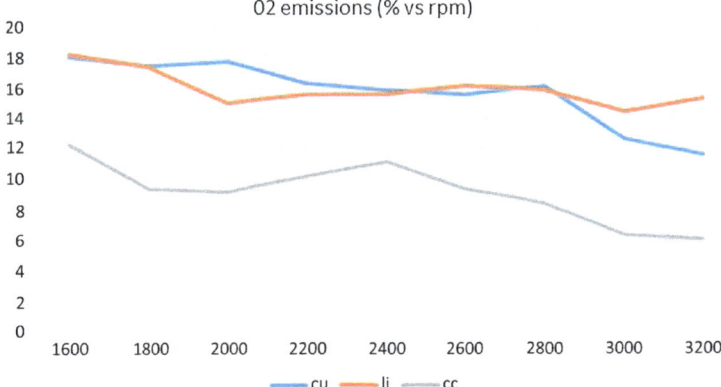

Fig. 7 O$_2$ emissions

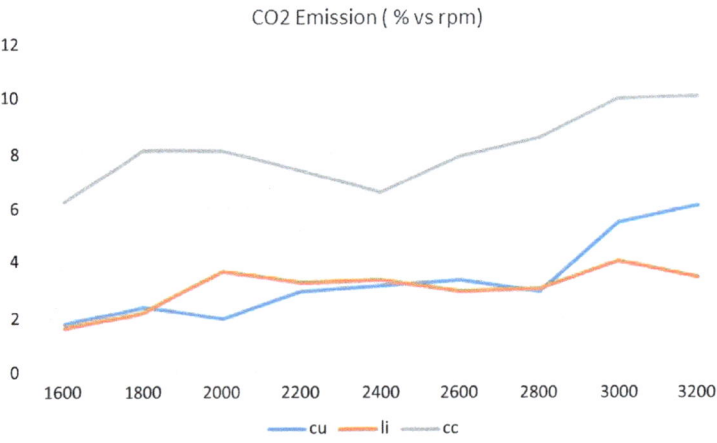

Fig. 8 CO$_2$ emissions

6 Conclusion

In this paper, it has been presented the effort of the authors to decrease the emission of CO$_2$ by adopting carbon capture techniques and storage mechanism, i.e., an adsorption technique is used to reduce the carbon emissions from the exhaust gas. Modified zeolite is used as the solid adsorbent. The absorbent blocks and seizes carbon molecule/particles in the exhaust.

This is the first attempt in the automobile sector for controlling CO$_2$ emission in the exhaust. To achieve this objective, the author had successfully designed a carbon storage model and analyzed its fluid flow inside the system. From the result

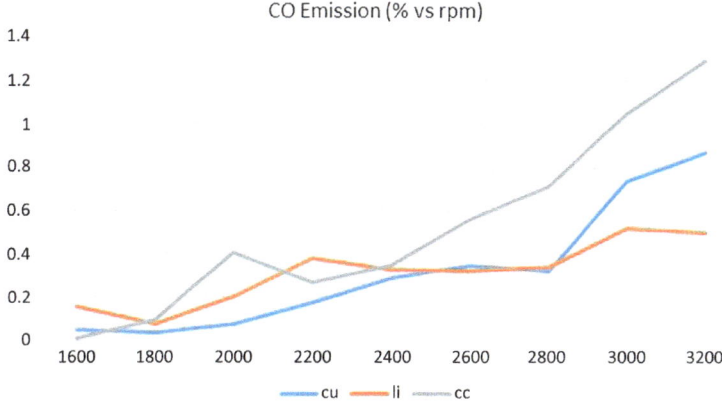

Fig. 9 CO emissions

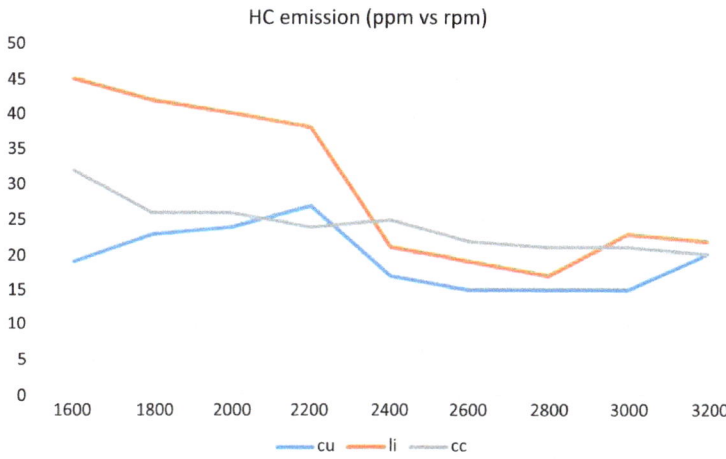

Fig. 10 HC emissions

of the CFD, analysis emphasizes that the back pressures are very less which means design is safe.

The outcome of the results shows the signification of the work done by the author.

442
P. Baskara Sethupathi et al.

Table 3 With only catalytic converter

RPM	CO (%)	HC PPM	CO_2 (%)	O_2 (%)	NO (%)	λ
1600	0.01	32	6.20	12.31	0	2.972
1800	0.10	26	8.10	9.35	0	2.13
2000	0.41	26	8.10	9.16	0	2.048
2200	0.27	24	7.40	10.20	0	2.307
2400	0.35	25	6.60	11.16	0	2.575
2600	0.56	22	7.90	9.38	0	2.072
2800	0.71	21	8.60	8.42	0	1.863
3000	1.05	21	10.1	6.36	0	1.526
3200	1.29	20	6.06	6.06	3	1.473

Table 4 With catalytic converter and Li reactor

RPM	CO	HC	CO_2	O_2	NO	λ
1600	0.16	45	1.6	18.25	6	
1800	0.08	42	2.2	17.39	1	8.462
2000	0.25	40	3.7	15.02	1	4.715
2200	0.38	38	3.3	15.62	0	5.160
2400	0.33	21	3.4	15.59	0	5.118
2600	0.32	19	3.0	16.21	0	5.814
2800	0.4	17	3.1	15.84	0	5.538
3000	0.52	23	4.1	14.45	0	4.059
3200	0.50	22	3.5	15.28	0	4.741

Table 5 With catalytic converter and Cu reactor

RPM	CO	HC	CO_2	O_2	NO	λ
1600	0.05	19	1080	18.10	0	
1800	0.04	23	2.40	17.45	0	8.097
2000	0.08	24	20.	17.45	0	9.449
2200	0.18	27	3.0	16.40	0	6.097
2400	0.29	17	3.20	15.87	0	5.489
2600	0.35	15	3.40	15.63	0	5.109
2800	0.32	15	3.0	16.13	0	5.794
3000	0.74	15	5.50	12.66	0	2.965
3200	0.87	20	6.10	11.67	0	2.607

Table 6 Maximum percentage change in CO_2, CO, and HC at specific RPM

Reactor	CO_2 reduction (%)	CO (increase)	HC (%)
Idle RPM			
Li reactor	72	20	26 (increase)
Cu reactor	70	15	45 (decrease)
Max RPM			
Li reactor	70	53	15 (decrease)
Cu reactor	40	38	No significant change

Fig. 11 Back pressure graph

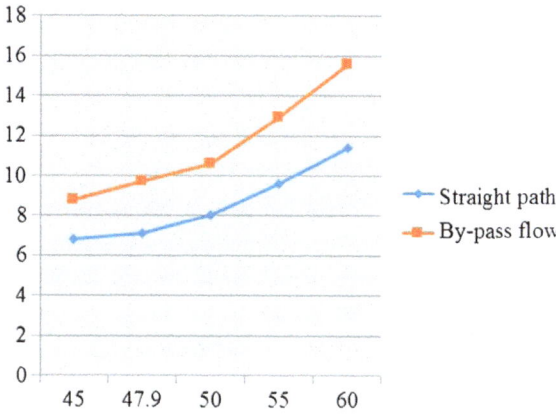

CFD ANALYSIS

Boundary Condition

Inlet condition		External condition
Mass flow rate (kg/h)	Temperature (°C)	Outlet pressure
47.63	704	1atm

➤ Inlet condition
✓ Mass flow rate : 47.63kg/h
✓ Temperature : 704°C

➤ Outlet condition
✓ Outlet pressure : 1atm

➤DOC-1
✓ Cell density : 400/4
✓ Porosity : 0.8136
✓ Inertial resistance : 2.64kg/m⁴
✓ Viscous resistance : 1833kg/m³ s

➤DOC-2
✓ Cell density : 400/4
✓ Porosity : 0.8136
✓ Inertial resistance : 2.64kg/m⁴
✓ Viscous resistance : 1833kg/m³ s

Fig. 12 Boundary condition

P. Baskara Sethupathi et al.

Uniformity Index

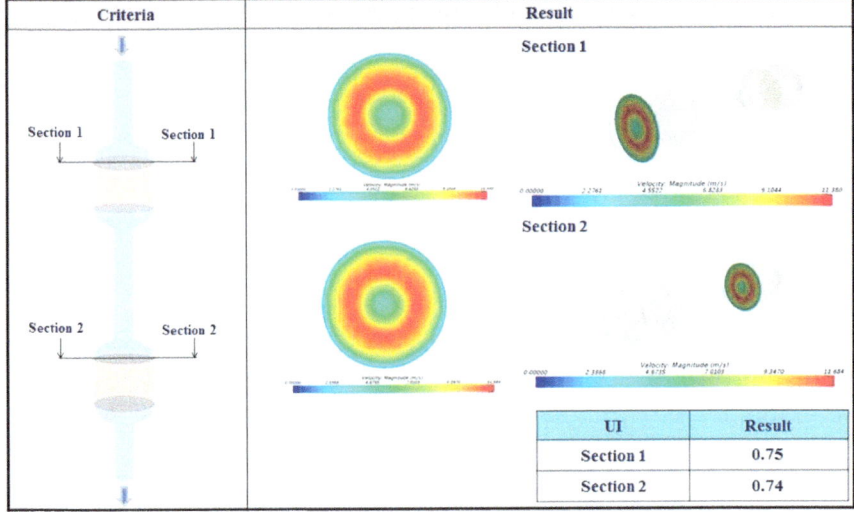

Fig. 13 CFD results

Fig. 14 Treated zeolite

Table 7 Bypass versus straight flow tabulation

Bypass flow			
S. no	MFR (kg/h)	Pressure (kPa)	Pressure (mbar)
1	45	0.888	8.8
2	47.9	0.97	9.7
3	50	1.06	10.6
4	55	1.29	12.9
5	60	1.56	15.6
Straight flow			
1	45	0.68	6.8
2	47.9	0.71	7.1
3	50	0.8	8
4	55	0.96	9.6
5	60	1.14	11.4

References

1. Sullivan, J.M., Sivak, M.: Carbon Capture in Vehicles: A Review of General Support, Available Mechanisms, and Consumer-Acceptance. UMTRI 2012–12
2. Jennifer, W.: Overview of Carbon Captures Methods
3. Recent Developments in CO_2 Removal Membrane Technology

Integration of Python-Based MDSPLUS Interface for ICRH DAC Software

Ramesh Joshi, Swanand S. Kulkarni and S. V. Kulkarni

Abstract Ion cyclotron resonance heating (ICRH) data acquisition control system (DAC) system for 100 kW, 45.6 MHz has been designed for RF ICRH system. This system is used for control and monitoring for 2, 20, and 100 kW RF amplifier stages, respectively. ICRH system consists of different power supplies for each RF amplifier stages. PLC-based DAC has been used for various power supplies monitor, control, and interlocks. Acquired data has been stored in MDSPLUS tree in binary format with defined signal list. Python is an open source scripting language which has been used for core development of software system. Python is very popular programming language for embedded as well as high-level programming. MDSPLUS is an open source package used for data archival and visualization. It is data archival and analysis tools used in various fusion experiments for data logging and visualization with configuration. It has standard developed program interface available with most of the popular languages like Fortan, C++, Python, Java, MATLAB, and IDL for graphical display of data and browsing. The paper introduces the interface of Python-based API for integration of Python and MDSPLUS for ICRH DAC software. Module allows user to read/write data directly in binary tree of MDSPLUS hierarchy. Java-based jScope tool is a part of MDSPLUS database which is used for data visualization.

Keywords PLC—programmable logic controller · DAC—data acquisition and control · ICRH—ion cyclotron resonance heating · Python

R. Joshi (✉) · S. V. Kulkarni
Institute for Plasma Research, Bhat Village Gandhinagar,
Ahmedabad 382428, India
e-mail: rjoshi@ipr.res.in

S. V. Kulkarni
e-mail: kulkarni@ipr.res.in

S. S. Kulkarni
Gujarat Technological University, Ahmedabad 382424, India
e-mail: kulkarnis1994@gmail.com

© Springer Nature Singapore Pte Ltd. 2018   447
K. Saeed et al. (eds.), *Progress in Advanced Computing and Intelligent Engineering*,
Advances in Intelligent Systems and Computing 563,
https://doi.org/10.1007/978-981-10-6872-0_42

1   ICRH DAC

High power ICRH system is an integral part of future fusion reactors and one of the important auxiliary heating systems utilized for tokamak heating experiment [1]. The DAC software with PLC-based hardware is design and developed for control and operation for ICRH on tokamak. There are several power supplies that have been installed and operated with the DAC system for monitoring and control. 2 and 20 kW stages use plate power supply each. For 100 kW system uses various power supplies installed which are screen grid, plate, filament, and control grid for each one stage. All these power supplies would have different voltage and current monitoring and control with different voltage settings and raise/lower functionalities. All these power supplies should be monitored and controlled by developed system with desired interlocks. Pre-ionization experiments are carried out in tokamak Aditya at different loop voltages, different magnetic fields to vary the position of the resonance layer, and also in the wide range of pressure range for the tokamak operation in ion cyclotron frequency range [2] (Fig. 1).

2   MDSPLUS Interface with Python

Model data system (MDSPLUS) is a data handling, log and analysis system used widely in the fusion experiments [3]. This data analysis and acquisition system has been adopted by various systems around the world [4]. It provides a powerful expression evaluator [4]. Python consists MDSPLUS as site package to import all available functionality to use data archival and analysis system for program interface. Python has ctypes functionality to integrate the C programming-based library implemented in MDSPLUS Python package. The numerical functionality within MDSPLUS has been provided by NumPy library implemented in Python for arrays and textual scalars. This module integrates Python and MDSPLUS using application program interface. Module allows user to put data directly in binary tree of MDSPLUS hierarchy.

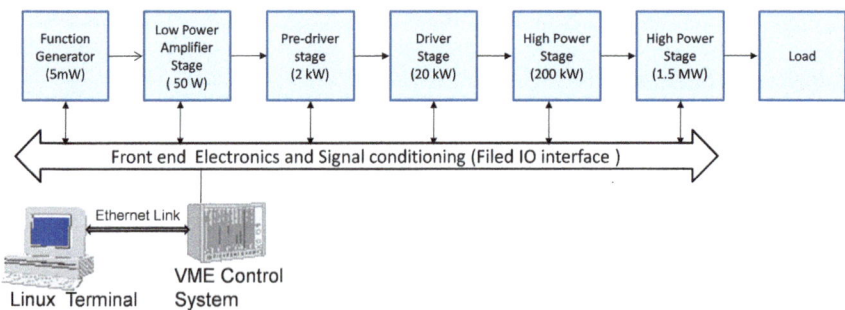

Fig. 1 Block diagram of ICRH generator DAC

The raw data has been acquired by the computer which is extracted from the data packet as per the channel number and type like analog or digital. Acquired data is stored in two main formats like binary or ASCII. These data are coming as data packet that would be stored mainly in file format not using RDBMS because it is in chunk of bytes, so it has been stored as hierarchical directory structure. The data is stored in the binary file format. There are three files which have been generated with each shot number named as characteristics, tree structure, and actual file data, respectively. The real-time data access layer of MDSPLUS is currently available for Windows, Linux, VxWorks, and RTAI [5, 6].

MDSPLUS provides the ability of having multiple concurrent writers and readers to the MDSPLUS data files [7]. Also multiple data acquisition applications can be writing to the same data file as they read each measurement. This functionality is quite powerful but does require the use of file locking to ensure that processes writing data do not corrupt data actively being written by other processes and that readers of the data do not attempt to read data when the data is in the process of being written. This file locking does slow down the data acquisition process somewhat, but when it occurs on local disk devices the performance degradation is minimal.

3 Advantages of Python Interface Module

Given the fact that Python is a high-level language and well suited for embedded systems, it is very easy to implement it with the required system of ICRH [8]. The following are the advantage points for using Python system.

- Well suited for embedded systems,
- High performance compared to other languages,
- Easy implementation with MDSPLUS package for data storage,
- Higher transparency rate between DAC and MDSPLUS storage system,
- Comparatively low complexity in case of coding script of API interface,
- Better integration with jTraverser and jScope for data visualization,
- More functionalities available in Python package.

Various new improvements and updates are available in the market for Python-MDSPLUS package, which makes implementation and usage of the system better and more resourceful.

4 Integration and Test Results

4.1 Application Programming Interface Code

The API code for ICRH DAC–MDSPLUS interface has been developed using Python, Shell IDE, on a Win 32-bit operating system.

Required software for implementation of the code:

- Python 2.7.9,
- MDSPLUS 7.0,
- NumPy 1.9.2 Win 32 superpack,
- SciPy 0.16.1 Win 32 superpack.

API Script: Increment Pulse number after each iteration.

```
From MDSplus import *
Import os,sys
TreeName = 'icgl456'
ShotNumber = -1

TreeObject1 = Tree(TreeName,ShotNumber)

#ICGL456 TREE NODES
TreeNodes = ['KW200_CG_C',.........        'KW200_RP_POW']

#.DATA SYSTEM FILES
DataFiles = ['kw200_cg_c.data',        'kw200_rp_power.data']

#DIRECTORY PATH SET ACCORDING TO THE INCOMING DATA
DirPath = "C:\ICGL456_DATA"


"""NEW PULSE CREATION SCRIPT"""
try:
TreeObject1.createPulse(PulseNumber)
    TreeObject1=Tree(TreeName,PulseNumber)
print TreeObject1
except:
print "ERROR: PULSE FAILED" #PULSE FAILED, SYSTEM EXIT.
sys.exit()

for j in range (0, len(FileObject4)):
FileData.append(float(FileObject4[j]))
try:
        Node = TreeNodes[i]
        TreeObject2 = TreeObject1.getNode(Node)
print "STATUS---> TREE NODE ACCESSED.\n"
print "CURRENT TREE---> ",TreeObject1
print "CURRENT NODE---> ",TreeObject2
        TreeObject3 = Float64Array(FileData)
TreeObject2.putData(TreeObject3)
        #print "DATA IN NODE--->",TreeObject3
```

Summary: The above script starts the pulse number of the Tree from has been stated from 1 and increment as per execution of the script. Pulse is created for the tree, and the pulse number is appended in the text file "ICGL456_PULSE.txt". As tree is accessed on the newly created pulse, data is acquired from in the specified path. The script accesses each text files and tree nodes simultaneously, in an ordered manner. Exception error has been implemented at each stage of the script to avoid system from implementing garbage value. Script has been using the MDSPLUS Python package import which provides the facility to have access of Tree object by which creation and manipulation with tree node data can be possible.

API Script: Reset and Delete Pulse

```
From MDSplus import *
Import os,sys

TreeName = 'icgl456'
ShotNumber = -1

TreeObject1 = Tree(TreeName,ShotNumber)
print TreeObject1

"""PULSE FILE OVERWRITE"""
print "\n\NPULSE FILE WILL BE OVERWRITTEN ON THIS NUMBER"
PulseNumber = input("\n\nPULSE RESTART NUMBER: ")
TreeObject1 = Tree(TreeName,PulseNumber)

"""DIRECTORY LIST"""
DirList = []
forDirName in os.listdir(DirPath):
DirList.append(DirName)

"""PULSE NUMBER EXTRACTION SCRIPT"""
PulseList = []
forListcount in DirList:
    Reverse = Listcount[::-1]
IndexNumber = Reverse.index('_')
RevPulse = Reverse[:IndexNumber]
Listcount = RevPulse[::-1]
IntDirPulse = int(Listcount)
    #print IntDirPulse
PulseList.append(IntDirPulse)

"""DELETING REST OF PULSE"""
NewPulseRestart = 1+PulseNumber
NewLength = 2+len(RestartArray)
for m in range(NewPulseRestart,NewLength):
TreeObject1.deletePulse(m)
```

```
"""DATA INSERTION IN ICGL456_PULSE.txt"""
FileIncrement = open("ICGL456_PULSE.txt","w")
NewIncrement = PulseNumber+1
NewIncrementArray=[]
for n in range (1,NewIncrement):
NewIncrementArray = n
FileIncrement.write(str(NewIncrementArray)+'\n')
FileIncrement.close()
```

Summary: Less frequent use of the resetting pulse value has been observed. Reset and delete Pulse script takes an input from the user; this input is considered as a pulse number from which the user wishes to reset the pulse. Specific pulse is searched and accessed by the tree; tree gets overwritten on that pulse so that new and correct data can be imported in the tree nodes. In the reset pulse execution, all the pulse files get deleted after the user defined pulse, and text file "ICGL456_-PULSE.txt" is updated with current pulse number. Thus, resetting the API to a new pulse number is customized as per user application (user defined). There are three functionalities that have been implemented as per the requirement of the script. Extraction, insertion, and deletion are the different operations of the node items that have been used for the proper operation during actual testing.

4.2   Result and Screen Shot

4.2.1   jTraverser (MDSPLUS)

As shown in Fig. 2, ICGL456 Tree; pulse file has been created, tree nodes accessed and data inserted in the node respective to the data file. Currently, 16 Tree nodes are visible of ICGL456 tree. Signal data type is created, which contains numeric data, as dimensional value or timestamp is known for each value (1 ms).

4.2.2   jScope (MDSPLUS)

MDSPLUS package comes with the jScope as an analysis tool used for graphical representation tool [9]. As shown in Fig. 3, jScope displaying eight different node data of Tree ICGL456, on shot number (pulse) 1 + 2; eight different waveforms in various colors are successfully visible for data analysis. More tree nodes can be added in the window on further implementation. The details given in the figure are dummy data which has been acquired during dummy testing.

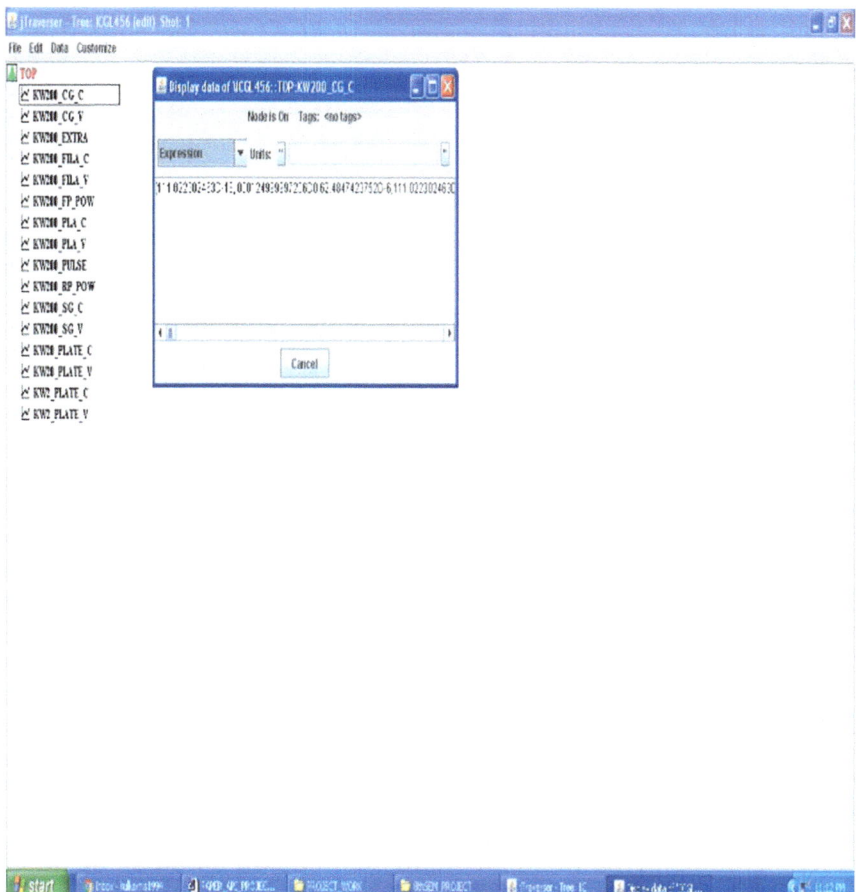

Fig. 2 jTraversus ICGL456 tree

4.2.3 Increment Script

As shown in Fig. 4, increment script increments the Pulse number on each iteration of execution and the Pulse numbers are successfully inserted in the text file, in vertical order (append). On next execution, Pulse number from file is analyzed, extracted, and incremented and executed. Pulse generated after the user-defined number is deleted instantly, high probability of them containing garbage data. Text file is updated back to the user-defined number. Increment script again reads the text file and according to that creates a new Pulse and increments the list in text file.

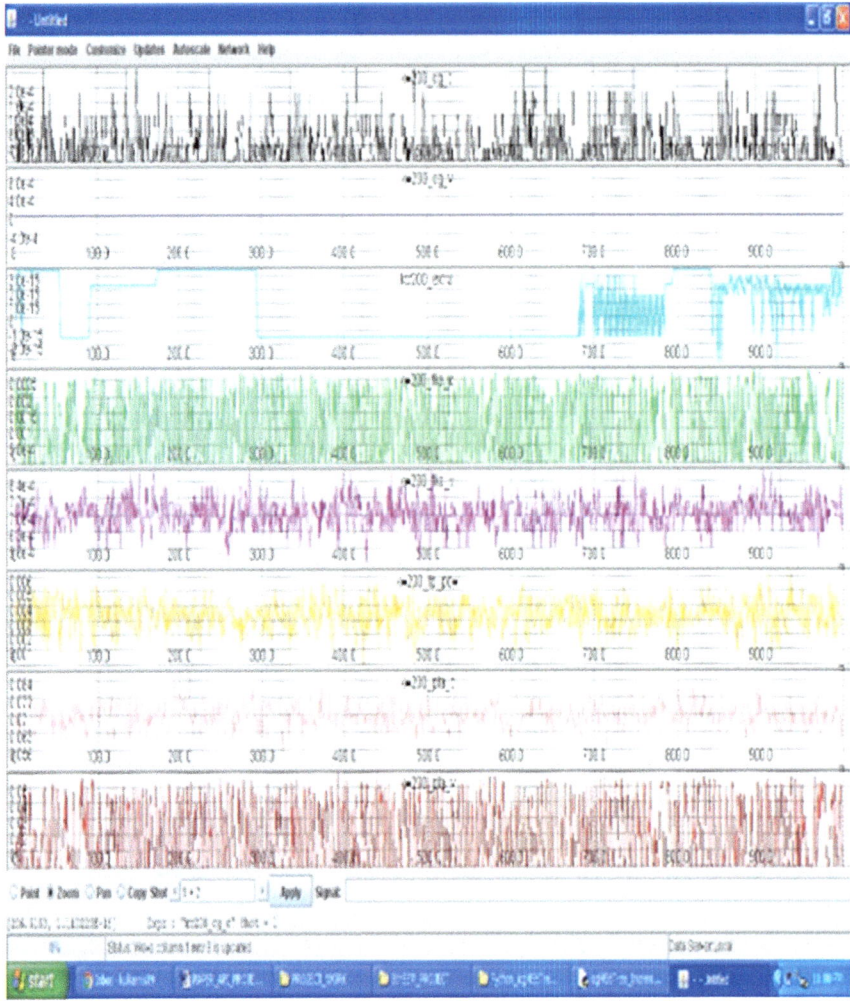

Fig. 3 jScope graph of ICGL456 tree

Finally, in this way new Pulse can be constantly being created at every instance of a time, and data can be successfully inserted into the desired Tree with correct Tree nodes. Unwanted Pulse can be deleted, and system can operate to its optimum potential.

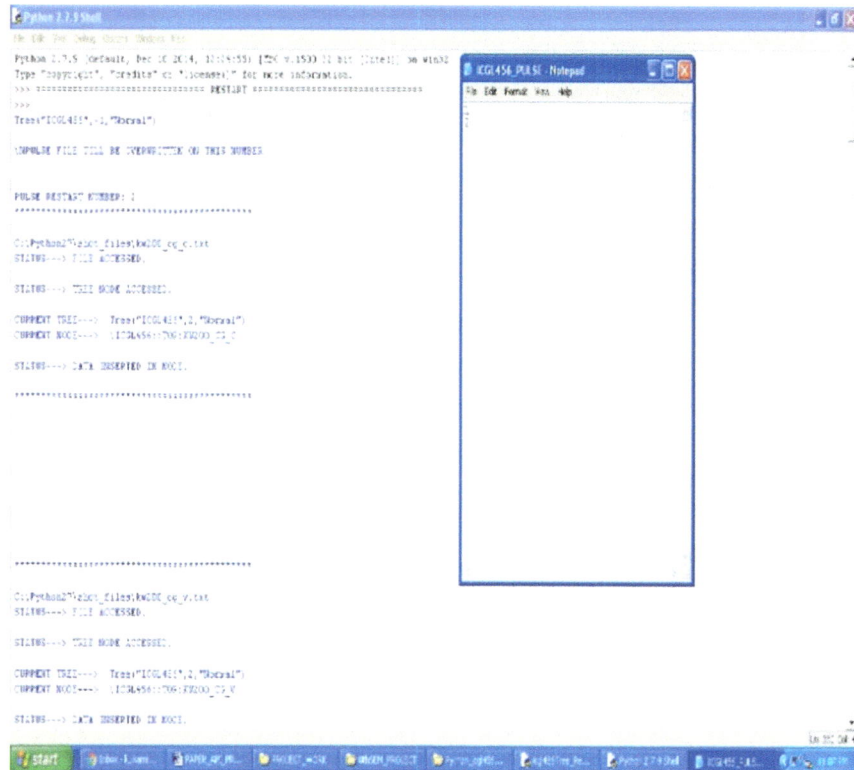

Fig. 4 Increment script and text file screen shot

5 Conclusion

This MDSPLUS software has successfully deployed on the system as per required functionality. Python-MDSPLUS integration module has been designed and deployed in the system. With such Python-based system, API accessing data, data storage, and data visualization are easily possible for MDSPLUS data storage system. System has been thoroughly tested on various Trees and subtrees with various data types as well as various data. Due to the development of the API, Python now works as a bridge between connecting the DAC data and MDSPLUS storage system and brings in total data transfer transparency with the DAC system and MDSPLUS. Python has great potential to work with this system on optimum level and bring in new improvements and updates for better user experience. This software is tested to execute the task of data archiving all required signals with analysis of acquired data that takes place in off-shot mode.

References

1. Ramesh, J. et al.: Design and development of PLC based DAC for 45.6 MHz, 100 kW ICRH system using EPICS and MODBUS/TCP IPR/TR-333/2015 (March 2015)
2. Kulkarni, S.V. et al.: Resonant and Non-resonant Type Pre-ionization and Current Ramp-up Experiments on Tokamak Aditya in the Ion Cyclotron Frequency Range. In: IAEA Fusion Energy Conference (2012)
3. MDSPLUS Online Reference. http://www.MDSPLUS.org
4. Fredian, T., Stillerman, J.: MDSPLUS current developments and future directions. Fusion Eng. Design **60**, pp. 229–233 (2002)
5. Fredian, T., Stillerman, J., Manduchi, G.: MDSPLUS extensions for long pulse experiments to appear in Fusion Engineering and Design
6. Barbalace, A. et al.: MDSPLUS real-time data access in RTAI. In: Proceedings of ICALEPCS07, Knoxville, Tennessee, USA, pp. 132–134, (2007)
7. RTAI Home page, [Online]. http://www.rtai.org
8. MDSPLUS 7.1.7. https://pypi.python.org/pypi/MDSPLUS
9. Fredian et al.: MDSPLUS Objects—Python Implementation, Plasma Science and Fusion Center Massachusetts Institute of Technology Cambridge MA 02139 USA
10. jScope. http://www.MDSPLUS.org/old/javascope/ReadMe.html

Part IV
Optical Networks, Wireless Sensor Networks, VANETs and MANETs

Parameter-Controlled Gas Sensor System for Sensor Modeling

Dipali Ramdasi and Rohini Mudhalwadkar

Abstract When a gas is passed over a chemically sensitive layer, its electrical properties change due to adsorption of gas molecules. This phenomenon is implied in thin film sensors. To improve the sensitivity and selectivity of sensors, a parametric modeling approach is preferred. In this, the parameters affecting sensor response are varied, and the sensor response is acquired for developing a model based on variation of parameters. The developed model suggests changes in sensor design and operating parameters, enhancing sensor performance for detecting explosives. A system in which the parameters of temperature, cycle time, and gas concentration can be varied is implemented using an embedded system approach. To facilitate the development of model, the sensor response is made available in comma-separated values. Also, a real-time plot of the sensor response is made available for identification of presence of a nitroaromatic explosive.

Keywords Nitroaromatic explosive · Parametric model · Real-time plot

1 Introduction

The use of various types of explosives by terrorists to spread violence and unrest among common people has motivated researchers to work on urgent and immediate detection of explosives. When explosives are packed in metal containers, they can be detected by metal detectors. Though this method is successful in case of landmine and weapon detection, [1, 2] this has a limited scope in explosives packaged intelligently. The timely detection of explosives and making them inactive is a challenging task. Sniffer dogs are considered as one of the most reliable tools for the detection of explosive vapors and compelled researchers to contribute in the area of electronic

D. Ramdasi (✉)
Cummins College of Engineering for Women, Pune 411052, India
e-mail: dipali.ramdasi@gmail.com

R. Mudhalwadkar
College of Engineering Pune, Pune 411005, India

© Springer Nature Singapore Pte Ltd. 2018   459
K. Saeed et al. (eds.), *Progress in Advanced Computing and Intelligent Engineering*,
Advances in Intelligent Systems and Computing 563,
https://doi.org/10.1007/978-981-10-6872-0_43

nose. Canines working as sniffer dogs need extensive training along with proper care and handling. Also, continuous monitoring with sniffer dogs is very difficult. The use of spectroscopy or chromatography for detection of explosives is very effective and accurate [3–5]. These techniques are very costly and time-consuming, while they can be used only for offline detection. To improve the sensitivity and selectivity of the explosive detection systems is a challenging task. Researchers worldwide are working on the performance improvisation of sensors and sensor system. To carry out this research, modeling of the sensor and sensor system is an approach used by researchers. This paper presents a microcontroller-based system which can assist the researcher in the development of parametric model of sensors for detection of explosives. Modeling the sensors by variation in physical parameters can help in the refinement of physical measuring techniques.

1.1 Explosives and Their Properties

The most commonly used explosives are Nitrogen based, typically dinitrotoluene (DNT), trinitrotoluene (TNT), cyclo-trimethylene trinitramine (RDX), pentaerythritol tetranitrate (PETN), and ammonium nitrate. The stored non-detonated explosives exhibit nitroaroma; hence, detection of nitroaromatic compounds like ammonia, nitrous oxide, and nitric oxide can lead to determination of the presence of explosives [6, 7]. As these compounds are at very low concentration, in a few parts per million (ppm) or parts per billion (ppb), detection of these compounds poses a challenge on the measurement technique.

1.2 Sensors for Explosive Detection

Though there are various sensors available, the thin film sensors based on polymers and metal oxides are one of the low-cost alternatives. In these, a gas-sensitive layer is deposited on a substrate with drop coating, spin coating, or electrochemical deposition methods. These sensors can be easily developed in laboratories, and extensive research can be performed with combination of materials and binding agents to enhance selectivity. When the thin film sensors are subjected to gas, the electrical properties of the sensor change due to adsorption of gas molecules on the gas-sensitive layer [8, 9]. The change in resistivity is measured in this paper, with which the concentration of the gas can be estimated.

1.3 Parametric Modeling

The literature survey suggests that the sensor response is dependent on parameters like the gas concentration, temperature of the sensor, humidity, and the time for which the sensor is subjected to the gas [10–12]. To develop a parametric model, the sensor array is subjected to a known concentration of target gas at predefined temperature and time. The response of the sensor is acquired by keeping temperature and time constant and varying the concentration by varying the pump speed for a single acquisition cycle. Thus, cycles are repeated for varying temperature and time. From the acquired sensor response, a model is developed considering the above parameters. The parameters like sensor material, sensor geometry is fixed for that particular sensor, hence are not considered while developing the model. From the developed model, unknown concentration and the presence of a gas can be estimated. This will help in identifying the presence of an explosive.

The developed system allows the researcher to set the parameters of temperature and time for which the sensor is exposed to gas. The speed of the volumetric flow meter can be set by the researcher, by which the parameter of concentration can be varied. A sensor array consisting of four thin film sensors can be connected to the system. The readings of all four sensors are transmitted to a computer along with the temperature and humidity. A graph of all the sensors is plotted in MATLAB environment, and the values of sensors are stored as comma-separated values, which can be used for further data analysis and parametric model development.

2 Elements of the Gas Injection System

To perform parametric modeling of sensors and evaluate the model, it is necessary to develop a hardware setup which enables the varying of parameters and observe the response of the sensor for the varying parameters. For this, an ATmega2560-based Arduino board at operating frequency of 16 MHz is used to develop the system. The general block diagram is illustrated in Fig. 1. The ATmega2560 is an 8-bit microcontroller from the AVR family [13, 14].

The input–output lines are used to interface the liquid crystal display (LCD) and the keyboard. A 20 × 4 LCD display is used to display messages, the current temperature, and humidity values. A 4 × 4 keyboard matrix is used to set parameters of temperature, cycle time, and motor speed. The ATmega2560 is used keeping in view, further expansion of the system if required. The programming is done in the open-source Arduino software, which allows writing code in C and uploads it on compiling to the board. Figure 2 shows the system setup. The measurement and control of physical parameters are described in the further subsections.

Fig. 1 General block
diagram of the system

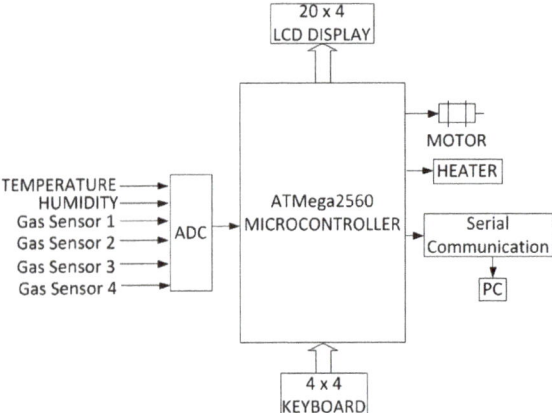

Fig. 2 System setup

2.1 Temperature Measurement and Control

As the sensor array will be exposed to nitroaromatic explosives, the sensor temperature cannot be exceeded beyond 80 °C. The temperature sensor LM35 is used for
measurement of temperature and connected to on-chip ADC channel 8 of the microcontroller. The reference voltage of ADC is set to 5 volts external. The temperature
reading in degree Celsius is displayed on the LCD as well as it is transmitted serially
to the computer along with the values of thin film sensors. The LM35 is mounted
centrally, at even distance from all the thin film sensors. To maintain the temperature
equal to the set temperature, a solder gun is used as a heater. It is mounted centrally,
equidistant from all the thin film sensors in the test enclosure. The solder gun is
driven by pin number 6 of the Arduino. An opto-isolator MOC3021 is used to connect pin 6 of the Arduino board to the triac BT136, which in turn drives the gun. The
solder gun is powered from a standard 230V AC source. The PID control algorithm

[15] is used to control the temperature of the sensors, with appropriate settings of the proportional, integral and derivative gains. The output of the PID control algorithm is then converted to pulse-width-modulated output which drives pin number 6.

2.2 Humidity Measurement

When the thin film sensors are subjected to a rise in relative humidity, it is observed that the resistance of the sensors reduces. To avoid misinterpretation of the results, it is necessary to observe the relative humidity of the environment in which the sensors are placed. The DHT11 is a humidity sensor, which gives a calibrated digital output that can be easily interfaced to an 8-bit microcontroller via a serial bus. The relative humidity of the measuring surrounding is displayed on the LCD as well as transmitted serially to the computer.

2.3 Cycle Time for which Sensors are Exposed to Gas

When the target gas passes over the sensors, the gas molecules get adsorbed to the gas-sensitive layer of the sensor, resulting in change of electrical properties of the sensors. The time for which sensors are exposed to the gas affects the response of the sensors. Increase in the cycle time results into an aggravation of adsorption of gas molecules, and hence there is an increase in the response of the sensors. The time for which the sensor array is exposed to gas can be set by the researcher using the keyboard. The on-chip 16-bit timer is used for this purpose. The gas is injected into the test chamber using a volumetric flow pump, driven by a motor. The timer starts along with the motor and turns the motor off after the desired time is elapsed.

2.4 Concentration of the Gas

The volumetric flow pump is driven by a DC motor. The volumetric flow rate when multiplied by the density of the target gas and the cycle time will give the mass of the target gas. When the speed of the motor changes, so does the mass, changing the concentration of the gas. The motor is interfaced to the Arduino board using the motor driver L293D. This is a high-current half-H driver typically used to drive motors.

2.5 Gas Sensors

The gas sensors used for testing the system are thin film sensors developed using spin coating technique. The gas-sensitive layer is polymer based with different doping levels for the four sensors in the sensor array for increased sensitivity and selectivity. After coating the substrate, the sensor is heated at 250 °C for 30 min, allowed to cool down and then used for gas sensing [16, 17]. All sensors are mounted in the test enclosure along with the heater, temperature sensor, and humidity sensor. The sensors are connected to the ADC channels 2 to 5 via buffers devised by operational amplifiers. The electrical parameter of change in resistance is measured to estimate the gas concentration. The target gas used for testing the system is nitric oxide with concentration varying from 5 ppm (parts per million) to 30 ppm. The sensor response is measured in terms of:

$$\frac{\Delta R}{R} = \frac{R_g - R_a}{R_a} \tag{1}$$

3 Sensor Interface and Real-time Plot

The response of the sensors needs to be analyzed for developing a parametric model. The digitized value of the sensor response is transmitted serially via a CP2102 (USB (Universal Serial Bus) to UART (Universal Asynchronous Receiver Transmitter) controller) at a baud rate of 9600 to the computer. A MATLAB program in the computer opens the serial port and accepts the data. Sensor 1 to Sensor 4 values are transmitted continuously to the computer/laptop till the motor is on and the timer is on. After each value transmission, the value is plotted in a graphical user interface (GUI) developed in MATLAB with a minimum delay of the transmission time. Thus, the trend of each sensor can be observed. As the thin film sensor response time depends on the adsorption rate, the typical response time is high and is stated by researchers as approximately 50 s. Thus, it can be assumed that the sensor response is being plotted in real time. Along with every response, the current temperature and humidity are also transmitted and displayed in the GUI. The complete sensor response is stored in the computer as a .xlsx file. This file can be later accessed by the researcher or a model building program to derive the model.

4 System Software

The software for the Arduino board is developed in C language. The temperature, cycle time, and motor speed in percentage are taken as inputs from the user. The keyboard and display connected to the microcontroller board facilitate this. Once all the input values are set by the user and the temperature is stabilized, the motor for

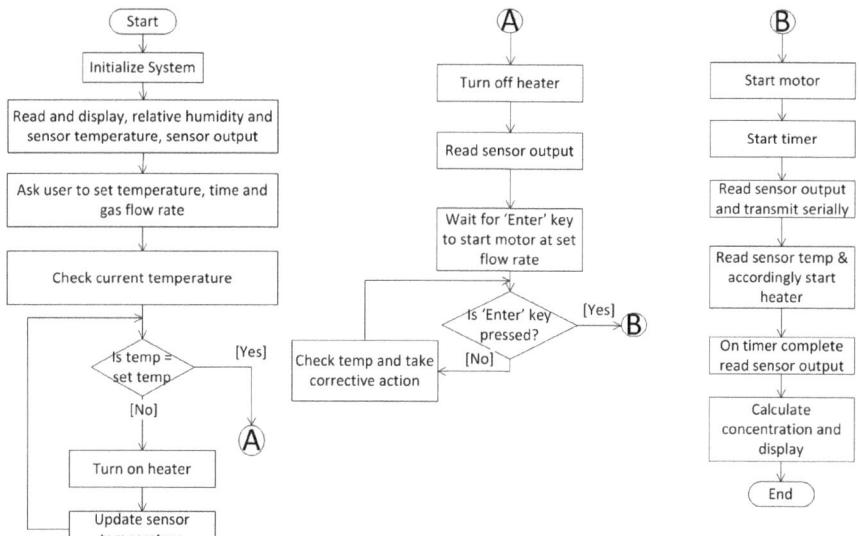

Fig. 3 Flow chart of the system

allowing gas to pass over the sensors is started. The timer and sensor response transmission also start immediately, and a real-time plot of the sensor array response is observed. The flow chart of the gas injection system pertaining to the microcontroller is illustrated in Fig. 3 and to that of the GUI and response plotting is illustrated in Fig. 4.

5 Results

The polymer-based thin film gas sensors were subjected to a known concentration of nitric oxide ranging from 5 to 30 ppm, and the response of the sensors was observed on a GUI developed in MATLAB. It was observed that the temperature control near the set value was achieved after a settling time of 6 min from the start. After the motor starts at the set speed, the real-time plots of the sensor response help in identifying the presence of a gas. A database of the sensor response is created by varying only one parameter and keeping the rest two constant out of the identified parameters of sensor temperature, cycle time, and concentration of gas. If there is a more than 20% change in the humidity of the sensing environment, the response of sensors is affected adversely. A screenshot of the real-time graph is shown in Fig. 5. The sensor response is stored in a .xlsx file. A screenshot of few entries in excel file is shown in Fig. 6. MATLAB considers these as comma-separated values, and a parametric model based on these values can be developed using a technique suitable to the researcher.

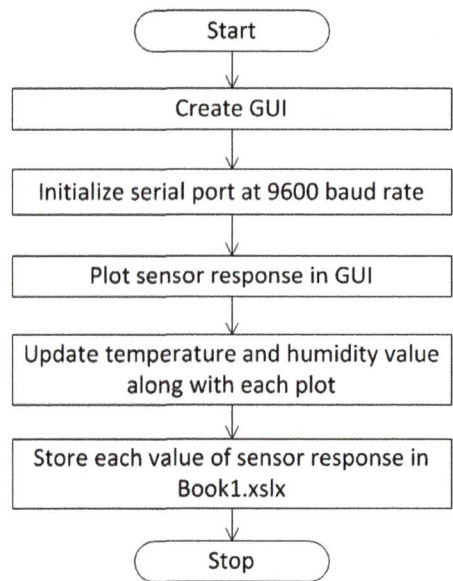

Fig. 4 Flow chart of the real-time sensor response plot

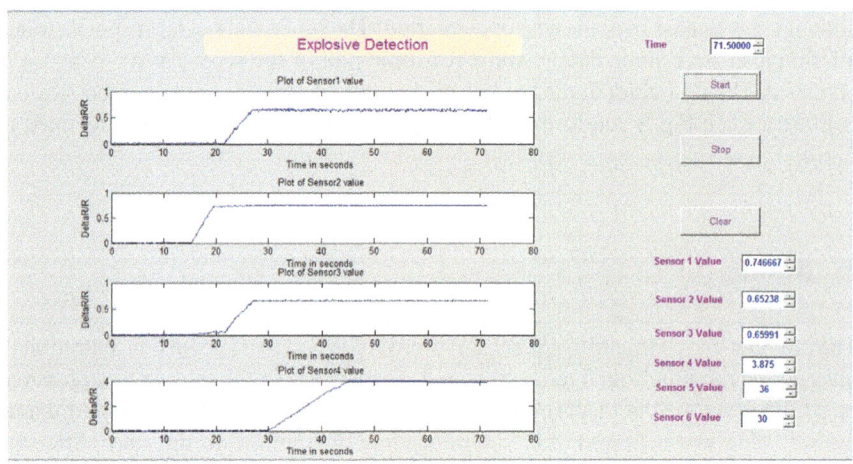

Fig. 5 Screenshot of the real-time sensor response plot

Fig. 6 Screenshot of the
sensor array response in
excel file

0.645238	0.753333	0.664414	0.009615
0.642857	0.746667	0.65991	0.009615
0.640476	0.74	0.653153	0.009615
0.630952	0.736667	0.655405	0.014423
0.619048	0.75	0.65991	0.009615
0.638095	0.763333	0.662162	0.004808
0.652381	0.743333	0.666667	0.019231
0.661905	0.75	0.666667	0.004808
0.630952	0.75	0.662162	0.024038
0.635714	0.75	0.655405	0.004808
0.638095	0.74	0.650901	0.019231
0.635714	0.743333	0.650901	0.004808
0.642857	0.743333	0.653153	0.004808
0.642857	0.746667	0.65991	0.009615
0.654762	0.746667	0.666667	0.014423
0.616667	0.743333	0.648649	0.024038
0.657143	0.75	0.65991	0.033654
0.642857	0.746667	0.671171	0.038462
0.65	0.75	0.65991	0.0625
0.659524	0.743333	0.664414	0.076923
0.652381	0.75	0.666667	0.072115
0.616667	0.736667	0.648649	0.105769
0.647619	0.743333	0.657658	0.139423
0.642857	0.743333	0.650901	0.144231
0.647619	0.746667	0.653153	0.1875

6 Conclusion

A gas injection system with provision for varying the parameters affecting thin film sensor response is developed using the ATmega2560 microcontroller. The polymer-based thin film sensors developed using the spin coating technique are used for acquiring the sensor response. By varying the parameters, a huge database is created and used for developing parametric model of the sensors for the detection of nitroaromatic explosives. The system exhibits real-time plots of the sensor response in a GUI developed in MATLAB. Thus, this system will assist researchers in the developing or enhancing sensor models based on experimentation. Using models developed, the optimized temperature and cycle time can be decided for enhanced sensitivity and selectivity. The same system can be extended for training and identification of volatile organic compounds using neural networks. Also, instead of thin film sensors, pellet sensors or commercially available sensors may also be used.

References

1. Moore, D.S.: Recent advances in trace explosives detection instrumentation. Springer Int. J. Sensing Imag. **8**(1) 9–38 (2007)
2. Gardner, J.W., Philip, N.B.: A brief history of electronic noses. Sensors Actuat. B Chem. **18**(1), 210–211 (1994)
3. Sanders, N.L., Kothari, S., Huang, G., Salazar, G., Cooks, R.G.: Detection of explosives as negative ions directly from surfaces using a miniature mass spectrometer. J. Anal. Chem. **82**, 5313–5316 (2010)
4. Primera-Pedrozo, O.M., Soto-Feliciano, Y.M., Pacheco-Londo, L.C., Herndez-Rivera, S.P.: Detection of high explosives using reflection absorption infrared spectroscopy with fiber coupled grazing angle probe/FTIR. J. Sensor Imag. **10**, 1–13 (2009)
5. Hatab, N.A., Eres, G., Hatzingerc, P.B., Gua, B.: Detection and analysis of cyclotrimethylenetrinitramine (RDX) in environmental samples by surface-enhanced Raman spectroscopy. J. Raman Spectr. **41**, 1131–1136 (2010) Infrastructure. Morgan Kaufmann, San Francisco (1999)
6. Meyer, R., Kohler, J., Homburg, A.: Explosives. Wiley (2008)
7. Crowl, D.A.: Understanding explosions, Vol. 16, Wiley (2010)
8. Brudzewski, K., Osowski, S., Pawlowski, W.: Metal oxide sensor arrays for detection of explosives at sub-parts-per million concentration levels by the differential electronic nose. Elsevier J. Sensors Actuat. B Chem. **161**, 528–533 (2012)
9. Bai, H., Shi, G.: Gas sensors based on conducting polymers. MDPI Sensor. J. **7**, 267–307 (2007)
10. Nenad, G., Igor, V., Kuzmanic, I.: Thin film gas sensor modeling and simulation. In: 48th International Symposium ELMAR-2006. Zadar, Croatia (2006)
11. Gardner, J.W., Bartlett, P.N., Pratt, K.F.E.: Modeling of gas-sensitive conducting polymer devices. IEEE Proc. Circ. Devic. Syst. **142–5**, 321–333 (1995)
12. Yong, L., Raymond, A.A., Thomas, J.M.: Parameter Identification and Simulation of a Thin Film Conducting Polymer Gas Sensor. Institute for Systems Research (2001)
13. Arduino Technical Center. http://www.arduino.cc/
14. Atmel Corporation. www.atmel.com/devices/atmega2560.aspx
15. Wang, Q-G.: PID tuning for improved performance. IEEE Trans. Control Syst. Technol. **7–4**, 457–465 (1999)
16. Ameer, Q., Adeloju, S.B.: Polypyrrole-based electronic noses for environmental and industrial analysis. Sensors Actuat. B Chem. **106–2**, 541–552 (2005)
17. Tomchenko, A.A: Semiconducting metal oxide sensor array for the selective detection of combustion gases. Sensors Actuat. B Chem. **93**(1), 126–134 (2003)

Analysis on Congestion Control Mechanism in Wireless Networks

Sivaprasad Abirami and Kumudhan Cherarajan

Abstract Nowadays, Congestion is proving to be a major issue in all kinds of network. Huge chunk of data transmission is the leading cause of congestion to take place. Hence, without increasing the capacity of the network our defined congestion control algorithms works dealing and managing efficient data transmission with minimum or no congestion through analysis. The algorithm is build on the basis of two parameters: The first parameter is the queue length, and the second parameter is the RTT [Round-Trip Time]. Before transmitting data to any destination, the source should check the available queue length of the destination based on some threshold. To understand the network congestion, RTT is calculated. Smaller the value of RTT, lesser is the congestion and vice versa. In this paper, two different algorithms based on the above parameters are defined, implemented, and analyzed.

Keywords Congestion control · Wireless network · Queue length based algorithm

1 Introduction to Congestion

When too many packets are there in the network, the performance degrades. This situation is called as congestion [1]. One of the biggest problems in the Internet world today is the huge amount of data transmission in the network beyond the capacity that has created congestion. Now, something should be done to control the problem of congestion. To control congestion, it is not possible to ask everyone to stop the data transmission between the networks.

When traffic in network increase beyond network capacity, then the situation becomes worse as shown in Fig. 1. Because of this situation, the router starts losing the packets. At very high traffic, performance degrades completely and almost none of the packets are delivered.

S. Abirami (✉) · K. Cherarajan
Sakec, University of Mumbai, Mumbai, India
e-mail: abi.lecturer@gmail.com

© Springer Nature Singapore Pte Ltd. 2018
K. Saeed et al. (eds.), *Progress in Advanced Computing and Intelligent Engineering*,
Advances in Intelligent Systems and Computing 563,
https://doi.org/10.1007/978-981-10-6872-0_44

Fig. 1 Performance graph of
network [1]

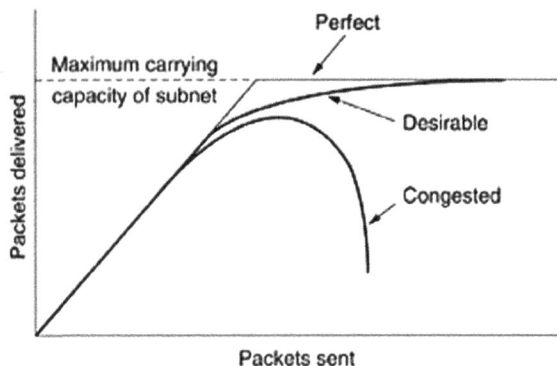

1.1 Routing

Routing is a method which is used to forward the packets. Routing algorithms are used to select the best path to transfer the packets. Different metrics will be utilized by different routing algorithms. The basic metrics are distance, delay, and hop count.

When congestion related information has to be transmitted it has to consider the shortest path to deliver the packets. Basically, the optimized algorithm used in this analysis is Ant-Based routing algorithm and Dijkstra's shortest path first routing algorithm. It was found that Dijkstra's shortest path first routing algorithm is optimized algorithm to select the best shortest path [2].

1.2 RTT (Round-Trip Time)

It is the total time calculated for a bit to travel between two most distant nodes on the network and return again. To calculate this time, the lengths of the cable segments are multiplied by the delay factor of the cable used along with the delay factor of the nodes and the other hardware components used and a safety buffer of 4 bit times.

Example:

$(150\,\text{meters} * 1.112\,\text{bits times}/\text{meter}) + 100\,\text{bit times} + (2 * 92\,\text{bit times}) + 4\,\text{bit times} = 454.8\,\text{bit times}.$

Figure 2 shows a sample network with two routers and two hubs. The distance between the hardware components is about 50 meters.

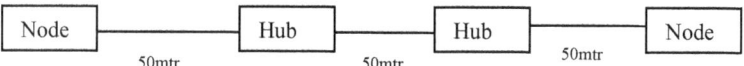

Fig. 2 Sample network [3]

2 Algorithm 1 [Congestion Control Algorithm Without RTT]

This algorithm deals with congestion control by transmitting data based on the available queue length of the destination. Initially, all the nodes are provided with the queue length of 100. For every data transmission, the queue length will be reduced by one. This updated queue length gets transmitted to all other nodes periodically. The duration of queue length update can be decided by the user in this simulation based on the type of network, the capability of the routers and the amount of data transmission. The data transmission happens only when the

Fig. 3 Flowchart of algorithm 1

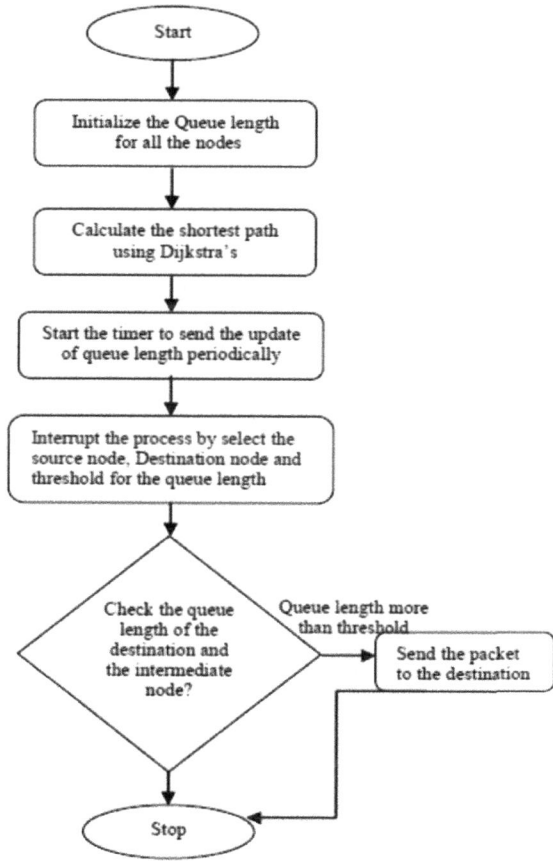

available queue length of the destination is more than the threshold value else the data transmission get stopped. By the concept of stopping the data transmission, the dropping of data gets eliminated (Fig. 3).

2.1 Steps in the Design of Algorithm 1

Step 1: Create the node and edges.
Step 2: Assign the initial queue length to all the nodes (100). Execute the Dijkstra's Routing algorithm to calculate the shortest path.
Step 3: Set periodic update timer and queue length threshold based on the network requirement.
Step 4: Start sending the updated queue length to all the nodes periodically. Interrupt the process by entering the source node, the destination node.
Step 5: Now for the destination node and the intermediate node it has to check the queue length, if the queue length greater than threshold then it has to select the best path and forward the packet.

3 Algorithm 2 [Congestion Control Algorithm with RTT]

In Algorithm 1, the queue length update is periodically updated without knowing the congestion status of the network. RTT is used to identify the congestion status of the network. Before the periodic update of the queue length, the congestion status of the network is identified. The congestion status of the network is identified by setting a threshold value of RTT. If the calculated RTT value is less than the threshold then, the time taken from the source to destination is very less, so the network is not congested and vice versa. The periodic update of the queue length and the data transmission is same as that of Algorithm 1 (Fig. 4).

3.1 Steps in the Design of Algorithm 2

Step 1: Create the node and edges.
Step 2: Assign the initial queue length to all the nodes (100). Execute the Dijkstra's Routing algorithm to calculate the shortest path. Set periodic update timer, queue length threshold and RTT threshold based on the network requirement.

Fig. 4 Flowchart of
Algorithm 2

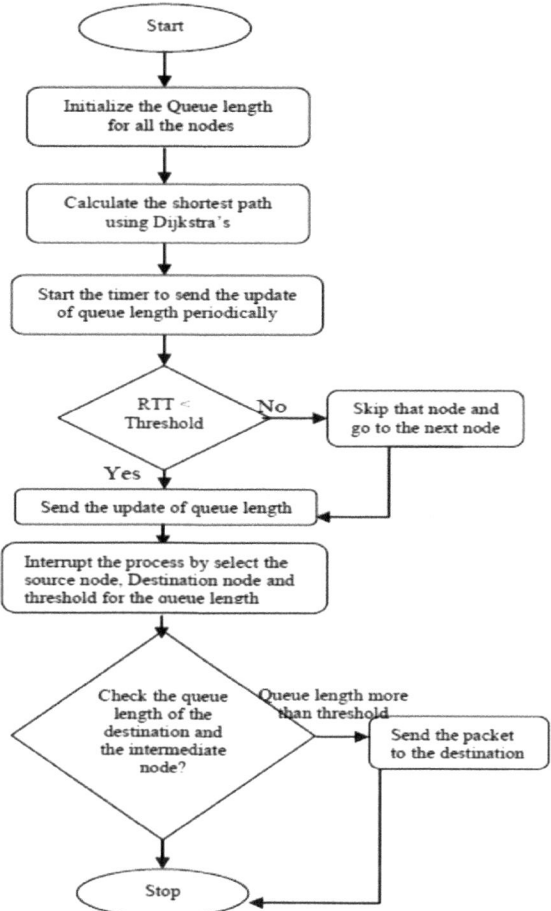

Step 3: Before sending the update calculate the RTT, if RTT is less than the threshold then send the update else skip and go to the next node. Interrupt the process by entering the source node, the destination node.

Step 4: Now for the destination node and the intermediate node it has to check the queue length, if the queue length greater than threshold then it has to select the best path and forward the packet.

4 Implementation

4.1 Edge Length Between the Nodes

The snapshot given below is used to enter the distance between the nodes which will be connected to form the network. The user can enter its own value to create the network by entering the distance between the nodes (Fig. 5).

4.2 Selection of Algorithm

The below given snapshot is used to select the algorithm which the user wants to execute (Fig. 6).

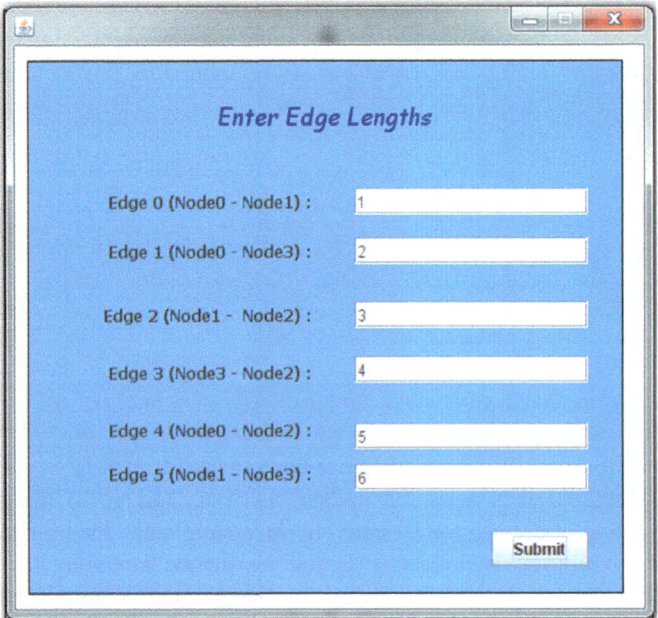

Fig. 5 Edge lengths between the nodes

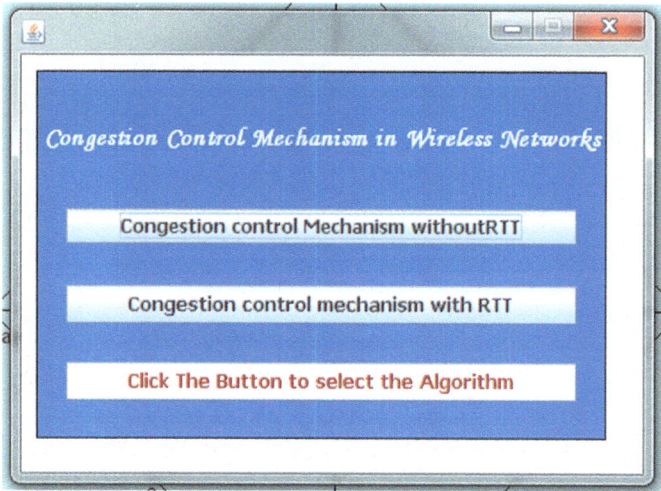

Fig. 6 Selection of algorithm

4.3 Algorithm 1 [Congestion Control Mechanism Without RTT]

The Algorithm 1's execution status is given below in which the "path table" is the output of the Dijkstra's routing algorithm in which all the shortest path between all the nodes is given. The "Set Timedelay" field asks the user to enter the periodic queue length update time. Queue status provides the current queue length of each and every node. "Set QueueLengthThreshold" asks the user to type the threshold value of the queue length. "Source" and "Destination" fields are used to provide the source node and the destination node to which the data transmission should be done (Fig. 7).

4.4 Algorithm 2 [Congestion Control Mechanism with RTT]

In this snapshot, two extra fields are there: The first one is "Set RTT threshold" which is used by the user to provide the RTT threshold based on the network, and the second field is "RTT values" which will provide the calculated RTT value of all the nodes to take decision whether the periodic update should be transmitted or not. (Fig. 8).

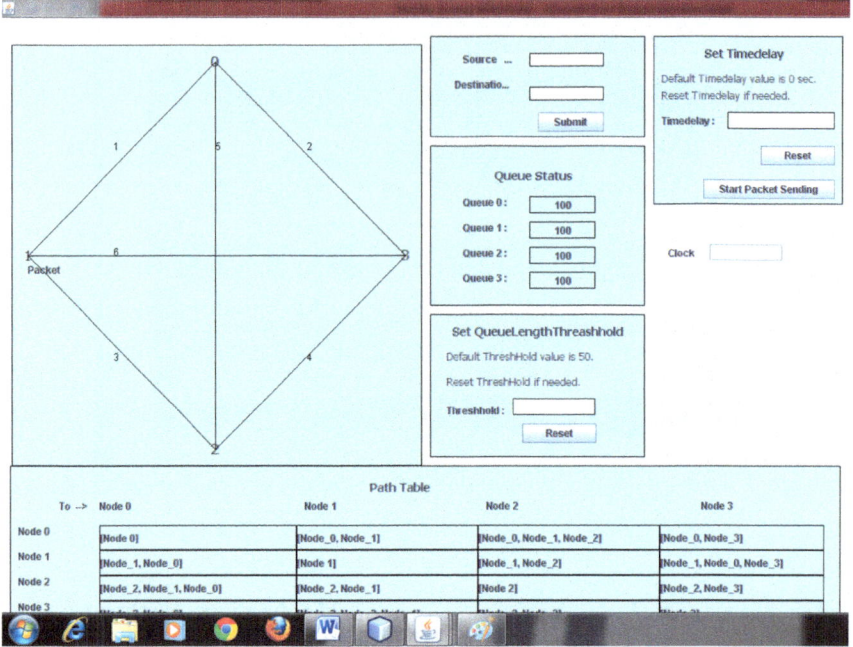

Fig. 7 Algorithm 1 [congestion control mechanism without RTT]

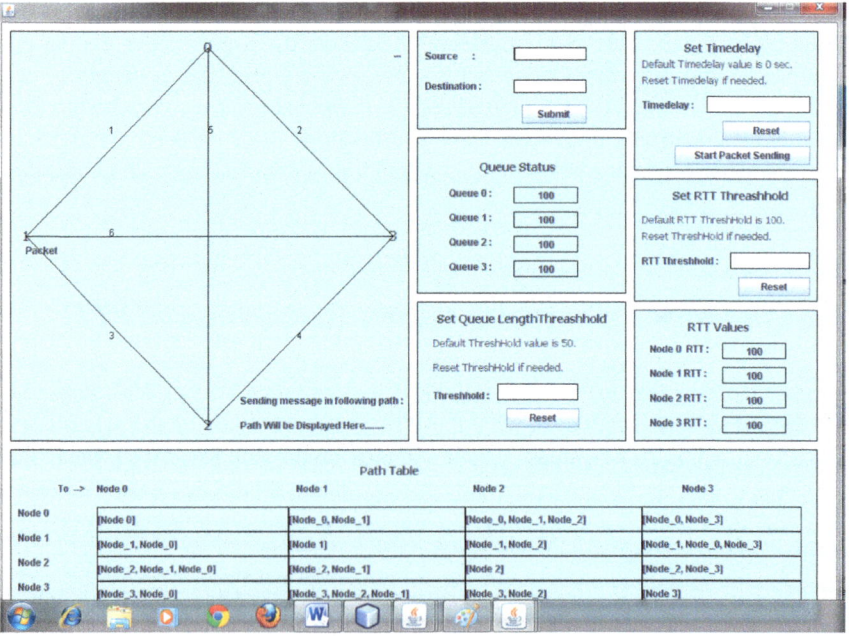

Fig. 8 Algorithm 2 [congestion control mechanism with RTT]

5 Result and Analysis

5.1 Analysis 1

Analysis 1 deals with the comparative study of throughput of the algorithms. The analysis is done by setting the same parameter like same time delay for the periodic update, the same queue length threshold, the same source, and the same destination in both the algorithm. The analysis is done with three different threshold value of queue length (Tables 1, 2, 3).

Table 1 Throughput when queue length threshold is 25

Parameters	Algorithm 1	Algorithm 2
Time delay	5 Sec	5 Sec
Q. Length threshold	25	25
Source	1	1
Destination	3	3
RTT Threshold	-	25
Data Transmission time	44 sec	50 Sec

Table 2 Throughput when queue length threshold is 50

Parameters	Algorithm 1	Algorithm 2
Time delay	5 Sec	5 Sec
Q. Length threshold	50	50
Source	1	1
Destination	3	3
RTT Threshold	-	50
Data Transmission time	44	54

Table 3 Throughput when queue length threshold is 75

Parameters	Algorithm 1	Algorithm 2
Time delay	5 Sec	5 Sec
Q. Length threshold	75	75
Source	1	1
Destination	3	3
RTT Threshold	-	75
Data Transmission time	46 sec	58 sec

5.1.1 Result

The Average Time delay for data transmission by Algorithm 1: $44 + 44 + 46 = 134/3 = 44.5$ s.

The Average Time delay for data transmission by Algorithm 2: $50 + 54 + 58 = 162/3 = 54.0$ s.

According to the above analysis, the time delay for data transmission is more for Algorithm 2 since before each and every update it has to calculate RTT.

5.2 Analysis 2

It is used to find the number of queue length updates forwarded by both the algorithms (Tables 4, 5, and 6).

Table 4 Update count when queue length threshold is 25

Parameters	Algorithm 1	Algorithm 2
Time delay	5 Sec	5 Sec
Q.Length threshold	25	25
Source	1	1
Destination	3	3
RTT Threshold	-	25
No. Of updates	18	3

Table 5 Update count when queue length threshold is 50

Parameters	Algorithm 1	Algorithm 2
Time delay	5 Sec	5 Sec
Q.Length threshold	50	50
Source	1	1
Destination	3	3
RTT Threshold	-	50
No. Of updates	18	9

Table 6 Update count when queue length threshold is 75

Parameters	Algorithm 1	Algorithm 2
Time delay	5 Sec	5 Sec
Q.Length threshold	75	75
Source	1	1
Destination	3	3
RTT Threshold	-	75
No. Of updates	18	13

5.2.1 Result

The average no. of updates done by Algorithm 1: $18 + 18 + 18 = 54/3 = 18$
The average no. of updates done by Algorithm 2: $3 + 9 + 13 = 25/3 = 8.33$
The No. of updates transmitted by Algorithm 2 is very less. Based on congestion the Algorithm 2 will work. If the network is not congested then the update will be transmitted to all the nodes. Algorithm 1 in any situation the update will be transmitted to all the nodes. The problem with Algorithm 2 is that based on the RTT threshold value the updates are transmitted. If the RTT is less than the threshold, the update is transmitted else it is not transmitted. So it is possible like according to Algorithm 2 the current status of the queue length may not be known to all the nodes.

6 Conclusion

Thus, the analysis of both the algorithms for dealing and managing the congestion brings us to a notion of carrying large chunks of data transmission by calculating the queue length of each node in the network. This defined and implemented algorithm can be challenged by increasing the number of nodes in the network and then analyzing the congestion traffic in the network.

References

1. Andrew, S.T.: Computer Network. Pearson Education. Fourth ed.
2. Leo, W.S., Alexander, S., Andre, K.P.: Performance analysis of Dijkstra, A* and Ant algorithm for finding optimal path. MICEEI. 2nd Makassar International Conference on Electrical Engineering and Informatics (2010)
3. Craig, Z.: The Complete Reference networking. McGraw Hill Education Private Limited (2001)
4. Forouzan.: Data Communication and Networking. McGraw Hill Education Private Limited. Fourth ed.
5. Jingyang, W., Xiaohong, W., Huiyong, W., Min, H., Lina, Ma., Zhengtao.: The research of active network congestion control algorithm based on operational data. International Conference on Communications and Networking in China. CHINACOM'07 (2007)
6. Liu, W., Chen, B.-C., Zou X.-L., Hangzhou.: A congestion control algorithms based on network measurement in Next-Generation Networks. Published in 2010 International Conference on Internet Technology and Applications
7. Musa, P., BahaŞen., Pınar, Y.: An efficient solving of the traveling salesman problem: the ant colony system having parameters optimized by the taguchi method. Turkish J. Elect. Eng. Comput. Sci.

A Distributed Transmission Power Efficient Fault-Tolerant Topology Management Mechanism for Nonhomogeneous Wireless Sensor Network

Manas Ranjan Nayak, Gyananjaya Tripathy and Amiya Kumar Rath

Abstract We propose a distributed and transmission power efficient fault-tolerant topology management technique, known as transmission power efficient disjoint path (TEDP), designed for nonhomogeneous wireless sensor networks (WSNs). A nonhomogeneous WSN is consisting of a substantially large count of energy-constrained and low-computational capability low-cost sensors and a small number of resource-extensive sensors with unrestrained battery power. The aim of our proposed algorithm is to designate every sensor node's communication range in order to ensure that every sensor node establishes k-vertex disjoint connectivity with resource-extensive nodes, and the overall transmission power utilization in the network is substantially minimized. With the employment of TEDP algorithm, the induced topologies can tolerate k − 1 node failures in the worst case. The experimental outcomes demonstrate that in contrast to existing solutions our algorithm attains the decline in overall transmission power substantially by 25% and the decline in utmost transmission power needed in a sensor node by 30%.

Keywords Heterogeneous network · Fault tolerance · Topology control
K-connectivity · Disjoint paths · Power efficiency

M. R. Nayak
SOA University, Bhubaneswar, India
e-mail: manas2nayak@yahoo.co.in

G. Tripathy · A. K. Rath (✉)
Department of Computer Science, VSSUT, Burla, Odisha, India
e-mail: amiyaamiya@rediffmail.com

G. Tripathy
e-mail: gyananjayatripathy@gmail.com

© Springer Nature Singapore Pte Ltd. 2018   481
K. Saeed et al. (eds.), *Progress in Advanced Computing and Intelligent Engineering*,
Advances in Intelligent Systems and Computing 563,
https://doi.org/10.1007/978-981-10-6872-0_45

1 Introduction

WSNs have been comprehensively deliberated for their wide ranging applications, for example, border or combat zone surveillance, space discovery, assessment of geological characteristics of a region, search and rescue operations, industrial process monitoring and management, and many others [1]. WSNs are usually consisting of substantially large set of miniature sensors typically restrained in their battery power, computation and communication reserves. These sensor nodes cooperate with each other in a disseminated and self-governing way to collectively achieve an assignment usually without having an infrastructural support in the environment. Ensuring energy efficiency and restoring network-wide connectivity in the event of failures are some of the indispensable characteristics for WSNs to make the network functionally sustainable in the case of sensor's battery power exhaustion, hardware/software malfunctioning, link failures, or hostile environmental circumstances [2]. Topology management is one of the significant techniques employed to substantially minimize battery power utilization and sustaining network-wide connectivity [3]. In the literature, numerous proactive and reactive topology management techniques have been propositioned to tolerate sensor breakdowns in WSNs [4].

In this paper, we propose a proactive fault-tolerant topology management algorithm which is to be employed in a heterogeneous WSN deployment. This is a two-layered sensor node deployment framework where the lower layer is composed of resource-constrained sensor nodes with restricted battery power, low communication and computational capabilities. The higher layer is composed of resource-extensive sensor nodes with enhanced battery power, storage, and high computational and communication capabilities. However, these resource-extensive sensor nodes (supernodes) are deployed in limited quantity owing to their high cost. These supernodes experience longer transmission ranges and higher transmission rates between them. In this heterogeneous WSN, deployed sensor nodes accumulate data from their environment and forward their accumulated data to the supernodes for further processing. Such category of WSN setup is considered to be more consistent and has enhanced network life span compared to homogeneous WSN setups.

This paper proposes an algorithm called the transmission power efficient disjoint path (TEDP) for instituting a fault-tolerant topology to trail accumulated data by sensor nodes to supernodes. In WSNs, assuring k-connectivity in the network topology is essential to ensure a definite degree of fault tolerance [4]. In the worst case, the ensuing topology can sustain up to $k - 1$ sensor failures. We present a distributed algorithm TEDP for resolving k-connectivity problem in an effective means with respect to total transmission power of the ensuing topologies, utmost dispensed transmission power of sensor nodes, and entailed transmissions of control messages. The experimental outcomes demonstrate that our TEDP algorithm realizes on an average 25% decline in total transmission power necessitated in the network based on the packet failure rate, and a 30% decline in utmost transmission power necessitated in a node contrasting to some of the extant solutions.

The effectiveness of power optimization is the outcome of our proposed scheme that we employ while ascertaining the disjoint routes. This scheme entails in accumulating the entire route information in preference to just having information about the next node on the routes and offers a large-scale discovery of the best routes all over the network except for the need of information about span of the network.

We present interrelated work on topology management techniques designed for WSNs in Sect. 2. Section 3 illustrates our proposed approach and TEDP algorithm and the experimental outcomes in Sect. 4. Finally, our proposed work is concluded in Sect. 5.

2 Related Works

Topology management is best described as managing a group of neighbor nodes in a WSN by means of fine-tuning the transmission range and/or deciding on distinct nodes in order to ensure the messages getting forwarded [5]. Topology management strategies may be split primarily into two categories, specifically, homogeneous as well as nonhomogeneous. In homogeneous strategies, all sensors do have identical transmission ranges in contrast to nonhomogeneous strategies where sensors may have diverse transmission ranges.

There are numerous topology management strategies recommended in literary works, and they may be categorized in line with the techniques they employ [6]. Several topology management techniques are designed on the transmission power regulation technique which in turn is dependent upon the capability associated with sensors to manage their transmission power [7, 8]. A number of algorithms employ sleep scheduling mechanism which is designed to diminish energy utilization while nodes are in the state of idle [9–11]. Some others employ symmetrical configurations, position and directional information [12–14]. The distinction among most of these studies and our proposed work is that we make an effort to lessen nodes' overall transmission power in two-tiered heterogeneous WSN setups while others resort to homogeneous setups. Moreover we give attention to connectivity amid a sensor node and supernodes, while they concentrate on ensuring connectivity amid any pair nodes.

Clustering may also be viewed as one more means of ensuring topology management in which the intention is to coordinate the network into a hierarchical organized structure to ensure load balancing among the nodes and enhancing the network life span [15]. Some techniques choose cluster heads based on numerous standardized criteria and build a layered structure [16, 17]. Nevertheless, in our proposed work we build a layered topology having supernodes and sustain fault-tolerant connectivity amid sensor nodes and supernodes.

Wireless sensor and actor networks (WSANs) typically have a two-tiered setup in which the lower layer is composed of inexpensive sensors and the higher layer is composed of resource-expensive nodes (often known as actors or supernodes) which usually undertake suitable measures [18]. WSANs essentially have two types of communication links: sensor-to-actor and actor-to-actor [19]. Numerous strategies have been suggested with regard to sustaining reliable sensor-supernode connectivity [20, 21]. However, these techniques do not make use of k-vertex disjoint connectivity amid inexpensive sensor nodes and supernodes, and therefore, the network may not withstand the failure of k − 1 sensor nodes. Though the technique prescribed in [19] ensures the k-supernode connectivity, it does not take into account the power effectiveness of the ensuing topologies. Our proposed technique varies from most of these efforts by sustaining k-connectivity as well as dealing with energy effectiveness in unison.

Cardei et al. [22] suggested a fault-tolerant topology control mechanism intended for heterogeneous WSNs having two-layer architecture ensuring both k-connectivity and energy effectiveness. This mechanism is designed to amend sensors' communication range to realize k-vertex disjoint connectivity to supernode as well as to minimize the utmost communication power level of inexpensive sensors. Authors recommend a centralized mechanism named global anycast topology control (GATC). This mechanism strives to minimize the communication power level of inexpensive sensors but achieves this having network-wide topology information. However, this may not be realistic considering its implementation in significantly large WSNs. Authors proposed a distributed approach named distributed anycast topology control (DATC) that offers k-vertex disjoint connectivity to supernode by means of augmenting the communication coverage of sensors by small raises. However, the distributed DATC algorithm provides a far more realistic means to ensure k-connectivity. DATC entails 1-hop neighborhood topology information, which may be stretched to n-hop. The aim of this algorithm is to make sure that any sensor node SNi is either directly reachable to any other sensor node SNj in its accessible region or presence of minimum k disjoint routes amid SNi and SNj.

The localized DATC starts with building a minimum set of neighbors with the least communication power level for every node. The power level of each node is augmented with little increments to discover other accessible nodes from the paths of neighbors. The path discovery scope may get restrained owing to this fact that several nodes outside the accessible neighborhood may be unknown to the node executing discovery. DATC is having this significant constraint for the reason that it has minimal possibility to uncover k disjoint paths for the 1- or 2-hop distance neighbor nodes from the node executing discovery. As opposed to DATC, our proposed algorithm ensures each node to uncover paths which includes nodes beyond its accessible region by storing entire path information from supernodes to sensor nodes. This information is stored in the local information table maintained by each sensor node. Like this, our proposed algorithm offers a lot more possibility to uncover improved k-disjoint paths and better power optimization in comparison with DATC. In the course of uncovering disjoint paths, sensor nodes serve possibly with the utmost transmission power as a consequence of improving the possibility of uncovering more disjoint paths. Our experimental outcomes are in compliance with our proposed discussion.

3 Transmission Power Efficient Disjoint Path

Our proposed fault-tolerant topology management algorithm operates in a two-tiered heterogeneous WSN architecture comprising resource-intensive supernodes and resource-constrained sensor nodes.

In order to accomplish fault-tolerant topology, we give attention to k-vertex disjoint connectivity to supernodes. This means every sensor has connectivity to a minimum of one supernode by means of k-vertex independent routes. In the worst case, the resultant topology is able to sustain the failure of $k-1$ sensor node failures. Our presented algorithm tends to eliminate the edges which do not belong to any one of the sensor-to-supernode k-disjoint paths. To make this happen, we figure out neighbor nodes that are on one such disjoint paths and that are not. This algorithm discovers a set of essential vertices to assure k-vertex supernode connectivity. Obtaining the set of essential vertices, each sensor node eliminates those edges not connected to an essential vertex. Subsequently, in order to save battery power of sensors, we reduce the transmission range of sensors nevertheless can reach the furthest node in the newly constructed neighbor set. Since this proposed algorithm is a localized and distributed algorithm, hence is implemented by every sensor node in the network. It does not require global topology information. Rather each node requires 1-hop topology information. The messages discover the disjoint paths from sensor nodes to supernodes incorporating path information from supernodes to inexpensive sensors.

3.1 Network Model

Our algorithm makes a consideration of a nonhomogeneous network comprising of SP_N resource-intensive nodes and S_N low-cost sensor nodes, where $SP_N \ll S_N$. Sensors are positioned at random, whereas supernodes are positioned at identified sites. We are concerned about the communications from sensor node to sensor node and from sensor node to supernode. We are least concerned about the communications from one supernode to another supernode for the reason that they are resource-intensive and can communicate directly employing their utmost transmission range. In the original WSN deployment structure, every sensor is having a maximum transmission range *TRmax*.

3.2 Problem Statement

We intend to build a two-tiered heterogeneous WSN topology to ensure k-vertex supernode connectivity in order to route data accumulated by sensor nodes to the supernodes. We strive to lessen the dispensed power required for transmission for each and every deployed sensor nodes and at the same time sustaining k number of disjoint routes from every sensor node to the subset of supernodes. Herein newly formulated topology, each sensor node ought to be connected to a minimum of one supernode having k number of vertex disjoint communication routes.

We characterize this problem in this way: provided a k-vertex disjoint supernode linked topology having SP_N resource-intensive nodes as well as SN resource-restrained nodes able to regulate the communication range to a predesignated constant *TRmax*, ascertain the communication range of every SN with the aim to substantially reduce the overall power required for transmission and the ensued topology continues to be having k-vertex supernode connectivity.

3.3   Algorithm for K-Vertex Supernode Connectivity

The proposed localized algorithm TEDP effectively dispenses diverse transmission power levels for nodes while sustaining k-vertex supernode connectivity. It operates in five phases. The first is the route information assortment phase which is undertaken by the supernodes via *Initialize* messages. This message can only be created and sent out by a supernode and encompasses supernode ID. All sensors deployed in the network receive this message and initiate the updation in their local information table based on the data encompassed in the *Initialize* message. Every sensor node does maintain a disjoint route list. If any updation takes place in the route list, then sensor nodes send out *RouteInfo* messages. When this *RouteInfo* messages are received by the sensor nodes, each sensor node makes a computation on the disjoint routes based on the entries in the information table and the information is incorporated in the received *RouteInfo*. If there is a reduction in the disjoint route cost after the computation is observed, then the message incorporating the revised information related to the route is forwarded. The highest cost of the route is the cost of the disjoint route in the set of disjoint routes maintained by a sensor node. The moment it is observed that no further reduction in the computation of routes lists, the first phase of the proposed algorithm comes to an end. The second phase ensures every sensor node computes its entailed neighbors by employing the locally computed disjoint routes list.

The third phase ensures every sensor node constructs a *Notification* message for each distinct disjoint route and forwards the message downward the successive nodes in the disjoint route. Each node in the distinct route tags each other as a neighbor for establishing k-vertex disjoint connectivity to supernode. The fourth phase eliminates those untagged neighbor nodes from the list of neighbors maintained at each sensor node. The concluding phase makes certain that each sensor node regulates its transmission power level to be within the coverage of the farthest neighbor in consistent with the newly formulated topology. The TEDP algorithm notations are introduced in Table 1. The route information assortment phase in TEDP is given in Algorithm 1.

Algorithm 1. Path Information Collection in TEDP

```
Input: M, K
Output: D

        Wake up all sensor nodes.
        Broadcast network message with supernode id.
        mc = 0,c = 0;
        T =∅;
        for all received message(p) at each sensor node do,
                if p ∈ T then
                        discard the entry
                else
                        D = DIS (T);
                        T = T ∪ P;
                        Sort(T);
                        T' = {Pᵢ ∈ T | i<= M};
                        Transmit PathInfo (T');
                end if
        end for
```

Table 1 TEDP Notations

T and T'	Set of all paths
D and D'	Disjoint path set
U	Union operation
M	Maximum paths
C and C'	Cost of paths
P and P_i	*RouteInfo* message
K	Degree of disjoint connectivity
Q	Subsets of size k of T
q	Single subset of T of size k

The initial phase commences with supernodes broadcasting *Initialize* messages throughout the network. Upon receiving the message, each node initiates to update its own route information table by creating a new entry on behalf of the newly instituted route to supernode. The entry information will consist of supernode ID and the link cost (span of the link) between the node which received the message and the supernode. The route cost is the span of the most elongated link in the route. The routes in the information table are arranged in accordance with cost in an increasing order.

When the information table is updated, each sensor constructs and relays a *RouteInfo* message containing its ID and information table to their accessible neighborhood employing utmost transmission power. Upon receiving *RouteInfo* message, each sensor node computes the union of the routes received through *RouteInfo* message and the existing in its local information table. The least cost for a set of disjoint routes for the recently computed union and the existing is estimated. The size of the estimated disjoint sets is no more than k since we provision k number of disjoint routes in the ensuing topology. The route information table gets updated when the newly computed cost is less than the existing route cost. Finding disjoint routes (DIS) is given in Algorithm 2.

Algorithm 2. Finding Disjoint Routes to Supernodes (DIS)

```
Input: T and k
Output: D

        D = Ø;
        If |T| < k then
                If P ∩ D = Ø then
                        D = D∪ P;
                end if
        else
                Q = {q ⊂ T | |q| = k}
                for all q
                        if q consists of disjoint paths then
                                C = cost(q);
                                qmin= q;
                        end if
                end for
        end if
```

If a *RouteInfo* message has not affected any update in the disjoint path list of the receiver, in that case no *RouteInfo* message is relayed by that node. Any update can be effect if a lesser cost disjoint route is ascertained through this message. The algorithm provisions an upper bound on the count of *RouteInfo* messages to be sent by the sensor for the duration of this phase. This restriction assures the convergence of this algorithm. It may happen in the worst case that there may be $|E|$ disjoint routes with distinct costs in the order of reducing costs. $|E|$ is the count of edges in the network. Then the sensor node will have to send maximum $|E|$ number of *RouteInfo* messages. Since we may have at most $O(n^2)$ edges given a graph having n number of nodes, we may come to this conclusion that a sensor node may send out $O(n^2)$ *RouteInfo* messages. This is an exceptional case when every node has an edge with every other node in the network. This upper bound may be articulated as $O(nd)$; here d indicates the utmost node degree.

It is proposed to restrict the count of hops a *RouteInfo* message has to travel for the duration of route information phase. The supernode specifies a time-to-live (TTL) field value for the *RouteInfo* communication and sends this value to sensor nodes through *Initialize* message. This TTL value is incorporated in every *RouteInfo* message and has the effect on the count of hops in the concluding established path. When a sensor node receives a *RouteInfo* message, it checks its TTL value in the message and if the TTL value has reached the preassigned value then no more *RouteInfo* message is initiated by the receiver node irrespective of whether this message has caused any update in its disjoint list. This is to ensure that the disjoint routes having the length exceeding certain prescribed restricted value are not contemplated in the computation of disjoint list.

The next phase of this algorithm ensures each and every node in a preferred route is to be tagged as essential neighbor nodes. In order to ensure this, every node sends out a *Notification* message to all its preferred routes. This message gets communicated alongside the route. Every neighbor node along the route tags one another as essential neighbors. If any pair of neighbors does not tag one another, then it is understood that the connected link amid these neighbors is not indispensable hence may be subsequently eliminated. Obtaining the decided list of essential neighbors, each node tends to regulate its transmission power to extend its coverage access to the furthest neighbor in the newly formulated topology.

The execution time complexity of this algorithm is $O(nd^2)$; here n is count of sensors deployed in the network, and d is the utmost node degree.

4 Performance Evaluation

4.1 Experimental Setup

In our simulation setup, the low-cost sensors are positioned indiscriminately in a 500 m × 500 m region and the least (5% of total deployed sensors) count of resource-rich supernodes are deployed uniformly in a deployment region. The utmost transmission range TRmax of the sensors is set to be 100 m, and the path attenuation exponent is set to be 2, and the degree of disjoint connectivity (k) is equal to 2. We assume the adaptable transmission range of supernodes is up to 300 m.

4.2 Overall Transmission Power

We take an account of our simulation outcomes presenting the power expended by the sensor nodes in the formulated topology (with k = 2) subsequent to the implementation of $DATC_{h = 1}$ (1-hop) and $DATC_{h = 2}$ (2-hop) local neighbor. In Fig. 1, the count of sensors enhanced from 100 to 500 and the count of supernodes is fixed at 5% of the count of sensors deployed in the network. This clearly indicates that the induced topologies by our TEDP algorithm incur less total transmission power than DATC algorithm for values of h = 1 and 2. This gives the explanation that the DATC approach encounters intricacies in finding the alternate disjoint routes to an elongated disjoint edge employing 1-hop or 2-hop neighbor region information. This leads to restricted disjoint path search scope; hence, the elongated disjoint paths cannot be replaced with shortened paths, in consequence, ensuing

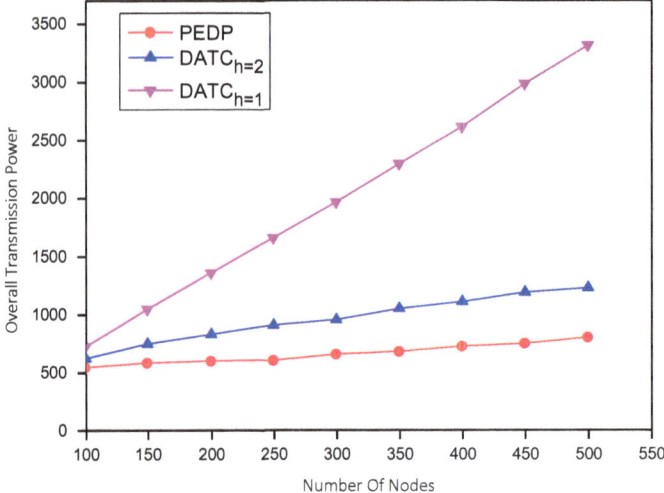

Fig. 1 Overall transmission power

long transmission ranges and thus expended transmission power. One more significant observation associated with the implementation of $DATC_{h=1}$ and $DATC_{h=2}$ is that $DATC_{h=2}$ having larger search scope offers considerably better results than $DATC_{h=1}$ with restricted search scope.

4.3 Utmost Transmission Power

Utmost transmission power is certainly an imperative performance metric for the induced topologies for the reason that this metric provides significance to ensuring equilibrium of battery power utilization among all nodes. Even though the total transmission power in the generated topologies is low, the topology may get disengaged if some of the sensors with large communication ranges employ utmost transmission power utilizing more battery energy than others leading to faster exhaustion of battery energy. In this section, we take an account of our simulation outcomes on utmost power among the entire nodes in the ensuing topologies induced by TEDP and DATC for the value of k = 2 and the count of supernodes which is 5% shown in Fig. 2.

The performance of our TEDP algorithm is considerably improved than that of DATC. This is because our algorithm can find k number of disjoint routes to the supernodes with the information available in the local path tables maintained by each sensor node, whereas DATC cannot find as many disjoint paths as our algorithm because DATC is having an access to information restricted to only its reachable neighborhood effecting utmost transmission ranges.

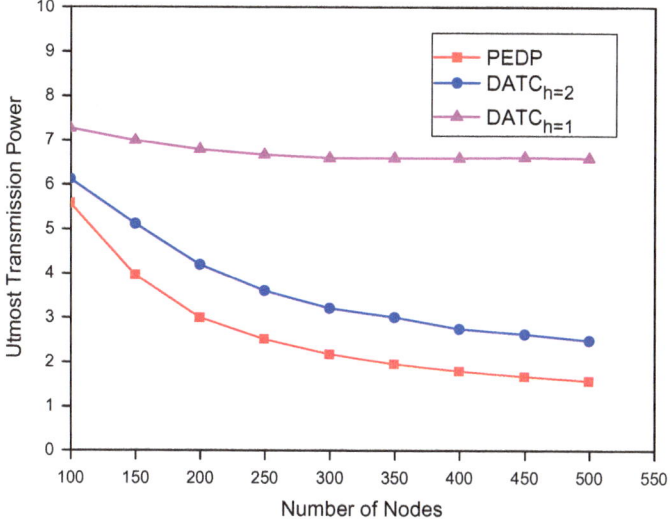

Fig. 2 Utmost transmission power

4.4 Overall Count of Control Message Transmissions

Here, we illustrate our simulation outcomes with respect to total count of transmissions of control messages for the duration of the implementation of TEDP and DATC. This is imperative to consider control message transmission as a performance metric as it is essential to take into account of energy utilization in the induced topologies as well as for the formation of those topologies. High message transmission cost may not be effective for ensuring energy efficiency in the formation of a stable topology. We observe that $DATC_{h=2}$ necessitates larger count of message transmissions as compared to TEDP or $DATC_{h=1}$. This is because $DATC_{h=2}$ necessitates each node to inform to all its 2-hop neighbor nodes. In contrast, both $DATC_{h=1}$ and TEDP entail a single message transmission when there is a path information table update. In $DATC_{h=2}$, each sensor node necessitates d number of message transmissions, where d is the node degree. Since $DATC_{h=2}$ entails larger count of message transmissions, hence, may not be realistic to put into practice. Transmission of too many messages for the duration of execution may trigger all the sensor nodes to exhaust their battery power more quickly, hence making the network ineffective.

4.5 Packet Loss Effect

In wireless medium, transmissions may suffer from intermittent packet losses as a result of communication link failures, collisions, deficiency of strong signal, high bit error, etc. In this paper, we intent to show a network situation where any communication link may suffer from packet losses with a known probability. We take a view that the overall transmission power increases with the increased value of PLR. This is because a large count of messages gets lost during transmissions owing to having a large PLR value causing the sensor node to have partial information about topology from the sensors in their neighborhood. Utilizing this limited information about the topology, algorithms may fail to obtain optimized disjoint paths for the reason that message encompassing the path information may be lost during its transmission due to transmission failures.

5 Conclusion

In this paper, we present a distributed fault-tolerant topology management algorithm for a nonhomogeneous WSN composed of resource-intensive supernodes and resource-constrained sensor nodes. Our algorithm ensures that every sensor node strives to establish k number of vertex disjoint connections with supernodes. The aim of our algorithm is to substantially reduce overall power utilization in WSM. Simulation outcomes show that our TEDP algorithm in comparison to existing DATC algorithm attains the decline in overall transmission power substantially by 25% and the decline in utmost transmission power needed in a sensor node by 30% under the supposition of no packet losses. With the consideration of PLR between 0.2 and 0.3, $DATC_{h = 2}$ performs considerably better compared to TEDP with respect to overall transmission power. In addition, TEDP entails fewer message transmissions and receptions in comparison to DATC. Our proposed distributed algorithm is appropriate for real WSN applications on account of its scalability to large networks.

References

1. Akyildiz, I.F., Su, W., Sankarasubramaniam, Y., Cayirci, E.: A survey on sensor networks. IEEE Commun. (Magazine) **40**(8), 102–116 (2002)
2. Liu, H., Nayak, A., Stojmenovic, I.: Fault-tolerant algorithms/protocols in wireless sensor networks. In: Guide Wireless Sensor Networks, pp. 261–291 (2009)
3. Bao, L., Garcia-Luna-Aceves, J.J.: Topology management in ad hoc networks. In: ACM International Symposium on Mobile Ad Hoc Networking and Computing. Annapolis (2003)
4. Younis, M., Sentruk, I.F., Akkaya, K., Lee, S., Senel, F.: Topology management techniques for tolerating node failures in wireless sensor networks: a survey. Comput. Netw. **58**, 254–283 (2014)

5. Wang, Y.: Topology control for wireless sensor networks. Wireless Sensor Netw. Appl., 113–147 (2008)
6. Gengzhong, Z., Qiumei, L.: A survey on topology control in wireless sensor networks. In: Proceedings of the 2nd International IEEE Conference Future Networks, pp. 376–380 (2010)
7. Li, N., Hou, J.C.: Localized fault-tolerant topology control in wireless ad hoc networks. IEEE Trans. Parallel Distrib. Syst. **17**, 307–320 (2006)
8. Blough, D.M., Leoncini, M., Resta, G., Santi, P.: The k-neigh protocol for symmetric topology control in ad hoc networks. In: Proceedings of the 4th ACM International Symposium on Mobile ad hoc Networking & Computing (2003)
9. Chen, B., Jamieson, K., Balakrishnan, H., Morris, R.: Span: An energy-efficient coordination algorithm for topology maintenance in ad hoc wireless networks. Wirel. Netw. **8**(5), 481–494 (2002)
10. Kumar, S., Lai T.H., Balogh, J.: On k-coverage in a mostly sleeping sensor network. In: Proceedings of the ACM Annual International Conference on Mobile Computing and Networking, pp. 144–158 (2004)
11. Baryshnikov, Y.M., Coffman, E.G., Kwak, K.J.: High performance sleep-wake sensor systems based on cyclic cellular automata. In: Proceedings of the International Conference on Information Processing Sensor Networks, pp. 517–526 (2008)
12. Wu, W., Du, H., Jia, X., Li, Y., Huang, S.C.-H.: Minimum connected dominating sets and maximal independent sets in unit disk graphs. Theor. Comput. Sci. **352**(1), 1–7 (2006)
13. Song, W.–Z., Wang, Y., Li, X.-Y., Frieder, O.: Localized algorithms for energy efficient topology in wireless ad hoc networks. In: Proceedings of the ACM International Symposium on Mobile Ad-Hoc Networking and Computing, pp. 98–108 (2004)
14. Haghpanahi, M., Kalantari, M., Shayman, M.: Topology control in large-scale wireless sensor networks: Between information source and sink. Ad Hoc Netw. **11**(3), 975–990 (2012)
15. Younis, O., Krunz, M., Ramasubramanian, S.: Node clustering in wireless sensor networks: recent developments and deployment challenges. IEEE Netw. **20**(3), 20–25 (2006)
16. Mamidisetty, K.K., Ferrara, M.J., Sastry, S.: Systematic selection of cluster heads for data collection. J. Netw. Comput. Appl. **35**, 1548–1558 (2012)
17. Azharuddin, M., Kuila, P., Jana P.K.: A distributed fault tolerant clustering algorithm for wireless sensor networks. In: Proceedings of the IEEE International Conference on Advances in Computing, Communication and Informatics, pp. 997–1002 (2013)
18. Akyildiz, I.F., Kasimoglu, I.H.: Wireless sensor and actor networks: research challenges. Ad Hoc Netw. **2**(4), 351–367 (2004)
19. Wu, J., Yang, S., Cardei, M.: On maintaining sensor-actor connectivity in wireless sensor and actor networks. In: Proceedings of the IEEE Conference Computer Communications, pp. 888–896 (2008)
20. Ozaki, K., Watanabe, K., Itaya, S., Hayashibara, N., Enokido, T., Takizawa, M.: A fault-tolerant model for wireless sensor-actor system. In: Proceedings of the 20th International Conference on Advanced Information Networking and Applications, vol. 2 (2006)
21. Melodia, T., Pompili, D., Gungor, V.C., Akyildiz I.F.: A distributed coordination framework for wireless sensor and actor networks. In: Proceedings of the 6th ACM International Symposium on Mobile Ad Hoc Networking and Computing (2005) 99–110
22. Cardei, M., Yang, S., Wu, J.: Algorithms for fault-tolerant topology in heterogeneous wireless sensor networks. IEEE Trans. Parallel Distrib. Syst. **19**(4), 545–558 (2008)

Dynamic Localization Algorithm for Wireless Sensor Networks

Meghavi Choksi, Saurabh K. Pandey, Mukesh A. Zaveri
and Sanjay Garg

Abstract Localization is required to maintain real-time position for processing in indoor and outdoor conditions. This may be utilized to support mobile computing and networking among nodes in wireless sensor networks. In this paper, a dynamic algorithm for location estimation is proposed for the wireless sensor networks. The algorithm is tested on the TinyOS-based emulation test bed for validation of the approach in real-time environment.

Keywords Localization · Wireless sensor networks · Distributed computing

1 Introduction

The rapid deployment and development of embedded sensors use the benefit of skilled processing, methodical use of technologies, adapt to fast changing environment, and responsive to cooperative movement. Localization plays an important role in coverage, deployment, routing and target tracking for safety critical applications. It may be used in infrastructure maintenance and defence applications. The implementation of localization process is governed by various characteristics. First, the accuracy of location computation depends on signal modality. For example, in

M. Choksi (✉) · S. Garg
Computer Science and Engineering Department, Nirma University,
Ahmedabad, Gujarat, India
e-mail: 14mcen03@nirmauni.ac.in

S. Garg
e-mail: sgarg@nirmauni.ac.in

S. K. Pandey · M. A. Zaveri
Computer Engineering Department, Sardar Vallabhbhai National
Institute of Technology, Surat, Gujarat, India
e-mail: ds14co006@coed.svnit.ac.in

M. A. Zaveri
e-mail: mazaveri@coed.svnit.ac.in

© Springer Nature Singapore Pte Ltd. 2018   495
K. Saeed et al. (eds.), *Progress in Advanced Computing and Intelligent Engineering*,
Advances in Intelligent Systems and Computing 563,
https://doi.org/10.1007/978-981-10-6872-0_46

humid environment, acoustic performs better than radio signal. Second, the range estimation phase of the scheme uses either range-based [2, 8] or range-free [1, 3] and anchor-based [6, 9] or anchor free [5, 11] approaches and then fed into next phase to calculate the positions. Third, the algorithms may use either absolute or relative localization techniques. Finally, the underlying computing architecture may be centralized, locally centralized or distributed. To support services using three-fold processing: sense, process and communicate, a robust localization scheme is needed that supports inevitable real-time computing architectures.

In this paper, we propose a dynamic localization algorithm to support location estimation in wireless sensor networks. The evaluation of proposed algorithm is done for average error in location estimation using TinyOS [7]-based emulation test bed with iris motes. The paper is organized as follows. Section 2 presents the proposed dynamic localization algorithm followed by experimental evaluation in Sect. 3. Finally, the paper is concluded in Sect. 4.

2 Proposed Dynamic Localization

In this section, we propose a dynamic localization algorithm for wireless sensor networks. Our algorithm analyses the network based on connectivity and density. In case of a sparse network, the centralized computing is used, while for a dense network, the distributed computing for position estimation is performed.

Initially, the anchor devices broadcast packet $(node_{ID}, current\_voltage, x, y)$, where (x, y) is position, $node_{ID}$ is unique identification and $current\_voltage$ is battery voltage remaining for the anchor nodes. A device whose location needs to be found, receives the packet and differentiate itself as *isolated*, *semi-isolated* and *normal* based on connectivity. The devices use received signal strength indicator (RSSI) that follows log-normal shadowing model [10] given by

$$p_i = p_o + 10n_p log\left(\frac{d_i}{d_o}\right) + x_\sigma \tag{1}$$

where p_i is received power with distance d_i, p_o is the power calculated at reference distance d_o, n_p is path loss exponent and x_σ is the standard normal distribution with zero mean.

The values, thus produced, need to be filtered for which we used two different methods. First, *feedback filter* [4] is used that ensures the smoothing of a large difference in signal strength values using

$$RSSI_{filter} = \alpha * RSSI_p + (1 - \alpha) * RSSI_{p-1} \tag{2}$$

where $RSSI_{filter}$ is the smooth value obtained, α is a constant ranging from 0.75 and above for p nodes. The second method used is *average based filtering* [4]. In this method, the average of all the signal strength measured is

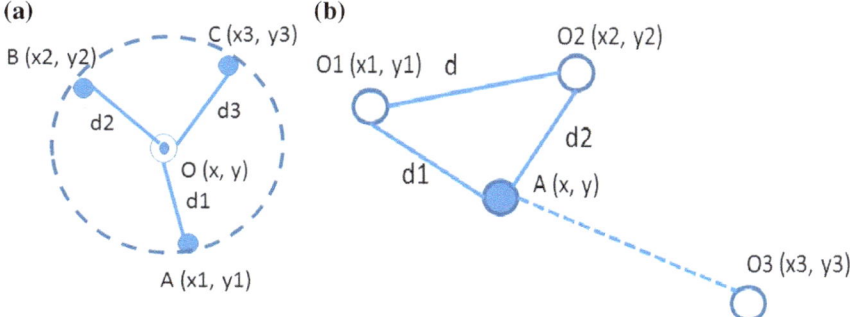

Fig. 1 a Distance calculation with one anchor; **b** Location estimation using three anchors

$$RSSI_{filter} = \frac{1}{n} \sum_{r=1}^{p} RSSI_i \tag{3}$$

where $RSSI_i$ is received signal strength and $RSSI_{filter}$ is filtered value of signal strength.

2.1 Isolated Scenario

A device is said to be isolated if it gets connected to only one anchor node. As shown in Fig. 1a, the distances from anchor node O for A, B and C are d_1, d_2 and d_3, respectively. In this case, only distance from the anchor node is calculated using centralized localization scheme.

2.2 Semi-Isolated Scenario

A device is said to be semi-normal when the number of anchor nodes are either 2 or 3. It uses centralized scheme for position estimation. As shown in Fig. 1b, device A receives signal strength from anchor nodes O_1, O_2 and O_3 and finds that the strength for O_1 and O_2 are better than O_3 for a standard threshold value. This threshold depends on the kind of application environment and service needed. Thus, the location estimation is done based on two anchor nodes O_1 and O_2, where the known distance between them is d. The distances d_1 and d_2 are calculated using Eq. 1 and then (x, y) is found by

$$x = \frac{d^2 - d_2^2 + d_1^2}{2d} \tag{4}$$

$$y = \sqrt{d_2{}^2 - (x_2 - x_1)^2} - y_2 \tag{5}$$

The mean error e is calculated using

$$e = \sqrt{(x_o - x)^2 + (y_o - y)^2} \tag{6}$$

where (x_o, y_o) is the actual position of deployment.

2.3 Normal Scenario

A device is designated as normal when the number of anchor nodes are equal to or greater than 4. Select those anchor nodes whose signal strength is better for a threshold value. The device whose location needs to be found checks its voltage. If voltage is less, then centralized scheme is used or else the distributed scheme is used. The weight of an anchor node is defined as

$$w_i = \frac{RSSI_i}{RSSI_i + RSSI_{i+1} + \cdots + RSSI_p} \tag{7}$$

where p denotes the selected number of nodes with known positions. The position (x, y) is estimated using

$$x = \frac{w_i * x_i + w_{i+1} * x_{i+1} + \cdots + w_p * x_p}{w_i + w_{i+1} + \cdots + w_p} \tag{8}$$

$$y = \frac{w_i * y_i + w_{i+1} * y_{i+1} + \cdots + w_p * y_p}{w_i + w_{i+1} + \cdots + w_p} \tag{9}$$

and mean error is calculated using Eq. 6.

The proposed algorithm is evaluated using standard parameters. The results and respective analysis are discussed in the following section.

3 Results

We present an analysis of emulation results for the proposed algorithm in this section. We consider a two-dimensional area $s \times s$ with $s = 10$ m. The motes are positioned using random approach. As shown in Fig. 2a, 10 iris motes are deployed in indoor conditions out of which the motes encircled are anchors while all motes in squared boxes have unknown locations. For all three cases: normal, semi-normal and isolated,

---

Algorithm 1 Dynamic Localization Algorithm (DLA)

1: **Input:** Deployed nodes k, anchor with coordinates (x_p, y_p).
2: **Output:** Error of estimation (e), Location of nodes
3: **procedure** LOCATION_DLA() /*procedure call*/
4:      Anchor node broadcasts packet with $(node\_ID, current\_position, current\_voltage)$
5:      **if** $anchornode = 1$ **then**
6:           use centralized scheme for distance calculation
7:           Calculate d_i: $p_i = p_o + 10 n_p log\left(\frac{d_i}{d_o}\right) + x_\sigma$
8:           Mean Error e: $(d - d_i)/d_i$
9:      **else if** $anchornode = 2$ or 3 **then**
10:          Use centralized scheme of localization
11:          $x = \frac{d^2 - d_2{}^2 + d_1{}^2}{2d}$
12:          $y = \sqrt{d_2{}^2 - (x_2 - x_1)^2} - y_2$
13:          Mean Error: $e = \sqrt{(x_o - x)^2 + (y_o - y)^2}$
14:      **else if** $anchornode \geq 4$ **then**
15:          Use centralized scheme as in step 13 for sparse networks.
16:          Use distributed scheme for dense networks using decision parameters.
17:          $w_i = \frac{RSSI_i}{RSSI_i + RSSI_{i+1} + \cdots + RSSI_p}$
18:          $x = \frac{w_i * x_i + w_{i+1} * x_{i+1} + \cdots + w_p * x_p}{w_i + w_{i+1} + \cdots + w_p}$
19:          $y = \frac{w_i * y_i + w_{i+1} * y_{i+1} + \cdots + w_p * y_p}{w_i + w_{i+1} + \cdots + w_p}$
20:          Mean Error: $e = \sqrt{(x_o - x)^2 + (y_o - y)^2}$
21:      **end if**
22: **end procedure**

(a)

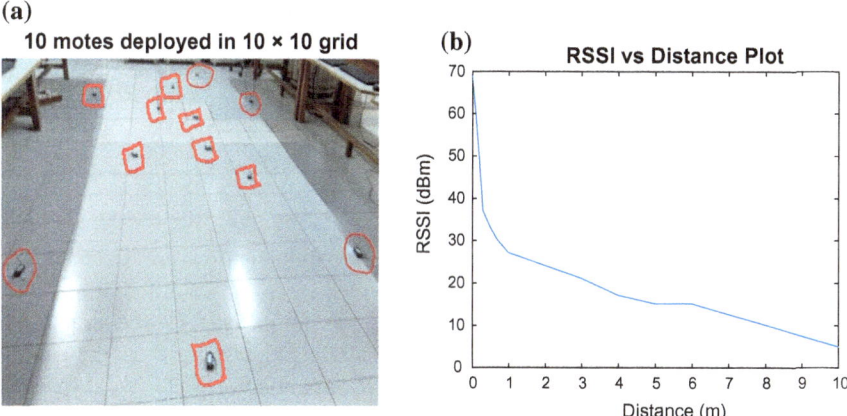

Fig. 2 a 10 motes deployed in 10×10 grid; **b** RSSI versus distance plot

in indoor conditions, $n_p = 2.7, x_\sigma = 10, d_o = 1$ m, and p_o is 27 dBm. The RSSI values for a mote are calculated at different distance measurements as shown in Fig. 2b. It can be evident that there is a decrease in signal strength with an increase in the distance of motes from the anchor nodes. As shown in Fig. 3a, the motes are kept at a

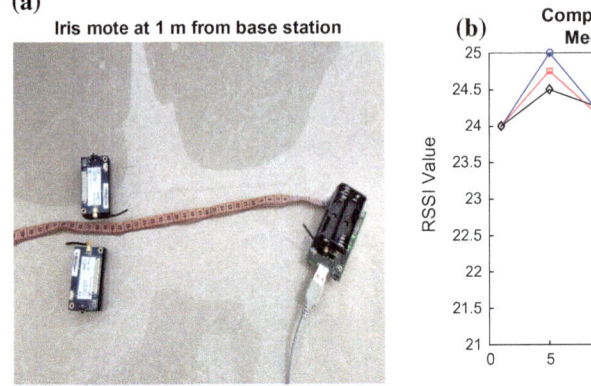

Fig. 3   **a** Iris motes at 1 m from base station; **b** Comparison between filtering method for $\alpha = 0.76$

Table 1   Calculation of mean error for 10 scenarios

Scenario	Original value	Measured value	Mean error
1	(1.0, 0.0)	(1.3, 0.462)	0.509
2	(0.2, 0.4)	(0.241, 0.86)	0.461
3	(0.5, 0.5)	(0.4, 0.320)	0.237
4	(0.5, 0.34)	(0.5, 0.14)	0.210
5	(0.5, 0.7)	(0.636, 0.69)	0.136
6	(0.5, 0.9)	(0.49, 0.95)	0.050
7	(1.0, 1.0)	(1.381, 1.381)	0.540
8	(0.0, 1.0)	(0.131, 0.918)	0.154
9	(1.0, 0.66)	(0.615, 0.96)	0.490
10	(1.0, 0.36)	(0.65, 0.32)	0.352

distance of 1 m from base station. In this case, a lot of fluctuations are observed. The values, thus produced, need to be filtered for which we used two different methods. For $\alpha = 0.76$, it is found that the average filtering technique performs better than the feedback filtering. Also, the feedback filter uses less data but increases the latency when calculating the new positions. A comparison among original signal strength received, values obtained from average filter and feedback filter is plotted in Fig. 3b.

The motes are static in nature and error is calculated for all the three schemes in the proposed algorithm. The error is calculated for ten different scenarios as tabulated in Table 1.

As can be observed from the table, in the distributed computing scheme, if the motes are nearer to anchor nodes, the localization error is more due to interference. The mean error is found to be least when the mote is placed at (0.5, 0.9) with value as 0.05. As depicted in Fig. 4, for the distributed hierarchy, when the number of

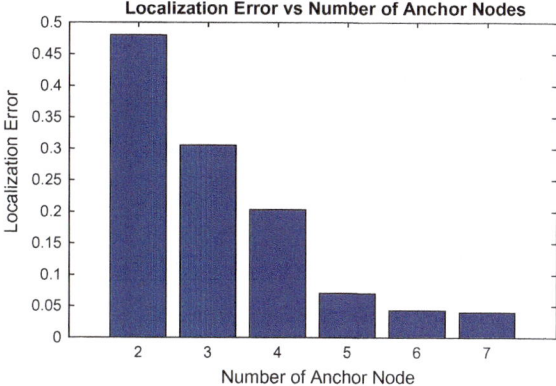

Fig. 4 Localization error versus number of anchor nodes

anchor nodes increase, the localization error is found to decrease significantly. For the set of motes deployed, the test bed is implemented with distributed, centralized and hybrid algorithms for ten different scenarios. The mean error obtained using centralized approach is 0.4, in case of distributed approach is 0.32. However, the proposed hybrid algorithm performs better than the previous two with mean error obtained as 0.26.

4   Conclusion

Localization for wireless sensor networks demands stringent quality measure for better communication. To achieve minimized energy consumption and maximized throughput for scalable applications, the basic need is to know the location of devices. The proposed dynamic localization method helps in implementing the algorithm a heterogeneous processing architecture. The algorithm is evaluated by real-time deployment of iris motes, and its evaluation confirms the validity of our approach. As part of future work, the evaluation of network needs to be done for a more complex environment.

References

1. Bulusu, N., Heidemann, J., Estrin, D.: Gps-less low-cost outdoor localization for very small devices. IEEE Pers. Commun. **7**(5), 28–34 (2000)
2. Cheng, X., Thaeler, A., Xue, G., Chen, D.: Tps: a time-based positioning scheme for outdoor wireless sensor networks. In: Proceedings of the Twenty-third Annual Joint Conference of the Computer and Communications Societies, pp. 2685–2696. IEEE, Mar 2004

3. Doherty, L., Pister, K.S., El Ghaoui, L.: Convex position estimation in wireless sensor networks. In: Proceedings of the Twentieth Annual Joint Conference of the Computer and Communications Societies, pp. 1655–1663. IEEE, Apr 2001
4. Halder, S.J., Giri, P., Kim, W.: Advanced smoothing approach of rssi and lqi for indoor localization system. Int. J. Distrib. Sens. Netw. **2015**(32), 1–11 (2015)
5. He, T., Huang, C., Blum, B.M., Stankovic, J.A., Abdelzaher, T.: Range-free localization schemes for large scale sensor networks. In: Proceedings of the 9th Annual International Conference on Mobile Computing and Networking, pp. 81–95. ACM, Sept 2003
6. Karalar, T.C., Rabaey, J.: An rf tof based ranging implementation for sensor networks. In: Proceedings of the International Conference on Communications, pp. 3347–3352. IEEE, June 2006
7. Levis, P., Lee, N., Welsh, M., Culler, D.: Tossim: accurate and scalable simulation of entire tinyos applications. In: Proceedings of the 1st International Conference on Embedded Networked Sensor Systems, pp. 126–137. ACM, Nov 2003
8. Niculescu, D., Nath, B.: Ad hoc positioning system (aps) using aoa. In: Proceedings of the Twenty-Second Annual Joint Conference of the Computer and Communications, pp. 1734–1743. IEEE, Mar 2003
9. Bahl, P., Padmanabhan, V.: RADAR: an in-building RF-based user location and tracking system. In: IEEE Computer and Communications Societies, vol. 2, pp. 775–784 (2000)
10. Rappaport, T.S., et al.: Wireless Communications: Principles and Practice. Prentice Hall, New Jersey (1996)
11. Shang, Y., Ruml, W., Zhang, Y., Fromherz, M.P.: Localization from mere connectivity. In: Proceedings of the 4th International Symposium on Mobile Ad Hoc Networking & Computing, pp. 201–212. ACM, June 2003

Improving Quality of Services During Device Migration in Software Defined Network

Raj Kumar, Abhishek Singh, Mayank Tiwary, Kshira Sagar Sahoo
and Bibhudatta Sahoo

Abstract Software Defined Networking (SDN) is a new approach of managing and programming networks enabled by OpenFlow. For load balancing, a migration among OpenFlowDevice (OFDevice) is needed from heavy-loaded controller to least-loaded controller. During migration of OFDevice, the response time, jitter and packet loss for the end user are high. To address this problem, we propose a method in which the response time, jitter and packet loss are minimized during device migration. In this approach for migrating the optimal OFSwitch, we use-*liveness, safety*, and *serializability*. The proposed approach focuses on selecting such a OFDevice which causes minimum load on the controller. The experimental results show that our proposed method improves response time, jitter and packet loss.

Keywords Software Defined Networking (SDN) · OFDevice · Controller

R. Kumar (✉) · A. Singh · M. Tiwary
Department of Information Technology, C. V. Raman College
of Engineering, Bhubaneswar 752054, Odisha, India
e-mail: rajkmrc710@gmail.com

A. Singh
e-mail: abhinavner26@gmail.com

M. Tiwary
e-mail: mayanktiwari09@gmail.com

K. Sagar Sahoo · B. Sahoo
Department of Computer Science and Engineering, National Institute
of Technology, Rourkela 769008, Odisha, India
e-mail: kshirasagar12@gmail.com

B. Sahoo
e-mail: bdsahu@nitrkl.ac.in

© Springer Nature Singapore Pte Ltd. 2018   503
K. Saeed et al. (eds.), *Progress in Advanced Computing and Intelligent Engineering*,
Advances in Intelligent Systems and Computing 563,
https://doi.org/10.1007/978-981-10-6872-0_47

1 Introduction

Software Defined Networking is the virtualization of traditional network in which control plane is separated from data plane. This decoupling enables us easy management and programming of networks with new innovation strategies based on Open-Flow protocols [1, 2]. Migration of OFDevice is needed for load balancing and to avoid any delay in processing of high-priority PACKET_INs. During migration of OFDevice among controllers, the response time for the particular OFDevice suddenly increases due to which there is an increase in jitter and packet loss [3].

Based on the specific configuration of the controller (CPU, memory and receiving buffer), the extent in the response time increases nonlinearly after a threshold number of *PACKET-IN* OFMessages per second. We proposed a method in which we migrate the OFDevice which causes least load in the heavy-loaded controller. During migration, the three standard properties—*liveness, safety*, and *serializability* [4] are taken under considerations.

1.1 *Contributions*

- We propose a migration method in Software Defined Network to improve the Quality of Service (QoS).
- We consider the response time, jitter and packet loss in the proposed method.
- We implement and evaluate the proposed method in mininet cluster and Java-based Floodlight controller.

2 Literature Survey

In [1], the load dynamically shifts among controllers according to the changing traffic patterns in the network. This removes the relevance of statically configured mapping between a switch and a controller and gives better response time when migrated dynamically. Game theoretic model was applied in [5] where an OFDevice was choosen for migration by a heavy-loaded controller (game sponsor) which invites its light neighbor controllers as game players to formulate the local game domain (GD) to acquire OFDevice which was taken as commodity under the constraints of zero sum game. A dynamic and adaptive load balancing algorithm for controller was developed based on distributed architecture for addressing flexible mechanism for load balancing among distributed controllers [3, 6]. This approach shows improved results in terms of throughput and interarrival *PACKET-IN* with respect to number of controller nodes.

3 System Model

Let $C = \{C_1, C_2, C_3, \ldots, C_n\}$ be the set of controllers and $S = \{S_1, S_2, S_3, \ldots, S_n\}$ be the set of OFSwitches, where C has many to many relationship with S and no element of C can be there which is not related to any element in S. For a OFSwitch S, there can be more than one controller also, where one acts as a master and the other acts a slave or equal. In OpenFlow 1.3 specifications [4, 7], a OFDevice can support the following controller roles:

- **ROLE_MASTER**: At this role, the controller processes every Packet_In message received and has full access to OFDevice and can assert changes to the OFDevice independently.
- **ROLE_SLAVE**: At this role, the controller does not receive any Packet_In messages from the OFDevices and has only read only access at a given time.
- **ROLE_EQUAL**: The default role of a controller is ROLE_EQUAL mode, where the controllers are having full access to the OFDevices.

For migrating a OFSwitch from one controller to another, not only these roles need to be changed but also properties of safe migration need to be preserved [1]. For a safe device migration from one controller to another, the following properties must be preserved:

- **Liveness**: There should be atleast one controller in an active mode at all time.
- **Serializability**: The controller would process the OFMessages as generated by OFSwitch in a synchronized fashion.
- **Safety**: Only one controller will process the message from the OFswitch to avoid any duplication in the flow table.

To preserve these properties and follow a safe migration [1], the source controller and the destination controller need to follow a specific flow order. The initial master controller, C_i sends the target master controller C_j, indicating the start of migration. The controller C_j after receiving the migration_start message from C_i, it sends a Role_Request message to the Switch for conversion of state from SLAVE to EQUAL mode. After receiving the successful reply of the Role_Change from the Switch, the target controller C_j communicates back to C_i for its role change success. Then, C_i fires a Flow_Add message to the Switch followed by a Barrier_Request message. After receiving the Barrier_reply (which depends upon the SDN application synchronization delay), the controller sends a Flow_Delete message for the same flow. Then, the controller again fires a Barrier_Request message and upon receiving the Barrier_Reply from the Switch, it indicates the target controller C_j for final Role change operations. Then, the source controller C_i fires a final Role_Request to change its role from Master to Slave and the target controller C_j fires a final Role_Request to change its role from Equal to Master.

4 Proposed Algorithm

In the proposed algorithm, we use two factors to select an optimal Switch in the network. And migration of this switch will reduce the load on the controller and hence increases the response time of the controller for OpenFlow Packets reply to the OFDevices. For this, we consider two controllers C_i and C_j as source and target controllers. And there a set of switches S_n are connected to both the controllers where one controller acts as a Master for a set of switches and the other controller acts as a Slave for the rest of the switches in the set.

Then, we try to find out the traffic flow statistics on individual switches during migration. When the controllers (C_i) CPU utilization exceeds a certain threshold, it starts the optimal switch selection module. This module selects a switch based on the flow of traffic on switch and then considers the CPU utilization caused by the switches over the controllers. The algorithm is defined as follows:

Algorithm 1: Switch Selection for Migration

    Inputs : The Switch set S.
             Source Controller C_i
    Outputs: A Optimal Switch s to be migrated from C_i

1 $RP_S \leftarrow NULL$
2 $CPU_S \leftarrow NULL$ Flow_Durations $\leftarrow NULL$
3 **for** $\forall s \in S$ **do**
4   **for** $\forall Flows \in s$ **do**
5   Received_Packets = (Packets) s.flowstats();
6   Max_Packets_s=MAX(Received_Packets);
7          $RP_S[s]$ = Max_Packets_s;
8          $CPU_S[s]$ = CPU_Util(s);
9   **end**
10 **end**
11 $Sort_{Desc}(RP_S)$;
12 $Sort_{Aesc}(CPU_S)$;
13 **for** $\forall v \in RP_S$ **do**
14   **if** $CPU_S[v]==min\_index\_threshold(CPU_S)$ **then**
15   return v;
16   **end**
17 **end**

In the initial steps of the algorithm (steps 3–10), we try to find out the maximum packets flow by the switches. Here s.flowstats() sends two STATS_REQUEST to switch in a very short interval and returns the difference of packet counts measured. Then, the two loops find out the max s.flowstats() for all switches connected to the controller and stores them in the set RP_S. Similarly, the set CPU_S holds the percentage of CPU utilization caused by the switches PACKET_INs to the controller. Then, we sort the set RP_S in descending order and the set CPU_S in ascending order. The final steps (13–17) find out an optimal switch to be migrated. Here we define a

$min\_index\_threshold(CPU_S)$, which is an index threshold for the array CPU_S. The loop in the steps 13–17 starts with the maximum in set RP_S and tries to find out whether the switch exists within the minimum range of CPU_S limited by the index threshold. In this process, the loop finds out such a switch which exists in the minimum range of CPU_S limited by the index threshold and having maximum value of RP_S.

5 Result Analysis

5.1 Experimental Setup

In the experimental setup, we used Java-based Floodlight SDN controller [8], Mininet [9] (installed on Ubuntu 14.04 environment). We created a topology with the help of MiniEdit comprises of two remote controllers c_1 and c_2 and 30 Switches. For simulating 30 switches, we created a Mininet Cluster [10]. All the OFSwitches are connected to both controllers. We deployed our proposed migration model in the above testbed and monitored the behavior using Iperf. To simulate the load on the controller, we considered 10 types of different PACKET_INs in the network and attached a calibrated number on a scale of 0–10 to each packet in based on the amount of CPU processing done for giving reply for the PACKET_IN by the controller. And during each run of simulation, these numbers are randomly assigned to the PACKET_INs (Figs. 1, 2, 3 and 4).

Fig. 1 Time versus Jitter with $\lambda = 0.56$

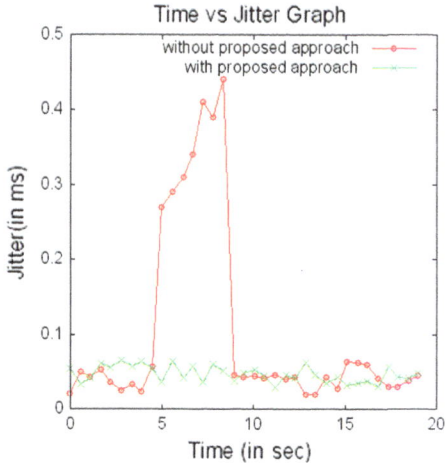

Fig. 2 Time versus Packet loss with $\lambda = 0.56$

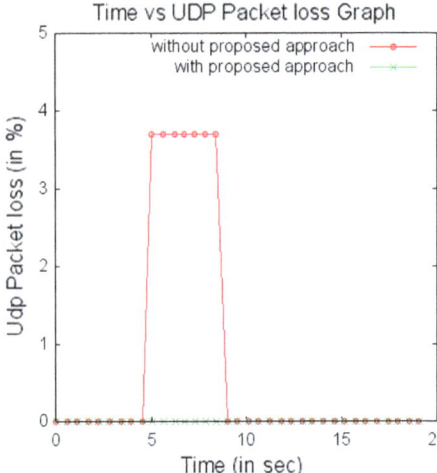

Fig. 3 Time versus Jitter with $\lambda = 0.56$

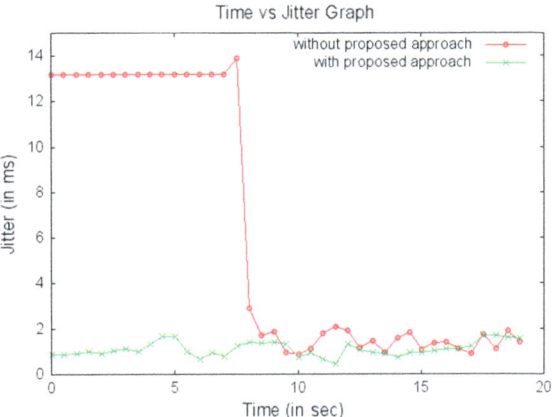

Fig. 4 Time versus Packet loss with $\lambda = 0.016$

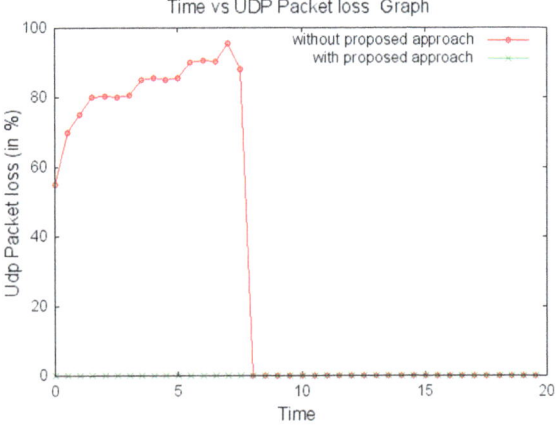

5.2 Experimental Results

- **Jitter** We perform two sets of experiments taking interval time, $i_1 = 0.56$ s and $i_2 = 0.016$ s and plot the jitter behavior of PACKET_IN messages with respect to the arrival time (Figs. 1 and 3), where there is sudden rise and fall of jitter during migration without proposed approach but with minimum variations in the jitter with our proposed approach.
- **Packet Loss** The packet loss is seen maximum in between 5 and 10 s (Fig. 2) for $i_1 = 0.56$ s and in the initial 8 s (Fig. 4) for $i_2 = 0.016$ s in the case of without proposed approach and with negligible packet loss in our proposed approach.

5.3 Conclusion

Through the proposed work, we have tried to increase the QoS performance of the loaded Switch when the network fires heavy load on the controllers in terms of PACKET_IN messages. The proposed methodology selects a Switch which in takes very less traffic in the network. We implemented the whole migration technique satisfying liveness, serializability, and safety properties. We performed our experiment on a Mininet testbed and monitored the results. In this way, we achieved better performance of QoS in terms of jitter and packet loss by selecting the optimal switch in the network for migration instead of selecting the switch which creates heavy load on the controller.

References

1. Dixit, A., Hao, F.: Towards an elastic distributed SDN controller. In: Proceedings of ACM HotSDN13, pp. 1–8 (2013)
2. McKeown, N., Anderson, T., Balakrishnan, H., et al.: OpenFlow: enabling innovation in campus networks. SIGCOMM CCR (2008)
3. Zhou, Y., Zhu, M., Xiao, L., Ruan, L., Duan, W., Li, D., Liu, R.: A load balancing strategy for SDN controller based on distributed decision. In: Proceedings of 13th International Conference on Trust, Security and Privacy in Computing and Communications, pp. 851–856 (2014)
4. Open Networking Foundation: OpenFlow management and configuration protocol OF-Config 1.3 (2012)
5. Cheng, G., Chen, H.: Game model for switch migrations in software-defined network. Electron. Lett. **50**(23), 1699–1700 (2014)
6. Cheng, G., Chen, H.: DHA: distributed decisions on the switch migration toward a scalable SDN control plane. In: Proceedings of IFIP Networking Conference (IFIP Networking), pp. 1–9 (2015)
7. Open Networking Foundation: OpenFlow Switch Specification Version 1.3.0, June 20
8. Floodlight. http://www.projectfloodlight.org/floodlight
9. Mininet. http://mininet.org
10. Lantz, B., OConnor, B.: A mininet-based virtual testbed for distributed SDN development. In: Proceedings of the Communications of SIGCOMM Review, vol. 45, pp. 365–366 (2015)

11. Levin, D., Wundsam, A., Heller, B., Handigol, N., Feldmann, A.: Logically centralized state distribution trade-offs in software defined networks. In: Proceedings of HotSDN (2012)
12. Tootoonchian, A., Ganjali, Y.: HyperFlow: a distributed control plane for OpenFlow. In: Proceedings of the INM/WREN (2010)

An Improved Mechanism to Prevent Blackhole Attack in MANET

Akhilesh Singh and Muzammil Hasan

Abstract Mobile node having ubiquitous characteristics, collectively form a group, generally known as mobile ad hoc network (MANET). It has a wide range of applications due to its easy installation, no any central authority-based failure point, no fixed infrastructure and many more. Openness of ad hoc network leads to the wide range of security issues that need to be investigated as well as resolved. In this paper, we are proposing the solution to the blackhole problem that overcomes this issue in more fruitful way. In general, attacker sends a forged RREP with high DSN and low hop count to fraudulently showing the source node having shortest path. When data are transmitted through that route all packets have been dropped. As per our approach, destination sequence number (DSN) is compared with a threshold value and fake messages are discarded. The simulation work has been carried out to compare the proposed approach with the standard AODV routing protocol on the basis of packet delivery ratio (PDR), and it shows positive results.

Keywords MANET · Blackhole attack · Packet drop attack
Routing attack

1 Introduction

At present, wireless networks are widely used for the communication purpose. They are of two types, one which has an infrastructure and present of access point and other which doesn't have fixed infrastructure also called ad hoc network. In ad hoc network, any node can leave or join the network at any point of time. Every mobile node has equipped with a wireless transmitter and a receiver. Every mobile node of the network is responsible for operation, creation and maintenance of the ad hoc

A. Singh (✉) · M. Hasan
Department of Computer Science and Engineering, MMMUT, Gorakhpur, India
e-mail: akhilesh840@gmail.com

M. Hasan
e-mail: muzammil_hasan@yahoo.com

© Springer Nature Singapore Pte Ltd. 2018   511
K. Saeed et al. (eds.), *Progress in Advanced Computing and Intelligent Engineering*,
Advances in Intelligent Systems and Computing 563,
https://doi.org/10.1007/978-981-10-6872-0_48

network. Confidentiality of data, availability of network can be achieved by removing the chances of attack on the network. MANET has feature like openness, varying topology, absence of central authority and no proper mechanism for defence, which cause security attacks on MANET.

Many types of security attacks are done on the MANET. One of them is blackhole attack in attacker node which sends forged information to the sender node and all the data packets are dropped by the attacker node. Many different mechanisms are used for the prevention of blackhole attack in MANET. Researchers use their different methodology to prevent this attack and secure the network from this attack. Some of them used destination sequence number which is known as DSN, some of them used cryptography mechanism to prevent the attack, some of them used trust-based mechanism and some of them used intrusion detection mechanism.

1.1 MANET

Ad hoc network is termed as independent basic service set (IBSS). In this type, communication between mobile nodes occurs without any access point. They directly send and receive messages from one node to another. Ad hoc network doesn't have any fixed infrastructure or central authority. Mobile nodes of the ad hoc network communicate with each other via wireless communication medium. In ad hoc network, nodes are free to roam, i.e. mobility is associated with the node, so such network is also called MANET that stands for mobile ad hoc network. Figure 1 shows the MANET with eight nodes [1].

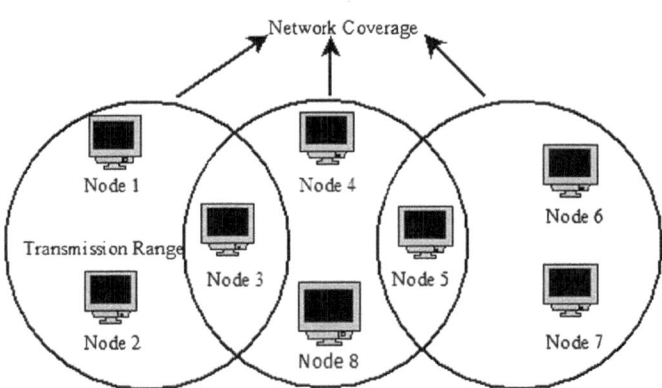

Fig. 1 MANET with eight nodes

1.2 MANET Characteristics

There are many characteristics which make MANET popular among the other communication network. These make MANET different from other networks.

Cooperative: If the source node wants to transmit data to the destination node and source and destination node are not in the range of communication. Then other nodes of the network cooperate with source and destination node for the communication.

Dynamic topology: Nodes in MANET are mobile in nature, so they roam throughout the network randomly and their location is unpredictable. This nature of the network creates complexity for the routing protocols.

Deficiency of fixed infrastructure: There is absence of central authority to monitor the ad hoc network. Traditional techniques of security are hardly applicable on MANET due to any presence of central authority, and it doesn't have any fixed infrastructure as the traditional networks have.

Resource constraints: MANET is a system of collective mobile nodes which have limited or low-power capacity. It also has limited memory, bandwidth, computation capacity, etc. So it is challenging to achieve a reliable and secure link of communication among the mobile nodes.

1.3 Routing Approaches Used in MANET

There are four types of routing approaches used in MANET. First is proactive routing, in which route are discovered before it is needed. It is also called table-driven routing. Some proactive routing protocols are given as follows: optimizes link state routing, destination-sequenced distance vector. Second is reactive routing in which routes are discovered when they are needed. Some reactive routing protocols are given as follows: dynamic source routing, ad hoc on-demand distance vector. Third is hybrid routing, which have some features of reactive protocol and some features of proactive protocol. Some hybrid routing protocols are given as follows: zone-based hierarchical link state routing protocol, order one MANET routing protocol. Fourth is hierarchical routing in which hierarchic level is maintained. Some hierarchical routing protocols are: cluster-based routing protocol, fisheye state routing protocol [1].

2 AODV Protocol

AODV is an important protocol among many routing protocols in MANET. In ad hoc network, every mobile node maintains a table which has routing information. That table information has the path from source to destination. When a node wanted

Fig. 2 AODV protocol

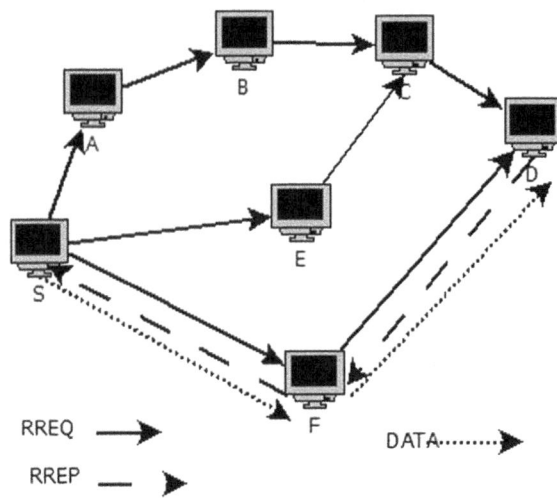

to transmit the data packet very first time, it checks the route in the route infor-mation table. If table contains the route for the desired destination then the node sends the data along with that route. If the node doesn't have the destination route then the node begins to discover a route by transmitting route request (RREQ) message to all its neighbours. All the nodes who get RREQ message check whether they have desired destination route, if they have route then send route reply (RREP) message to source node otherwise they broadcast RREQ message on behalf of source node [2] (Fig. 2).

Packet format of RREQ message is
<S_Add, SSN, B_Id, D_Add, DSN, Hop>

Packet format of RREP message is:
<S_Add, D_Add, DSN, Hop, Life_Time>

2.1   Blackhole Attack

Route discovery is a main task of AODV routing protocol where an attacker gets chance to attack on the network. When an attacker gets RREQ message and gives a prompt response of fake RREP message. After getting a RREP message source node forward data through that attacker node. When attacker node gets data from the source node, it drops all the data without forwarding data to destination node. This is considered as blackhole attack or packet drop attack [2].

Figure 3 shows an attack of blackhole attack on AODV network. In this net-work, S defines the source and D defines destination. When source node S wants to send data to the destination node D then it broadcasts RREQ messages to all its

Fig. 3 Blackhole attack

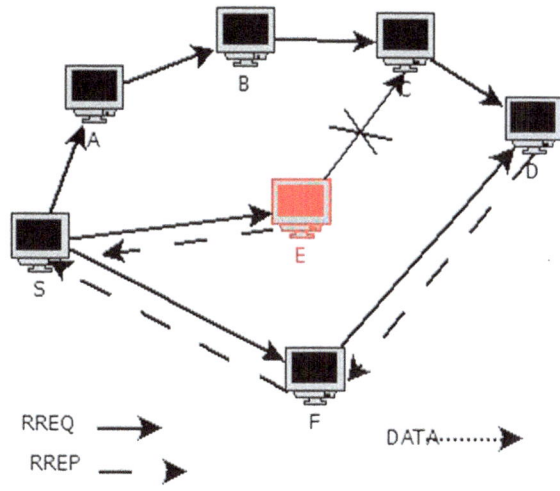

RREQ ⟶

RREP — ▶

DATA·········▶

neighbours. When attacker node E gets RREQ message, it immediately unicast fabricated RREP message with highest DSN and lower hop count. When S gets that fabricated RREP then it sends data through node E and node E drops all packets which come to this node.

3 Literature Review

Pooja Jaiswal and Rakesh Kumar [3] proposed their mechanism to prevent network form blackhole attack. In their approach, difference between SSN and DSN is considered. If difference is very high then it comes from an attacker node.

Deng et al. [4] proposed their mechanism to prevent network form blackhole. In their mechanism, sender node searches for alternative path of destination node. If path exists then there is node attack. This approach does not prevent from cooperative blackhole attack.

Tarun Varshney et al. [5] proposed their mechanism to prevent network form blackhole. Their approach is called watchdog mechanism. In their approach, when node sends data, it sets a watchdog. This watchdog monitors that whether the forwarded packet is also forwarded from next node in the route.

Anand A. Aware and Kiran Bhandari [2] proposed an approach to prevent blackhole attack. In this approach, first RREP is ignored as it assumes that it is from attacker. When the source node sends the data packet they use SHA 1 hash function for the message digest. This solution has problem that it is not necessary that first RREP is from attacker, and this solution doesn't prevent cooperative blackhole attack.

4 Proposed Mechanism

In order to prevent the above-described problem, we propose a new mechanism to prevent the attack. This new mechanism prevents the blackhole attack.

4.1 Calculation of Threshold

Sequence number has min value 0 and maximum value is 32-bit arithmetic (2^{32}).

$$DSN_{min} = 0 \quad DSN_{max} = 4294967295$$

In the proposed approach, we are defining a threshold value for the elimination of malicious node. Malicious node sends very high sequence number, nearer to the DSN_{max}. So we are defining the threshold by the calculation of following formula:

$$Th = DSNmax \times 97\%$$

where Th is the defined threshold. By defining this threshold, actually we are eliminating 3% of maximum sequence number, because attacker used sequence number nearer to maximum sequence number.

4.2 Flow Diagram for Additional Processing

Once threshold is defined, RREP message is verified by using that threshold value. When source node gets RREP message for the RREQ message which the source node generates it verifies those RREP messages. Figure 4 shows the processing flow chart at the source node.

4.3 Improved AODV Mechanism

In AODV, source node broadcasts RREQ and this message is forwarded until it reaches to the destination. Destination generates RREP for that RREQ. Improved mechanism can prevent attack on network. Improved mechanism is shown below:

Symbolisation	
S	Source Node
D	Destination Node
I	Intermediate Node
Th	Threshold
DNS	Destination Sequence Number

(continued)

(continued)

Improved_AODV (input: RREP)	
1.	Begin
2.	S broadcast RREQ to all its neighbours
3.	I receive RREQ reply if have fresh route otherwise forward
4.	D receive that RREQ
5.	D generate RREP and unicast it through the route RREQ came
6.	When S receives RREP, S checks DSN of RREP
7.	If DSN less than Th
8.	Route established
9.	Else
10.	Discard RREP
11.	End if
12.	Encrypt data and transmit it to D
13.	D receives data and Decrypt
End	

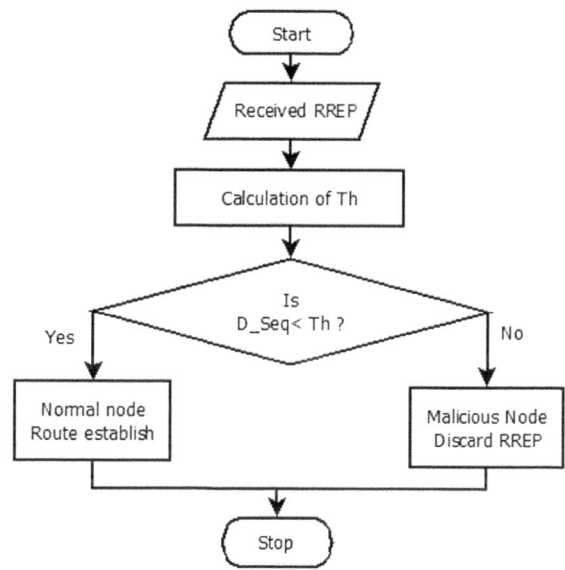

Fig. 4 Flow chart for additional process at source node

5 Simulation and Analysis

We have used network simulator 2 (NS2) for the simulation purpose. NS2 is a tool which provides implementation to different protocols. At the physical and data link layer usages IEEE 802.11 and network layer usages AODV routing protocol to route the packet. We are using 10–75 mobile nodes with transmission rate of 0.2 Mbps. The results of simulation are shown below (Figs. 5, 6 and Table 1).

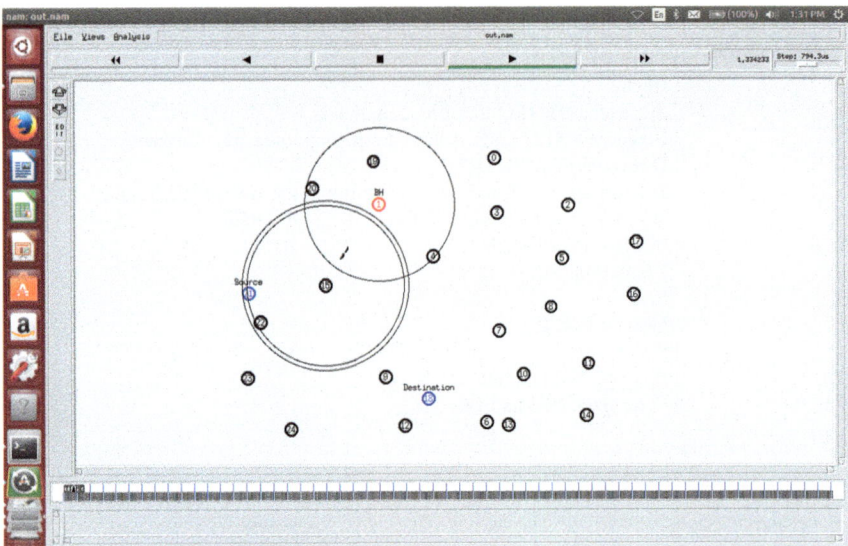

Fig. 5 Screenshot of blackhole attack

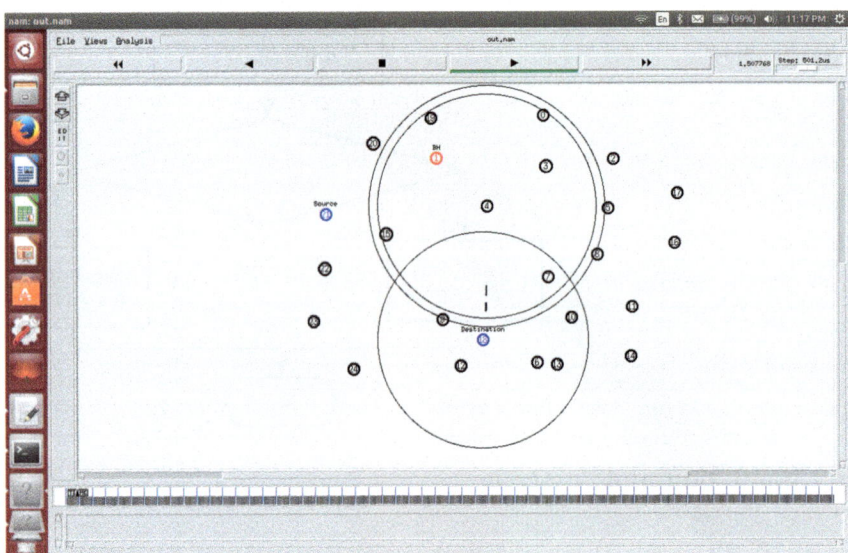

Fig. 6 Screenshot of prevention in case of single blackhole

Table 1 Simulation parameters

Constraint	Value
Simulator	NS2
MAC layer protocol	IEEE 802.11
No. of nodes	10–75
Routing protocol used	AODV
Simulation time	100 s
Traffic model	CBR
Terrain area	1000 m × 1000 m

5.1 Simulation Result

Simulation results are shown by the help of following graphs. These graphs show packet delivery ratio, throughput and delay with change of number of nodes in case of single blackhole and PDR with change of number of nodes in case of multiple attacker nodes are present means under cooperative blackhole attack.

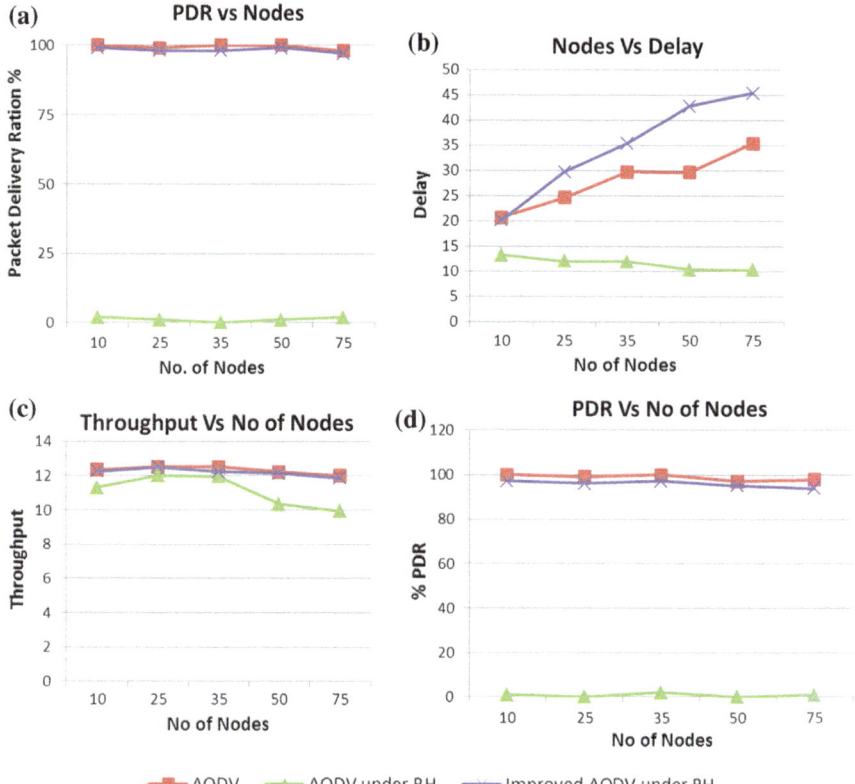

Fig. 7 **a** Effect of PDR with number of nodes. **b** Effect of time delay with number of nodes. **c** Throughput of the network with number of nodes. **d** Effect of PDR with number of nodes in case of multiple blackhole nodes

6 Conclusion and Future Work

MANET has a dynamic infrastructure and battery constraint. Complex computation consumes more battery power. Improved mechanism is capable for prevention of blackhole attack and increases the PDR and throughput of the network, and it does not have high complex computation. This improved mechanism is also capable for preventing cooperative blackhole attack.

In future, we are going to mitigate the time delay. We think there are some other directions which mitigate the effect of blackhole attack in a better way.

References

1. Singh, A., Hasan, M.: An analysis of prevention mechanism of blackhole attack. In: 2016 International Conference on Innovations in information Embedded and Communication Systems (ICIIECS'16), Tamilnadu, vol. 1, pp. 117–122 (2016)
2. Aware, A.A., Bhandari, K.: Prevention of black hole attack on AODV in MANET using hash function. In: 2014 3rd International Conference on Reliability, Infocom Technologies and Optimization (ICRITO) (Trends and Future Directions), pp. 1–6. IEEE (2014)
3. Jaiswal, P., Kumar, R.: Prevention of black hole attack in MANET. Int. J. Comput. Netw. Wirel. Commun. **2**(5) (2012)
4. Deng, Hongmei, Li, Wei, Agrawal, Dharma P.: Routing security in wireless ad hoc networks. IEEE Commun. Mag. **40**(10), 70–75 (2002)
5. Varshney, T., Sharma, T., Sharma, P.: Implementation of watchdog protocol with AODV in mobile ad hoc network. In: 2014 Fourth International Conference on Communication Systems and Network Technologies (CSNT), pp. 217–221. IEEE (2014)

Efficient Placement of ONUs via Ant Colony Optimization Algorithm in Fiber-Wireless (FiWi) Access Networks

Uma Rathore Bhatt, Aakash Chhabra, Nitin Chouhan and Raksha Upadhyay

Abstract FiWi is a fusion of optical network and wireless network. It gives high bandwidth, high data rates, high mobility, and low cost. The connection interface between the PON and WMN is established by the use of an optical network unit (ONU). ONU is an important component of communication as well as it affects the overall cost of system, hence its placement in the network becomes an important issue. Various research works have already been done to provide the solution. We aim to present a new optimizing algorithm for ONU placement and its minimization; this will result in the enhancement of network performance and reduction in the cost of overall network.

Keywords Fiber-Wireless (FiWi) · Optical network unit (ONU)
Ant colony optimization (ACO)

1 Introduction

Fiber-Wireless (FiWi) access network provides an impressive platform to support many powerful emerging future applications and services in "anytime–anywhere" approach, simultaneously it strengthens our society for future as it is merging

U. R. Bhatt (✉) · A. Chhabra · N. Chouhan · R. Upadhyay
Department of Electronics and Telecommunication, Institute of Engineering
and Technology, Devi Ahilya University, Indore, MP, India
e-mail: umarathore@rediffmail.com

A. Chhabra
e-mail: aakash.chhbr@gmail.com

N. Chouhan
e-mail: nitinchouhan03@gmail.com

R. Upadhyay
e-mail: raksha_upadhyay@yahoo.co.in

© Springer Nature Singapore Pte Ltd. 2018
K. Saeed et al. (eds.), *Progress in Advanced Computing and Intelligent Engineering*,
Advances in Intelligent Systems and Computing 563,
https://doi.org/10.1007/978-981-10-6872-0_49

together the capacity and merits of both optical network and wireless network [1, 2]. It has tree–mesh architectural topology such that the optical back end has tree topology, whereas wireless front end has mesh topology. Figure 1 describes FiWi network architecture; the feeder fiber is deployed from OLT to optical splitter; distribution fiber is deployed between optical splitter and each ONUs; and the ONUs connect wirelessly with the wireless routers. The users can access Internet by wireless and optical devices. For Internet access in upstream, users communicate to nearest router, and then router precedes information to its primary ONU through multihop wireless pathways. Now, user data transferred to the OLT and then finally to the Internet backbone. In downstream, the Internet provides provision to the user in the opposite manner [3, 4].

Since ONU is connecting interface between the PON and WMN, which has the capability to handle Internet traffic in addition to p2p traffic [5, 6]. Hence, its optimal placement in the network is a major concern, and it also affects the overall cost. Therefore, it is necessary to decrease the number of ONUs while ensuring connectivity among wireless routers. In this work, we implement a competent optimizing algorithm for ONU placement which is based on ant colony optimization (ACO) algorithm.

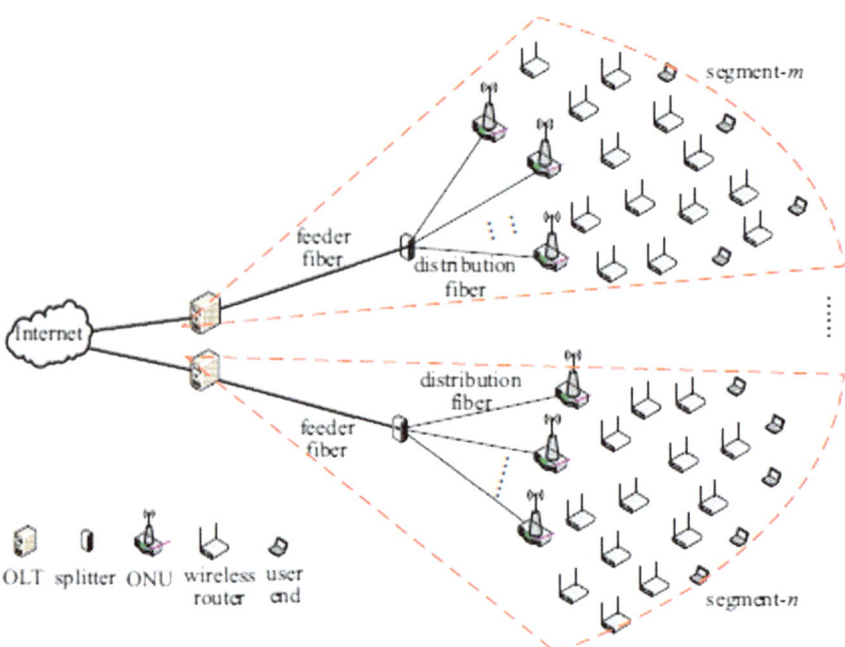

Fig. 1 FiWi architecture [4]

2 Related Work

The number of algorithms already been implemented for ONU placement in FiWi. Suman Sarkar et al. [7–9] propose ONU placement algorithms, viz. Random approach, deterministic approach, greedy algorithm, and simulated annealing (SA) algorithm. In random approach, the ONUs are placed randomly although in deterministic placing of ONU at the middle of each section. The greedy algorithm places the ONUs in deterministic way firstly, and after that the distance of each ONUs from user is found out. The least distance ONU is considered as primary ONU, and this process repeats for every user. For optimal solution, simulated annealing (SA) algorithm is evolved which provides global optimization. In [10], author proposed primal model for optimum ONU placement which considers Internet traffic and p2p traffic for network performance evaluation.

The tabu search algorithm [6] is proposed to reduce the total wireless hop count from routers to ONUs. With the help of this algorithm, the optimal ONU placement is done to maximize the network throughput. In [5], LBOP algorithm is proposed to reduce the number of ONUs and load balancing among ONUs during traffic hours. In [11], author proposed a hybrid algorithm for ONU placement which works in two stages. After the implementation of these two stages, the number of essential ONUs gets reduced and hence results in the cost-efficient network.

3 Proposed Work

In this work, we proposed an algorithm for the optimum placement of ONUs. The work is divided into two parts, i.e., the placement of ONU and its minimization with the help of ACO algorithm. We consider a network grid area of $L \times L$, where L is length of network, and grid size of $G_s \times G_s$, where G_s is grid dimension. Initially, the ONUs are placed in the middle of each grid, and the wireless routers are positioned randomly in the network grid under consideration.

Ant Colony Optimization algorithm is a heuristic algorithm which works on the real ant system [12, 13]. The process starts with the touring of ants from one position to different paths and deposits the pheromone and coming back to the original position. This involves the feedback about the pheromone concentration so that the best path can be opted. The pheromone concentration evaporates exponentially with respect to distance and time of contour which can be shown in Eq. (1).

$$P = e^{-\lambda dt} \tag{1}$$

Here, P denotes pheromone concentration, λ is constant parameter, d is distance of contour, and t is time to cover the contour.

Our proposed work aims toward the minimization of ONUs in the network. The work is done in phases listed below.

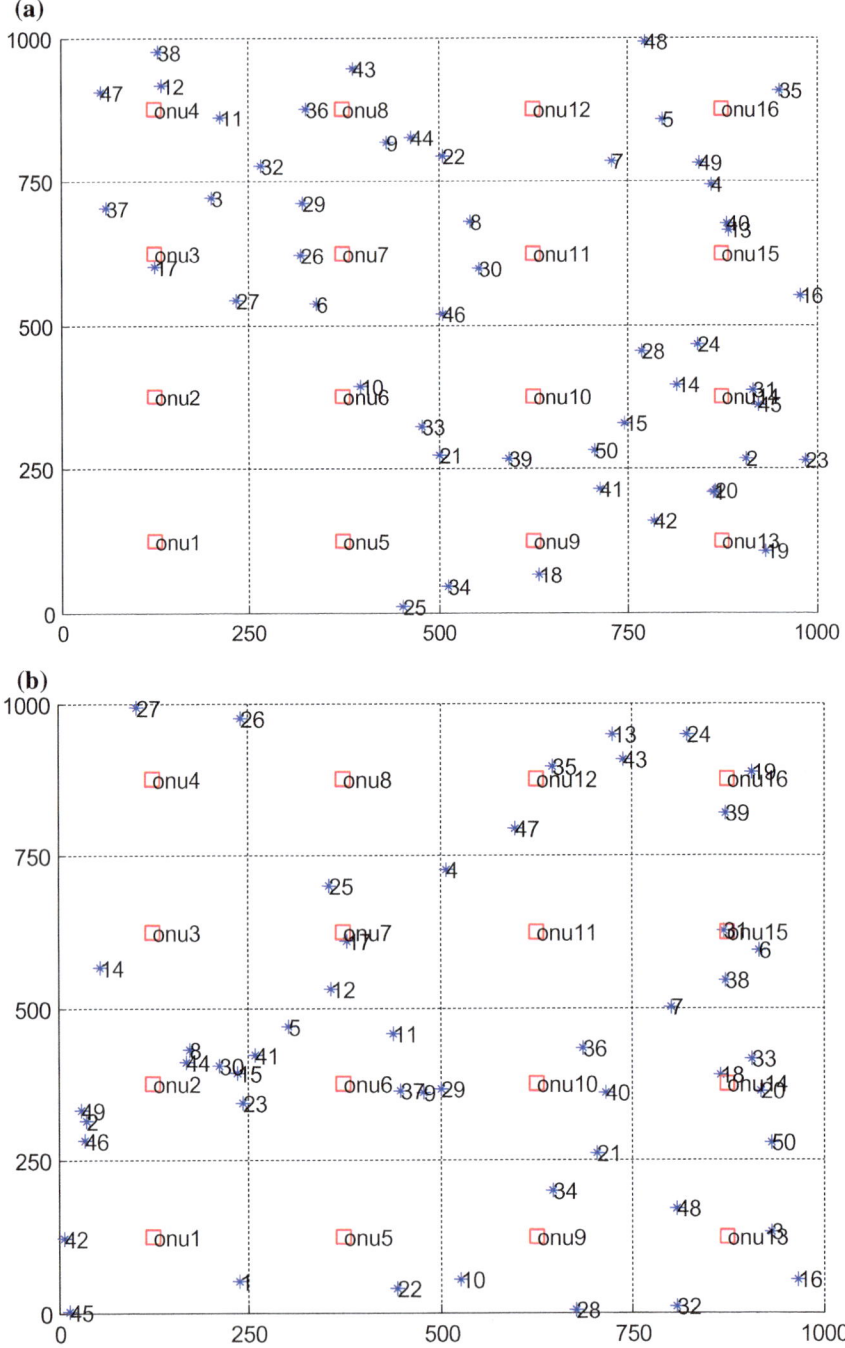

Fig. 2 **a** Initial placement of ONUs (□) with wireless routers (∗) for scenario 1. **b** Initial placement of ONUs (□) with wireless routers (∗) for scenario 2

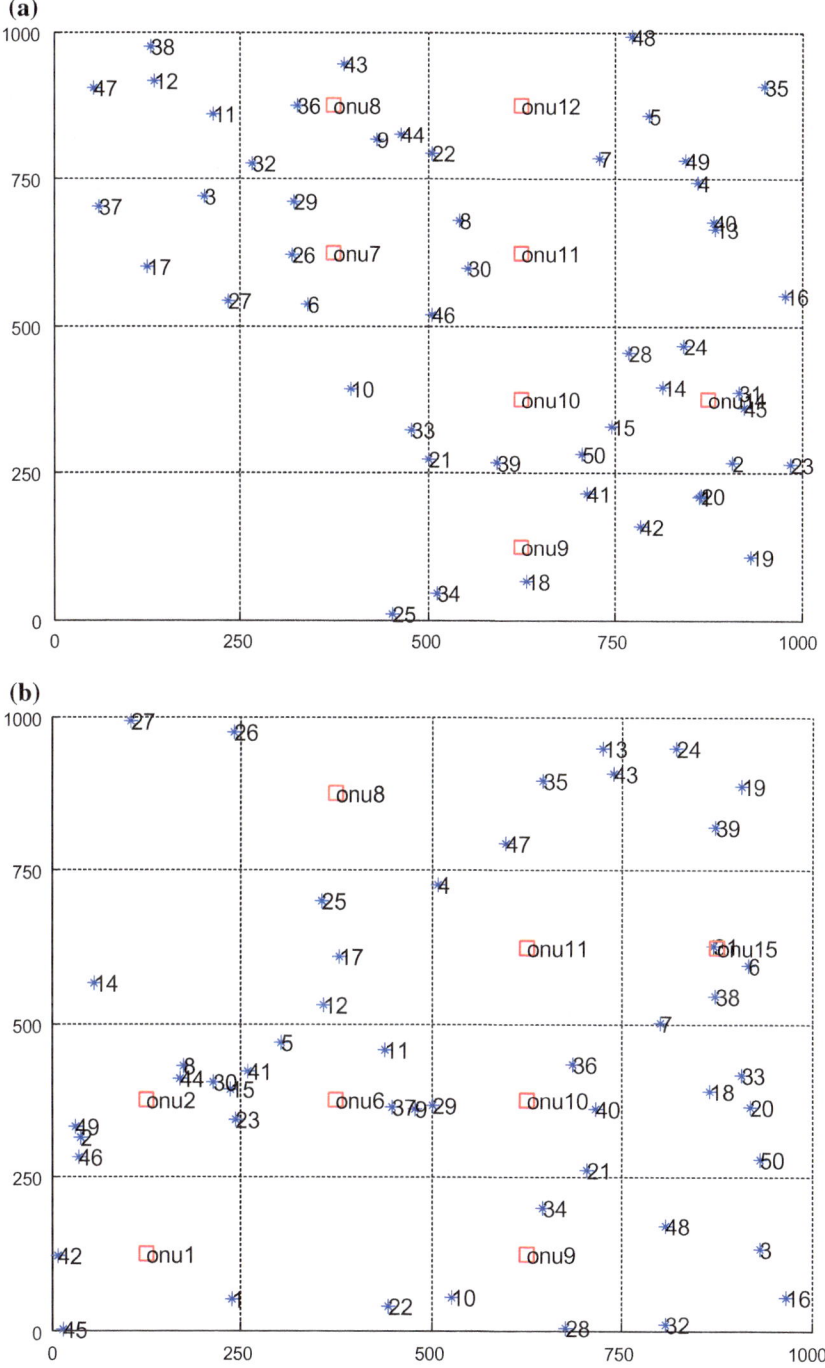

Fig. 3 **a** Final placement of ONUs (□) with wireless routers (✳) for scenario 1. **b** Final placement of ONUs (□) with wireless routers (✳) for scenario 2

Phase I: In the first phase, those ONUs are removed from the grid which do not have any router that means no communication of ONU with any router within the group in one hop way.

Phase II: In the second phase, for further minimization, we apply the ACO algorithm with the increase in the transmission range, i.e., communication of ONU with router in predefined hop ways. In this, we start ants from each router to each ONU, and the ant comes back with feedback knowledge of pheromone. The particular wireless path of router to ONU is selected which has higher concentration.

Phase III: Select and remove those ONUs from network, having minimal pheromone concentration or least wireless connectivity.

Figure 2a, b shows the initial placement of ONUs for two network scenarios. In these, 4 × 4 grid area is considered, and 50 wireless routers are randomly placed. The ONUs are positioned in the middle of grid, so total required ONUs are 16. After the implementation of the proposed algorithm, only seven ONUs are required for scenario 1 shown in Fig. 3a and eight ONUs for scenario 2 shown in Fig. 3b with new position of ONUs from where they connect with all wireless routers in predefined hop ways.

4   Results

4.1   Simulation Settings

The simulation has been carried out with the consideration of following settings:

- Network area of FiWi under consideration is 1000 × 1000 square units.
- Considering 4 × 4, 5 × 5, and 6 × 6 grid with wireless routers randomly from 20 to 100 in two different scenarios.

4.2   Simulation Results

The presentation of proposed algorithm can be analyzed by the simulation results obtained for two different scenarios which provide better understanding. Figure 4 represents the number of ONUs versus different grid sizes (i.e., 4 × 4, 5 × 5, and 6 × 6) for two network scenarios in which 50 wireless routers are randomly placed. It shows the reduction in the number of initially placed ONUs after execution of algorithm such that all wireless routers can communicate. Figure 5 depicts the number of ONUs for 4 × 4 grid size with wireless routers varying from 20

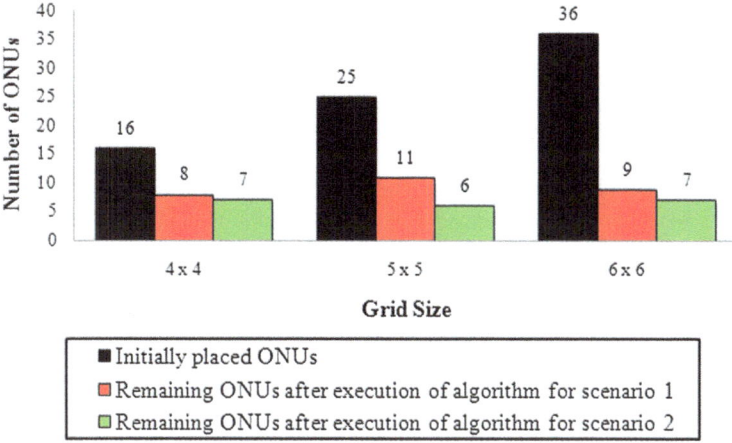

Fig. 4 Number of ONUs versus grid size for randomly placed 50 wireless routers in the network

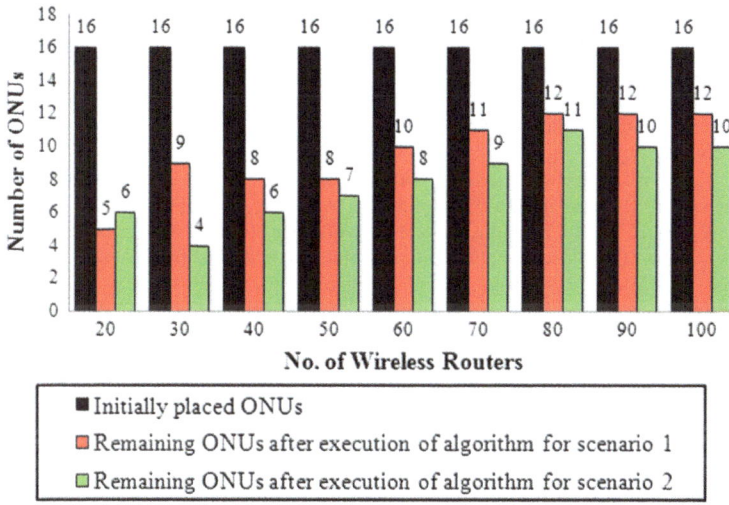

Fig. 5 Number of ONUs versus number of wireless routers for 4 × 4 grid in network

to 100 placed randomly in network. It is witnessed that the number of ONUs is reduced by the use of proposed algorithm in a way that wireless routers communication may not be hampered. Since the ONU has limitation to handle the load of wireless router, so as the number of routers rises the essential number of ONUs also rises.

5 Conclusion

The paper presents an algorithm which is based on ACO with the aim of optimal minimization of the number of ONUs essential in the network. Besides, the proposed work also ensures the connectivity among ONUs and wireless routers. The proposed algorithm has been tested on two different network scenarios. For both scenarios, we observe that proposed algorithm results in the less number of ONUs essential for the deployed network. The minimization of the number of ONUs results in the minimization of the cost of network which is highly desirable. Hence, the proposed work efficiently provides solution with respect to optimal ONU placement as well as reduction in deployment cost for the FiWi networks.

References

1. Ghazisaidi, N., Maier, M.: Fiber-wireless (FiWi) access networks: a survey. IEEE Commun. Mag. (2009)
2. Bhatt, U.R., Chhabra, A., Upadhyay, R.: Fiber-wireless (Fi-Wi) architectural technologies: a survey. In: International Conference on Electrical, Electronics and Optimization (2016) (In press)
3. Liu, Y., Guo, L., Gong, B., Ma, R., Gong, X., Zhang, L., Yang, J.: Green survivability in fiber-wireless (FiWi) broadband access network. Opt. Fiber Technol. **18**, 68–80 (2012). ELSEVIER Inc
4. Liu, Y., et al.: Protection based on backup radios and backup fibers for survivable fiber wireless (FiWi) access network. J. Net. Comp. App. **36**, 1057–1069 (2013)
5. Liu, Y., et al.: Load balanced optical network unit (ONU) placement in cost-efficient fiber-wireless (FiWi) access network. Optik—Int. J. Light Electron Opt. (2013). Elsevier
6. Zheng, Z., Wang, J., Wang, X.: ONU placement in fiber-wireless (FiWi) networks considering peer-to-peer communications. In: Proceedings: GLOBECOM IEEE, pp. 1–7 (2009)
7. Chowdhury, P., Mukherjee, B., Sarkar, S.: Hybrid wireless-optical broadband access network (WOBAN): prototype development and research challenges. IEEE Netw. 41–48 (2009)
8. Sarkar, S., Dixit, S., Mukherjee, B.: Hybrid wireless-optical broadband-access network (WOBAN): a review of relevant challenges. J. Lightwave Technol. **25**, 3329–3340 (2007)
9. Sarkar, S., Yen, H., Dixit, S., et al.: Hybrid wireless-optical broadband access network (WOBAN): network planning and setup. J. Sel. Areas Comm. **26**, 12–21 (2008). IEEE
10. Sarkar, S., Yen, H., Dixit, S., et al.: Hybrid wireless-optical broadband access network (WOBAN): network planning using Lagrangian relaxation. IEEE/ACM Trans. Netw. **17**(4), 1094–1105 (2009)
11. Bhatt, U.R., Chouhan, N., Upadhyay, R.: Hybrid algorithm: a cost efficient solution for ONU placement in fiber wireless (FiWi) access network. J. Opt. Fiber Tech. **22**, 76–83 (2015). ELSEVIER
12. Nanda, B.K., Das, G.: Ant colony optimization. A computational intelligence technique. Int. J. Comput. Commun. Technol. **2**, 105–110 (2011)
13. Selvi, V., Umarani, R.: Comparative analysis of ant colony and particle swarm optimization techniques. Int. J. Comput. Appl. **5** (2010)

Performance Analysis of Different PON Standards up to 10 Gbps for Bidirectional Transmission Employing Square Root Module

Anukul Sharma, Amit Kumar Garg and Vijay Janyani

Abstract This paper analyzes the performance of 10 GPON for 50 km bidirectional fiber link using travelling wave semiconductor optical amplifier (TSOA) with square root module (SRM). In the proposed network, the system utilizes 1550 and 1300 nm wavelength for downlink and uplink transmission, respectively. With the use of TSOA-SRM combination in the architecture, we have observed the improvement in bit error rate, quality factor and SNR for data rates up to 10 Gbps with 64 users in downlink and 40 users in uplink, improvement in SNR further helps us to reach extended GPON.

Keywords Gigabit passive optical network · Square root module
Travelling wave semiconductor optical amplifier

1 Introduction

Rapid increases in demand for bandwidth and multimedia services in recent years have drawn our attention toward passive networks. The focus is in achieving high data rate transmission, reliability of the network, and network coverage [1]. Passive optical network based on wavelength division multiplexing provides one such approach that increases the transmission capacity [2]. Researchers aim at PON for providing the bidirectional communication which includes downstream and upstream transmission over a single bidirectional fiber, reach extension so as to reduce network interfaces that result in reducing cost [3]. In downstream trans-

A. Sharma (✉) · A. K. Garg · V. Janyani
Department of Electronics and Communication Engineering,
Malaviya National Institute of Technology, Jaipur 302017, Rajasthan, India
e-mail: anukulsharma.07@gmail.com

A. K. Garg
e-mail: amitgroup1988@gmail.com

V. Janyani
e-mail: vjanyani.ece@mnit.ac.in

© Springer Nature Singapore Pte Ltd. 2018
K. Saeed et al. (eds.), *Progress in Advanced Computing and Intelligent Engineering*,
Advances in Intelligent Systems and Computing 563,
https://doi.org/10.1007/978-981-10-6872-0_50

529

mission data are transmitted from optical terminal in transmitter side to network units at the receiver, while in the upstream direction, vice versa.

In a fiber optical transmission, the strength of signal transmitted degrades as increasing the fiber's length. To compensate the loss introduced during transmission, the transmitted signal is being amplified using optical amplifiers. The semiconductor optical amplifier (SOA) is utilized in access networks to extend the splits ratio and reach of networks [4]. Optical amplification is provided by SOA to overcome the losses in reach extended networks. By providing amplification, the length of fiber increases without increasing interfacing nodes results in cost benefits. Amplification option using SOA includes post-amplification, pre-amplification, and in-line amplification [3]. Laser diode with the small bias current and low reflecting facets are treating as travelling wave amplifier which provides polarization independent pre-amplification [5]. For SOA to behave as travelling wave amplifier reflectivity of Fabry-Perot cavity must be small, so that reflectance is negligible. Travelling wave semiconductor optical amplifier is used because of its higher optical bandwidth and low polarization sensitivity [6].

The transmitted signal through fiber suffers from linear dispersion such as chromatic dispersion, polarization mode dispersion, which limits the fiber length for distribution and it increases with increase in fiber length [7], we use linear equalizer at receiver to compensate its effect. But optical communication is intrinsically nonlinear because of square law introduced by the use of photodiode at receiver, because of which linear equalizer which is designed to decrease the effect of dispersion, now gives the sub-optimal solution. A nonlinear square root operator after Photodetector is used to compensate the square effect introduced by the photodiode, and hence improves the performance of linear equalizer at the receiver [8]. This nonlinearity happens because square root module itself is nonlinear in nature. Thus, the output of the square root device is less nonlinear compared to its input.

Lee et al. [9] proposed a GPON system reach extended to 60 km using Raman amplification. The system uses symmetric bidirectional traffic at the data rate of 2.5 Gbps with 32 subscriber units. Downstream and upstream transmissions are taking place at 1490 nm and 1350 nm, respectively. The system achieved error-free operation with received power of −28 dBm.

Zhu and Nesset [10] describe a PON having split ratio of 1:64 and reach extension up to 60 km is possible using Raman amplification. The system operates at 2.5 Gbps bidirectional transmission using the downstream and upstream wavelengths of 1490 nm and 1310 nm, respectively.

GPON architecture for 60 km length with the split ratio of 1:64 using Raman amplifiers and square root module at 2.5 Gbps bidirectional transmission has been described. The system utilizes wavelength of 1490 nm and 1310 nm for downstream and upstream transmissions, respectively. Architecture shows an improved performance in Q factor, received signal power, and SNR using square root module [11]. For higher split ratio, power budget issue is resolved by using various modulation formats [12]. Another architecture based on hybrid dense wavelength division multiplexing reach extended up to 40 km with 128 split ratios has been demonstrated [13]. Analysis of cost and energy consumption for different hybrid multiplexing techniques for different fiber reach and data rates in given [14].

A bidirectional model for extended reach PON link operates up to 10 Gbps over the fiber length of 50 km using travelling wave semiconductor optical amplifier has been described. The system uses 1550 nm wavelength for downlink transmission with 64 ONUs and 1300 nm wavelength for uplink transmission with 40 users. The system results in error-free transmission at 10 Gbps for 50 km fiber length [15].

This paper proposed an architecture which utilizes TSOA and square root module for the fiber length of 50 km at 10 Gbps for 64 users in downlink and 40 users in uplink. The operating wavelength for upstream and downstream communication is 1300 nm and 1550 nm, respectively. The architecture with TSOA-SRM combination for different data rates, fiber lengths has been analyzed and compared to the structure which utilizes TSOA only. This paper comprises of Sect. 2 which discusses the simulation setup of proposed architecture, Sect. 3 which includes results and discussion, while Sect. 4 concludes the paper.

2 Simulation Setup

Block diagram for proposed architecture is shown in Fig. 1. The system transmits for both downstream and upstream direction at 1550 nm and 1300 nm, respectively, at the fiber length of 50 km for 64 ONUs at the bit rate of 10 Gbps.

The simulation setup has been inherited from [11, 15]. However in [15], they have considered a conventional ONU (without Square root module), and in [11], they have considered Raman amplification instead of TSOA and that too only up to 2.5 Gbps. In our design, we propose the use of TSOA along with ONU's

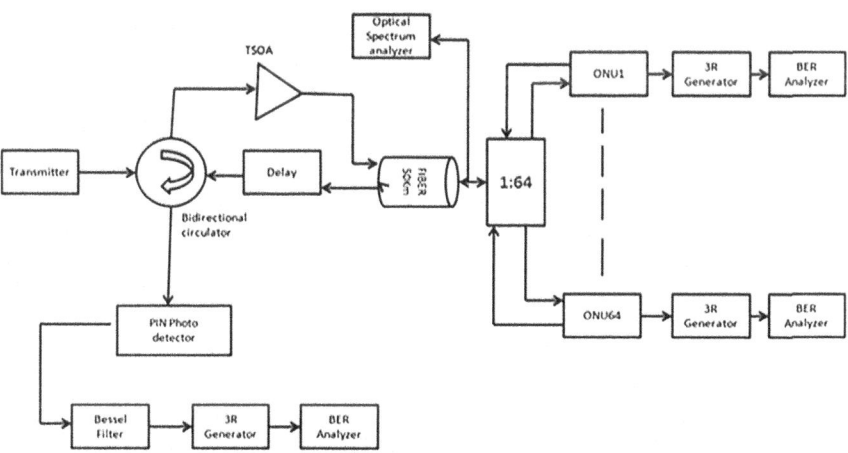

Fig. 1 Block diagram of simulation setup

Fig. 2 Optical network unit
(ONU) structure

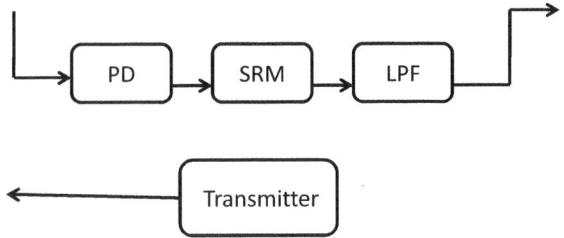

incorporating Square root module, capable of working up to 10 Gbps. The transmitted signal is generated by data source, NRZ rectangular driver, continuous wave
(CW) laser source, and optical amplitude modulator. The bit rate of 10 Gbps with
pseudorandom sequence (PRBS) is generated which is given to NRZ pulse generator block to generate coded NRZ signal. The continuous wave (CW) laser along
with generated NRZ signal is given to the optical amplitude modulator. The bidirectional circulator provides wavelength-dependent isolation of downlink and
uplinks optical modulated signal. TSOA placed after circulator in downlink having
0.15 A injection current to provide optical amplification. The signal is passed
through the bidirectional optical fiber with 0.2 dB/Km attenuation losses. For
broadcasting, optical splitter with the split ratio of 1:64 is used. Finally, the signal is
received by each ONU at receiver.

The ONU is shown in Fig. 2 which consists of photodetector PIN to transform
the signal into electrical form from optical form. The electrical signal then passed
through square root module, which is simulated in MATLAB, and then after
passing it through low pass filter signal is received by 3R generator. BER analyzer
calculates the system performance such as BER, Q factor, eye aperture. Delay
signal is used to provide null signal at the output port in bidirectional transmission.
An optical spectrum analyzer is used at various locations to observe the characteristic of the signal.

3 Results and Discussions

The performance of 10 Gbps bidirectional transmission GPON over a fiber length
of 50 km using TSOA and TSOA with square root module has been analyzed.
Downstream and upstream transmission takes place at the wavelength of 1550 nm
and 1300 nm, respectively. Obtained simulation-based results are discussed below.

3.1 Bit Error Rate

Figure 3 shows the received BER at 10 Gbps for a different number of users in
downlink using TSOA and TSOA with SRM. We obtained a BER of 1.4×10^{-13}

Fig. 3 BER versus different number of users in downlink

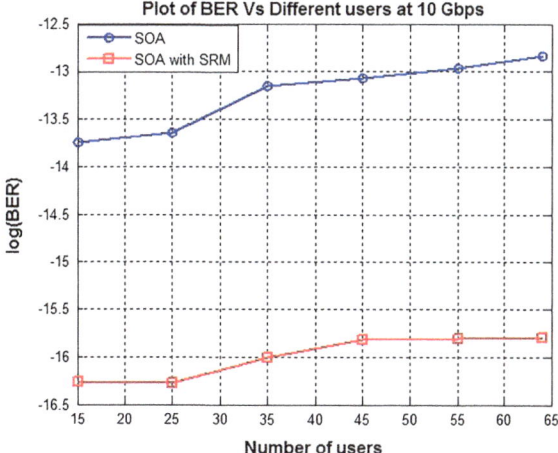

for the system using only TSOA and got an improved BER of 1.59×10^{-16} for the system using TSOA with SRM for 64 users.

3.2 Quality Factor

We observe from the result as shown in Fig. 4 that there is an improvement in Q factor for all data rates up to 10 Gbps for the system utilizing TSOA with SRM compared to TSOA only. We get Q factor of (20.88, 9.19, and 7.27) and (25.71, 11.88, 8.165) for the system using TSOA only and TSOA-SRM combination for 2.5 Gbps, 5 Gbps, and 10 Gbps data rates, respectively. Figure also shows that quality factor decreases as we increase data rates.

Fig. 4 Q factor versus bit rates for 64 users in downlink

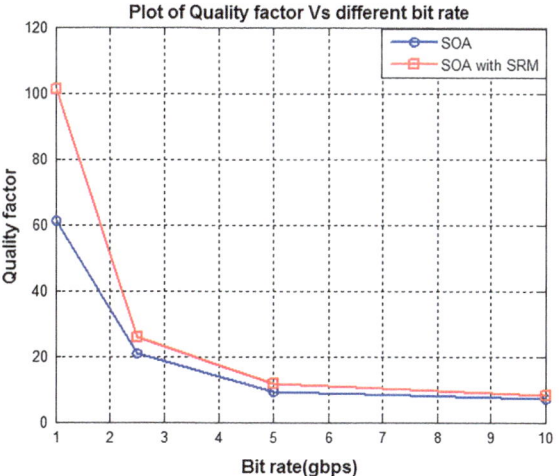

Fig. 5 Q factor versus fiber
length in downlink

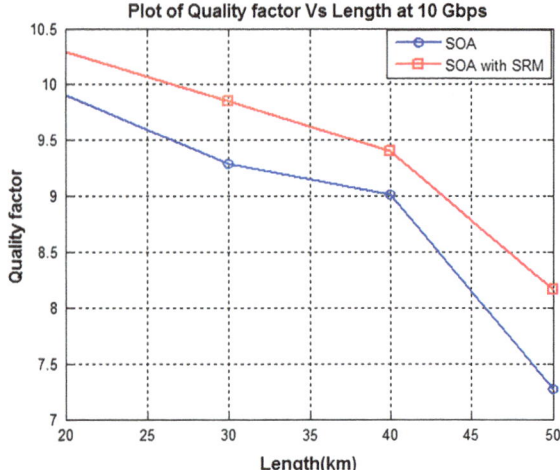

It has been seen from Fig. 5 that there is an improvement in Q factor over the
length greater than 20 km up to 50 km using TSOA with SRM as compared to only
TSOA. The figure also shows that as distance increases, quality factor decreases.
Figure 6 corresponds to the eye diagram for 40 users in uplink; Q factor obtained is
15.07. Figure 7 corresponds to eye diagram for 64 users in downlink, obtained Q
factor is 8.16.

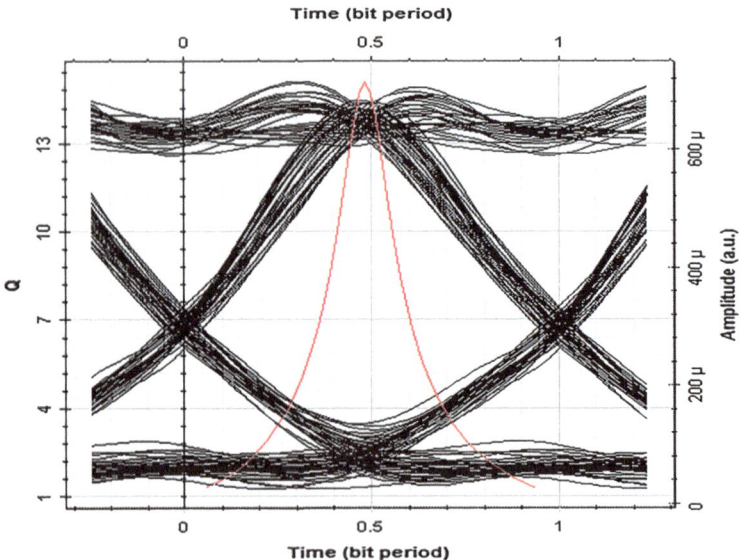

Fig. 6 Eye diagram for 40 users in uplink

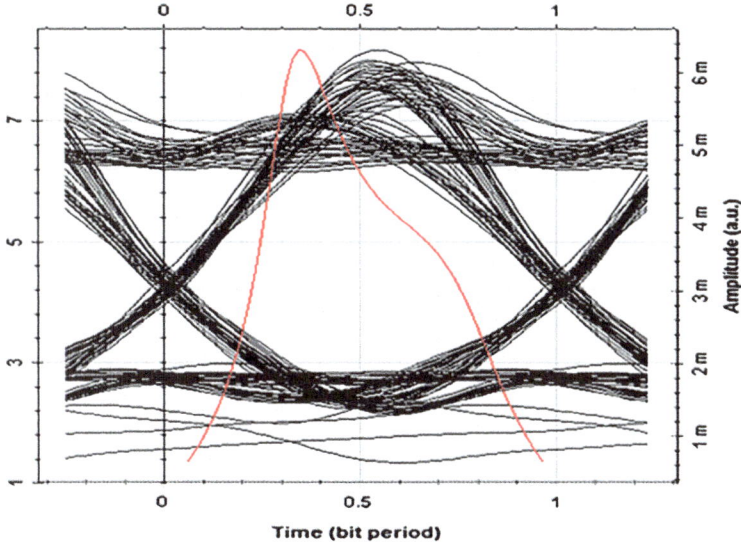

Fig. 7 Eye diagram for 64 users in downlink

3.3 Received Signal Power

It is seen from Table 1 that an improvement of 43 dB in received signal power is obtained for 50 km length at 10 Gbps with 64 users in downlink using TSOA-SRM. For various lengths, the increase in received signal power is shown in Table 1.

3.4 Signal-to-Noise Ratio

We achieved an SNR of 59.72 dB using TSOA-SRM combination as compared to 16.21 dB obtained by using only TSOA as shown in Fig. 8. The figure also shows that over a span of 20–50 km length of the fiber, a decrease in SNR for TSOA is 11.99 dB and for TSOA-SRM combination decrease in SNR is 6.22 dB. So, the SNR is improved by around 5 dB when fiber length is increased by 20–50 km when TSOA-SRM combination is used. Since the power at the input of the SRM is

Table 1 Received signal power for various fiber lengths using TSOA-SRM combination and TSOA only

Length (km)	20	40	50
Received signal power using TSOA (dBm)	−71.79	−79.78	−83.78
Received signal power using TSOA-SRM (dBm)	−34.05	−38.19	−40.27

Fig. 8 Signal-to-noise ratio
versus length

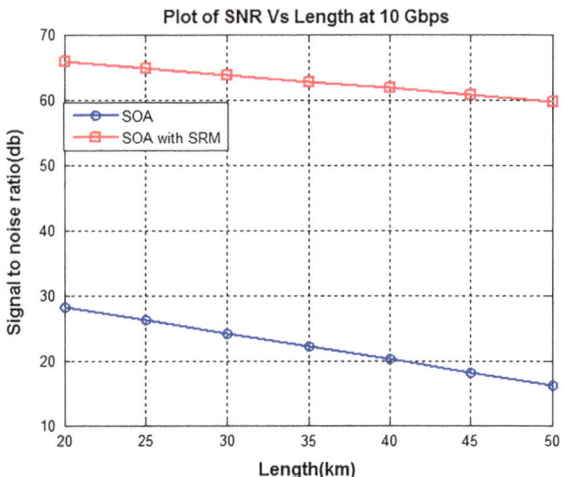

of the order of picowatts, after passing through a module, the power is of the order of microwatts. Hence, the received power increases which ultimately results in improvement of SNR. Improved power helps in extending the length of GPON. Moreover, the BER is also improved as received power increased.

4 Conclusion

We demonstrated a 10 GPON bidirectional system having 1:64 split ratio using travelling wave semiconductor amplifier with square root module at 1550 nm and 1300 nm for downlink and uplink, respectively, over a fiber length of 50 km. We obtained a Q factor and BER of 8.16 and 1.59×10^{-16}, respectively, for 64 subscribers at downlink, and for 40 subscribers in uplink, Q factor obtained is 15.07. We found an improvement in Q factor, BER, and SNR with the use of TSOA with SRM for all bit rates up to 10 Gbps over using the only TSOA, which further helps in improving the length of GPON and useful in high data rate passive optical network applications.

References

1. Schrenk, B., Chatzi, S., Bonada, F., Lazaro, J.A., Tomkos, J., Prat, J.: Dual waveband remote node for extended reach full duplex 10 Gb/s hybrid PONs. J. Lightwave Tech. **28**(10), 1503–1509 (2010). IEEE
2. Moon, J.H., Choi, K.M., Lee, C.H.: Overlay of broadcasting signal in a WDM-PON. In: Optical Fiber Communication Conference and National Fiber Optics Engineers Conference, pp. 3. IEEE, California (2006)

3. Michie, C., Kelly, T., Andonovic, I., McGeough, J.: Reach extension of passive optical networks using semiconductor optical amplifiers. In: 10th International Conference on Transparent Optical Networks, vol. 1, pp. 194–197. IEEE, Athens (2008)
4. Spiekman, L.H.: Semiconductor optical amplifiers in access networks. In: 14th Opto-Electronics and Communication Conference, pp. 1–2. IEEE, Hong Kong (2009)
5. Kashima, N.: Analysis of laser diode as transmitter and preamplifiers in time compression multiplexing system. J. Lightwave Technol. 10(3), 323–329 (1992). IEEE
6. Rani, A., Dewra, S.: Semiconductor optical amplifier in optical communication-review. IJERT Int. J. Eng. Res. Technol. 2(10), (2013)
7. Antonelli, C., Aquilla, L., Brodsky, M., Shtaif, M.: Distance limitation on the entanglement distribution over optical fiber due to chromatic and polarization mode dispersion. In: CLEO: 2011 Laser-Science to Photonics Applications, pp. 1–2. IEEE, Baltimore (2011)
8. Prat, J., María, M., Santos, C., Omella, M.: Square root module to combat dispersion-induced nonlinear distortion in radio-over-fiber systems. IEEE Photonics Technol. Lett. 18(18), 1928–1930 (2006). IEEE
9. Lee, K.L., Riding, J.L., Tran, A.V., Tucker, R.S.: Extended-reach gigabit passive optical networks for rural area using distributed Raman amplifiers. In: Optical Fiber Communication Conference, pp. 1–3. IEEE, San Diego (2009)
10. Zhu, B., Nesset, D.: GPON reach extension to 60 km with entirely passive fibre plant using Raman amplification. In: 35th European Conference On Optical Communication, pp. 1–2. IEEE, Vienna (2009)
11. Kumar, N.: Improved performance analysis of Gigabit passive optical networks. Optik—Int. J. Light Electron. Opt. 125(7), 1837–1840 (2014). Elsevier
12. Garg, A.K., Janyani, V.: Power budget improvement in energy efficient long reach hybrid passive optical network. In: 2015 International Conference on Microwave and Photonics (ICMAP), pp. 1–2. IEEE, Dhanbad (2015)
13. Kim, K.O., Doo, K.H., Lee, S.S.: Design of hybrid PON system for GPON reach extension on the basis of colorless DWDM-PON and 3R generator. In: Global Telecommunications Conference (GLOBECOM 2010), pp. 1–4. IEEE, Miami (2010)
14. Garg, A.K., Janyani, V.: Identification of cost and energy efficient multiplexing techniques for LR-PON for different networks scenario. In: IEEE 2015 Workshop in Recent Advances in Photonics (WRAP). IEEE, IISC Bangalore (2015) (In press)
15. Kocher, D., Kaler, R.S., Randhawa, R.: 50 km bidirectional FTTH transmission comparing different PON standards. In: Optik—Int. J. Light Electron. Opt. 124(21), 075–5078 (2013). Elsevier

A Review of Recent Energy-Efficient Mechanisms for Fiber-Wireless (FiWi) Access Network

Vijendra Mishra, Raksha Upadhyay and Uma Rathore Bhatt

Abstract Fiber-wireless (FiWi) access network is a category of networks which utilize optical network's bandwidth and wireless network's flexibility. This last mile technology will serve all needs of access networks for many years to come. However, its energy consumption has also been increased rapidly due to the development of high bandwidth application like Internet of Things (IoT). Therefore, it has become critically important to discuss the energy-related issue for FiWi. Based on the literature, this chapter provides an exhaustive survey of recent progress on energy efficiency mechanisms such as ONU sleeping mechanisms, power saving modes (PSM), and some cooperative-based energy saving schemes. Furthermore, it also provides future directions for better energy efficiency in FiWi network.

Keywords FiWi · Optical network · Wireless network · Access network
Energy efficiency · ONU (optical network unit) and PSM (power saving mode)

1 Introduction

Hybrid wireless optical broadband access network (WOBAN) which is also mentioned as fiber-wireless access networks was proposed in early 1990s to make an efficient "last mile" access technique [1, 2]. FiWi integrates wireless and optical access technologies for their respective merits, such as stable transmission, high capacity, and large bandwidth from optical part, and easy deployment and

V. Mishra (✉) · R. Upadhyay · U. R. Bhatt
Department of Electronics and Telecommunication, Institute of Engineering
and Technology, Devi Ahilya University, Indore, MP, India
e-mail: vijendramishra88@gmail.com

R. Upadhyay
e-mail: raksha_upadhyay@yahoo.co.in

U. R. Bhatt
e-mail: umarathore@rediffmail.com

© Springer Nature Singapore Pte Ltd. 2018   539
K. Saeed et al. (eds.), *Progress in Advanced Computing and Intelligent Engineering*,
Advances in Intelligent Systems and Computing 563,
https://doi.org/10.1007/978-981-10-6872-0_51

flexibility from wireless part. However it may not be economical in most of the cases to deploy fiber to every end user from central office (CO); and on the other hand, deploying wireless links from center office to end user may lead to the problem of limitation of spectrum [3]. Hence, we select an optimum solution between fiber and wireless by using fiber to the maximum possible distance till end user and then adopting wireless technology for further extension [4]. This integration of two different access technologies has attracted many researchers and also shown a rapid development in associated domain over the recent years, ranging from multidimensional communication to telecommunication to sensor network [5].

Typical FiWi network has been divided into two parts, i.e., at the front end of network wireless access is placed and deployed with optical access at backhaul [2, 5], as shown in Fig. 1. In this type of network, usually a wireless end user (WEU) transfers data to its neighbor wireless mesh routers (WMR). After that, packets are been transferred to the mesh, through hops to its respective gateways and then to optical network unit-base station (ONU BS) [4], and then finally injected to the optical part of FiWi network and then to the optical line terminal (OLT). The flow of traffic can be either in upstream direction or in downstream direction [6].

As FiWi access networks are supposed to be the future of access networks for wide variety of applications, it becomes important to target various issues such as ONU placement, survivability, architectures, and energy consumption. In modern era, energy efficiency has already become the selecting criteria for different access technologies, and hence, a deep study of energy-efficient technique in FiWi is also

Fig. 1 Structure of FiWi access network [2]

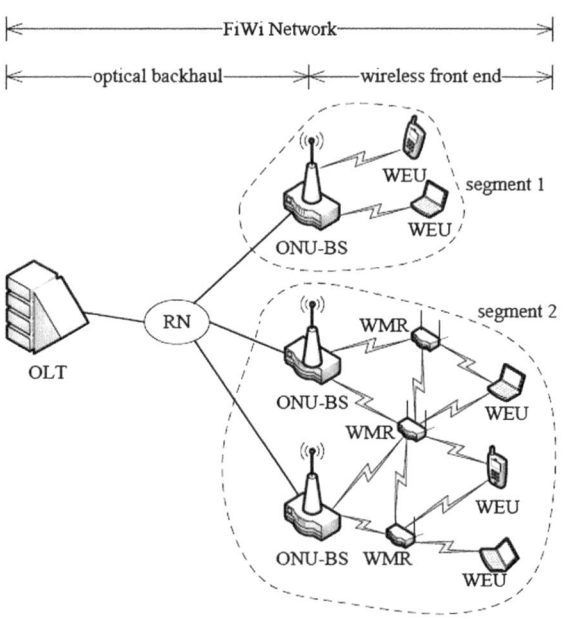

required. This chapter presents a survey on different energy-efficient mechanisms of FiWi.

2 Motivation

Till today, there are many survey papers that are published on different issues of FiWi networks, like different architectures, routing techniques, and optimization placement of devices. In 2007, Dixit et al. [4] presented an overview of WOBAN which mentioned different architectures, locating and relocating different devices and routing techniques. Ghazisaidi et al. [2] in 2009 provided a state of art on access technologies and their standards through his survey article, targeting on FiWi network integration architectures. To underline the eventual growth on FiWi network, Maier [1, 2, 4, 5] presented few survey papers in 2014 and 2013, where focus is on network planning and QoS. This chapter enlightens the energy-related aspects of fiber-wireless access network, which have not been found in other surveys. Literature of this chapter shows that energy efficiency can be analyzed through adapting to different ONU sleeping mechanisms and shifting to different power saving mode standards.

Content of this chapter is as follows: In Sect. 3, literature associated with effect of ONU sleeping on energy efficiency has been cumulated. Power saving mode at wireless end and its cooperative approach are discussed in Sect. 4. Then, in Sect. 5, we briefly explore the different issues affecting energy efficiency in FiWi. Finally, in Sect. 6, future scope and conclusion are mentioned.

3 ONU Sleeping Mechanism

Power consumption in any access network largely depends on efficient use of high-power devices in network. In FiWi networks, efficient use of ONU controls the energy efficiency of complete network as ONU is responsible for consuming most of the energy of FiWi network. Hence, efficient use of high-power devices is gaining interest to ensure power consumption in access networks [6]. ONU sleeping mechanism is one of the most popular techniques of reducing the active or working duration of high-power devices which ultimately leads to power saving in FiWi networks. Figure 2 shows the timing diagram of ONU sleep process, which indicates the dependency of sleep duration on data available and sleep request [7]. Figure 2 shows the transition between active and sleep state. At starting of each interval, OLT sends a sleep request to ONU, which contains information about the presence of data. If ONU finds sleep request with data information, it provides an ACK and then data are transferred to ONU.

In recent FiWi network, various sleeping mechanisms for passive optical network (PON) are already explored. Rohde and Schwarz [8] demonstrated an energy

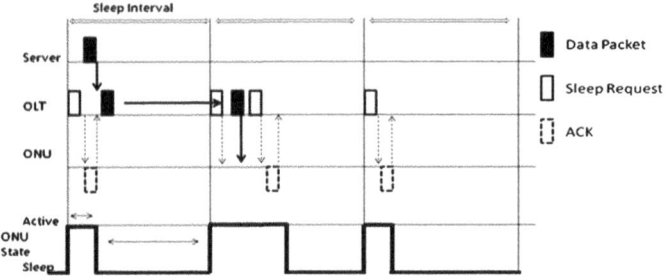

Fig. 2 ONU sleep process

management mechanism (EMM) whose main motive is to enhance the sleeping duration of high-power devices and compute scheduling at OLT using different algorithm. Tamura et al. [9] showed a power consumption technique which uses high-frequency transition method between active and sleep modes which can be obtained by allocating extra period. Kubo et al. [10] proposed a variable sleep and active period mechanism, which assigns ONU sleep period with reference to its interval, queue length, and the classes of service. Over the years, there had been a lot of research done on PON for improvement in energy efficiency, since PON is an integral part of FiWi; therefore, all those energy-efficient mechanisms can also be applied directly to FiWi in future.

4 Power Saving Mode

As mentioned in [11], wireless access network interfaces can either be in one of the two states at one instant of time, i.e., active or sleep. In the active state, the access point (AP) is made ON and it can perform communication, or may remain idle, while in sleep state, access point is turned OFF and it cannot detect or sense the traffic. Wireless interface in active state consumes more power than in sleep state. Hence, one always try to make network interface to remain more in sleep state, which can be achieved by using power saving mode (PSM) that is defined in [12]. They are classified into two different operational categories, namely PSM for wireless end and PSM cooperative based.

4.1 PSM/APSM for Wireless End

PSM is one of the primary energy saving techniques at wireless end. In this, station transmitter is turned off periodically when it does not have any data to be transmitted or received. Traffic assigned for the stations in sleeping duration has to be buffered which induced a significant delay in transmission. Hence, techniques associated with PSM/APSM have to optimize between reducing power consumption and delay.

Yuan et al. [13] discussed IEEE 802.11 PSM for WLAN, which shows the effect of various factors like traffic indicating message period and beacon interval onto performance of power saving mode. Rosenberg et al. [14] in 2006 presented an access point centric PSM, in which problem of power consumption had been addressed and for the same suggested two heuristic solutions, i.e., Longest Processing Time in selecting stations (LPTSS) and Shortest Processing Time in scheduling packets (SPTSP). Liu et al. [15] mentioned that optimization of beacon period reduces energy consumption and delay through proper scheduling.

4.2 Cooperation-Based Energy Saving Scheme

Power saving mode at wireless end and ONU sleeping mechanism at optical end had been studied separately but separate designing may not be able to fully utilize advantages of both energy saving schemes. When these two approaches are studied together, it is referred as cooperative approach. However, cooperative approach faces big issues of increase in delay and latency. It is due to the fact that total latency is cumulative latency of both sleeping mechanisms. Therefore, most of the literature associated with the cooperative approach, propose methods which reduce energy consumption while decreasing delay as well.

Nishiyama et al. [16] showed power saving mechanism used by STA in WLAN that can be controlled by ONU. It also analyzes the ONU energy consumption and latency. In 2013, Nishiyama et al. [16] continued his work by proposing a cooperative efficient scheme that includes ONU synchronization and dynamic sleeping at every fixed interval, which computes optimal sleeping of backend devices. Yang et al. [15] presented an energy-efficient technique that collaborates ONU sleeping with saving techniques at wireless end, through which change in topology can be produced by adding ONU into sleep/active state. Miyanabe et al. [17] suggested a power saving mechanism that is synchronized and provides energy consumption and delay, which functions on the basis of synchronization between the initialization of ONU sleep and saving modes. As PSM specifically targets the optimization between delay and sleeping interval, different unexplored optimization approaches can be used in PSM to improve energy efficiency in FiWi networks.

5 Other Issues for Energy Efficiency in FiWi

Other than issues mentioned in previous sections, there are many other directions where energy efficiency can be analyzed. Some of them are as follows.

In [18], Peng et al. had presented energy comparison between two different architectures of FiWi, which shows different methodology of integration between radio and fiber [18]. Radio over fiber (RoF) consumes lesser energy as compared to radio and fiber (R&F) hence it is suitable for mobile broadband communication. With reference to PON, Chowdhury et al. in [19] have discussed that WDM makes use of array waveguide grating that ultimately provides much larger power margin. In comparison with point-to-point communication system, energy-efficient problem is better addressed in PONs as OLT is being shared among multiple ONU which is well described by [20]. Fernando et al. [21] have clearly mentioned in their article that parameters of sleeping duration and end of sleep mode are outside the mentioned standards. Nishihara et al. [22] have mentioned the details of a new standard for power saving EPON IEEE 1904.1.

6   Conclusion

This chapter presents a comprehensive survey of recent energy-efficient mechanisms on FiWi access networks. Work included in this chapter specially emphasizes on the different energy-efficient mechanisms. Studies of literature assure openness of various directions in FiWi for energy efficiency. One of the future aspects is to combine various schemes to optimize different parameters. Few new trends in research emphasize on integrating FiWi networks with emerging technologies like Smart Grid, DAS (distributed antenna system), and NG-PONs (next-generation PON). After reviewing several energy efficiency-related schemes, we can say that energy efficiency is a major concern in FiWi.

Acknowledgements This research work is supported by Ministry of Communications and IT, DeitY, Government of India, under Visvesvaraya PhD Scheme.

References

1. Maier, M.: Fiber-wireless (FiWi) broadband access networks in an age of convergence: past, present, and future. Adv. Opt. **2014**, 1–23 (2014)
2. Ghazisaidi, N., Maier, M., Assi, C.: Fiber-wireless (FiWi) access networks: a survey. IEEE Commun. Mag. **47**, 160–167 (2009)
3. Sarkar, S., Yen, H.H., Hsin, S., Dixit, S., Mukherjee, B.: A novel delay-aware routing algorithm for WOBAN. IEEE Netw. **22**, 20–28 (2008). University of California
4. Dixit, S., Sarkar, S., Mukherjee, B.: Hybrid wireless-optical broadband-access network (WOBAN): a review of relevant challenges. J. Lightwave Technol. **25**, 3329–3340 (2007)
5. Ghazisaidi, N., Maier, M.: Fiber-wireless (FiWi) access networks: challenges and opportunities. IEEE Netw. **25**, 36–42 (2011)

6. Martin, M., Ghazisaidi, N., Martin, R.: The Audacity of Fiber-Wireless (FiWi) Networks, vol. 6, pp. 16–35. AccessNets, Springer Lecture Notes LNICST (2009)
7. Coimbra, J., Schütz, G., Correia, N.: Energy efficient routing algorithm for fiber-wireless access networks: a game approach. Comput. Netw. **60**, 201–216 (2014)
8. Rohde, Schwarz: 802.11ac Technology Introduction. White Paper (2012)
9. Tamura, H., Yahiro, Y., Fukuda, Y., Kawahara, K., Oie, Y.: Performance analysis of energy saving scheme with extra active period for LAN. In: Proceedings of the IEEE GLB, pp. 198–203. (2007)
10. Kubo, R., Kani, J., Ujikawa, H., Sakamoto, T., Fujimoto, Y., Yoshimoto, N., Hadama, H.: Study and demonstration of sleep and adaptive link rate control mechanisms for energy efficient 10G-EPON. J. Optical Commun. Netw. **2**, 716–729 (2010)
11. IEEE Computer Society LAN MAN Standards Committee: IEEE Std 802.11: Wireless LAN Medium Access Control (MAC) and Physical Layer (PHY) Specifications (1999)
12. IEEE 802.11: Wireless LAN Medium Access Control (MAC) and Physical Layer (PHY) Specifications, IEEE Standards Association Std. (2012)
13. Yuan, R., He, Y., Ma X., Li, J.: The IEEE 802.11 power saving mechanism: an experimental study. In: IEEE Conference WCNC, pp. 1362–1367 (2008)
14. Rosenberg, C., Lee, J., Chong, E.K.P.: Energy efficient schedulers in wireless networks: design and optimization. Mobile Netw. App **11**, 377–389 (2006)
15. Liu, Y., Guo, L., Yang, J.: Energy-efficient topology reconfiguration in green fiber-wireless (FiWi) access network. In: WOCC Conference, pp. 555–559 (2013)
16. Nishiyama, H., Togashi, K., Kato, N., Ujikawa, H., Suzuki, K.I., Yoshimoto, N.: On the effect of cooperation between power saving mechanisms in WLANs and PONs. In: IEEE International Conference on Communications, pp. 6225–6229 (2013)
17. Miyanabe, K., Nishiyama, H., Kato, N., Ujikawa, H., Yoshimoto, N.: Synchronized power saving mechanisms for battery powered mobile terminals in smart FiWi networks. In: VTC Fall (2014)
18. Peng, P.C., Lin, C.T., Chen, J., Peng, C.F. Peng, W.R., Chiou, B.S., Chi, S.: Hybrid optical access network integrating fiber-to-the-home and radio-over-fiber systems. IEEE Photonics Technol. Lett. **19**, 610–612 (2007)
19. Chowdhury, A., Chang, G.K., Jia, Z., Chien, H.C., Huang, M.F., Yu, J. Ellinas, G.: Key technologies of WDM-PON for future converged optical broadband access networks. IEEE/OSA J. Opt. Comm. Netw. **1**, C35–C50 (2009)
20. GPON Power Conservation, ITU-T Supplement G. Supplement 45 (2009)
21. Fernando, D., Milosavljevic, M., Kourtessis, P., Senior, J.: Cooperative cyclic sleep and doze mode selection for NG-PONs. In: ICTON, pp. 1–4 (2014)
22. Nishihara, S., Hajduczenia, M., Mukai, H., Elbakoury, H., Hirth, R., Kimura, M., Kato, M.: Power-saving methods with guaranteed service interoperability in ethernet passive optical networks. IEEE Commun. Mag. **50**, 110–117 (2012)

Impediments in the New Establishment of Grid Computing Architectures in Low and Lower-Middle Income Countries

Prashant Wankhede

Abstract A sustainable grid computing infrastructure is characterized by high-speed computer network, a suitable number of computing devices, and the software modules that are capable of establishing a reliable communication framework between the networked components. The number of issues and their extent of impact in the low and lower-middle income countries vary significantly depending upon the type of the grid application and economic modalities of the nation. This paper rectifies such major and minor pitfalls arriving in the fabrication process of a typical grid architecture and their extent of hurdles in such countries. The pitfalls after a deep analysis also result the paper in suggesting worthy resolution methods possible to overcome these issues. An efficient grid thus consists of a proper cache of suitable mechanisms to expound the problems listed in this paper to achieve a goal of a novel grid infrastructure.

Keywords Grid computing · Issues in grid computing · Grid computing pitfalls in low and lower-middle income countries

1 Introduction

Grid computing [1, 2] is a disseminated design [3] of huge quantities of PCs associated together in determining a solution for a perplexing issue which are too convoluted to be in any way to be endeavored by any individual. In the grid computing model, servers or personal computers run unbiased duties and are loosely coupled by means of the Internet or low-pace networks. Computer systems can also connect via immediate networks or via scheduling structures.

P. Wankhede (✉)
Department of Computer Science and Engineering, National Institute of Technology, Hamirpur 177005, Himachal Pradesh, India
e-mail: prashantsday@gmail.com

© Springer Nature Singapore Pte Ltd. 2018
K. Saeed et al. (eds.), *Progress in Advanced Computing and Intelligent Engineering*,
Advances in Intelligent Systems and Computing 563,
https://doi.org/10.1007/978-981-10-6872-0_52

Grid computing occurs on a network of computing assets managed with superior styles of middleware [4] technology, which are mediated through grid services [5]. Facilities provided by grid services includes network access control, grid and personal security, access to the shared or independent data, including digital libraries and complex Web databases, and access to long-term (for instance, digital archives) and large-scale interactive (for instance, server farms) storage services. Grid services are being replaced with the aid of open supply Internet services, which carry out the similar obligations and were evolved over a sufficiently longer period of time.

The motive for the increase of grid computing is the capability for an organization to reduce the capital and operating costs [6] of its computing assets, at the same time preserving the computing caliber it requires. Because the computing sources of most agencies are usually underutilized [7], but they may be essential for positive operations. Thus, the return of interest on computing investments can result in substantial growth through lively participation in grid computing methods despite the fact that networking and service costs might also boom to some extent.

2 Requirements of Grid Computing

To make a fully functional computer grid consisting of various connected networked devices, the middleware, and central control, following requirements are identified.

- A persistent access to underlying network utilized by an organization in order to achieve organization goals which may need to have the capability of handling immense transfer data which in turn varies with the application, the type of network and type of devices. The network is selected as such which are known to have some functionalities (e.g., latency) which function under the threshold specified by the application, and it also determines fair inclusion of networked devices without compromising the grid performance.
  Such networks are selected after evaluating their reliability [8] for the said task. Background information such as the specifications of the underlying network and the past history about the network reliability becomes critical in this regard.
- Another most desideratum factor in setting up grid infrastructure is the availability of middleware technologies that are going to mediate the services fetched by the central server system to end devices which are ultimately going to perform the part of computation.
  Designing such systems requires fair amount of knowledge [9, 10] about the application and the background of the area of the concern Internet on which it will be employed. Extant technologies which are already developed for grid computing infrastructure currently operating worldwide can form the basis of idea on which such incipient organizations can absorb some lessons on designing new approaches for their organization. Albeit, the operating

conditions in different countries in which such systems will be employed may produce varying foibles which also may require locally developed subsystems or techniques to bolster up the underlying middleware framework.

In addition to the above system, fully capable grid services software [11] is also required which would perform according to the expectation set by the application of an organization. This software may be a functional portable application or an underlying system-dependent module, as they may have to carry only the specific jobs designated without relying on the type or nature of the jobs.

- Grid computing systems are designed to perform by taking into consideration of available CPU cycles at the connected machines or devices. Therefore, statistical data projected from the survey of network components should elucidate such availability and their reliability for performing the required task and the CPU capability it needs.

  Other resources such as available memory and storage requirement also play crucial role in determining the suitability of the devices on the network for the designated job. Additionally, a combination such a computer having large idle CPU power [12, 13] and low storage availability should be selected so that it can only perform CPU intensive jobs and not the jobs whose computations result in occupying significant computer storage. A similar implementation would require to choose proper combination of resources based on the threshold set by the application of the grid.

- Since the advent of Globus [14, 15], grid computing has become a sinecure with ease of scavenging for the necessary resources with the consent provided by the end user for those resources through the Globus toolkit to execute complex CPU intensive tasks at their machine. Hence, there should be conceding agreement between the parties which are going to avail the use of grid infrastructure to achieve a particular organization goal [16–22].

3 Characteristics of GRID Computing Infrastructure

3.1 Circumscribing a Possibly Large Area

Grid computing infrastructure should be designed such that it encompasses a sufficient quantity of computing resources [23] such as PC's, PDA, cellphones, or even the components connected to a sophisticated Internet of things applications. Therefore, the grid engages a wide range of devices available in the network without limiting the performance of the device for the other user applications.

As the need for the more computer power grows, the grid can sometimes expand to large geographical area which may involve devices located at distant places but are capable of performing at the same capacity of local devices. Thus, significance of such devices can be ignored when the boundaries of application cannot be determined beforehand.

3.2   Vicissitudes of Machines

Widely varying processing devices further add to problems in designing software's, drivers, and supporting middleware [24] technologies and their implementations. Therefore, both hardware and software system should at least be designed in a way that takes into account the architecture, specification and operating environment of different devices that are going to be involved in the computation.

Furthermore, resources should be allowed to be used in such a way that their local or nonlocal presence does not cause any performance drops or reliability discrepancies.

Problems on grid computing infrastructure thus can be separated on the basis of resources available, and their locality with respect to the server which is nearest to the device in concern. This server may be the central server or the machine which acts as a mediator for the local subnetwork in overall network of the grid computing infrastructure. Locality can be further expanded to include problems which require urgent computations or results.

3.3   Legal Understanding Between the Users and Administration

Toolkit such as Globus when installed on the computer system requests authorization from the user to fetch the computing resources available while the machine is in ideal state or at minimal use. Thus, with the use of such tools, it becomes easy to do regulatory task which may require significant human intervention. Furthermore, if an organization is interested in grid framework which requires unorthodox permissions that organization is required to make such agreements with the users involved in the grid beforehand.

4   Issues Faced by the Low and Lower-Middle-Income Localities in the Establishment of a Typical Grid Computing Infrastructure

Based on the various surveys [25–28] carried out globally both in grid and cloud computing practices, following issues have been identified to have to cause inveterate problems in the construction of an efficient grid in the pecuniary countries.

Although, the number and types of issues can be of significantly vary in case of difficulties faced by organizations, the issues listed here are the chief pitfalls in the grid establishment and the resolution of these issues should be enough to construct a typical grid but not the highly customized application-specific version of the grid.

Because there can be numerous qualms in the fabrication process of next generation grids, the specificity of the issues cannot be delineated given the extent of the study such as this which results in giving only brief yet terse agendas necessary for new adventures in the grids.

4.1 Coordinated Resource Sharing

The most hackneyed problem in grid computing is coordinated sharing of computing resources [29] available at the user end, and the techniques which are used to mitigate this problem have still not acquired the capability of engaging heterogeneous devices to be used by the grid infrastructure.

4.2 Difficulty in Designing Software Model

As the nature of burgeoning organization becomes more and more dynamic, the software components are becoming less and less capable of performing normal tasks on the grid. The software components should be designed in such a way that problem-solving in the multi-institutional and dynamic virtual organizations should be fairly efficient.

Software venders such as IBM, Sun are continually devolving new applications for the use of grid computing but their utilization by the specific organization depends ultimately upon the application, supporting software components and the overall nature of the grid.

4.3 Unavailability of Reliable Scavenging Agents

An efficiently operating grid computing infrastructure necessarily involves an ubiquitous processing cycles scavenging application which can fairly detect suitable ideal CPU cycles among the devices connected in grid. Heterogeneity and variable computer performance play crucial role in designing such application, as they define the extent to which current problem will be exposed to computational systems and the temporal availability of results.

4.4 Unreliable Network Backbone

In scarcely developed geographical surroundings, it is not always possible to have computational devices diligently connected to their network and their reliability is sometimes questionable. Therefore, software components have bear the loss of improper or partial results from the end devices arising due communications failures, software conflicts due high loads and power failures.

4.5 Complex Mathematical Models and Their Inability to Adapt as a Software Technique

Grid computing scheduling algorithms such as back-filling and roundtable have promulgated some problems [30] such as inadvertent theoretical evaluation because their objective function is too confounding to comprehend.

Therefore, strategies involving suitable simulation models should be employed prior to developing framework for the grid computing projects as it prone to entail complex computational algorithms.

4.6 Inadequate Regulatory Laws and Unavailability of Sufficient Funds

Some grid systems such as recently devised Smart Grid [31, 32] system, which are basically employed on the underlying power network grid [33] may require expensive components, for example, power system stabilizers (PSS) installed on the generator which are required for the power supply to work [34] are not cheap and thus limiting the existing grid functionality.

Because of this, grid infrastructure should identify the networks which are currently performing under similar working restrictions which inhibit their ability to perform at a required level along with extant established high-caliber grid networks. Furthermore, these grid computing infrastructures are forced to perform on the networks which do not support performance metrics necessity identified for the grids in concern.

Regulatory authorities [35] further exacerbate the problem by not expanding the laws or rules made for digital technology that could potentially make the extant grid systems to perform at better yet more efficient than the current working model. For example, protocols and software are although obsolete, are still implemented because of lack up to date governance, and/or political pressure.

4.7 Grid Security

Cyber criminals around the world are always exploring new horizons to attack which brings forth new challenge to be tackled by newly established grids.

Therefore, it is imminent that to have the software resources to immune against attacks such as denial of service (DOS) attacks (even sometimes distributed denial of service attacks, popularly called DDOS attacks), malware and collaborated attacks [36–38].

5 Other Issues Affecting the Grid Computing

Following issues are responsible for pitfalls against good grids that are based mainly the moral or social perception against the technology and its utilization. There differs country wise extensively depending upon the social environment and regulatory laws.

1. One major concern by the user coming under grid purview is the privacy [39, 40] but the concern against the same is diminishing day by day as the independent software components are made without hampering users workspace and its privacy.
2. Power theft [34] in case of Smart Grid's plays an ineluctable role in case of lower income and large geographical area occupying countries where long distances and inadequate security infrastructure result in unreliable and inefficient grid network.
3. Social concern [41] over the fair availability of the electricity where the large electronic components are already exposed to improper resource allotment.
4. Some concerns such as government is utilizing the electricity for the other purpose; instead, it was made essential for in the first place.
5. There also such concerns that the organization is using the grid infrastructure for the task that is not previously intended.
6. Some components are known for their RF emissions [42] which might put off some users.

6 Possible Solutions and Future Work

Most problems in grid computing are typically mitigated by establishing a promising backbone network which would sustain the ever-growing demand of grid thus the functionality of the grid. Although the number of computing devices such as PC's or cellphones is growing at exceptional rate in the world, the number of servers is not increasing at the same rate. This can be overcome by employing local server concept. On the other hand, software agents which are designed for scavenging processing cycles or other resources should be more versatile to take advantage of growing number of devices and their heterogeneity.

A singular regulatory authority should be followed if possible to some extent so as to avoid clumsy or inflexible rules and regulations in case of communication protocols, software methodologies and different technologies adopted by the authorities. This can easily be avoided by designing applications which experience very few hurdles in case of different working environments. For example, tools such as Globus toolkit is continually adopting support for wide range of operating environments and architectures.

The design of Middleware technologies should predetermine utilization of available computing resources efficiantly and at the same time, yield minimum viable satisficatory results according to the requirement analysis.

Furthermore, public awareness about the grid computing technology may become useful against altering the social perception. This could occupy a leading role in case of Smart Grids as the scope for development or innovation in this area is incomprehensible.

7 Conclusion

This paper elucidates some important yet incisive pitfalls in the fabrication process of efficient grid frameworks in low and lower-middle income countries. The paper presents all such issues that are involved while designing a sophisticated or state of the art of grid computing infrastructure. It points out the factors associated with the process, and their public awareness which leads the citizens to show support for grid technology and to be an integral part of complex problem-solving process that nation or an organization is interested to have answer for.

In retrospect, the poor digital infrastructure is the main bane to proper grid computing network because required hardware and software resources are inadequate to achieve such organization goals. Secondly, improper legislation and their inadvertent implementation lead to incoherent grid infrastructure resulting in an inefficient and an unreliable grid computing network.

Although grid networks such as Teragrid [43], NASA's [44] grid or SETI's [45] grid in the USA are exceptional in their class, they can act as a basic architectural model and ideal to look for gradual inspiration for any organization in the adventure of building a successful grid.

References

1. Foster, K.: The Grid: Blue Print for a New Computing Infrastructure. Morgan Kaufmann Publications (1999)
2. Foster, I., Kesselman, C., Nick, J.M., Tuecke., S.: The Physiology of the Grid: An Open Grid Services Architecture for Distributed Systems Integration. Grid Forum white paper (2003)
3. Pardeshi, S.N., Patil, C., Dhumale, S.: Grid computing architecture and benefits. Int. J. Sci. Res. Publ. **3**(8) (2013). ISSN 2250-3153
4. Schulze, B., Buyya, R., Porto, R.: Middleware for clouds and e-Science. In: 8th International Workshop on Middleware for Grid, Clouds and e-Science (2010)

5. Arnin, K., Hategan, M., von Laszewski, G., Zaluzec, N.J.: Abstracting the grid. In: 2004 12th Euromicro Conference on Parallel, Distributed and Network-Based Processing, USA (2004)
6. Doninger, C., Cary, N.C.: Maximize IT Flexibility and Lower Costs with Grid Computing on Windows. Paper 227-29. http://www.oracle.com/us/027587.pdf
7. Integrated Defense Staff, India. http://www.ids.nic.in/tnl_jces_Jun_2011/PDF/pdf/2.%20Grid_computing.pdf
8. Laxmi, K.V., Krishnaiah, R.V.: A reliable grid service infrastructure. Int. J. Comput. Eng. Appl. III(III) (2013)
9. Yu, J., Buyya, R.: A taxonomy of workflow management systems for grid computing. J. Grid Comput. Springer (2005)
10. Yu, J., Buyya, R.: A Taxonomy of Scientific Workflow Systems for Grid Computing. ACM Sigmod Record (2005)
11. Foster, I., Kesselman, C.: The Grid 2: Blueprint for a New Computing Infrastructure. Elsevier (2003)
12. Goldchleger, A., Kon, F., Goldman, A., Finger, M., Bezerra, G.C.: InteGrade: object-oriented grid middleware leveraging the idle computing power of desktop machines. In: Concurrency and Computation Practice and Experience. Wiley (2004)
13. Bhavsar, M.D., Pradhan, S.N.: Scavenging idle CPU cycles for creation of inexpensive supercomputing power. Int. J. Comput. Theory Eng. 1(5) (2009)
14. Globus@Home. http://www.globus.org
15. Foster, I., Kesselman, C.: Globus: a metacomputing infrastructure toolkit. Int. J. Supercomput. Appl. 11(2), 115–128 (1997)
16. Abbas, A.: Grid Computing: A Practical Guide to Technology and Applications. Charles River Media (2003)
17. Berman, F., Fox, G., Hey, A.J.G.: Grid Computing: Making The Global Infrastructure a Reality. Wiley, Chichester (2003)
18. Foster, I., Kesselman, C.: The Grid 2: Blueprint for a New Computing Infrastructure. Morgan Kaufmann Publishers, San Francisco (2003)
19. Foster, I., Kesselman, C., Tuecke, S.: The anatomy of the grid: enabling scalable virtual organizations. Int. J. High Perform. Comput. Appl. 15(3), 200–222 (2001)
20. Joseph, J., Fellenstein, C.: Grid Computing. Prentice Hall PTR, Englewood Cliffs (2003)
21. Minoli, D.: A Networking Approach to Grid Computing. Wiley Interscience, Hoboken (2004)
22. Plaszczak, P., Wellner Jr., R.: Grid Computing: The Savvy Manager's Guide. Morgan Kaufmann, San Francisco (2005)
23. Dongarra, J., Lastovestsky, A.: An overview of heterogenous high performance and grid computing. In: Engineering the Grid: Status and Perspective. American Scientific Publishers (2006)
24. Berman, F., Fox, G., Hey, A.J.G.: Grid Computing: Making the Global Infrastructure a Reality. Wiley (2003)
25. https://www.idc.com/getdoc.jsp?containerId=prUS25558515
26. http://www.evansdata.com/press/viewRelease.php?pressID=65
27. Ahuja, S.P., Myers, J.R.: A survey on wireless grid computing. J. Supercomput. 37, 321 (2006)
28. Bichhawat, A., Joshi, R.C.: A survey on issues in mobile grid computing. Int. J. Recent Trends Eng. Technol. 4(2) (2010)
29. Foster, I., Kesselman, C., Tuecke, S.: The anatomy of the grid: enabling scalable virtual organizations. Int. J. Supercomput. Appl. (2001)
30. Uwe, S.: Evaluation of Scheduling Algorithms for Grid Computing. Computer Engineering Institute, University of Dortmund (2001)
31. Rusitschka, S., Eger, K., Gerdes, C.: Smart grid data cloud: a model for utilizing cloud computing in the smart grid domain. In: First IEEE International Conference on Smart Grid Communications (SmartGridComm) (2010)

32. Gungor, V.C., Sahin, D., Kocak, T., Ergüt, S.: Smart grid technologies: communication technologies and standards. IEEE Trans. Industr. Inf. **7**(4) (2011)
33. Hammons, T., Wong, K.P., Lai, L.L.: Transmission System Operation and Interconnection. National Transmission Grid Study, United States Department of Energy, IEEE General Meeting (2007)
34. Grundvig, J.: Detecting Power Theft by Sensors and the Cloud: Awesense Smart System for the Grid. Huffington Post (2013)
35. Buyya, R., Abramson, D., Giddy, J.: Economic models for resource management and scheduling in grid computing. In: Concurrency and Computation Practice and Experience. Wiley (2002)
36. Kaufman, L.M.: Data security in the world of cloud computing. IEEE Secur. Priv. **7**(4) (2009)
37. Balakrishnan, M.: Freescale Trust Computing and Security in the Smart Grid. Freescale Semiconductor, Inc. (2013)
38. Kar, S.: An anomaly detection scheme for DDoS attack in grid computing. PhD Dissertation (2009)
39. Chaitanya, N.S., Ramachandram, S., Padmavathi, B., Skandha, S.S., Kumar, G.R.: Data privacy for grid systems. In: Communications in Computer and Information Science, pp. 70–78. Springer (2011)
40. Li, M., Yao, H., Guo, C., Zhang, N.: Privacy protection mechanism in grid computing environment. In: Proceedings of the First European Conference on Computer Network Defence School of Computing, University of Glamorgan, Wales, pp. 33–39. Springer, UK (2006)
41. Akoumianakis, D.: Virtual community practices and social interactive media: technology lifecycle and workflow analysis. Inf. Sci. Ref. **1** (2009)
42. RF Emissions and the Smart Grid. http://www.networktasman.co.nz/Advanced_Meters/Radio%20Frequency%20Safety.pdf
43. The TeraGrid Project. http://www.teragrid.org/
44. NASA@National Grid Computing Forum. http://www.nasa.gov/centers/ames/news/releases/2001/01_92AR.html
45. SETI@Home. http://setiathome.ssl.berkeley.edu/

Fuzzy A-Star Based Cost Effective Routing (FACER) in WSNs

Arabinda Nanda and Amiya Kumar Rath

Abstract The routing is one of the major operations in WSNs. In our proposed work distance, energy and the angle formed by the node with the base station are taken as input parameters. The Mamdani fuzzy inference system is used to select the chance of a sensor node to become a cluster head (CH). A-Star search algorithm is used to locate an optimum route from source to sink node. The data packets are routed on the selected path.

Keywords Fuzzy inference system · A-Star search · Optimal route Hierarchical routing

1 Introduction

In WSNs, data packets are sending from source sensor node to base station through many intermediate nodes. Sensor nodes are not possible to recharge after deployment. Routing is an important issue in WSNs, to solve the energy problems some extent. The network structure based routing protocol in WSNs are hierarchical routing, flat routing and routing based on location. In case of flat routing, the entire nodes are treated equally. Once a node needs to send data, it computes the optimum path from source to sink node, then it sends the data packets through the path. In case of hierarchical routing, sensor nodes are divided into many number of unequal or equal size clusters. A CH is selected from each cluster. The CH receives the sensed data from its member nodes, aggregates it and sends the aggregated data to

A. Nanda (✉)
Department of Computer Science & Engineering, Siksha 'O' Anusandhan University,
Bhubaneswar, Odisha, India
e-mail: aru.nanda@rediffmail.com

A. K. Rath
Department of Computer Science and Engineering, Veer Surendra Sai University
of Technology, Burla, Odisha, India
e-mail: amiyaamiya@rediffmail.com

© Springer Nature Singapore Pte Ltd. 2018
K. Saeed et al. (eds.), *Progress in Advanced Computing and Intelligent Engineering*,
Advances in Intelligent Systems and Computing 563,
https://doi.org/10.1007/978-981-10-6872-0_53

the base station. In location-based routing, every sensor node knows the location of other sensor nodes. We proposed fuzzy A-Star-based technique for hierarchical routing. Unlike the hard computing methods, a soft computing fuzzy system is capable to work fine with diverse and incorrect inputs. It does not need complex calculations, the speed of processing and not needed large amount of energy.

2 Literature Review

Several techniques of routing are used for optimal energy handling in WSNs. Fuzzy is one of the efficient techniques, which attempts for energy utilization. Our proposed method FACER is compared with FEAFR proposed by Pagi et al. [1] because its approach is nearly similar to our approach. The comparative analyses of both the models are given below (Table 1).

The comparative study of various fuzzy based routing is given below (Table 2).

Table 1 Comparative study of FEAFR model with FACER

Models	Routing type	Method	Input parameters	Speed and memory uses
FEAFR [1]	Flat routing using Dijkstra's algorithm $F(n) = g(n)$	Fuzzy	Distance, angle, energy	Dijkstra's algorithm is an uninformed search used for routing that is less speed and more memory uses
Our proposed model FACER	Hierarchical routing using A-Star algorithm $F(n) = g(n) + h(n)$	Fuzzy	Distance, angle, energy	A-Star algorithm is a heuristic search used for routing, that is more speed and less memory uses

Table 2 Comparative study of various fuzzy based routing

References	Routing type	Methods	Input parameters
Ran et al. [2]	Energy optimized routing	Fuzzy approach	Remaining energy, energy consumption rate, transmission energy, queue size, distance and weight
Zarafshan et al. [3]	Directed diffusion	Fuzzy diffusion approach	Weight, energy and traffic load
Arabi et al. [4]	Hybrid energy efficient routing	Fuzzy method	Node centrality, energy and concentration
Gharajeh and Khanmohammadi [5]	Static three-dimensional fuzzy routing	Fuzzy	Number of neighbors and distance

(continued)

Table 2 (continued)

References	Routing type	Methods	Input parameters
Mahmood et al. [6]	Online maximum lifetime routing	Fuzzy algorithms	Level of battery
Iman and Saadeh [7]	Energy-aware routing protocol	Fuzzy based	Distance, residual energy and depth
Bushehr and Kavian [8]	Clustering based routing	Fuzzy logic	Concentration and energy
Vakili et al. [9]	Novel fault-tolerant routing protocol	Fuzzy logic	Residual energy
Tamene and Rao [10]	Distributed cluster based routing	Fuzzy	Energy, distance
Mohamed [11]	Optimal route path	Fuzzy routing technique	Distance to base, residual energy

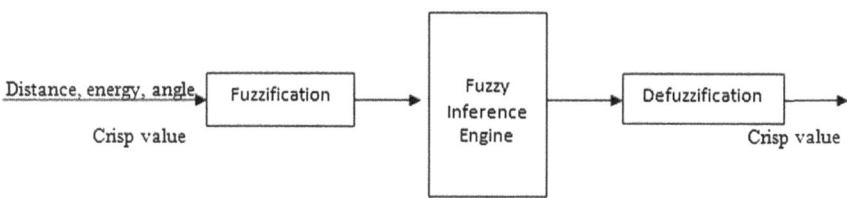

Fig. 1 Fuzzy inference system (Mamdani)

3 Proposed Work

In our proposed model, a whole network is partitioned into many clusters using fuzzy clustering algorithm. Then, the fuzzy system used to select a chance of a node to have a CH. We have used three input variables, which are distance, angle and energy, and output is a chance of a node to have a CH (Fig. 1).

We are using Mamdani fuzzy inference system due to its simple rule-based method that does not need any complex calculations. The Mamdani fuzzy inference system has following steps:

1. *Fuzzification*: It is finished by assigning membership functions to the input values. We require three membership functions to fuzzify distance, energy and angle of a node. We are using triangular membership because of their suitability for real-time operations (Table 3).

Table 3 Input parameters and membership function

Input parameters	Membership function ($\mu(x)$)		
	Low (LW)	Medium (MM)	High (HH)
Distance (m)	0–150	150–200	200–250
Energy (j)	0–5	5–10	10–12
Angle (°)	0–90	90–180 and 270–360	180–270

Fig. 2 Angle between source sensor node and the sink node

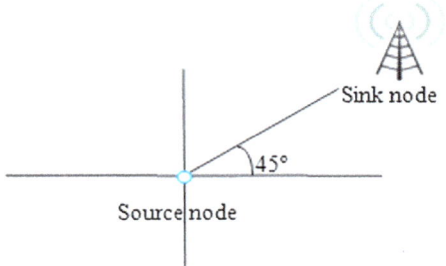

Let us assume the nodes are present at the origin of 2D. The sink is also present in the first quadrant of 2D. The angles 45° is considered as high vicinity. The angles between 90°–180° and 270°–360° are considered as a medium, whereas the angles between 180° and 270° are not chosen because it is opposite to the first quadrant and the nodes present in this quadrant are a large distance from the sink node (Fig. 2).

2. *Rule generation and Evaluation*: In this phase 27, fuzzy if... then rules are generated.

If the antecedents of If...Then rules are connected thought OR operation, then the consequences calculated as: $\mu_{A \cup B(x)} = maximum \left[\mu_{A(x)}, \mu_{B(x)} \right]$

If the antecedents of If...Then rules are connected thought AND operation, then the consequence calculated as: $\mu_{A \cap B(x)} = minimum \left[\mu_{A(x)}, \mu_{B(x)} \right]$ (Table 4).

3. *Defuzzifications*: It is the opposite of fuzzification. Different methods are COA, COG, MOM, etc. We use COG method for defuzzification

$$COG = \frac{\sum_{i=1}^{n} x_i * \mu_{i(x)}}{\sum_{i=1}^{n} \mu_{i(x)}}$$

The output of defuzzification is a number. If that number is less than and equal to the pre-assigned threshold, then the chance of selection of CH nodes are success. After the end of the fuzzy process, the A-Star search is used to find the shortest path from source node to sink node.

Table 4 Fuzzy information system

Rules number	Distance	Angle	Residual energy	Chance
R1	LW	LW	LW	MM
R2	LW	LW	MM	MM
R3	LW	LW	HH	HH
R4	MM	LW	LW	LW
R5	MM	LW	MM	MM
R6	MM	LW	HH	MM
R7	HH	LW	LW	LW
RS	HH	LW	MM	LW
R9	HH	LW	HH	MM
R10	LW	MM	LW	LW
R11	LW	MM	MM	MM
R12	LW	MM	HH	MM
R13	MM	MM	LW	LW
R14	MM	MM	MM	MM
R15	MM	MM	HH	LW
R16	HH	MM	LW	LW
R17	HH	MM	MM	LW
R18	HH	MM	HH	MM
R19	LW	HH	LW	LW
R20	LW	HH	MM	LW
R21	LW	HH	HH	LW
R22	MM	HH	LW	LW
R23	MM	HH	MM	MM
R24	MM	HH	HH	LW
R25	HH	HH	LW	LW
R26	HH	HH	MM	LW
R27	HH	HH	HH	MM

4 Results

Result analysis of FACER with FEAFR performed using MATLAB 7.0. The analysis is performed from the aspect of avg. energy consumption, throughput and routing time.

The result of Fig. 3 shows avg. energy consumption. The avg. energy consumption of FACER is lesser than FEAFR as number of nodes increases.

The result of Fig. 4 shows throughput. The throughput of FACER is higher than FEAFR as number of nodes increases.

The result of Fig. 5 shows time of routing. The timing of routing of FACER is lesser than FEAFR as number of nodes increases.

Fig. 3 Avg. energy
consumption

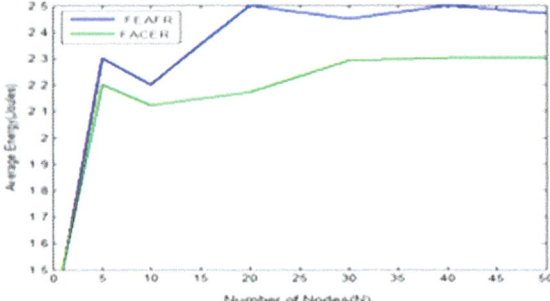

Fig. 4 Throughput

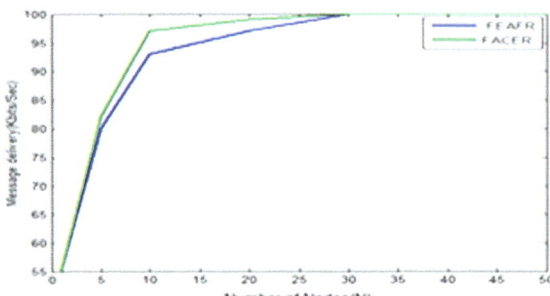

Fig. 5 Routing time

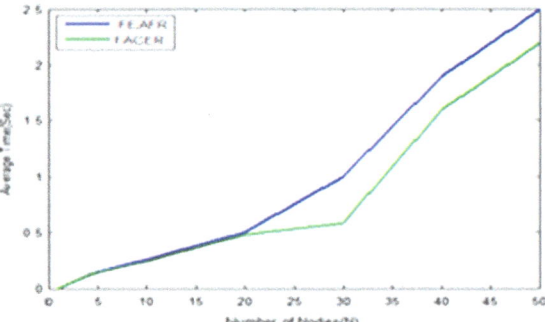

5 Conclusions

The proposed method FACER used to find an optimum shortest route from source
sensor node to sink node. Routing is approved through the nodes on the recognized
path. The fuzzy system uses to find the chance of a node to become CH. The A-Star
search algorithm finds the shortest route path from source node to sink node. The
proposed approach is admissible, complete and optimal one as it uses A-Star
algorithm.

References

1. Pagi, A.B., Budyal, V.R., Sataraddi, M.J.: Fuzzy based energy aware flat routing (FEAFR) in wireless sensor networks. Int. J. Emerg. Technol. Adv. Eng. **4**(7), 108–115 (2014)
2. Ran, G., Zhang, H., Gong, S.: Improving on LEACH protocol of WSNs using fuzzy logic. J. Inf. Comput. Sci. **7**(3), 767–775 (2010)
3. Zarafshan, F., Karimi, A., Al-Haddad, S.A.R.: A novel fuzzy diffusion approach for improving energy efficiency in WSNs. Int. J. Mach. Learn. Comput. **2**(4), 506–509 (2012)
4. Arabi1, Z., Khodaei, Y.: A hybrid energy efficient routing using a fuzzy method in WSNs. Int. J. Distrib. Parallel Syst. **1**(1), 01–11 (2010)
5. Gharajeh, M.S., Khanmohammadi, S.: Static 3D fuzzy routing based on the receiving probability in WSNs. Computers **2**, 152–175 (2013)
6. Minhas, M.R., Gopalakrishnan, S., Leung, C.M.: Fuzzy algorithms for maximum lifetime routing in WSNs. IEEE Wireless Commun. **11**(6), 6–28 (2004)
7. Iman, M.A., Saadeh, M.K.: Fuzzy-based energy aware routing protocol for WSN. Int. J. Commun. Netw. Syst. Sci. 403–415 (2011)
8. Bushehr, M.T., Kavian, I.Y.S.: Energy efficient clustering algorithm for WSNs using fuzzy logic. Int. J. Comput. Appl. **89**(14), 01–05 (2014)
9. Fard, M.V., Mazinani, S.M., Hoseini, S.A.: Introducing a novel fault tolerant routing protocol in WSNs using fuzzy logic. Int. J. Comput. Sci. Inf. Technol. **5**(5), 171–176 (2013)
10. Tamene, M., Rao, K.N.: Fuzzy based distributed cluster formation and route construction in WSN. Int. J. Comput. Appl. **140**(5), 21–27 (2016)
11. Mohamed, A.H.: Fuzzy modeling for WSNs. Int. J. Comput. Appl. **138**(13), 29–33 (2016)

Low Delay Routing Algorithm for FiWi Access Network

Raksha Upadhyay, Shweta Pandey and Uma Rathore Bhatt

Abstract Fiber wireless FiWi access network is an integration of two networks. The first part of network is called passive optical network (PON), used at backend. The second part of network, wireless mesh network (WMN) is used at front end. This combination facilitates high data rates, stability, and flexibility. Routing is very important part of wireless mesh network (WMN), and therefore of FiWi access network. It directly affects the network parameters such as throughput, delay, packet delivery ratio which in turn affects the network performance. In this paper, a routing algorithm is proposed for FiWi network, on the basis of minimum angle criteria. The algorithm results in lesser delay as compared to standard routing algorithm.

Keywords Passive optical network (PON) · Wireless mesh network (WMN)
Optical line terminal (OLT) · Optical network unit (ONU)

1 Introduction

Two types of network PON and WMN are the basis of FiWi network as shown in Fig. 1. PON covers optical part and it provides high bandwidth accessibility. It is a dominating technology which is supported by different components such as OLT, splitters, and ONUs. OLT is responsible for injecting packets in internet backbone.

R. Upadhyay (✉) · S. Pandey · U. R. Bhatt
Department of Electronics and Telecommunication, Institute of Engineering
and Technology, Devi Ahilya University, Indore, MP, India
e-mail: raksha_upadhyay@yahoo.co.in

S. Pandey
e-mail: shwetaimagin21@gmail.com

U. R. Bhatt
e-mail: umarathore@rediffmail.com

© Springer Nature Singapore Pte Ltd. 2018
K. Saeed et al. (eds.), *Progress in Advanced Computing and Intelligent Engineering*,
Advances in Intelligent Systems and Computing 563,
https://doi.org/10.1007/978-981-10-6872-0_54

Fig. 1 FiWi network

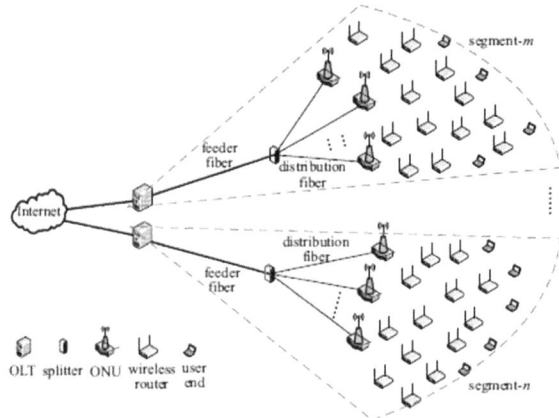

Splitters are used to connect different ONUs to OLT. Signal is converted into wireless and vice versa with the help of ONU [1]. Feeder fiber is deployed from OLT to splitter, and after that distribution fiber is deployed between splitter and ONUs. Each ONU drives multiple wireless routers. Front end consists of WMN, which is carried by wireless gateways, wireless routers and end users. When user needs to access the internet, packet goes to the OLT through primary ONU and nearby routers. Finally then injected to the backbone of the network.

For connecting the radio nodes in WMN, mesh topology is used. If any node fails to operate, the path can be established among the rest of the nodes, through intermediate nodes. Therefore, at front end routing is an important phenomenon which also affects the performance of the overall network. There are different parameters such as throughput, delay, energy efficiency which decide the network performance [2]. This paper focuses on delay incurred in routing process at front end of FiWi network.

2   Related Work

At the front end of FiWi, the wireless routers are used to inject the packets into the wireless mesh of network. The packet travels through different nodes to find destination which is also called multi-hop communication. As the packet travels through several routers or intermediate nodes in the mesh, so packet delay is the major problem and it increases as the mesh network becomes large. Packet loss may also occur due to multiple failures. Thus, to handle these problems different algorithms are proposed.

A well-known algorithm called as Delay Aware Routing Algorithm (DARA) has been implemented in FiWi network which calculates delay in terms of propagation delay, transmission delay, slot synchronization delay, and queuing delay [3].

Another algorithm is called Capacity and Delay Aware Algorithm (CADAR) which calculates delay on the basis of capacity. It assigns capacity to each link. Link-state advertisement (LSA) is responsible for advertising the states of all nodes. CADAR gives more accurate results than DARA [4]. Risk and Delay Aware Routing Algorithm has also been implemented in FiWi network. A path which contains minimum delay is considered, and on the basis of this a risk list table is maintained. If failure occurs, RL is updated, and packets are sent again [5].

In literature, routing algorithms are also available for WMN requiring lesser delay and fewer hop counts. Most common routing algorithm for WMN is minimum distance algorithm works on the shortest path algorithm which does not consider traffic demands. This algorithm works on the basis of minimum distance. If a user wants to communicate, then it first finds the intermediate node which is at minimum distance and then packet is transferred to that particular node. Now that particular intermediate node works as source, and it again repeats the process until destination is reached. However, this algorithm suffers with some issues like increased delay, poor load balancing, and high congestion. Another algorithm called Angle Based Multicast Routing Algorithm (AMRA) works for WMN [6]. This algorithm helps in reducing delay on the basis of calculating angle. In this, a node selects next node of the route on the basis of specified angle criteria. Nodes which do not satisfy the criteria cannot be a part of route to the destination. In order to further improve the performance of FiWi network, a different algorithm called Minimum Algorithm can be implemented on FiWi network, which helps in reducing number of hops by calculating minimum angle.

3   Proposed Work

A simple routing algorithm called Minimum Angle Routing Algorithm is proposed for FiWi network in this section. This algorithm is based on minimum angle criteria. For example, if there are n number of routers which want to communicate with ONU then each router will draw a reference line to ONU and with this line, angle will be calculated for every intermediate nodes. That intermediate node which draws minimum angle from source is considered, packet will be delivered to it and now this node will work as source. This process will be continued until destination is reached [7, 8].

On the basis of this algorithm, delay is calculated. In the minimum distance algorithm, as the number of routers increases, hop count also increases which causes increase in delay while proposed algorithm is based on minimum angle with reference to ONU in which hop counts are minimized. Delay is directly proportional to number of hop counts. As the number of hops will increase, it increases delay also.

4   Results and Analysis

A random scenario of routers and ONU is shown in Fig. 2 for FiWi network. Path is calculated on the basis of minimum distance algorithm. There are five number of routers in which three routers will communicate in multiple hops while others will in single hop. If router no. two wants to communicate with ONU then possible path on the basis of minimum distance is through router no. $2 \rightarrow 4 \rightarrow 3 \rightarrow 1 \rightarrow ONU$. Number of hops will be four.

On the other hand, if hop count is calculated on the basis of proposed algorithm as shown in Fig. 3 then for router no. two, path will be through routers $2 \rightarrow 6 \rightarrow ONU$ and hop count will be two. Hence, delay is reduced.

Fig. 2 Random scenario of ONU and routers for minimum distance algorithm

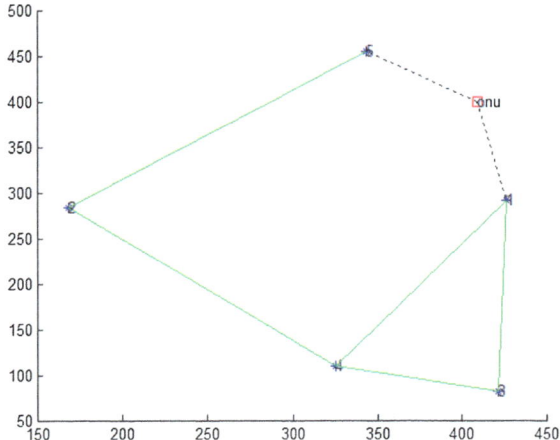

Fig. 3 Random scenario of ONU and routers for improved minimum angle algorithm

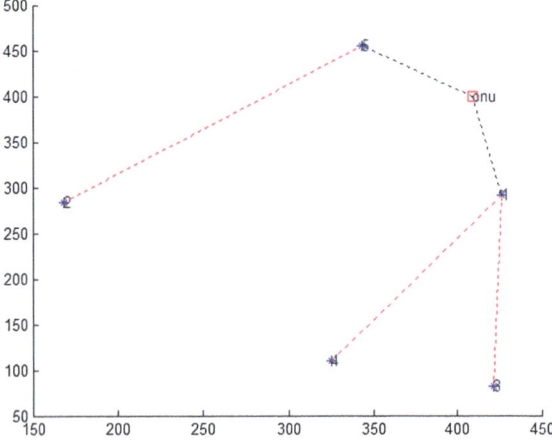

Fig. 4 Hop Count versus
Routers

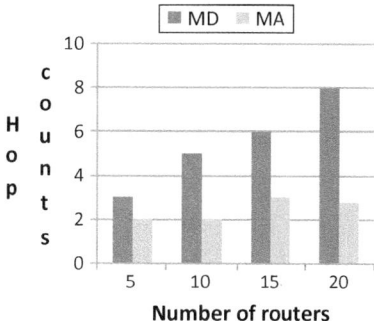

Fig. 5 Delay versus Routers

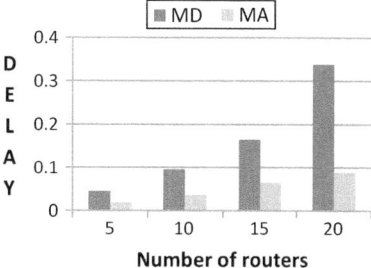

Delay is calculated assuming 100 packets are transmitted at 10 Mbps for every router which communicates in multiple hops. Figure 4 compares hop count versus wireless routers for both algorithms minimum distance and minimum angle. It is shown that as the number of routers increases, hop count also increases in both cases but minimum angle algorithm has better result than minimum distance algorithm in terms of reduced hop counts.

Figure 5 shows the graph of delay (in sec) versus wireless router for both cases. As the number of routers increases, delay also increases in both cases but proposed algorithm gives lesser delay as compared to minimum distance algorithm.

5   Conclusion and Future Work

Front end of FiWi network consists of WMN. This paper implements minimum angle routing algorithm for FiWi network to get the reduced delay. The route is established on the basis of minimum angle criterion. It results in fewer hop counts as compared to traditional minimum distance algorithm. Using delay other parameters such as throughput can be computed. Further the combination of minimum distance and minimum angle can also be applied to FiWi network for further improvement in network performance.

References

1. Sarkar, S., Yen, H., Dixit, S., Mukherjee, B.: Hybrid wireless-optical broadband access network (WOBAN): network planning using Lagrangean relaxation. IEEE/ACM Trans. Netw. **17**, 1094–1105 (2009)
2. Liu, Y., Guo, L.: Green survivability in FiWi access network. Opt. Fiber Technol. **18**, 68–80 (2012)
3. Sarkar, S., Yen, H., Dixit, S., Mukherjee, B.: A novel delay-aware routing algorithm DARA for a hybrid wireless-optical broadband access network (WOBAN). IEEE Netw. **20**, 20–28 (2008)
4. Reaz, A., Vishwanath, R., Sarkar, S., Dixit, S., Mukherjee, B.: CADAR an efficient routing algorithm for wireless-optical broadband access network. In: IEEE Communications Society Subject Matter Experts for Publication in the ICC proceedings, pp. 5191–5195 (2008)
5. Sarkar, S., Yen, H., Dixit, S., Mukherjee, B.: RADAR risk and delay aware routing algorithm in a hybrid wireless optical broadband access network (WOBAM). In: Optical Fiber Communication and the National Fiber Optic Engineers Conference, pp. 1–3 (2007)
6. Thernal, B., Thirunandan, K.: Angle based multicast routing algorithm (AMRA). Indian J. Sci. Technol. **8** (2015)
7. Chakraborty, I., Hussain, A.: A simple joint routing and scheduling algorithm for a multihop network. In: Computer Systems and Industrial Informatics (ICCSII), pp. 1–5 (2013)
8. Chakraborty, I., Agarwal, A.: A simple routing algorithms based on energy consumption. In: IEEE Sponsored 9th International Conference, pp. 1–9 (2012)

A Novel Approach to Avoid Selfish Nodes During Allocation of Data Items in MANETs

Satyashree Samal, Manas Ranjan Mishra, Bibudhendu Pati,
Chhabi Rani Panigrahi and Joy Lal Sarkar

Abstract Performance of mobile ad hoc networks (MANETs) increases by the cooperation of mobile nodes among themselves. In MANETs, mobile nodes share their resources such as battery power, memory, and bandwidth. But, in the real world some nodes do not want to contribute their resources either fully or not at all and thereby storing their resources for future use. So these nodes behave as selfish due to their resource constraint. In this work, we propose an approach for detection of the selfish nodes in MANET and avoid them during the data distribution process when there is sybil attack by considering the routing optimality, checking optimality, and forwarding optimality. The analysis of the simulation results indicate that the network performance has been increased as compared to the traditional network in terms of data accessibility and hence decreasing overall selfishness alarm after detecting the selfish nodes.

Keywords MANET · Sybil attack · Selfish node · Data accessibility

S. Samal (✉) · M. Ranjan Mishra · B. Pati · J. Lal Sarkar
Department of Computer Science and Engineering, C.V. Raman
College of Engineering, Mahura, India
e-mail: sony.samal123@gmail.com

M. Ranjan Mishra
e-mail: manascvrce@gmail.com

B. Pati
e-mail: patibibudhendu@gmail.com

J. Lal Sarkar
e-mail: joylalsarkar@gmail.com

C. Rani Panigrahi
Department of Information Technology, C.V. Raman
College of Engineering, Mahura, India
e-mail: panigrahichhabi@gmail.com

© Springer Nature Singapore Pte Ltd. 2018   571
K. Saeed et al. (eds.), *Progress in Advanced Computing and Intelligent Engineering*,
Advances in Intelligent Systems and Computing 563,
https://doi.org/10.1007/978-981-10-6872-0_55

1 Introduction

Recently, the rapid advancement of wireless technology has given light for new research directions [1–8]. The emergence of the temporary network leads to the development of MANETs which is constructed by nodes that are mobile in nature and lack of base stations. MANETs have several applications such as emergency redemption, military environment as well as natural disaster areas due to quick deployment. Each host in the network acts as a router or a relay node [9]. They communicate and transfer data using one-hop or multi-hop techniques. The mobile nodes in a MANET are mainly classified as three types [9] and are genuine, partially selfish, and fully selfish nodes. The nodes which share the data items allocated by others within its limit are known as genuine nodes. The memory space which are not shared by the nodes easily while allocating the data item in the network are termed as partially selfish nodes, and they share minimum amount of memory with other nodes. On the other hand, the nodes which refuse completely to hold or allocate the data items from the others for benefit of the network are termed as selfish nodes. Due to this highly dynamic and autonomy property of the nodes, frequent network partition occurs [10, 11]. Hence, some data items in one network are unavailable to some other nodes in another network which results in reducing data availability in the network. This availability of data is known as data accessibility which is one of the important parameters of the MANETs. To overcome this problem, several researchers proposed different techniques to replicate the data items and then allocating the individual node to maximize the data availability [10, 11]. Due to resource constraints in this network, as time passes, some nodes behave in selfish manner. For example, the data packets may not be forwarded by these selfish nodes to conserve their energy. The selfish nodes allocate a data item by their own and hence the memory space does not remain free for other's use and is called as selfish replica allocation [9]. But, the data items are allocated by these nodes for their own purpose. In reality, this misbehavior of the selfish nodes creates several issues in the network. The concept of selfishness has been proposed in Gnutella [12] in which the resources are not shared to others, and they are only used by the users for benefit of their own.

MANETs can be categorized as two types such as open MANETs and closed MANETs [13, 14]. Open MANET is a collection of nodes in which all nodes participate voluntarily, whereas in closed MANETs, all nodes are having different objectives such as some of them may behave as selfish nodes in order to conserve their resources. In this work, we assume a closed MANET which corresponds to a real environment. So, the network is constructed as a collection of genuine as well as selfish nodes. If the selfish nodes will not be detected, then they may create several problems to the network. So, detecting selfish nodes is highly necessary for the networks.

The organization of the paper is as follows. In Sect. 2, we provide the related work and our design mechanism is briefly discussed in Sect. 3. Section 4 describes the cooperation-optimality analysis. Section 5 describes the simulation model along with the analysis of results. Finally, we conclude the paper in Sect. 6.

2 Related Work

This section reviews the literature related to this work. For detection of the misleading nature of nodes in the network, several techniques have been proposed and are described as follows:

A. *Credit-Based Scheme*
A scheme of providing incentives in terms of virtual currency called nuglets for forwarding of data among nodes in wireless ad hoc network was suggested by Buttyan et al. [15]. Nuglets are earned by nodes by furnishing services to others and have to pay in the form of nuglets to get services from other nodes. Chen et al. [16] proposed a mechanism called iPass which is an auction-based incentive scheme enabling the cooperative packet forwarding scheme in MANET. When one node provides data to others, each will have to pay the price of packet sending the services to the intermediate nodes. So, a joint result of incentive engineering and flow controlling in an open MANET is iPass.

B. *Reputation-Based Scheme*
Buchegger et al. [17] proposed a protocol named CONFIDANT. The objective of CONFIDANT is to make misbehavior of nodes unattractive and uncooperative and are excluded from the network.

C. *Acknowledgment-Based Scheme*
This scheme includes the response of an acknowledgment from receiver to sender to confirm about the forwarded packet. Balakrishnan [13] defined the TWOACK scheme to detect misbehavior of nodes, and the problems have been resolved by broadcasting the routing protocol to keep away them in future route. From the above literature, it can be summarized that the above three techniques considered only the network-related issues like not forwarding the data packets to the target node to conserve the energy as well as other resources of the network .

3 Proposed Mechanism

The main objective of the proposed mechanism is to detect the presence of selfish nodes when the networks are associated with the sybil attack.

- Let the set of nodes in an open MANET is MN $= MN_1, MN_2, \dots, MN_n$ where n is the total number of nodes and MN_i ($1 \le i \le n$) is a specific mobile node.
- The set of data items is represented as DN $= \{DN_1, DN_2, \dots, DN_m\}$ where m is the total number of data items, and we assume that the data items are not updated for simplicity and is given in Fig. 1.

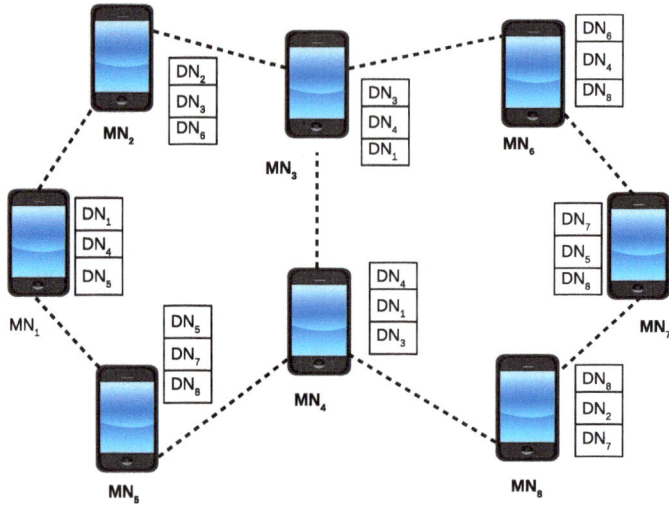

Fig. 1 Example of selfish replica allocation [10]

- Equal size of data items has been taken where original data items of a node MN_i is denoted as DN_i, and others are replica of data item. Each node MN_i in MN has fixed number of memory space and represented by M_i which is limited for replica and original data items.
- Each node MN_i in MN has its own access frequency to each data item DN_i in DN and is denoted as AF. Access frequency is the rating of the specific data item accessed by one node.

To detect the selfish nodes we have used the concept as described in [18]. In addition to that the sybil attack also has been considered when the same nodes have different identities. In this case, the proposed technique is followed by routing optimality, checking optimality, and forwarding optimality. A node is considered as selfish due to its resource constraint and may have multiple identities. The problem in such a case is that same node may take large amount of memory space. To work with sybil attack, following mechanism needs to be considered.

The sybil mechanism is important especially to identify the sybil nodes which are particularly under harmfull attack. A sybil mechanism is defined as $s_y = (R_t, R_{eg}, P_v)$ where R_t denotes the radio resource testing, R_{eg} denotes the registration, P_v indicates the position verification. These three parameters are used in sybil mechanism to prevent the sybil attack. In case of R_t, we have taken two assumptions. Firstly, only one radio is used for any physical device and there is no radio that simultaneously can send or receive on more than one channel. To send the data packets, a sybil mechanism helps a node to ensure that there are no neighboring nodes that have sybil identities. A node can choose a channel randomly. If the neighboring nodes that were assigned that channel is legitimate, it should hear the message. Let us consider that z of the identifier's n neighbors that are actually sybil nodes. So, the probability to detect the sybil nodes is $\frac{z}{n}$ and the probability of not detecting the sybil nodes is $(\frac{n-z}{n})^r$ where r indicates that test is repeated for r rounds. Assume that out of N neighbors, a node can test n neighbors at a time. At that time among these n neighbors, let q be the sybil nodes out of Q, w be the correct nodes out of W, and m be the number of malicious nodes out of M. Now, the probability to detect the sybil nodes denoted as P_{detect} and is computed by using Eq. (1) [19]:

$$P_{detect} = \sum_{allq,w,m} \frac{Q(q)W(w)M(m)}{N(n)} \frac{q - (M - m)}{n} \tag{1}$$

Now, if this test is repeated for r rounds then Eq. (1) can be modified as:

$$P_{detect} = 1 - \left(1 - \sum_{allq,w,m} = \frac{Q(q)W(w)M(m)}{N(n)} \frac{q - (M - m)}{n} \right) \tag{2}$$

R_{eg} be the next tuple where sybil mechanism prevents the sybil attack. In case of wireless networks, there may be one central authority to manage the whole network. The whole network must belief the central authority. The central authority disseminates the information to the whole network because it has the knowledge about the deployed nodes. To detect sybil attacks, an entity could poll the network and compare the results to the known deployment and any node can check the identities to validate another node as legitimate. P_v determines the position verification, and we consider that the wireless network is immobile once it is deployed. Sybil nodes can be detected using this approach because they will appear to be at exactly the same position as the malicious node that generates them.

4 Cooperation-Optimality Analysis

In this section, we analyze the technique as routing optimal (R_o), checking optimal (C_o), and forwarding optimal (F_o). A routing protocol (R_p) can be cooperation optimal if it satisfies these three optimality conditions. If $R_p \in R_o, C_o, F_o$ then only $R_p \in CO_o$. To show that $R_p \in R_o, C_o, F_o$, we have to prove $R_p \in R_o$, $R_p \in C_o$, and $R_p \in F_o$.

4.1 Routing Optimality

In this section, we have shown that the proposed technique is routing optimal.

Theorem 1 *Byzantine nodes cannot disrupt the routing stage of the protocol.*

Proof During the routing stage, Byzantine nodes have the following possible actions [20]:

- Byzantine nodes do not forward the *RREQ/RREP* messages.
- Byzantine nodes report the *link costs* which may be false.
- Byzantine nodes may be harmfull in the forwarding stage but may act honestly in the routing stage.

Between the source node s and the destination node d, the *RREQ/RREP* messages must be delivered successfully. Byzantine nodes do not stop the paths which have been computed by the protocol. Therefore, in all of the three cases there is no chance that Byzantine nodes can disrupt the routing stage of the protocol.

Theorem 2 *The proposed technique is routing optimal.*

Proof As Theorem 1 states that Byzantine nodes cannot disrupt the routing stage of the technique and according to Su et al. [20], generalized second price (GSP) satisfies Nash equilibrium and their actions maximize their utilities. Therefore, proposed technique is routing optimal.

4.2 Checking Optimality

In this section, it has been shown that the proposed technique is checking optimal.

Theorem 3 *The sybil nodes can not disrupt the correct nodes as they are detected in the checking stage.*

Proof During the packet delivery, we consider that sybil nodes have the following possible actions:

- Direct communication and indirect communication.
- Fabricated identities and stolen identities.

When a legitimate node sends its radio packet to a sybil node, one of the malicious device listens to that packet. Packets sent from a sybil node which are actually sent from one of the malicious devices indicates as direct communication. Whereas in case of indirect communication, a legitimate node does not communicate directly with the sybil nodes. There are two ways that the sybil node can get their identities: one is to fabricate a new identity and another is that it can steal an identity from a legitimate node. During the checking stage, sybil mechanism works with two conditions as mentioned in R_t, R_{eg}, P_v. So, there is no chance that sybil nodes disrupt the correct nodes.

4.3 Forwarding Optimality

Su et al. proved that Byzantine nodes cannot disrupt the forwarding protocol and they also proved that their proposed protocol that unifies GSP [21] as well as FORBID mechanism is forwarding optimal [22]. The proposed technique also follows these protocols. So, the proposed technique is also forwarding optimal.

5 Results and Analysis

Simulation has been done using network simulator (NS2) for both proposed and existing approach to evaluate the performance of the network. The parameters used in our approach has been given in Table 1. Each node selects another node as destination randomly among the network and starts moving while simulation starts.

The result of data accessibility is shown in Fig. 2a. From Fig. 2a, it is observed that when the percentage of selfish nodes increases, data accessibility decreases in the network.

But, it has been seen that after detection of selfish nodes, data accessibility increase as compared to the network scenario without selfish node detection because it avoids selfish node in the path of allocating the replica. Figure 2b shows the communication cost with increase in percentage of selfish nodes. Figure 2b indicates that communication cost is increasing. This is because even if it is present nearer to the allocating node, the data is not allocated to the selfish node. The cost of communication is defined in terms of hop counts to send or receive the data. Figure 3a shows that the average delay between the requesting nodes to the serving node has increased after the detection of selfish node. This is due to the fact that data item is not served or allocated to selfish node even if it is present closer to the serving node. Hence, it

	Simulation parameter	Type/Values
Table 1 Parameters used for proposed approach	Number of nodes	Random
	Percentage of selfish nodes (%)	(0–100)
	Number of Memory	3
	Topology	Random
	Mac type	802.11
	Queue type	Drop tail/Priority
	Antenna type	Queue
	Routing protocol	AODV
	Simulation time	100 s
	Size of the network	100 * 100

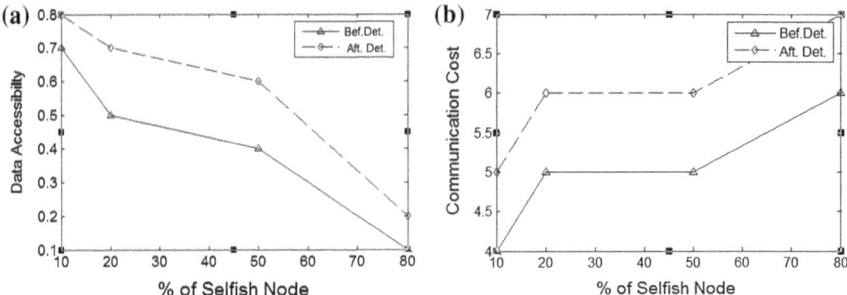

Fig. 2 **a** Data accessibility versus % of selfish node. **b** Communication cost versus % of selfish nodes

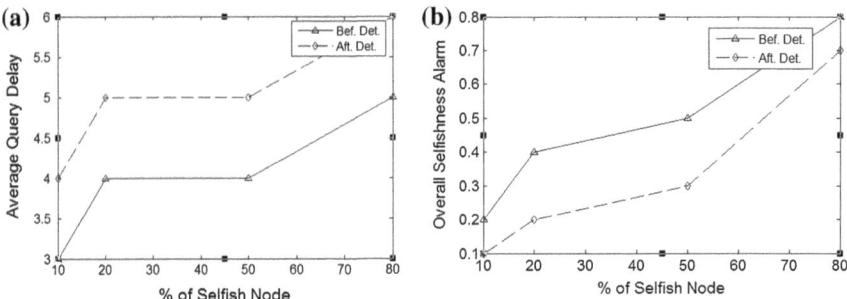

Fig. 3 **a** Average query delay versus % of selfish nodes. **b** Overall selfishness alarm versus % of selfish nodes

will take more time to deliver the data item to the destination. With increase in the number of selfish nodes, overall selfish alarm is increased but after detection overall selfishness alarm is also reduced which excludes the selfish node and is shown in Fig. 3b.

6   Conclusion

In this work, authors emphasize on the network issues related to the selfish property of the nodes and to detect them in the presence of sybil attack. The simulation results indicate higher performance in terms of data accessibility which is an important aspect of the network, since providing data to the end users effectively is the ultimate aim of this work. In future, we plan to reduce the risk factor of the identified selfish nodes which may improve the performance of the network, and we would also like to consider other parameters such as packet delivery ratio, jitter, and throughput in order to compute the performance of network.

References

1. Panigrahi, C.R., Pati, B., Tiwary, M., Sarkar, J.L.: EEOA: improving energy efficiency of mobile cloudlets using efficient offloading approach. In: IEEE International Conference on Advanced Networks and Telecommunications Systems, pp. 1–6 (2015)
2. Pati, B., Sarkar, J.L., Panigrahi, C.R., Tiwary, M.: ECHSA: an energy-efficient cluster-head selection algorithm in wireless sensor networks. In: Proceedings of 3rd International Conference on Mining Intelligence and Knowledge Exploration, pp. 184–193 (2015)
3. Sarkar, J.L., Panigrah, C.R., Pati, B., Das, H.: A novel approach for real-time data management in wireless sensor networks. In: Proceedings of 3rd International Conference on Advanced Computing, Networking and Informatics, vol. 2, pp. 599–607 (2015)
4. Panigrahi, C.R., Sarkar, J.L., Pati, B., Das, H.: S2S: a novel approach for source to sink node communication in wireless sensor networks. In: Proceedings of 3rd International Conference on Mining Intelligence and Knowledge Exploration, pp. 406–414 (2015)
5. Pati, B., Sarkar, J.L., Panigrahi, C.R., Verma, R.K.: CQS: a conflict-free query scheduling approach in wireless sensor networks. In: Proceedings of 3rd International Conference on Recent Advances in Information Technology, pp. 13–18 (2016)
6. Mishra, M., Panigrahi, C.R., Sarkar, J.L., Pati, B.: GECSA: a game theory based energy-efficient cluster-head selection approach in wireless sensor networks. In: International Conference on Man and Machine Interfacing, pp. 1–5 (2015)
7. Mishra, M., Panigrahi, C.R., Pati, B., Sarkar, J.L.: ECHS: an energy aware cluster head selection algorithm in wireless sensor networks. In: Proceedings of International Conference on Man and Machine Interfacing, pp. 1–4 (2015)
8. Pati, B., Sarkar, J.L., Panigrahi, C.R., Tiwary, M.: ARTQS: an advanced real-time query scheduling approach in wireless sensor networks, pp. 219–224 (2015)
9. Choi, J.-H., Shim, K.-H., Lee, S., Wu, K.-L.: Handling selfishness in replica allocation over a mobile adhoc network. IEEE Trans. Mobile Comput. 11(2) (2012)
10. Hara, T.: Effective replica allocation in ad hoc networks for improving data accessibility. In: Proceedings IEEE INFOCOM, pp. 1568–1576 (2001)
11. Hara, T., Madria, S.K.: Data Replication for improving data accessibility in ad hoc networks. IEEE Trans. Mobile Comput. 5(11), 1515–1532 (2006)
12. Adar, E., Huberman, B.A.: Free riding on Gnutella. First Mon. 5(10), 1–22 (2000)
13. Balakrishnan, K., Deng, J., Varshney, P.K.: TWOACK: preventing selfishness in mobile ad hoc networks. In: Proceedings of IEEE Wireless Communication and Networking Conference, pp. 2137–2142 (2005)
14. Miranda, H., Rodrigues, L.: Friends and foes: preventing selfishness in open mobile ad hoc networks. In: Proceedings of IEEE International Conference on Distributed Computing Systems Workshops, pp. 440–445 (2003)
15. Buttyan, L., Hubaux, J.P.: Nuglets: a virtual currency to stimulate cooperation in self-organized mobile ad hoc networks. Technical Report No. DSC/2001/001. Swiss Federal Institute of Technology (EPFL) (2001)
16. Chen, K., Nahrstedt, K.: iPass: an incentive compatible auction scheme the enable packet forwarding service in MANET. In: Proceedings of the 24th International Conference on Distributed Computing Systems, pp. 534–542 (2004)
17. Buchegger, S., Boudec, J.-Y.L.: Performance analysis of the CONFIDANT protocol (cooperation of nodes: fairness in dynamic ad-hoc networks). In: Proceedings of the 3rd ACM International Symposium on Mobile Ad Hoc Networking and Computing, pp. 226–236 (2002)
18. Choi, J.-H., Shim, K.-H., Lee, S., Wu, K.-L.: Handling selfishness in replica allocation over a mobile ad hoc network. IEEE Trans. Mobile Comput. 11(2), 178–291 (2012)
19. James Newsome, J., Shi, E., Song, D., Perrig, A.: The Sybil attack in sensor networks: analysis and defenses. In: 3rd International Symposium on Information Processing in Sensor Networks, pp. 259–268 (2004)
20. Su, X., Peng, G., Chan, S.: Multi-path routing and forwarding in non-cooperative wireless networks. IEEE Trans. Parallel Distrib. Syst. 25(10), 2638–2647 (2014)

21. Su, X., Chan, S., Peng, G.: Auction in multi-path multi-hop routing. IEEE Commun. Lett. **13**(2), 154–156 (2009)
22. Su, X., Chan, S., Peng, G.: Generalized second price auction in multi-path routing with selfish nodes. In: Proceedings IEEE GLOBECOM, pp. 3413–3418 (2009)

Part V
Cloud Computing
and Distributed Systems

Minimizing Ciphertext in Homomorphic Encryption Scheme for Cloud Data

Manish M. Potey, C. A. Dhote and Deepak H. Sharma

Abstract Data security is a major concern in cloud computing. It must satisfy the three goals of security in computing—integrity, confidentiality, and availability. Homomorphic encryption is a technique in which user or cloud service provider (CSP) can perform operations on cloud data without performing decryption. Many algorithms are available for homomorphic encryption. But these algorithms generate large size ciphertext. This paper focuses on homomorphic encryption which generates small size ciphertext. It is a variant of scheme proposed by Dijk et al. In an experimentation of this scheme, encrypted data are stored in DynamoDB of Amazon Web service (AWS) public cloud. When user requires data, it can be downloaded on users machine and then decrypted.

Keywords Data security · Cloud computing · Homomorphic encryption
AWS · DynamoDB · CSP · Security goals

1 Introduction

In cloud computing, it is necessary to address security at all levels like application, host, and network. The information available at all these levels must ensure confidentiality, integrity, and availability. The data have several states from its creation to it destruction. These states include create, transfer, use, share, store, archive, and

M. M. Potey (✉) · D. H. Sharma
Department of Computer Engineering, K.J. Somaiya College of Engineering,
Mumbai, India
e-mail: manishpotey@somaiya.edu

D. H. Sharma
e-mail: deepaksharma@somaiya.edu

C. A. Dhote
Department of Information Technology, PRMIT&R, Amravati, India
e-mail: vikasdhote@rediffmail.com

© Springer Nature Singapore Pte Ltd. 2018
K. Saeed et al. (eds.), *Progress in Advanced Computing and Intelligent Engineering*,
Advances in Intelligent Systems and Computing 563,
https://doi.org/10.1007/978-981-10-6872-0_56

destroy. This paper focuses on security of cloud data at rest. Data security in cloud computing must also cover security of data in transit and following [1] aspects.

Data Lineage—In cloud computing, the data in cloud are moving from one location to other. Following the path of data is called as data lineage.

Data remanence—It is an issue when data get exposed after deletion to the unauthorized party for an auditor's assurance.

Data provenance—Provenance means data must be computationally accurate, and it also possesses integrity. For ensuring confidentiality generally encryption is used.

Normally, to perform computation, it is required to decrypt the data. But homomorphic encryption (HE) allows computations on encrypted data. So the data remain in its encrypted state in most of the processing stages on the cloud. The results can be verified by decrypting ciphertext.

Homomorphic encryption technique allows user to perform multiple types of operations on encrypted data. Only one kind of operation is allowed in a partially homomorphic encryption technique. Figure 1 shows the proposed system.

This paper is arranged in six parts. Section 2 gives brief outline about related work carried out for homomorphic encryption algorithms. Section 3 discusses about existing scheme by Van Dijk, and Sect. 4 explains the proposed low size ciphertext homomorphic encryption scheme. The implementation of proposed work is described in Sect. 5; the conclusion of the paper is given in Sect. 6.

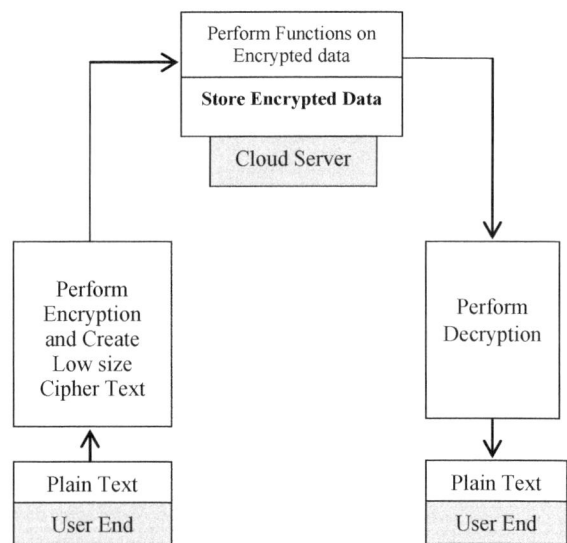

Fig. 1 Proposed homomorphic encryption scheme

2 Related Work

Initially, homomorphic encryption was suggested by Gentry [2, 3] in his paper and thesis. Public key encryption is used in this scheme. It consists of encryption, decryption, key-generation, and evaluation phases.

Rivest et al. [4] in 1978 proposed privacy homomorphism for bank data application. The hardware configuration is suggested and encryption functions perform operations without decrypting. But it has limited applicability.

Paillier has proposed partial homomorphic scheme. The author in [5] surveys various homomorphic encryption schemes. This paper introduces basics of encryptions to homomorphic encryption. He discussed various schemes with respect to parameters like security and efficiency.

The need of fully homomorphic encryption and other security issues in cloud computing is given by Aderemi et al. [6], Goldwasser, and Micali [7] proposed a new encryption model based on probability. In this new model, trapdoor is replaced by unapproximable predicate.

Improvement over Gentry's model is proposed in [8–11]. Y. Govinda Ramaiah [12] proposed efficient homomorphic algorithm over integer. It is also a variant of Van Dijk scheme.

Most of these researchers proposed a fully or partially homomorphic encryption scheme. The authors in [8, 10, 11] have proposed improved homomorphic scheme with comparatively low size ciphertext and small key. The scheme proposed by Gentry was not efficient [8, 9, 13] due to its large key size and high evaluation time. Most of the researchers used additional noise and bootstrapping while performing homomorphic encryption. It makes the ciphertext very large than its plaintext. Due to this space complexity increases. Whenever there is need to perform some advance operations like sorting, indexing, searching etc. on such encrypted data, the size of ciphertext plays a major role. So there is need to provide homomorphic encryption scheme which generates low size ciphertext which can be used for practical application. Such scheme is proposed in this paper.

3 Existing Scheme

The FHE scheme is proposed in [8] uses various parameters. These parameters are
- y—Size of public key in bits
- x—Size of secret key in bits
- w—Noise in bits
- v—Number of integers in public key.

Security parameter of polynomial is s. The noise w is considered as $\omega(\log s)$. The secret key x should be greater than $w \cdot \Theta(\log^2 s)$. To protect against lattice-based attacks v is considered as $\omega(x^2 \log s)$. v is taken as $y + \omega(\log s)$. It also uses additional noise $w' = w + \omega(\log s)$. It is considered as f. For better homomorphism, parameters are suggested as $w = s$, $w' = 2s$, $x = \tilde{O}(s^2)$, $y = \tilde{O}(s^5)$, and $v = y + s$. This homomorphic scheme contains of following algorithms.

Generate_Key (s): Select secret key KS with odd y bit integer from open interval $[2^{x-1}, 2^x)$. For $i = 0, 1, ..., v$. Select random number T_i from the interval $[0, 2^y/KS)$ and another number U_i from interval $(-2^w, 2^w)$ and calculate $N_i = KS \cdot T_i + U_i$. The public key is $(T_0, T_1, ..., T_y)$, and secret key is KS.

Encryption (PKey, Msg $\in \{0, 1\}$):
Select a number D from $(-2^w, 2^w)$ for adding noise. Select a subset $I \subseteq \{1, 2,..., v\}$. Calculate sum $= \Sigma_{i \in I} \cdot T_i$ and Ciphertext $CT = (Msg + 2(I + Sum)) \bmod T_0$.

Decrypt (KS, CT): Calculate Message $= (CT \bmod KS) \bmod 2$.
In order to get enough homomorphism with this set of parameters, the complexity is given as $\tilde{O}(s^{12})$, key generation complexity is $\tilde{O}(s^{12})$, encrypt and decrypt complexity $\tilde{O}(s^{10})$.

4 Proposed Homomorphic Encryption Scheme

The following scheme is used for low size ciphertext homomorphic encryption. This scheme consists of Key generation, encryption, and decryption phases. The parameters J and K form a secret Key. The parameters $P0$ and $P1$ form a public key. The security parameter is considered as 's'. Number N to be encrypted is accepted from user. The primitives used in this scheme are given as follows.

Key Generation (s):

1. Select secret key J with 64 or 128 bit and K-32 or 16 bit prime number

2. Select D and F as 128-bit prime number

3. Choose s bit random integer K'

4. Generate public key $P0 = J * D$ and $P1 = J * F + K * K'$.

Encrypt (N, P0, P1):

1. Select $T1$, $T2$ as s bit random numbers

2. Compute $P2 = [T1 * P1] \bmod P0$

3. Output—Ciphertext $CX = [N + T2 * P2] \bmod P0$.

Decrypt (CX, J, K):

1. Output N = *(CX mod J) mod K.*

The homomorphic addition and multiplications are performed using following mathematical operations.

$$Addition\ (N1,\ N2)$$
$$Output\ Add = N1\ + N2$$

$$Multiplication\ (N1,\ N2)$$
$$Output\ Mult\ = N1 * N2$$

The outcome of above scheme is given below.

Consider *J = 8446413785904601499, K = 4100490077*, number to be encrypted *N = 27*.

The random value of *D* and *F* in 128 bit is considered as

$$D = 31187299836494291758136689800901245640 3$$
$$F = 226929990875395563252430422634127344009$$

Four-bit random number *K = 15* and compute *P0* and *P1*.

$$P0 = 26342083928410571016159281387795498164069434989268259480 97$$
$$P1 = 191674460336514651269710690522636708559108988807393742064 6$$

Encrypted value of input *N = 27* is

$$CX = 105234942881548267407508472536394102464407773987040423327$$

Decrypted value of *N* is *27*.

Proposed scheme complexity is $\tilde{O}(s^3)$, 's' as security parameter. In encryption, multiplication is main operation. The s number of bits is multiplied by s^2 number of bits then the bit level operation complexity comes out to be $\tilde{O}(s^3)$, Decryption complexity depends on ciphertext size. The size of ciphertext at bit level is $\tilde{O}(s^3)$. For performing integer operation, the complexity is comes out be $\tilde{O}(1)$. So the overall complexity is $\tilde{O}(s^3)$.

5   Implementation

This part explains about the working model of low ciphertext size homomorphic encryption. The Eclipse IDE for Java EE developers is used to connect to the Amazon Web Services (AWS) DynamoDB service. Here, one simple bank example is considered. On DynamoDB, two tables are created namely *Bank* and *Keys*. Bank table contains the *username, balance* (in homomorphic encrypted form), and *password*. Key table contains *username, P_key* (*J* parameter used in scheme), and *R_key* (*K* parameter used in scheme).

In this application, four options are provided namely—addition, subtraction, check balance, and exit. By using Eclipse, user can execute this application and logged into users account and can perform addition and subtraction operation on encrypted balance amount field. Once all tasks are performed user can logout using exit option.

The following steps [14] are performed for the implementation.

Step 1: Creation of DynamoDB instance on Amazon Web Services.
Step 2: Database tables are created with proper schema.
Step 3: Get the credentials from AWS and perform access controls.
Step 4: Install Java SDK and Eclipse Kepler version at user end.

After the installation of AWS SDK on Eclipse framework, the user is available with all the needed packages.

Step 5: Follow the steps as given in AWS SDK.
Step 6: Developing the code in Java for this homomorphic encryption scheme.
Step 7: Finally execute the code using Eclipse.

5.1   Results

The experimentation is carried out with a plaintext as 111. The experimental results of this scheme by varying size of parameters are provided in the following Table 1.

In Table 1, the proposed scheme generates ciphertext of size 58–77 bytes. This scheme is not inserting additional noise. This scheme is similar to the scheme in [8].

Table 1 Experimental results

Parameters values in bits used in the algorithm				Size of ciphertext
J	K	D	F	
64	16	128	128	58 bytes
128	16	128	128	77 bytes
128	32	128	128	77 bytes

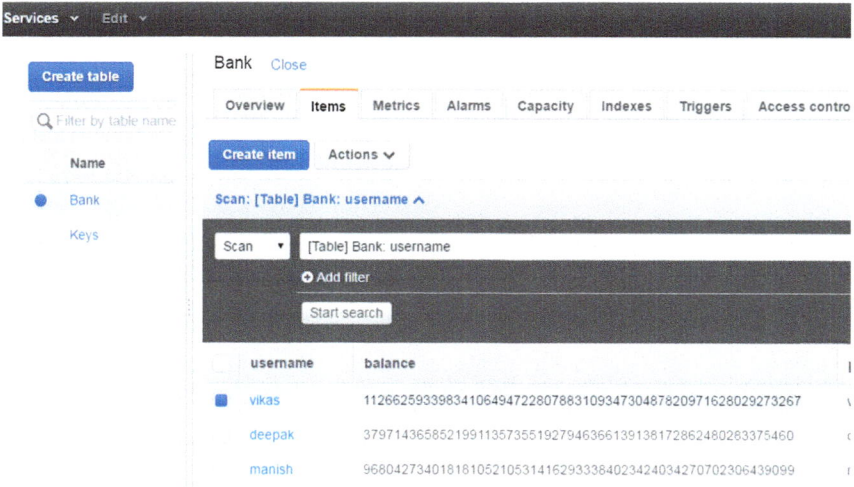

Fig. 2 Bank table on DynamoDB

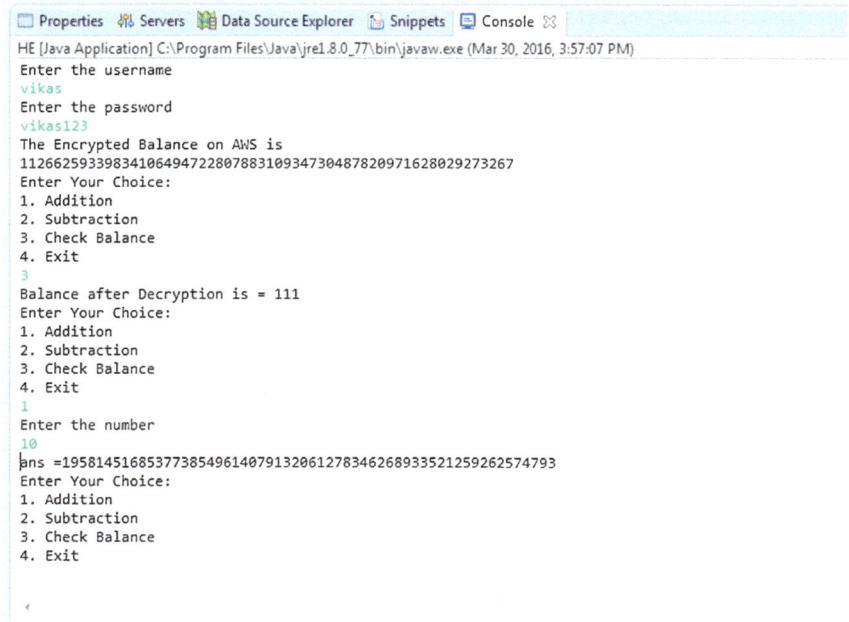

Fig. 3 Execution at client

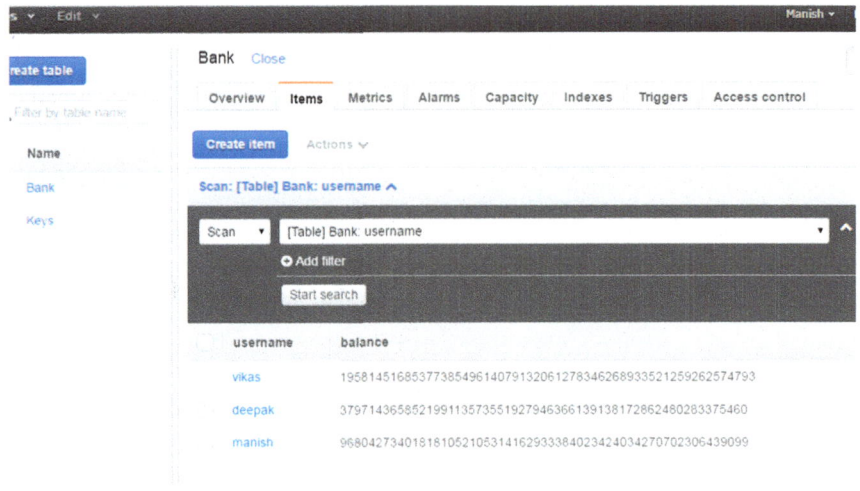

Fig. 4 Updated table on AWS after performing operations

5.2 Sample Execution Results

User is allowed to check balance in plaintext. The data on DynamoDB database on AWS are shown in Fig. 2. Here, the balance of particular *username* is checked.

The application is executed at client by giving particular *username's* credentials and performed sample operations. It is shown in Fig. 3.

Once code is executed at client, the data on AWS DynamoDB table are updated. As shown in Fig. 4, balance of first user (*vikas*) is updated.

6   Conclusion and Future Work

At no stage, the data are exposed in plaintext using homomorphic encryption. It ensures confidentiality of the data. The ciphertext created through this approach using homomorphic encryption is small in size as compared to Van Dijk scheme. It is an efficient and practically applicable homomorphic encryption suitable for application on cloud. The message expansion is low. The overall complexity of this scheme is $\tilde{O}(s^3)$.

There is need of security analysis in such low size ciphertext algorithm. There is also a need to evolve various algorithms for searching and querying operations on encrypted data under this scheme.

References

1. Kumaraswamy, S., Mather, T., Latif, S.: Cloud security and privacy: an enterprise perspective on risks and compliance. O'Reilly Media, Inc. (2009)
2. Gentry, C., Halevi, S.: Implementing gentry's fully-homomorphic encryption scheme. In: Annual International Conference on the Theory and Applications of Cryptographic Techniques. Springer, Berlin, Heidelberg (2011)
3. Gentry, C.: A fully homomorphic encryption scheme. Doctoral Dissertation. Stanford University (2009)
4. Rivest, R.L., Adleman, L., Dertouzos, M.L.: On data banks and privacy homomorphisms. Found. Secure Comput. **4** (1978)
5. Fontaine, C., Galand, F.: A survey of homomorphic encryption for nonspecialists. EURASIP J. Inf. Security 2007 (2009)
6. Atayero, A.A., Feyisetan, O.: Security issues in cloud computing: the potentials of homomorphic encryption. J. Em. Trends Comput. Inf. Sci. **2**(10), 546–552 (2011)
7. Goldwasser, S., Micali, S.: Probabilistic encryption. J. Comput. Syst. Sci. **28**(2), 270–299 (1984)
8. Van Dijk, M., Vaikuntanathan, V., Halevi, S., Gentry, C.: Fully homomorphic encryption over the integers. In: Annual International Conference on the Theory and Applications of Cryptographic Techniques. Springer, Berlin, Heidelberg (2010)
9. Smart, N.P., Vercauteren, F.: Fully homomorphic encryption with relatively small key and cipher text sizes. In: International Workshop on Public Key Cryptography. Springer, Berlin, Heidelberg (2010, May)
10. Stehlé, D., Steinfeld, R.: Faster fully homomorphic encryption. In: International Conference on the Theory and Application of Cryptology and Information Security. Springer, Berlin, Heidelberg (2010, December)
11. Coron, J.S., Mandal, A., Naccache, D., Tibouchi, M.: Fully homomorphic encryption over the integers with shorter public keys. In: Annual Cryptology Conference. Springer, Berlin, Heidelberg (2011, August)
12. Govinda, R.Y., Vijaya, K.: Efficient public key homomorphic encryption over integer plaintexts. In: International Conference on Information Security and Intelligence Control (ISIC), 2012. IEEE (2012)
13. Brakerski, Z., Vaikuntanathan, V.: Fully homomorphic encryption from ring-LWE and security for key dependent messages. In: Proceedings of the 31st Annual Conference on Advances in Cryptology, pp. 505–524. Springer (2011, August)
14. AWS Toolkit For Eclipse. http://docs.amazonaws.cn/en_us/AWSToolkitEclipse/latest/GettingStartedGuide/aws-tke-gsg.pdf?

Performance Analysis of Cloud Resource Provisioning Algorithms

Shilpa Kukreja and Surjeet Dalal

Abstract People have different opinion on Cloud Computing. Cloud Computing is changing the way of work by deploying the applications and files in the Cloud. People can easily access the applications and documents from anywhere in the world by just assembling the number of computers and servers via the Internet. So hosting the applications in the Cloud makes easy for the group members to cooperate. We can define Cloud Computing as the utility computing, which delivers the rental services on demand and scalable infrastructure over the Internet, and the virtualized datacenter that supports those services. Cloud Computing is the big change for the professionals and academia in the computing world. Cloud is different from the desktop computing since it provides online tools, guidance, and advancement to the customers. Cloud Computing does not bind the software to our personal computer rather it provides the efficient way to store them on the servers which can be accessed via the Internet. In case our PC fails to deliver the services, nothing to worry because you have your software in the Cloud. Cloud not only provides the above service, but also bears flexibility, better quality, long-term deployment, and many more.

Keywords Cloud Computing · Load balancing · Resource provisioning
Cost-effective · SLA

1 Introduction

Cloud Computing is the growing trend in the world of Information Technology. Cloud provisions the dynamically scalable infrastructure as a service through the Internet. We can use this service as that of electricity, water, gas, and telephone

S. Kukreja · S. Dalal (✉)
Department of CSE, SRM University, Sonipat, Haryana, India
e-mail: profsurjeetdalal@gmail.com

© Springer Nature Singapore Pte Ltd. 2018
K. Saeed et al. (eds.), *Progress in Advanced Computing and Intelligent Engineering*,
Advances in Intelligent Systems and Computing 563,
https://doi.org/10.1007/978-981-10-6872-0_57

without intention of where they are hosted [1, 2]. The customer has to pay as per the usage of the "renting" resources. This kind of computing is called utility computing.

In utility computing, the resources can be leased or released through Internet on demand basis. Users use the resource and pay as per the usage. This model is referred to as a Cloud. The Cloud infrastructure manages the platform and leases the resources. The Cloud pricing model is not fixed rather it is prepared according to the consumption of the resource [3]. That is the positive point, why Cloud is the famous technological trend in IT. Amazon, Google, and Microsoft are the top Cloud service providers in the world. Due to their amazing features, the customers are also earning the benefit of maximum resource utilization and minimized consumption cost.

Cloud Computing is an integration of grid computing and cluster computing. When the servers in the datacenters at the different geographical locations are connected in a single system through the main server, it is called grid. Each datacenter has a number of servers which is called cluster [4]. Cloud is basically a datacenter which has interconnected servers. The companies who own worldwide business, they need this kind of infrastructure, e.g., Toyota, Walmart, Citibank, and Tata-group. These companies have their own Cloud. Their network is of 100–1000 servers connecting to the different locations. The infrastructure of these companies is handled by IBM, Dell, Intel, Amazon, Salesforce.com, Sun, and HP [5].

Summarizing the major characteristics why Cloud is famous are:

- No extra software is required. Users access the services like using the resources, storing data, and many others with the help of Internet without knowing the location.
- The Cloud billing system is based on the general utility. Consumption of the resource decides the payment.
- Security is an important concern for data confidentiality. This is the task of service provider to retain the security level. If the user can pay some more money for securing their data, the provider can devote the resource with higher security [4].

1.1 Cloud Resource Management

The aim of Cloud Computing is to gain the effective management of the resources on a large scale in a cost-effective manner. Cloud provider has to focus on the allocation and optimal resources utilization because if any failure occurs, it will affect the system performance [6]. So resource management is a most challenging job for the Cloud provider. By having the successful Cloud resource administration, the client can exploit the Cloud administrations, and the punishment cost for the supplier is decreased.

2 Resource Provisioning Issue in Cloud Computing

Resource provisioning can be achieved by providing the availability of resources to the customers from anywhere in the world. An enterprise can earn the maximum profit by leveraging the availability of resources based on the service-level agreement (SLA) [7, 8]. This important feature benefits to the organizations and enterprises in the form of cost-effectiveness, scalability, performance, and operating efficiency. All these goals require the effective resource provisioning. Cloud Computing offers two plans for provisioning the resources. These are:

2.1 Advance Reservation Plan

Reservation plan starts with the signed contract between the Cloud provider and the customer. In this contract, resources can be reserved before utilization but the payment for the reserved resources has to be done early in the contract. Once the resources are reserved in advance, the customer cannot acquire more resources later on and may not completely meet the customer's requirement. This is a disadvantage of this scheme. But this plan is advantageous in terms of cost because the service provided by this plan is cheaper. Moreover, the best of reservation plan is hard to achieve due to the unpredictable customer's future resource requirement.

2.2 On-Demand Plan

According to this plan, the provider dynamically provisions the resources to the customer when needed. There is no need for the advance reservation of the resources. Cost of this scheme is charged as per the resource usage. How much of the resource is used depends on the customer and charged on pay-per-use basis. This is the best mechanism to provide the required number of resources at the right time. Due to the high resource availability, on-demand plan improves the performance of the system, customer satisfaction, and reduces the SLA violations meeting the QoS objectives. Since the plan offers on-demand provisioning of resources that meets the actual demand of the customer, the customer has to pay high due to the expensive service of the plan.

3 Prior Work

Different authors offered various provisioning algorithms and considered a number of parameters as the objective such as cost, response time, execution time, service time, time-limit, and customer satisfaction. As system performance depends upon the mentioned parameters, we categorize previous resource provisioning algorithms into the following categories:

3.1   Cost-Based Resource Provisioning Techniques

Henzinger et al. [9] proposed a flexible Provisioning of Resources in Cloud Environment (FlexPRICE) in which the Cloud provider and the user developed a balanced relationship through which they could work together. FlexPRICE was implemented through PRICES, the pricing model, that computed the cost on the basis of computational cost, data transfer cost, setup cost, and time discount factor. The time discount factor allowed discounting the delayed execution of the job.

Robust Cloud Resource Provisioning (RCRP) algorithm has been proposed by Chaisiri et al. [10] that minimized the overprovisioning as well as underprovisioning cost which leads to the consumer's benefit. It is observed that RCRP algorithm has provided the optimal solution which is near to the minimum total provisioning cost. The formulated solution has not only met the constraints but also reduced the oversubscribed and on-demand provisioning cost.

Wang et al. [11] have addressed the problem of service availability to support the long-term applications on minimum lease cost in SpotCloud. On the basis of the results, it has been examined that the algorithm truly promised the performance gain in service availability at minimum cost.

In this paper, Chaisiri et al. [7] have provisioned the resources by the proposed optimal virtual machine placement algorithm. Optimal Stochastic Integer Programming (SIP) has been used by the authors to implement OVMP algorithm to rent the resources to the customers. From the result, it has been found that using OVMP algorithm, the resource provisioning cost has been minimized.

Wang et al. [12] have proposed SpotCloud, a real working system that enabled the customers to utilize their local resources by integrating them into the Cloud. They also have suggested the distributed market design to the sellers to have better knowledge of their resource.

3.2   SLA-Based Resource Provisioning Algorithms

Nguyen Van et al. [13] introduced the autonomic resource manager that controlled the virtual environment in which provision of resources separated from the placement of virtual machines. Both VM provisioning and VM placement problems were formulated by constraint satisfaction. The aim of the resource manager was to optimize the utility function by considering the degree of SLA fulfillment and operating cost with regard to the customer satisfaction and performance objectives. Utility function and constraint programming approach both were combined to achieve self-optimization.

The authors Buyya et al. [14] aimed to minimize the provider's cost and SLA violations for dynamically sharable resources. All the goals such as dynamic change of customers, mapping customer's requests, and handling heterogeneity of virtual machine could be managed by the SaaS provider by using the proposed algorithm. The results showed that the ProfminVMminAvaiSpace algorithm has optimized the cost as compared to the other algorithms. The aim of the proposed work was to efficiently allocate the right amount of resources to fulfill the client's requests.

Elprince [15] proposed autonomic resource controller that effectively allocated the sufficient amount of resources needed for a given application without violating the service-level agreement. In case of SLA violation, large penalties had to be paid by the datacenter to the customer. The Cloud framework in has adopted the automated SLA negotiation mechanism and WLARA algorithm that has properly allocated the resources for service provisioning to guarantee SLA fulfillment. Both workload and geographical location directly affected SLA parameters.

Versteeg have offered SLA-based provisioning algorithms that minimized the number of resources and improved customer satisfaction level (CSL) by minimizing the SLA violations. Both of the algorithms have performed well in terms of quality of service which was higher than the level specified in SLA.

3.3 Deadline-Based Resource Provisioning Algorithms

The authors Le et al. in [16] have designed resource provisioning model in which changes in workload adapted by contemplating the queue model and average interval time. Three job scheduling algorithms were adopted: Shortest-job-first (SJF), Nearest Deadline First (NDF), and First-come-first-serve (FCFS). The results showed that the provisioning model provided elasticity in provisioning the resources for the dynamic workloads, and FCFS came out with better performance than the other algorithms, and SJF worked fast for the shorter deadline than the others.

Calheiros et al. [17] have considered the whole organization's workload and proposed dynamic provisioning and scheduling architecture that enabled the applications to complete within the deadline in a cost-effective way. The accounting mechanism has been introduced to know the cost measured by the use of public Cloud resources assigned to the users. The results showed that the approach made an efficient use of the public Cloud resources and enabled deadlines to be met at reduced cost.

The resource provisioning approach in minimized the cost of VM for executing mapreduce applications while keeping the deadline as objective. Hwang et al. have proposed two provisioning algorithms: One was deadline-based, and other was for minimizing the cost of resource usage. For *DTP-cost* algorithm, a single virtual machine performed multiple jobs. So the number of virtual machines required was less, thereby reducing the cost as compared to LFF-cost algorithm.

The work in presented mapreduce Cloud service model, Cura, that automatically created the configuration cluster for the mapreduce jobs and also provided deadline-aware mechanism that reduced the cost on delay of execution of the job. Cura also introduced a VM allocation technique that ensured the fast response time guaranteed for short-time jobs while using few number of servers as compared to other techniques. The experimental results showed that Cura effectively managed the deadline-awareness to achieve better resource management and resulted in a 80% cost reduction.

Buyya have addressed the case of scientific applications where the applications required less data transfer and also the size of these applications was less than 1 MB. The authors have presented Aneka's features shown in the experiment demonstrated that the resources were efficiently allocated from different sources to meet the deadline while reducing the application execution time.

Zhu have presented the framework supporting the dynamic adaptation for applications in Cloud environment in which applications needed to maximize QoS metric having the deadline and budget constraints. The proposed dynamic resource provisioning algorithm was based on feedback-control policy. The algorithm was effective and met the deadline performing the parameter adaptation.

3.4 Profit-Based Resource Provisioning Algorithms

The proposed optimal resource provisioning algorithm in [18] is divided into two levels: The first one was application-level which was implemented on SaaS provider and SaaS users, and second was resource-level implemented on SaaS provider and Cloud provider. SaaS provider met the QoS requirements such as budget and deadline of SaaS users.

Choi et al. [19] aimed to reduce the penalty cost while meeting the SLA constraint in order to maximize the profit. In this paper, SLA constraint was a deadline that met the customer satisfaction and reduced the SLA violation. As the deadline decreased, provider's profit increased. It means lower the deadline and SLA violations, higher the profit.

The authors in [20] have addressed the problem of maximizing the revenue through SLA by proper resource allocation and pricing schemes. Feng et al. have implemented resource allocation problem using queuing theory and considered the quality of service (QoS) parameters to find the optimal solution. This paper has given the optimal results that made the Cloud environment beneficial in facing the resource shortage.

The proposed policies in [8] helped the IaaS Cloud service providers to earn the profit when the provider was the member of Cloud federation. Federation also helped the underutilized providers to sell their unused resources to the other members of federation in order to take the revenue.

4 Conclusion

In the last section, we have summarized various resource provisioning algorithms and have examined that the only few parameters are considered by the above techniques which are not enough [21–31]. Most of the algorithms throw light on minimizing the cost and earning the benefit. Additionally, the algorithms have some disadvantages that make the negative impact on the performance of the technique and need improvement. The table with the cons is as follows (Table 1).

Table 1 Resource provisioning techniques

Year	Algorithm/technique	Parameters used	Disadvantages
2010	FlexPRICE	Time and cost	1. Fast computation requires high cost 2. User's overhead increased
2010	Robust Cloud resource provisioning	Customer' demand, price, resource availability	1. High computational complexity
2013	Optimal provisioning with dynamic resources	Minimized lease cost, service availability	1. Higher cost has to be paid by the buyers for more VMs
2009	Optimal virtual machine placement algorithm	Cost and number of virtual machines	1. SIP reserves large number of VMs in first stage which is close to the total VMs available 2. EVF cannot adapt to the changes in price
2015	SpotCloud: a real working system	Resource utilization, service availability	1. Delay of 27 min in connecting to the customer is very high 2. Increased migration cost
2009	Autonomic virtual resource management	SLA fulfillment, operating cost	1. The average solving time for VM packing phase is worst 2. Multiple objective conflicts with one another
2011	SLA-based profit maximization algorithm	Response time, service initiation time, penalty cost	1. ProfminVio algorithm is costly due to the large number of VMs initiated 2. The accounts occupy maximum available space
2013	Autonomous resource controller	Waiting time, response time, latency time	1. Implementation process is time-consuming 2. Dataset requires 2 years record which is hard to find
2013	SLA negotiation mechanism and workload	Price, time slot, response time	1. It cannot execute too many requests beyond datacenter's capacity

(continued)

Table 1 (continued)

Year	Algorithm/technique	Parameters used	Disadvantages
	& location-aware resource allocation		
2014	Customer-driven SLA-based resource provisioning algorithms	service initiation time, penalty rate, arrival rate	1. The first algorithm, BFResvResource, causes more SLA violations 2. Complexity of the algorithm increases with the increased number of VMs
2013	Deadline-driven resource management scheme	Submitted time, deadline, total resource demand	1. Longer processes have to wait in case of SJF 2. NDF algorithm slows down the job execution leading to more SLA violations
2012	Resource management system	Specific deadline, budget	1. Public Cloud executes the task which could not meet their deadline with in-house resources
2012	LFF-based resource provisioning and DTP-based resource provisioning	Minimized cost, time-limit	1. LFF algorithm has slow acceptance rate and incurs high cost
2012	Aneka's deadline-driven provisioning mechanism	Execution time, usage cost	1. Expected small delay in task processing
2011	Federated resource provisioning for IaaS provider	Elasticity, cost, profit, QoS, resource utilization	1. It is observed that FAPO policy has low VM utilization 2. Higher spot price of VM increases the chances of request rejection
2012	Optimal SLA-based resource allocation algorithms	Price, service time, available resources	1. Revenue decreases with the increased number of servers

We have analyzed from the above table that in the paper [20, 11], users have to pay high cost whether it is due to the number of virtual machines or due to the fast computation requirement. The same problem has been found in [14] where minimized cost is considered as the target. Yes, we agree with the point given in [11] that acquiring more virtual machines will ease the task and help faster computation, but there is another negative impact that the system will get more complex with the increased number of virtual machines.

References

1. Phatak, M., Kamalesh, V.N.: On cloud computing deployment architecture. In: IEEE International Conference on Advances in ICT for Emerging Regions, pp. 11–14. IEEE, Bangalore (2010)
2. Buyya, R., Vecchiola, C., Selvi, T.: Mastering Cloud Computing. McGraw Hill Education Pvt. Ltd. ISBN 978-1-25-902995-0 (2013)
3. Zhang, Q., Cheng, L., Boutaba, R.: Cloud computing: state-of-the art and research challenges. J. Internet Serv. Appl. **1**, 7–18 (2010)
4. Jadeja, Y., Modi, K.: Cloud Computing concepts, architecture and challenges. IEEE Conference on Computing, Electronics and Electrical Technologies, pp. 877–880. IEEE, Kumarc Oil (2012)
5. Rittinghouse, J.W., Ransome, J.F.: Cloud Computing: Implementation, Management and Security. CRC Press Taylor and Francis group. ISSN 978-1-4398-0680-7 (2010)
6. Nair, T.R.G., Vaidehi M.: Efficient resource arbitration and allocation strategies in cloud computing through virtualization. In: IEEE International Conference on Cloud Computing and Intelligence Systems (CCIS), pp. 397–401. IEEE, Beijing (2011)
7. Chaisiri, S., Lee, B.S., Niyato, D.: Optimal virtual machine placement across multiple cloud providers. In: IEEE Asia-Pacific Conference on Service Computing, pp. 103–110. IEEE, Singapore (2009)
8. Toosi, A.N., Calheiros, R.N., Thulasiram, R.K., Buyya, R.: Resource provisioning policies to increase IaaS provider's profit in federated cloud environment. In: 13th IEEE International Conference on High Performance Computing and Communications, pp. 279–287. IEEE, Banff (2011)
9. Henzinger, T.A., Singh, A.V., Singh, V., Wies, T., Zufferey, D.: FlexPRICE: flexible provisioning of resources in cloud environment. In: 3rd IEEE International Conference on Cloud Computing, pp. 83–90. IEEE, Miami (2010)
10. Hwang, U., Kim, K.H.: Minimizing cost of virtual machine for deadline-constrained mapreduce applications in cloud. In: 13th ACM/IEEE International Conference on Grid Computing, pp. 130–138. IEEE, Beijing (2012)
11. Palanisamy, B., Singh, A., Liu, L., Langston, B.: Cura: A cost-optimized model for mapreduce in cloud. In: 27th IEEE International Symposium on Parallel and Distributed Processing, pp. 1275–1286. IEEE, Boston (2013)
12. Wang, H., Wang, F., Liu, J., Wang, D., Groen, J.: Enabling customer-provided resources for cloud computing: potentials, challenges, and implementation. IEEE Trans. Parallel Distrib. Syst. **26**, 1874–1886 (2015)
13. Van, H.N., Tran, F.D., Menaud, J.M.: SLA-aware virtual resource management for cloud infrastructures. In: 9th IEEE International Conference on Computer and Information Technology, pp. 357–362. IEEE, Xiamen (2009)
14. Wu, L., Buyya, R., Garg, S.K.: SLA-based resource allocation for software as a service provider in cloud computing environments. 11th IEEE/ACM International Symposium on Cluster, Cloud and Grid Computing, pp. 195–204. IEEE, Los Angeles (2011)
15. Elprince, N.: Autonomous resource provision in virtual data centers. In: IFIP/IEEE International Symposium on Integrated Network Management, pp. 1365–1371. IEEE, Ghent (2013)
16. Le, G., Xu, K., Song, J.: Dynamic Resource Provisioning and Scheduling with Deadline Constraint in Elastic Cloud. In: IEEE International conference on Service Sciences (ICSS), pp. 113–117. IEEE, Shenzhen (2013)
17. Naeimi, H., Natarajan, S., Vaid, K., Kudva, P., Natu, M.: Cloud Atlas-unreliability through massive connectivity. In: 31st IEEE Test Symposium on VLSI, p. 1. IEEE, Berkeley (2013)
18. Vecchiola, C., Calheiros, R.N., Karunamoorthy, D., Buyya, R.: Deadline-driven provisioning of resources for scientific applications in hybrid clouds with Aneka. J. Future Gener. Comput. Syst. **28**, 58–65 (2012)

19. Zhu, Q., Aggarwal, G.: Resource provisioning with budget constraints for adaptive applications in cloud environment. In: IEEE Transactions on Services Computing, vol. 5, pp. 497–511. IEEE (2012)
20. Feng, G., Garg, S., Buyya, R., Li, W.: Revenue maximization using adaptive resource provisioning in cloud computing environments. In: 13th ACM/IEEE International Conference on Grid Computing, pp. 192–200. IEEE, Beijing (2012)
21. Wang, J., Varman, P., Xie, C.: Avoiding performance fluctuation in cloud storage. In: IEEE International Conference on High Performance Computing, pp. 1–9, IEEE, Dona Paula (2010)
22. Marinescu, D.C.: Cloud Computing: Theory and Practice. Elsevier. ISBN 978-0-12404-627-6 (2013)
23. Hadley, B., Hume, A., Lindberg, R., Obraczka, K.: Phantom of the cloud: towards improved cloud availability and dependability. In: 4th IEEE International Conference on Cloud Networking (CloudNet), pp. 14–19. IEEE, Niagara Falls (2015)
24. Kovalchuk, S.V., Smirnov, P.A., Maryin, S.V., Tchurov, T.N., Karbovskiy, V.: Deadline-driven resource management within urgent computing cyber infrastructure. In: International Conference on Computational Science, pp. 2203–2212. Elsevier, Russia (2013)
25. Calheiros, R.N., Buyya, R.: Cost-effective provisioning and scheduling of deadline-constrained Applications in Hybrid Cloud. In: 13th International Conference on Web Information System Engineering, pp. 171–184. Springer, Heidelberg (2012)
26. Chaisiri, S., Lee, B.S., Niyato, D.: Robust cloud resource provisioning for cloud computing environment. In: IEEE International Conference on Service-Oriented Computing and Applications (SOCA), pp. 1–8. IEEE, Perth (2010)
27. Wang, H., Wang, F., Liu, J., Xu, K., Wu, D., Lin, Q.: Resource provisioning on customer-provided clouds: optimization of service availability. In: IEEE International Conference on Communication (ICC), pp. 2954–2958. IEEE, Budapest (2013)
28. Li, C., Li, L.Y.: Optimal resource provisioning for cloud computing environment. J. Supercomput. 62, 989–1022 (2012)
29. Choi, Y., Lim, Y.: Resource management mechanism for SLA provisioning on cloud computing for IoT. In: IEEE International Conference on Information and Communication Technology Convergence, pp. 500–502. IEEE, Jeju (2015)
30. Son, S., Jung, G., Jun, S.C.: SLA-based cloud computing that facilitates resource allocation in the distributed data centers of a cloud provider. J. Supercomput. 64, 606–637 (2013)
31. Versteeg, S., Wu, L., Garg, S.K., Buyya, R.: SLA-based resource provisioning for hosted software-as-a-service applications in cloud computing environment. IEEE Trans. Serv. Comput. 7, 465–485 (2014)

Optimized Approach to Electing Coordinator Out of Multiple Election in a Ring Algorithm of Distributed System

Rajneesh Tanwar, Krishna Kanth Gupta, Sunil Kumar Chowdhary, Abhishek Srivastava and Michail Papoutsidakis

Abstract In a distributed system, multiple processes on multiple nodes can execute at same instance of time and may want to enter into the critical section. Several distributed algorithms require that there be a node with the coordinator responsibilities that perform some kind of coordination activities within the system. In election algorithm, two or more processes may identify coordinator had crashed simultaneously but actually this does not cause any issue to the election process except waste of network bandwidth. In this paper, I have proposed a solution for electing a coordinator in the distributed system simultaneously to help in reducing the network bandwidth by dropping out the extra network messages generated during election process.

Keywords Coordinator · Multiple · Election · Network · Ring

1 Introduction

A distributed system basically an accumulation of self-ruling PCs, associated through a system and dissemination middleware that run simultaneously, and coordinate their exercises through message disregarding the correspondence system.

R. Tanwar (✉) · K. K. Gupta · S. K. Chowdhary · A. Srivastava
Amity University, Noida, Uttar Pradesh, India
e-mail: rajneeshtanwar15@gmail.com

K. K. Gupta
e-mail: krishnaknth143@gmail.com

S. K. Chowdhary
e-mail: skchowdhary@amity.edu

A. Srivastava
e-mail: abhishek.sri13@gmail.com

M. Papoutsidakis
Piraeus University of Applied Sciences, Athens, Greece
URL: http://islab.teipir.gr

© Springer Nature Singapore Pte Ltd. 2018   603
K. Saeed et al. (eds.), *Progress in Advanced Computing and Intelligent Engineering*,
Advances in Intelligent Systems and Computing 563,
https://doi.org/10.1007/978-981-10-6872-0_58

The primary target of distributed system is, however, there are heterogeneous nodes in the system, and it makes a single system image or uniprocessor image to the client, through different transparencies. The correspondence between the procedures is accomplished by trading messages. The product of the distributed system is firmly coupled, and the procedures of the framework coordinate with each other. It is prudent to share the system assets among simultaneous running procedures; however, there are loads of assets in like manner, and common prohibition calculation like mutual exclusion algorithm is utilized to deal with asset distribution. On the off chance that they continue sitting tight for the allotment of basic assets, they may struck in the stop. For the same, we require a supervisor/organizer to arrange the exercises in the whole framework to sort out framework errands. If there should arise an occurrence of facilitator disappointment or accident, the hub who distinguished the non-accessibility of the organizer starts the election process. There are numerous calculations that are utilized for race of organizer, for example, bully, ring calculation, and some are adjusted calculations taking into account prior methodology.

Election algorithm is a procedure of choosing a process as coordinator among the conceivable nodes inside of the distributed system. This methodology in view of the idea of the priority number and hub/process having the most noteworthy priority number among the currently active processes can be chosen as a facilitator. At whatever point organizer neglects to react to process demand, process accepts that facilitator is down/slammed and begins the new race process.

In this paper, I modified the existing ring election algorithm and reduced the number of messages communicated during the election of coordinator in assumption that multiple nodes get to know about the crashing of coordinator and initiating their own election procedures.

Prerequisite:

1. All the processes or nodes in the ring are logically interconnected and assigned with unique priority number (UID) from 1 to N.
2. Each process or nodes know the structure of the ring.
3. No packet should be lost in between nodes, and delivery time will be variable and finite.

2 Existing Ring Election Algorithm

Election algorithm is used for selecting the coordinator process among N number of processes. The role of the coordinator is to maintain consistency within the system by synchronization of the activities performed within the system. In existing ring election approach, all the processes are logically arranged in unidirectional ring pattern and each process has a communication channel to the next process. Each process is assigned with the Unique Identification (UID) priority number and knows the structure of the ring. In case, any process is down then sender skips over that process and forwards message to the next successor until the active process is located.

The working of algorithm is as follows:

Step 1: When any process (say Pi) forwards a request message to current coordinator and does not receive a reply within prescribed time frame, then sender assumes coordinator is crashed or down.

Step 2: Now process Pi can initiate the election process by sending a message (contains UID) to its successor.

Step 3: On receiving the election message from process Pi, the successor appends his UID information in the message and forwards to his active successor.

Step 4: In the same manner election message processed by each process and finally message received by process Pi which contains details of all active processes with their UID's.

Step 5: Process Pi checks for processes having highest UID, elects process (suppose Pj) as a coordinator and circulates coordinator information message with other processes over the ring (Fig. 1).

In a normal scenario, existing algorithm takes $N - 1$ messages to fetch information of the existing nodes and more $N - 1$ messages to communicate the coordinator's

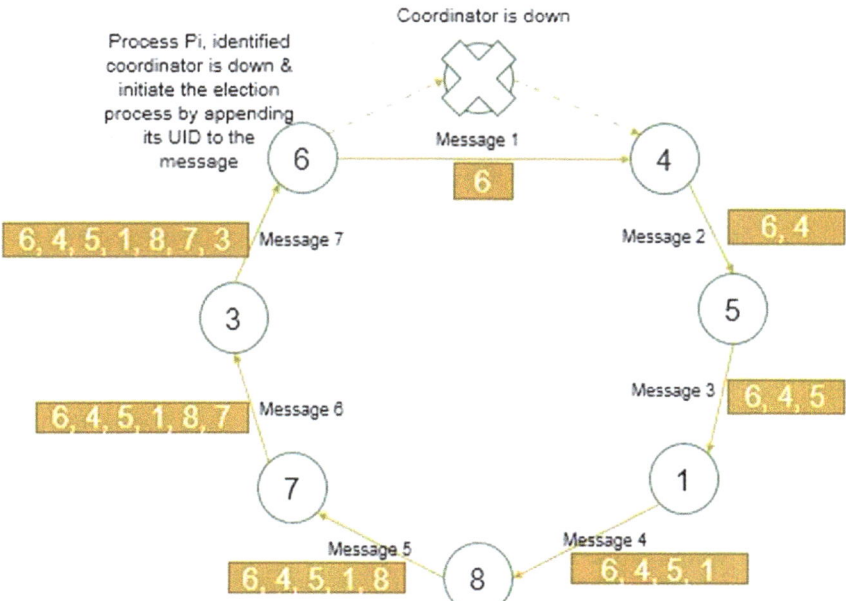

Fig. 1 Existing ring algorithm

election messages to all nodes. So, it consumes $(N − 1) + (N − 1) = 2(N − 1)$ number of messages for selection for coordinator among existing nodes when election process is initiated by the single process only.

In case, if failure of coordinator is identified by multiple processes and each process will initiate their election process for coordinator selection. As a result, total number of messages exchanged during the selection of the coordinator will increase exponentially. For example, suppose ring consists of N processes in the system and out of 15 one is the coordinator. So only $N − 1$ remaining processes can identify and initiate election process. Therefore, total number of messages exchanged will be $N − 1 * 2(N − 1) = (N − 1)2$ i.e., $0(n2)$ messages. Initiation of multiple elections by multiple processes just add-on to the complexity of the system and increasing the count of the number of unnecessary messages which can led to network congestion also. The result of all the election processes remains the same.

3   Proposed Ring Election Algorithm

3.1   Leader Election: In Case, if Single Process Initiate the Election Process

The proposed algorithm is as follows:

When a Process (Pi) sent some request to the coordinator and coordinator failed to reply to the request of Process Pi, then Pi assumes that coordinator is crashed or down. Now process Pi can initiate the election process.

Step 1:   When any process Pi (process 3 in Fig 2a) forwards a request message to current coordinator and does not receive a reply within prescribed time frame, then sender assumes coordinator is crashed or down.

Step 2:   Now Process Pi generates message in which process claims itself as a coordinator with his UID and forwards the message to its successor.

Step 3:   When successor process Pm (process 4 in Fig 2a) received the message, first it checks for the UID mentioned in the message and compares with own UID. If UID of the successor is less, then it accepts Pi as the coordinator and simply forwards the message from process Pi to its successor otherwise successor Pm drops the message from the Pi and generates the same message as generated by Pi but claims himself as a coordinator with his UID and forwards to its successor.

Step 4:   The same message keeps on propagating further in a ring until process with higher UID intercepts and drops that message or message received by the process that initiates that message itself.

Step 5:   Once election message received by initiator process say Pi (process 8 in Fig 2b), then Pi swore itself as a coordinator.

Step 6:   In case of initiation of the election process by multiple processes, then.

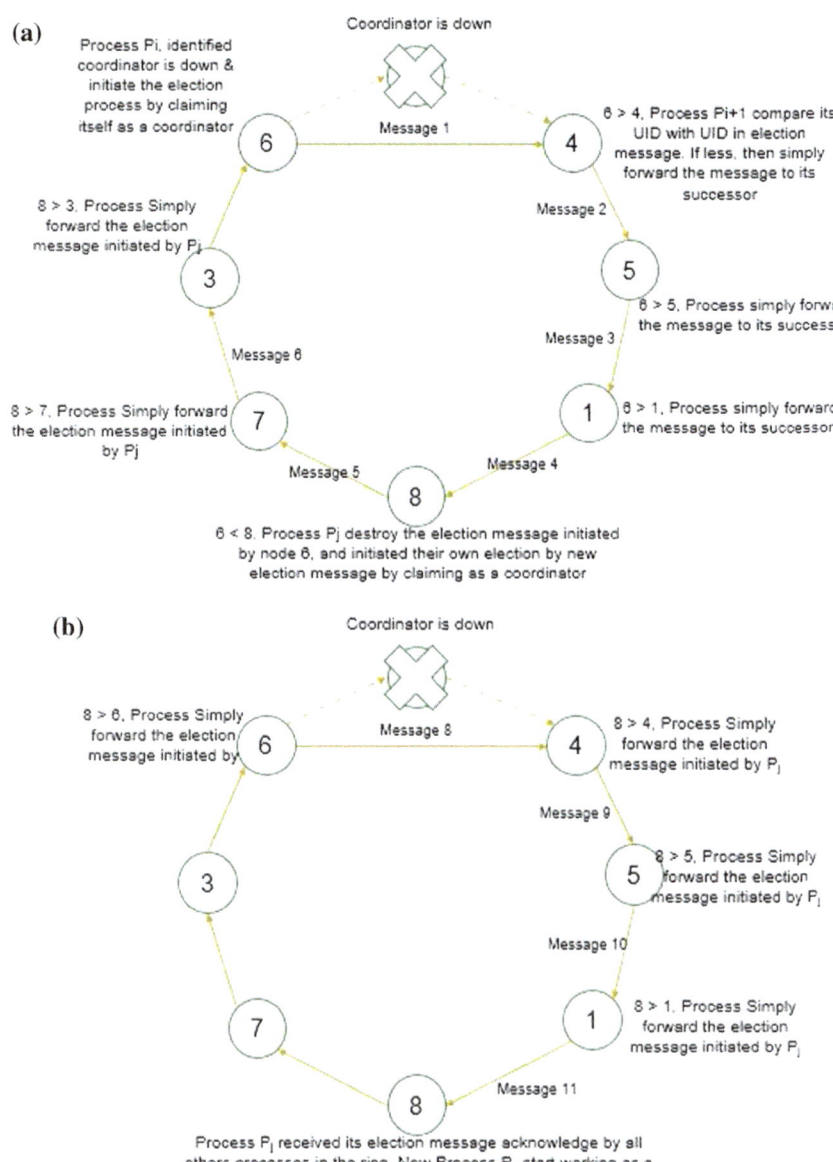

Fig. 2 **a** Proposed working of algorithm, **b** proposed working of algorithm

3.2 Leader Election: In Case, if Multiple Process Initiate the Election Process

The proposed algorithm is as follows:

When a Process (Pi) sent some request to the coordinator and if coordinator failed to reply to the request of Process Pi, then Pi assumes that coordinator is crashed or down. Now multiple process P1, P2 … Pn − 1 can initiate their own election process. As per the proposed algorithm, message of any election process only forwarded when the priority of the process is less than the priority of process mentioned in the message and request is not duplicated. If requested message is duplicated, then forwarding process will simply drop the request message or new message with different priority or UID.

For example: process 5, process 3, and process 8 get to know about the crashing of the coordinator and initiate the election process at same time and follow the same approach as per the proposed algorithm of leader election (refer below figure).

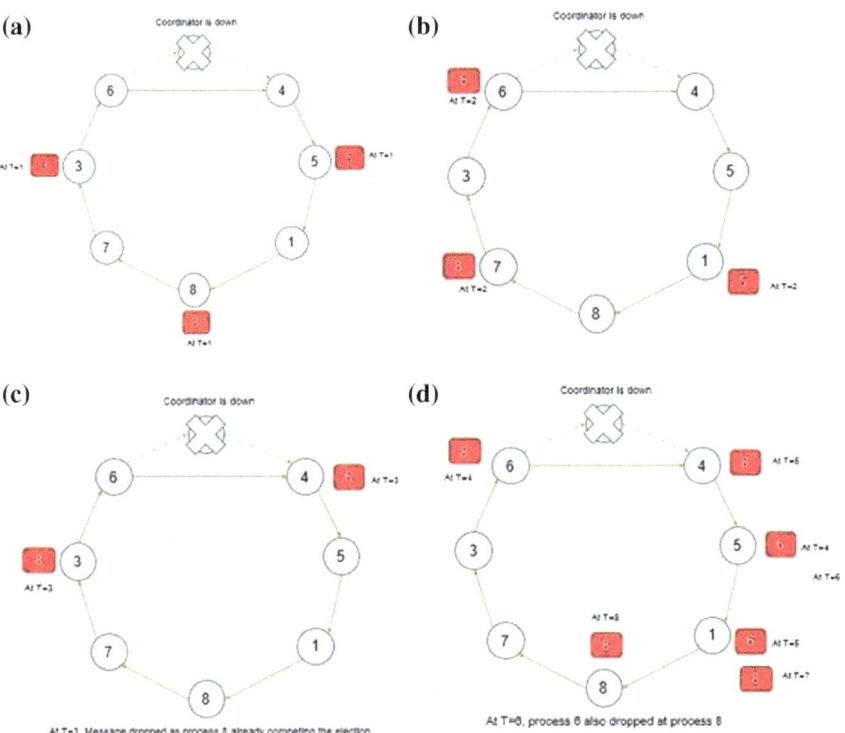

After initiating election by process 5, at T = 3 i.e., at process 8 the election message dropped by the process 8, as process 8 already itself involved with the election of the coordinator. Similarly, the election message of process 6 also gets dropped at process 6 but at the later stage, and only single process would complete the election process and elect the new coordinator. So, as per the process of dropping the message by the other processes would result in reducing the number of network packets that circulate around the network for electing a coordinator and improve the network efficiency by reducing the network congestion.

4 Performance Analysis

Best case:

Existing Ring Algorithm: If N is the number of the processes in the ring, the number of messages required for election of coordinator by a single process is 2 $(N - 1)$ i.e., for example: for 8 processes election cost $2 * (8 - 1) = 14$ messages.

As per the mentioned example above, if the same process is initiated by multiple processes (suppose m), then that count becomes $m * 2 * (N - 1)$ i.e., $m = 3$ and $N = 8$, total cost will be $3 * 2 * (8 - 1) = 42$ messages.

Proposed Algorithm: As per the mentioned example and as per proposed approach, the total number of messages consumed during the whole election process is as follows:

Number of messages consumed by node or process $5 = 2$.
Number of messages consumed by node or process $3 = 5$.
Number of messages consumed by node or process $8 = 7$.
Total number of message exchanged is 14 which is just equal to existing algorithm available for electing a coordinator by the single process only.

5 Conclusion

As per as my proposed election algorithm, I have reduced the number of messages used or exchanged for coordinator election nearly by half of the original ring algorithm. Hence, my proposed algorithm is a better approach than ring algorithm.

References

1. Chowdhary, S.K.: Improved and efficient ring election algorithm in distributed system. In: Proceedings of Computer Application and Signal Processing. ISBN: 978-981-09-2579-6 (2014)
2. Basu, S.: An efficient approach of election algorithm in distributed systems. Int. J. Comput. Sci. Eng. (IJCSE) (2013)

3. Beaulah Soundarabai, P., Thriveni, J., Venugopal, K.R., Patnaik, L.M.: An improved leader election algorithm for distributed systems. Int. J. Next-Generat. Netw. (IJNGN) **5**(1) (2013)
4. Vasudevan, S., Kurose, J., Towsley, D.: Design and analysis of a leader election algorithm for mobile ad hoc networks. In: Proceedings of the 12th IEEE International Conference on Network Protocols (ICNP) (2004)
5. Arghavani, A., Ahmadi, E., Haghighat, A.T.: Improved bully election algorithm in distributed systems. In: Proceedings of the 5th International Conference on IT and Multimedia at UNITEN (ICIMU). Malaysia (2011)
6. Katwala, H., Shah, S.: Study on election algorithm in distributed system. IOSR J. Comput. Eng. (IOSRJCE) **7**(6) 34–39. ISSN: 2278-0661, ISBN: 2278-8727 (2012)

Access Control Framework Using Dynamic Attributes Encryption for Mobile Cloud Environment

Neha Agrawal and Shashikala Tapaswi

Abstract Unified communication allows the users to share data between multiple devices having different platform. Previously, data was stored on local servers and accessed by employees through their desktops that have security policies inbuilt. In the age of integrated communication, users can access stored data from anywhere, anytime via smart devices and 4G wireless technologies. Traditional security protocols like access control are not well suited for mobile environment. Dynamic features of mobile devices can incorporate the conventional access control scheme to escalate the security of traditional protocols. In this work, dynamic attributes like spatial or temporal attribute, application usage, unlock failures, location and proximity of mobile devices are employed to curb the access of data. Using pairs of mobile agents, the work deals with the constraints of mobile devices and provides an efficacious solution. The proposed scheme can identify the authenticity of data owner and preserve the anonymity of the user.

Keywords Mobile cloud computing · Mobile agent · Unified communication
Access control · Dynamic attributes · Encryption

1 Introduction

Unified communication enables sharing of information among multiple devices working on diverse platforms. Traditionally, organizations employ access control mechanisms to protect the data stored on local servers and users access them

N. Agrawal (✉) · S. Tapaswi
Atal Bihari Vajpayee-Indian Institute of Information Technology and Management,
Gwalior 474015, MP, India
e-mail: nehaiiitm345@gmail.com

S. Tapaswi
e-mail: stapaswi@iiitm.ac.in

© Springer Nature Singapore Pte Ltd. 2018   611
K. Saeed et al. (eds.), *Progress in Advanced Computing and Intelligent Engineering*,
Advances in Intelligent Systems and Computing 563,
https://doi.org/10.1007/978-981-10-6872-0_59

through their secured desktops. Cloud computing, bring-your-own-device (BYOD), outsourcing and high bandwidth mobile broadband made it possible to access the data using mobile devices anywhere, anytime. After 2012, BYOD turned into a live field and according to the Cisco, 95% of the employees within the organizations are permitted to utilize their mobile device to access the stored data [1]. Access controls used for desktop systems are not desirable for mobile devices, since they are dependent solely on static attributes.

Cloud computing allows the users to store their data on public data storage, which enhances the flexibility, reduce consumption and maintenance cost and charge users based on service usage. The data shared using smart devices on public data storage increases productivity, but on the flip side increases security vulnerabilities as well. Traditionally it has been assumed that storage server, data owner and users belong to the same domain, hence the server is fully trusted. Data confidentiality is not guaranteed in cloud computing, BYOD and outsourcing environments as the data is stored and processed in third-party storage. Personal data may be leaked by a third party if data is stored in its original configuration. Due to this reason, cloud computing is not as much lucrative as it should be since the users are reluctant to store their sensitive data on public cloud. To overcome this personal leakage, plethora of researches has been done to store encrypted data in public servers.

Access control through encryption is employed to achieve confidentiality as well as impede the malevolent users from unauthorized access to personal data. To ensure access control and confidentiality simultaneously, an attribute-based encryption (ABE) cryptographic method is applied. Data owners are able to store encrypted data using fine-grained access policies; hence, the users who fulfil the access constraints can only access the data.

The classical access control, security technique used was role-based access control (RBAC), where data is encrypted using static attribute (roles) and functions like AND and OR. Hence, a user belongs to the specific role can only decrypt the data. Based on the number of the authority responsible for monitoring the attribute, the ABE scheme can be a single authority (SA)-based or multiple authorities (MA)-based. Only one authority is involved in SA-ABE, whereas multiple authorities are involved in MA-ABE to monitor disjoint set of attributes. Role-based access control ABE depends on predefined static attributes, therefore cannot be used directly in the era of BYOD. Since no real-time verification is involved in the above scheme, it is highly vulnerable to attack. Hence, dynamic attributes which are collected by mobile device are used to strengthen the access control scheme.

The rest of the paper is formed as follows: Sect. 2 discusses related work done so far; system architecture, novelties of the new scheme, workflow and proposed algorithm are explained in Sect. 3. Conclusion and future line of work are described in Sect. 4.

2　Related Work

Access control is a traditional security issue. In the literature, different access control methods have been proposed. In 1990s, many researches were based on role-based access control (RBAC) scheme. With this system, only authorized users are permitted to access the system information. The authorization and authentication of employees are associated with their roles. To alter the access control privileges, the security administrator needs only to lift the existing use and assign new role membership. Due to its simplicity, various extended RBAC models have been offered.

In [2], author proposed context-aware dynamic access control scheme. Access privileges have been allotted to users based on their functions and context data. Similar works have been proposed in [3, 4]. In [3], authors proposed event-based RBAC, where one can consider any aspect of context other than time and space which is measurable. They employed a great number of events and environmental conditions to resolve the user's system access. A temporal RBAC model is suggested in [4], where the author used temporal conditions. The shortcoming of [3, 4] is that they are not suited for mobile devices because it is not possible to restrict data and users at the same place.

Various RBAC models for mobile systems have been proposed which were based on location. According to the proposed work discussed in [5], near field communication (NFC) receiver has been provided in mobile devices for user's location verification. In [6], permission sets are granted and/or revoked using spatial information. Based on location, the access permissions are dynamically assigned to users. A more complex location-aware RBAC model is discussed in [7]. Two types of associations have been used: in particular location specific role is assigned, and in some specific locations some roles can only be activated. The flaws associated with [5–7] are they restrain the range of user and do not consider important mobile environment attributes.

The work done so far have not intended to provide confidentiality, since they assume the data is stored in trusted server. But to make the unified communication possible, data may be stored in others servers also. Hence, encryption of data became a focal point for all researchers. To ensure access control and confidentiality simultaneously, the ABE cryptographic method is applied. Using fine-grained access policies, data owner stores their encrypted data. A static location-based data encryption technique is proposed in [8], where a predefined latitude and longitude is associated with each static location.

Some of the groundbreaking works in ABE are being talked about here. The author of [9] gave fuzzy identity-based encryption (IBE). In the proposal, the message encrypted with identity λ can only be decrypted using a private key for identity μ if both identities λ and μ are close to each other. The positive points of the scheme are it is secure against collusion attack as well as error tolerant. Two main types of ABE have been proposed, namely key-policy attribute-based encryption (KP-ABE) and ciphertext-policy attribute-based encryption (CP-ABE). In paper [10], the author proposed KP-ABE, where set of attributes are associated with ciphertext and access permissions are linked with private keys which control the users for accessing the ciphertext. CP-ABE is proposed in [11], where access policies have been incorporated in the ciphertext. For accessing a particular ciphertext, the user needs to satisfy the attributes associated with that ciphertext.

Single authority ABE (SA-ABE) suffers from a single point of failure; to overcome this flaw, multiple authorities-based ABE (MA-ABE) has been proposed. In MA-ABE, various independent authorities are involved for monitoring attributes and distributing private keys. Similar work was proposed in [12], where the owner can select a set of attributes and can decide a number d_k. The user possessing minimum d_k number of attributes can decrypt the message. The author of [13] improves the scheme of [12] by removing the central authority. In [13], the author introduced anonymous key issuing protocol, which is responsible for maintaining the privacy of the users. The author of [14] proposed fully decentralized ABE scheme which removes the necessity of trusted servers. In this, user can possess any number of attribute from each authority and the flexibility to join or leave the system has been provided to the authorities.

The literature discussed so far is not considering other important dynamic attributes which are needed for a mobile computing environment; thus, access control in mobile devices is still a formidable challenge. A most pioneering work in this field was done in [15], where number of dynamic attributes in mobile device has been considered. The authors used application usage, unlock failure, location and proximity of smart devices which make the access control scheme robust. The algorithm maintains the security of conventional ABE schemes and supports single authority as well as multiple authorities-based encryption.

According to Pitoura and Samaras [16] constraints in mobile computing environment are:

1. Poor resources,
2. Reliability,
3. Low bandwidth and
4. Frequent disconnection.

The work of Li et al. [15] handles resource poor constraint in terms of battery power by using elliptic curve cryptography (ECC). ECC uses 224 bits size key which is lower than 2048 bits size key used in RSA. In the above-discussed literature study, there is no provision for checking the authenticity of the data owner. Due to this, there may be a case that an attacker can upload malicious data on the cloud storage to perform some unlawful act. Hence, still there is a need for research in this field to concentrate on other constraints of mobile computing and check the authenticity of data owner as well. In this paper, a new algorithm is proposed to handle these research gaps.

3   Proposed Approach

3.1   Preliminaries

3.1.1   Attribute-Based Encryption

In ABE, owner encrypts the data using credentials corresponding to necessary attribute. The user satisfying those attributes can only decipher the data. The trusted authorities preserve the necessary credentials for each attribute and provide decryption credentials to users after verification of his/her attributes. The data owner obtains encryption credentials from attribute authority(ies) (AA) for the attributes after being verified by the certification authority (CA). The owner uploaded encrypted data into the cloud storage and the users having decryption credentials can only decrypt the data. Figure 1 shows the ABE components used in this scheme.

3.1.2   The Client/Intercept/Server Model

To address the constraint of frequent disconnection, the proposed work deploys a client-side agent (CSA) and server-side agent (SSA). CSA will run at the user's mobile device, while SSA will run within the wired network. The CSA acts as a local proxy server on the client side, and SSA acts as a local client server at server side. For the reason of virtual insertion of the pair of agents between client and server, it diminishes the effects of wireless network. The client request is intercepted by the CSA, and together with the SSA, CSA performs various optimization techniques to reduce data transmission over wireless network. Hence, the proposed scheme handles low bandwidth constraint, improves data availability and sustains mobile computation even in disconnection-prone mode.

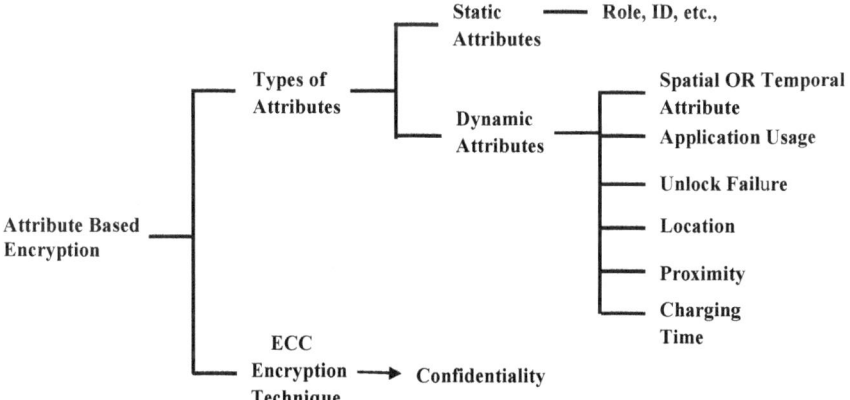

Fig. 1   ABE technique used in the proposed work

3.1.3 Bilinear Pairing

Let M_1, M_2 and M_T are multiplicative cyclic groups of prime order p. The generators of M_1 and M_2 are P_1 and P_2, respectively. $M_1 \times M_2 \rightarrow M_T$ is a bilinear pairing map e that satisfies below three properties.

a. *Bilinearty*: There exists $e(x^a, x^b) = e(x, y)^{ab}$, $\forall\ a, b \in Z_q$ and $\forall\ x \in M_1$, $\forall\ y \in M_2$.
b. *Non-degeneracy*: There exists $e(x, y) \neq 1$, $\forall\ x \in M_1$, $\forall\ y \in M_2$.
c. *Computability*: There exists a proficient algorithm to evaluate $e(x, y)$, $\forall\ x \in M_1$, $\forall\ y \in M_2$.

3.1.4 Elliptic Curve Cryptography

It is an elliptic curve theory-based public key encryption scheme used to generate smaller, faster and efficient cryptographic keys. Since it requires less computing power and battery usage, it is widely used in the field of mobile computing. The security of ECC depends on the discrete logarithm problem, i.e. to find x when $Q = xP$, where Q and P are two points on elliptic curve E.

3.2 System Architecture

The paper extends the framework given in [15] by the inclusion of certification authority and pair of agents. Figure 2 shows the proposed framework.

The system architecture consists of five entities: attribute authority (AA), certification authority (CA), mobile hosts (which contains user and CSA), SSA and cloud service provider (CSP). The roles of each of the components are given below.

Certification Authority is considered as a trusted authority responsible for verifying users' and owners' identities and issues certificates to them. CA maintains the list of static attributes like role, ID for all users and owners. The users' and owners' static attributes must be in the list for receiving the certificates from the CA. After getting certificates, the user/owner requests AA for the credentials.

Attribute Authorities provide encryption and decryption credentials to owner and user, respectively. The data owners choose a set of attributes from independent authorities (or from single authority), and the corresponding authority provided the encryption credentials to them. Using the obtained credentials together with dynamic attributes (received from a mobile device), the owner encrypts the data and uploads on cloud storage. The dynamic attributes used in this scheme are spatial and temporal attribute, application usage, unlock failure, charging time, location and proximity of mobile device.

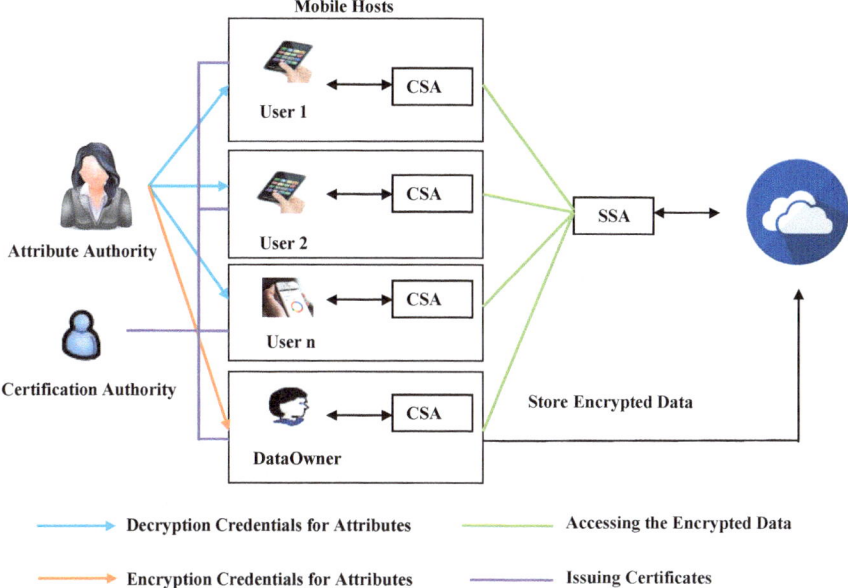

Fig. 2 System architecture

Mobile Hosts represent user requesting CSA for a cloud service. CSA intercepts the user request and transmits to the CSP through the SSA. CSP response sends back to the CSA through SSA. During disconnection, the requests of the clients and responses from the cloud server have been queued into the cache of the corresponding agents. Hence, the system works even in disconnected mode.

Server Side Agent is placed within the wire-line network acting as a surrogate of the client on the server side. SSA together with CSA performs various optimizations like differencing, caching, header reduction to reduce the traffic over a wireless network.

Cloud Service Providers offer the facility of storage and computational energy to users and owners. Owner uploads the encrypted data on the available storage, and users avail the computational power for searching the data on the cloud.

3.3 Novelties of the Proposed Scheme

1. The proposed scheme deals with the aforementioned constraints of mobile computing.
2. The proposed scheme can identify the authenticity of the data owner.

3. The proposed scheme helps in preserving the anonymity of the user.
4. Provides the comparable performance.

3.4 System Workflow

Figure 3 explains the systematic flow of the proposed architecture which explains by the given algorithm.

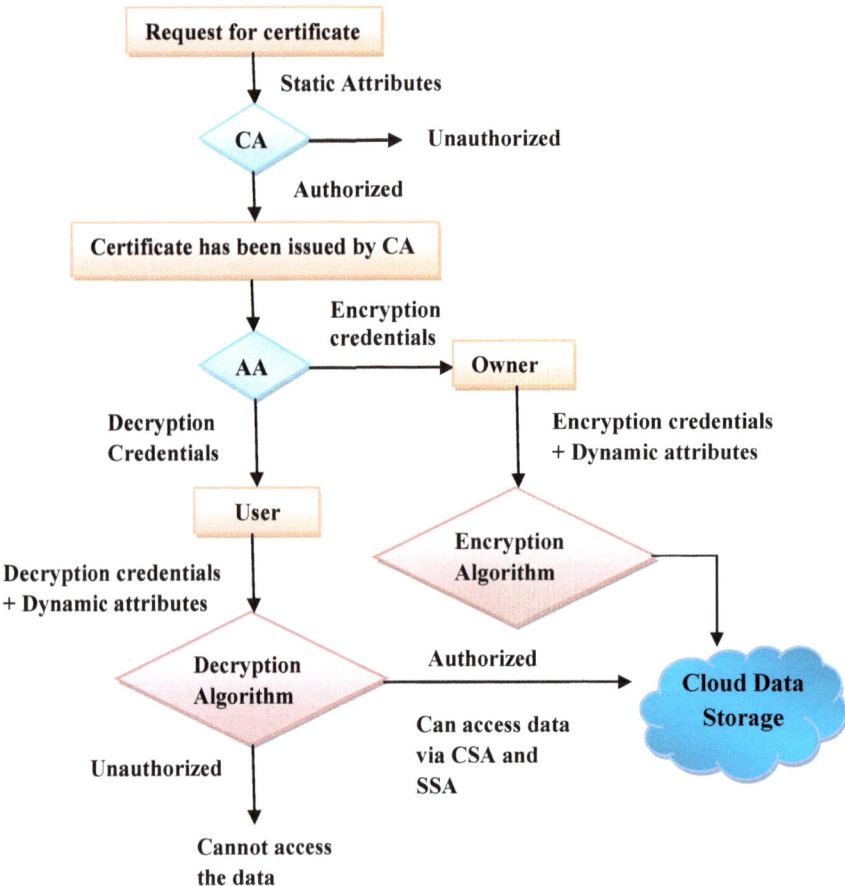

Fig. 3 System workflow

Proposed Algorithm

If Clients and data owner requests for certificates then;
 	
    CA verifies their static attributes;
 	
        If static attributes are in the list then;
 	
            Issue certificate;
 	
        Else
 	
            Request declined;
 	
Authorized clients and data owner receives encryption and decryption credentials from AAs;
 	
Data owner encrypts data using dynamic attributes and encryption credentials;
 	
If clients posses the those dynamic attributes then;
 	
    Access the data via CSA and SSA;
 	
Else
 	
    Request declined;

3.5 Scheme

Let U, O and A represent finite sets of users, owners, attribute authorities, respectively. Static attributes of the owner and user are represented by finite set, $static\_attr = [role, ID, degree]$.

Certificate Issuing. Each certificate authority (CA_i) has a pair of keys SK_{Ci} and PK_{Ci}, where SK_{Ci} is the secret key and PK_{Ci} is the corresponding public key (*where* $i \in A$). CA issues certificates to owners and users, which are encrypted with its secret key SK_C.

$$h\left[CA_{info} \,\|\, static\_attr_j\right] \rightarrow Certificate_j, \; j \in U \; or \; j \in O \qquad (1)$$

$$E\left[Certificate_j\right]_{SKCi} \rightarrow Certificate'_j \qquad (2)$$

Setup. It takes a parameter as input says λ and generates a bilinear group and a set of parameters as output. Generated parameters p, q and G_1 are public parameters. Using these parameters, public and private keys are generated for each attribute.

$$S\left(1^\lambda\right) \rightarrow p, q \text{ and } G_1 \qquad (3)$$

Key Issuing. Attribute authority(ies) verifies(y) certificates by decrypting with the CA's public key PK_{Ci} and then issues encryption and decryption credentials to owner and user, respectively.

$$D\left[Certificate'_j\right]_{PKCi} \rightarrow Certificate_j \quad [verified] \tag{4}$$

$$AA \ randomly \ generates \ secrets \ \alpha \ and \ \beta \ for \ each \ attribute \ i \ \epsilon \ static\_attr \tag{5}$$

$$AA \ generates \ the \ corresponding \ public \ key \ and \ private \ key \leftarrow \{p, q, \alpha, \beta\} \tag{6}$$

Encryption. The encryption credentials obtained from AA(s) together with owner's mobile device's dynamic attributes is the input to the encryption algorithm which produces ciphertext as output. Elliptic curve Diffie-Hellman (ECDH) key exchange is used to generate a shared secret using a public key which in this case is a point P on the elliptic curve E.

$$The \ Owner \ selects \ a \ secret \ random \ number \ \gamma \ \epsilon \ Z_q \tag{7}$$

He takes encryption credentials, dynamic attributes and P to generate his public key and transfer to user

$$Q_{owner} = \{A \cup B \cup \gamma\}P \rightarrow user \tag{8}$$

Decryption. The decryption credentials obtained from AA(s) together with user's mobile device's dynamic attributes are the input to the decryption algorithm which produces plain text as output. ECDH is used to generate a shared secret. The user can retrieve the correct data if he is authenticated.

$$User \ selects \ a \ secret \ random \ number \ \delta \ \epsilon \ Z_q \tag{9}$$

He takes decryption credentials, dynamic attributes and P to generate his public key and transfer to owner

$$Q_{user} = \{C \cup D \cup \delta\}P \rightarrow owner \tag{10}$$

$$Owner's \ shared \ secret \ key \rightarrow \{Q_{user} \cup A \cup B \cup \gamma\} \tag{11}$$

$$User's \ shared \ secret \ key \rightarrow \{Q_{owner} \cup C \cup D \cup \delta\} \tag{12}$$

4   Conclusion

Cloud is not a utopian environment, so security in the cloud is paramount aspect. In this work, an access control framework for mobile cloud computing network has been proposed, which incorporates the dynamic attributes of mobile device.

Data owner uses dynamic attributes along with static attributes for the encryption of data for storing it on a public cloud. The proposed scheme provides an additional layer of security by identifying malicious data owner and also effective for mobile environment. Future work of the paper includes detection of attacks and reduces the delay which occurs due to the involvement of attribute and certification authority.

References

1. Cisco Study: IT Saying Yes to BYOD, Cisco. http://tinyurl.com/d8fv2uj (2012). Accessed 15 July 2015
2. Zhang, G., Parashar, M.: Context-aware dynamic access control for pervasive applications. In: Communications Networks and Distributed Systems Modelling and Simulation Conference, pp. 21–30 (2004)
3. Bonatti, P., Galdi, C., Torres, D.: ERBAC: Event-driven RBAC. In: 18th ACM Symposium Access Control Models and Technologies, pp. 125–136. ACM (2013)
4. Bertino, E., Bonatti, P.A., Ferrari, E.: TRBAC: a temporal role-based access control model. ACM Trans. Inf. Syst. Secur. (TISSEC) 4(3), 191–233 (2001)
5. Kirkpatrick, M.S., Bertino, E.: Enforcing spatial constraints for mobile RBAC systems. In: 15th ACM Symposium Access Control Models and Technologies, pp. 99–108 (2010)
6. Hansen, F., Oleshchuk, V.: SRBAC: a spatial role-based access control model for mobile systems. In: 7th Nordic Workshop on Secure IT Systems (NORDSEC), pp. 129–141 (2003)
7. Ray, I., Kumar, M., Yu, L.: LRBAC: a location-aware role-based access control model. In: Information Systems Security, Lecture Notes in Computer Science, vol. 4332, pp. 147–161. Springer, Berlin, Heidelberg (2006)
8. Liao, H.C., Chao, Y.H.: A new data encryption algorithm based on the location of mobile users. Inf. Technol. J. 7(1), 63–69 (2008)
9. Sahai, A., Waters, B.: Fuzzy identity-based encryption. In: Advances in Cryptology—EUROCRYPT, Lecture Notes in Computer Science, vol. 3494, pp. 457–473 (2005)
10. Goyal, V., Pandey, O., Sahai, A., Waters, B.: Attribute-based encryption for fine-grained access control of encrypted data. In: 13th ACM Conference on Computer and Communication Security, pp. 89–98. New York, USA (2006)
11. Bethencourt, J., Sahai, A., Waters, B.: Ciphertext-policy attribute-based encryption. In: IEEE Symposium on Security and Privacy, pp. 321–334 (2007)
12. Chase, M.: Multi-authority attribute based encryption. In: Theory of Cryptography, Lecture Notes of Computer Science, vol. 4392, pp. 515–534. Springer, Berlin, Heidelberg (2007)
13. Chase, M., Chow, S.S.M.: Improving privacy and security in multi-authority attribute-based encryption. In: 16th ACM Conference on Computer and Communication Security, pp. 121–130. New York, USA (2009)
14. Lewko, A.B., Waters, B.: Decentralizing attribute-based encryption. In: Advances in Cryptology—EUROCRYPT 2011, Lecture Notes in Computer Science, vol. 6632, pp. 568–588 (2011)
15. Li, F., Rahulamathavan, Y., Conti, M., Rajarajan, M.: Robust access control framework for mobile cloud computing network. J. Comput. Commun. Elsevier 68, 1–12 (2015)
16. Pitoura, E., Samaras, G.: Data management for mobile computing. In: Springer Science + Business Media, LLC. ISBN 978-1-4613-7526-5, ISBN 978-1-4615-5527-8 (eBook). https://doi.org/10.1007/978-1-4615-5527-8

Improving Energy Usage in Cloud Computing Using DVFS

Sambit Kumar Mishra, Priti Paramita Parida, Sampa Sahoo, Bibhudatta Sahoo and Sanjay Kumar Jena

Abstract The energy-related issues in distributed systems that may be energy conservation or energy utilization have turned out to be a critical one. Researchers worked for this energy issue and most of them used Dynamic Voltage and Frequency Scaling (DVFS) as a power management technique where less voltage supply is allowed due to a reduction of the clock frequency of processors. The cloud environment has multiple physical hosts, and each host has several numbers of virtual machines (VMs). All online tasks or service requests are scheduled to different VMs. In this paper, an energy-optimized allocation algorithm is proposed where DVFS technique is used for virtual machines. The fundamental idea behind this is to make a compromise balance in between energy consumption and the set up time of different modes of hosts or VMs. Here, the system model that includes different sub-system models is explained formally and the implementation of algorithms in homogeneous as well as heterogeneous environment is evaluated.

Keywords Cloud computing · DVFS · Task allocation · Energy consumption
VM · Virtualization

Please note that the LNCS Editorial assumes that all authors have used the western naming convention, with given names preceding surnames. This determines the structure of the names in the running heads and the author index.

S. K. Mishra (✉) · P. P. Parida · S. Sahoo · B. Sahoo · S. K. Jena
National Institute of Technology, Rourkela, India
e-mail: skmishra.nitrkl@gmail.com

P. P. Parida
e-mail: pritiparamita.parida@gmail.com

S. Sahoo
e-mail: sampaa2004@gmail.com

B. Sahoo
e-mail: bibhudatta.sahoo@gmail.com

S. K. Jena
e-mail: skjena@nitrkl.ac.in

© Springer Nature Singapore Pte Ltd. 2018
K. Saeed et al. (eds.), *Progress in Advanced Computing and Intelligent Engineering*,
Advances in Intelligent Systems and Computing 563,
https://doi.org/10.1007/978-981-10-6872-0_60

1   Introduction

In the recent time, the exceptional demand for computational power by recent researchers and business applications needs a large number of data centers which consumes extensively more energy. It was estimated that the expense of consumption of energy by IT industries in the USA as 4.5 billion dollars and the approximation shows double cost by 2011 [1]. Generally, there are other critical environmental issues that emerge from high power utilization like emissions of huge amount of CO_2. One of the estimations is that the IT hardware equipment is responsible for around 2% of emissions of global CO_2 [2]. Therefore, proper utilization of computing resources or reduction of energy consumption becomes a critical research topic. Cloud computing is a model that can increase the resource utilization of the system and therefore reduces numerous of IT hardware equipment. In other words, the cloud computing is a shared model in which the resources are shared among different components of the system to achieve better efficiency and performance. Moreover, data center in the cloud needs reliable Internet through which a large number of physical hosts or servers are connected. There are several service providers in the cloud called Cloud Service Provider (CSP) who is capable of managing all services to the end user. The cloud computing system has several open challenges and security issues [3].

In order to reduce the electrical power for the whole system, the resource scheduling should be proper or optimized one. The scheduling of tasks to have different resources is a known NP-complete problem because of large solution space. A scheduler has schedules different resources to all the tasks. Therefore, an efficient scheduler can manage less computing resources to execute all the tasks in parallel. Most of the researcher works on static as well as dynamic energy efficient scheduling procedure for the allocation of resources [4–6]. In dynamic scheduling, processors consume approximately less than 30% of power consumption as compared to static scheduling [5]. One of the powerful techniques of the cloud computing is the virtualization. This virtualization technique provides an abstraction of hardware resources. It is applied to the IaaS layer (where VM management, VM deployment, etc., are done) to virtualize the hardware resources.

Researchers give more attention toward minimizing energy consumption and corresponding cost of data centers. To balance the distribution of power management, different techniques come into picture. Dynamic Voltage and Frequency Scaling (DVFS) technique is one of such technique through which electrical power can be minimized. Broadly, there are two different methods for the reduction of power consumption and those are (a) Dynamic Component Deactivation (DCD) and (b) Dynamic Performance Scaling (DPS). Enabling and disabling of different electrical components is done by using DCD mechanism. Whereas DPS allows reduction of a certain frequency and the voltage supply moderately, DVFS is one of the DPS techniques. Here, the research is applied to develop a heuristics for the allocation of tasks to decrease the energy utilization or energy consumption by applying the DVFS approach. Before submission of tasks to the cloud by user, an agreement has to be made between the user and the CSP and the agreement is known as service level agreement (SLA). If the CSP can provide services according to the negotiation or SLA, the user submits the task or service request. The primary contributions of this work are outlined as follows.

- We have clearly explained the models that include cloud system model, task allocation model, and DVFS-oriented energy model.
- We have used DVFS mechanism to the system for the reduction of energy consumption, and this technique is applied to the set of VMs.
- We have proposed Energy Efficient DVFS-based Task-Scheduling Algorithm (*EEDTSA*) for service or task allocation in the cloud and compared with the random allocation algorithm.

We organized the rest of the paper as follows. Prerequisites for our work are in Sect. 2 as related work. Section 3 describes the system model for cloud infrastructure. Section 4 presents our scheduling algorithm and its description. Simulation results from experimental assessment and observations are reported in Sect. 5. We have concluded our work of this paper in Sect. 6.

2   Related Work

The resource allocations in cloud environment can differentiate in two views. First one is from user points of view where the user has two main objectives as minimizing the execution time and also minimizing the cost. Second one is from Cloud Service Provider (CSP) points of view where the CSP has objectives as resource should be utilized optimally and maintenance cost should be less. The DVFS technique is a common technique that is used to minimize the consumption of power of different electrical devices like mobiles, computers. An energy-aware resource allocation heuristic algorithm is proposed for data center that also delivered Quality of Service (QoS) [7]. They also provide an autonomic mechanism for managing the cloud resources efficiently and effectively which satisfies certain level of SLAs. They mapped virtual machines to satisfactory resources; however, there is some degradation in the system performance. In [8], the authors have explained the need of a good job scheduler, which schedules all the jobs provided the running time of the jobs should be less and used fewer resources. They provide several mapping definitions with the functionalities of DVFS controller and MMS-DVFS. They have proposed an algorithm for mapping VMs with jobs, and the same is simulated in CloudSim. They have experimented for multiple times, and the result shows better.

There are huge applications such as social networks [9], sensor networks [10], and mobile nodes [11] that always require the source of the data to be computed in the cloud. In [11], the authors have described four steps for scheduling problem in Mobile Cloud Computing (MCC) environment, and these steps are (a) finding the set of offloading tasks onto the cloud, (b) mapping of tasks to local cores, (c) frequency determination, and (d) scheduling of tasks. They have used different methodologies for minimizing the consumption of energy and lastly used DVFS technique for further reduction of energy utilized in mobile devices. Their system model constitutes a Directed Acyclic Graph (DAG), where nodes represent tasks and an edge represents the sequence of execution of tasks. They have proposed MCC task-scheduling algorithm having three steps. The first step is for minimizing the execution time, the second step is for minimizing the energy consumption, and the third step is for further reduction of energy consumption using DVFS. Huang et al. [12] have also modeled the cloud system with the help of DAG. They have proposed parallel algorithm for reduction of energy consumption in heterogeneous environment which obeying certain SLAs.

The authors in [5] aim to schedule applications for the reduction of consumption of power for the execution of the parallel task with DVFS technology and that to maintain a relationship between task execution time and energy consumption of the system. They have also focused on two basic issues: One is Best Effort Scheduling (BES), and another is energy performance trade-off scheduling. To use the two models: DVFS-oriented performance model and power consumption model, they have conducted multiple experiments to solve the task allocation problem efficiently and developed an algorithm named as Power-aware Threshold Unit (PTU) algorithm. Here, they have considered energy as SLA. In [13], various security issues in different layers are described. In [14], Calheiros and Buyya have proposed a methodology for the execution of a set of urgent tasks and a set of CPU-intensive tasks on the cloud frameworks that consume minimum energy.

3   Models

In this section, the cloud system model, DVFS-oriented energy model, and the task allocation model are described in this study.

3.1   Cloud System Model

An energy-aware allocation for the execution of the task in the context of E_{SLA} (SLA for less energy consumption). Users can define the performance specifications for the computation of services as well as can specify the maximum amount of energy for executing their tasks. Requirement of energy for processing the tasks is an SLA between the cloud user and the CSP. The E_{SLA} has several metrics (like execution time of a task, deadline of the task, amount of energy consumption, CO_2 emission rate). An example of E_{SLA} is: Execute these tasks for x time units if the total energy consumption of the service is below w watt. When an SLA is met, then the tasks with their E_{SLA} are submitted to the task queue. After that, the scheduler (which has a complete information about all the VMs) schedules all the tasks from the task

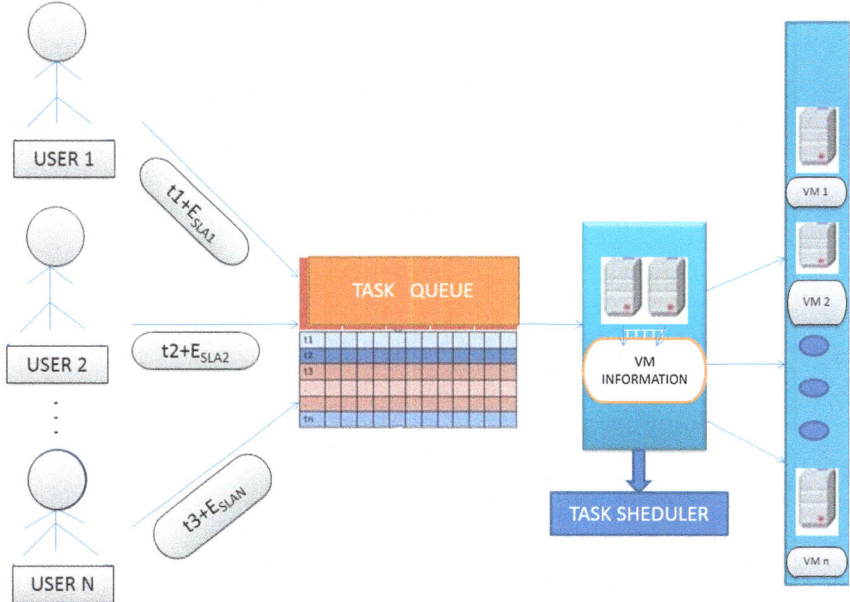

Fig. 1 Framework for task scheduling with SLA-based energy consumption

queue to different virtual machines batch-wise. Then, the tasks are executed within the specified energy by providing the computing resources or virtual machines. Each VM has a local queue to place the allocated tasks (Fig. 1).

3.2 Task Allocation Model

A task can be modeled as four tuples (t_i, D_i, ER_i, L_i). Here, t_i is the task identification, D_i is the deadline of task, ER_i is the energy requirement of task, and L_i is the length of the ith task in million instruction (MI). Task is represented by a directed acyclic task graph $G = (V, E)$ as shown in Fig. 2. Each vertex in the graph accounts for a task, and each directed edge represents the precedence of task execution. In Fig. 2, it is shown that task 1 and task 2 are entry tasks (i.e., the task with no parent). Similarly, task 8 is the exit task (i.e., the task with no child). Independent tasks can execute in parallel (task 1 and task 2 are independent tasks).

Fig. 2 An example of a task graph

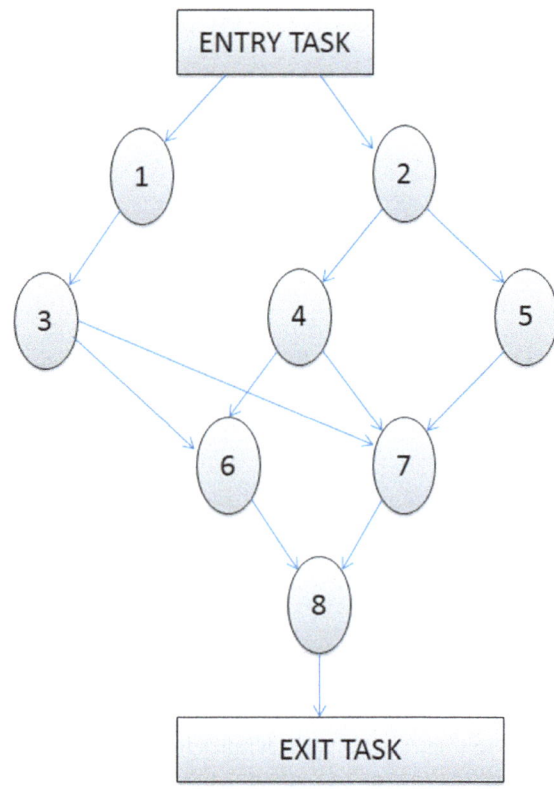

3.3 DVFS-Oriented Energy Model

The consumption of power by the heterogeneous nodes in the cloud is calculated from the CPU, main memory, secondary storage, and bandwidth usage. Since this resource utilization affects the power consumption of the system, we should optimize it. CPU consumes more energy as compared to other parts of a system. Therefore, we focus on optimizing the energy consumption and the CPU utilization. Present-day CPUs support Dynamic Voltage and Frequency Scaling (DVFS) methods to change its frequency dynamically to improve the energy problems. The CPU utilization is typically proportional to the overall system load. The energy consumption (E) of the cloud system is the sum of static energy consumption (E_s) and the dynamic energy consumption (E_d) as in Eq. 1 [15].

$$E = E_s + E_d \tag{1}$$

Here, the dynamic energy consumption is calculated as follows in Eq. 2.

$$E_d = \sum_{t_p} P_d.t_p = \sum_{t_p} (k.v^2.f.t_p) \qquad (2)$$

Here, t_p is the time period, k is a constant related to the device, P_d is the dynamic power consumption with voltage (v) and frequency (f).

Since static energy consumption is directly proportional to the dynamic energy consumption [16], we can have the total energy consumption as follows in Eq. 3.

$$E = \sum_{t_p} (k.v^2.f.t_p) \qquad (3)$$

4 Algorithm

This section discusses the details of proposed scheduling algorithm, Energy Efficient DVFS-based Task-Scheduling Algorithm (*EEDTSA*). Here, the purpose is to minimize energy consumption overally by changing various voltage–frequency pair for each task. Inputs to the algorithm are all tasks along with their length (in million instructions), DAG task graph, deadline of all tasks, all VMs, and the voltage–frequency pairs of processors of all VMs as listed in Table 1. The output of the algorithm contains the allocation vector (A) and the total energy consumption (E) of all task. Here, *WinSize* is same as the number of VMs. So, select a set of tasks T from total task and $|T| = WinSize$. Here, the Heterogeneous Earliest Finish Time (HEFT) list-scheduling algorithm [17] is applied to map all the tasks in the DAG to different processors. It results in a task allocation to calculate the time slots for the execution of tasks and information interchange among virtual machines. Here, a matrix is found as output noted as ET_{ij}, which is the execution time of ith task on jth VM.

5 Experimental Result

For the experimental purpose, each VM has a single processor among the listed processors in Table 1. Here, we perform the comparative evaluation of our heuristic algorithm (EEDTSA) with the random algorithm. The processors mentioned in the Table 1 have 1 MB cache, and 35 W is the TDP (thermal design point) value that is the maximum volume of heat made by the processor. For the experiment, various combinations of processors are chosen from Table 1 for each VM.

In scenario-1, the number of tasks is 2000 which is fixed, and the number of virtual machines increases from 50 to 150 in the gap of 10. The comparison graph for the calculation of energy using the random allocation technique and our *EEDTSA* is shown in Fig. 3. From the figure, the consumption of energy value in *EEDTSA* is less. In scenario-2, the number of virtual machines is 100 which is fixed, and the number of tasks increases from 1000 to 10000 in the gap of 1000. The comparison

Table 1 Voltage–Frequency pair

Processor model	Frequency (GHZ)	Voltage (V)
Turion 64 ML-30	1.6	1.15
Turion 64 ML-34	1.8	1.20
Turion 64 ML-37	2.0	1.50
Turion 64 ML-40	2.2	1.62
Turion 64 ML-44	2.4	1.68

Algorithm 1 : *EEDTSA*: Energy Efficient DVFS-based Task-Scheduling Algorithm

---

Input: *Task*: set of tasks sorted in descending order of their length; DAG task graph; D: Dead-line of all tasks; V: set of VMs; E_{SLA}, F: the voltage-frequency (vf) pairs of processors of all VMs;

Output: Energy efficient Allocation result (A and $|A| = |Task|$) and the total energy consumption (E)

1: Initialize, $E = 0$, $n = |Task|$, $count = 0$;
2: Sort all VMs, V in descending order of their processing speed;
3: Select a set of tasks T form set Task and $|T| = WinSize$;
4: Map the DAG to V using HEFT algorithm;
5: **for** each task $t_i \in T$ **do**
6:     **for** each VM $v_j \in V$ **do**
7:         **if** Dead-line and E_{SLA} satisfies **then**
8:             *Continue*;
9:         **else**
10:             **if** j == 1 **then**
11:                 Reject the task t_i and *Break*;
12:             **end if**
13:             Allocate t_i in v_{j-1};
14:             $A[count \times WinSize + i] = j - 1$;
15:             $count = count + 1$;
16:             **for** each $vf_k \in F$ **do**
17:                 E_k = Calculate energy consumption of t_i using Eq. 3;
18:             **end for**
19:             $E_{Min} = Min_{k \in F} E_k$;
20:             $E = E + E_{Min}$; and *Break*;
21:         **end if**
22:     **end for**
23: **end for**
24: **if** Execution of all task not completed **then**
25:     Goto Step-3;
26: **end if**
27: Return A and E;

Fig. 3 Comparison of energy consumption in *EEDTSA* and random allocation algorithm when number of VM is fixed and task varies

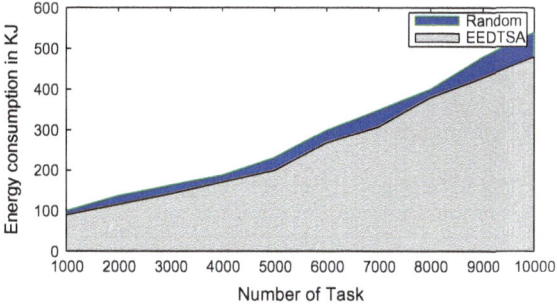

Fig. 4 Comparison of energy consumption in *EEDTSA* and random allocation algorithm when number of task is fixed and VM varies

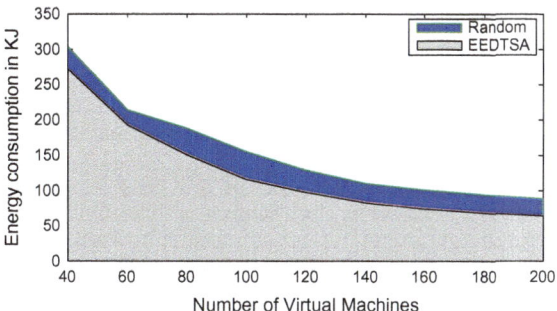

graph for the calculation of energy using the random allocation technique and our *EEDTSA* is shown in Fig. 4. From the figure, the consumption of energy value in *EEDTSA* is less.

6 Conclusion

In this paper, we are interested in energy-aware allocation result in the cloud and propose a novel scheduling algorithm *EEDTSA*. In *EEDTSA*, DVFS is applied to allocate tasks in the cloud resource. We have explained elaborately different heuristic algorithms where researchers used DVFS technique. We provide a cloud system model for the conservation of energy by applying E_{SLA} as a service level agreement. We have simulated the *EEDTSA* and also made comparison graphs with random allocation algorithm. The comparison graph shows that *EEDTSA* consumes less energy as compared to the random allocation. There is a plan to implement some recent existing algorithm to compare with *EEDTSA*.

References

1. Brown, R.: Report to congress on server and data center energy efficiency. In: Lawrence Berkeley National Laboratory, pp.109–431 (2008)
2. Gartner, Gartner estimates ICT industry accounts for 2 percent of global CO_2 emissions. http://www.gartner.com/it/page.jsp?id=503867. Accessed 11 Jan 2011
3. Puthal, D., Sahoo, B.P.S., Mishra, S., Swain, S.: Cloud computing features, issues, and challenges: a big picture. In: IEEE International Conference on Computational Intelligence and Networks (CINE), January 2015, pp. 116–123
4. Mishra, S.K., Deswal, R., Sahoo, S., Sahoo, B.: Improving energy consumption in cloud. In: 2015 Annual IEEE India Conference (INDICON), December 2015, pp. 1–6
5. Huai, W., Huang, W., Jin, S., Qian, Z.: Towards energy efficient scheduling for online tasks in cloud data centers based on DVFS. In: IEEE 9th International Conference on Innovative Mobile and Internet Services in Ubiquitous Computing (IMIS), pp. 175–186 (2015)
6. Sahoo, S., Nawaz, S., Mishra, S.K., Sahoo, B.: Execution of real time task on cloud environment. In: 2015 Annual IEEE India Conference (INDICON), December 2015, pp. 1–5
7. Beloglazov, A., Abawaj, Y., Buyya, R.: Energy-aware resource allocation heuristics for efficient management of data centers for cloud computing. Futur. Gener. Comput. Syst. **28**(5), 755–768 (2012)
8. Wu, C.M., Chang, R.S., Chan, H.Y.: A green energy-efficient scheduling algorithm using the DVFS technique for cloud datacenters. Futur. Gener. Comput. Syst. **37**, 141–147 (2014)
9. Puthal, D., Nepal, S., Paris, C., Ranjan, R., Chen, J.: Efficient algorithms for social network coverage and reach. In: IEEE International Congress on Big Data, pp. 467–474 (2015)
10. Puthal, D., Nepal, S., Ranjan, R., Chen, J.: DPBSV-an efficient and secure scheme for big sensing data stream. In: Trustcom/BigDataSE/ISPA, IEEE, vol. 1, pp. 246–253 (2015)
11. Lin, X., Wang, Y., Xie, Q., Pedram, M.: Task scheduling with dynamic voltage and frequency scaling for energy minimization in the mobile cloud computing environment. IEEE Trans. Serv. Comput. **2**, 175–186 (2015)
12. Huang, Q., Su, S., Li, J., Shuang, K., Huang, X.: Enhanced energy-efficient scheduling for parallel applications in cloud. In: Proceedings of 12th IEEE/ACM International Symposium on Cluster, Cloud and Grid Computing (ccgrid), IEEE Computer Society, pp. 781–786 (2012)
13. Puthal, D., Nepal, S., Ranjan, R., Chen, J.: Threats to networking cloud and edge datacenters in the internet of things. IEEE Cloud Comput. **3**(3), 64–71 (2016)
14. Calheiros, R.N., Buyya, R.: Energy-efficient scheduling of urgent bag-of-tasks applications in clouds through DVFS. In: IEEE 6th International Conference on Cloud Computing Technology and Science (CloudCom), pp. 342–349 (2014)
15. Kim, K.H., Buyya, R., Kim, J.: Power Aware Scheduling of Bag-of-Tasks Applications with Deadline Constraints on DVS-enabled Clusters. In: CCGRID, pp. 541–548 (2007)
16. Li, J., Martnez, J.F.: Dynamic power-performance adaptation of parallel computation on chip multiprocessors. In: HPCA, pp. 77–87 (2006)
17. Topcuoglu, H., Hariri, S., Wu, M.: Performance-effective and low complexity task scheduling for heterogeneous computing. IEEE Trans. Parallel. Distrib. Syst. **13**(3), 260–274 (2002)

Modeling of Task Scheduling Algorithm Using Petri-Net in Cloud Computing

Suvendu Chandan Nayak, Sasmita Parida, Chitaranjan Tripathy and Prasant Kumar Pattnaik

Abstract Task scheduling is an NP-hard problem. In earlier, different task scheduling algorithms are proposed for cloud computing environment. By adapting new and modified task scheduling algorithms, better resource utilization can be obtained. In real time, most of the tasks are deadline-based tasks. The deadline-based task has different parameters. The backfilling algorithm is used to schedule these types of tasks in Haizea. In this paper, we modeled the existing backfilling algorithm for scheduling deadline-based task using Petri-Net. The paper presents the design model of the existing backfilling algorithm. The model specifies real-time challenges of backfilling algorithm using Petri-Net. The work also comes forward with some design issues of backfilling algorithm using Petri-Net.

Keywords Task · Deadline · Scheduling · Backfilling · Petri-Net

S. C. Nayak (✉) · S. Parida
Department of Computer Science and Engineering,
C.V. Raman College of Egineering, Bhubaneshwar, India
e-mail: suvendu2006@gmail.com

S. Parida
e-mail: sasmitamohanty5@gmail.com

S. C. Nayak · C. Tripathy
Department of Computer Science and Engineering,
Veer Surendra Sai University of Technology, Burla, India
e-mail: crt.vssut@yahoo.com

P. K. Pattnaik
School of Computer Engineering, KIIT University, Bhubaneshwar, India
e-mail: patnaikprasantfcs@kiit.ac.in

© Springer Nature Singapore Pte Ltd. 2018   633
K. Saeed et al. (eds.), *Progress in Advanced Computing and Intelligent Engineering*,
Advances in Intelligent Systems and Computing 563,
https://doi.org/10.1007/978-981-10-6872-0_61

1   Introduction

Cloud computing is the next generation of Internet Technology. From the last few years, it comes forward as challenging area for researchers where task scheduling is one of them. The basic concept of task scheduling is to schedule all the tasks within the deadline with limited resources where more resource utilization can be achieved. In an optimal way, the task scheduling problem is known as NP-hard problem [1]. The elasticity is one of an important properties of cloud computing in which flexible resource provisioning is provided by clouds called auto-scaling nature. Resources are allocated or released according to the requirement of user applications at any time and charged. However, resource under-provisioning usually decreases the system's performance, whereas, resource over-provisioning always leads to idle resources [2]. Therefore, it is a challenge for cloud providers to provide the exact amount of cloud resources to the user with minimum cost.

The OpenNebula is one of the open source cloud platforms which provides a flexible platform for resource allocation and computation for both ends of cloud users and cloud service providers. The OpenNebula supports basically three types of leases: (1) Advance reservation, (2) Best effort leases, and (3) Immediate leases [3]. The user request is considered as task or lease in OpenNebula. In real a time constraint is associated with the best effort lease, called deadline sensitive leases [4, 5].

Different algorithms are used to schedule deadline sensitive leases, like swapping and backfilling algorithm [4, 6]. Backfilling algorithm performs better than swapping. In case of swapping algorithm, leases are non-preemptive whereas in backfilling algorithm, leases are preemptive in nature. It provides better resource utilization than swapping by allocating more number of leases.

It is very essential to model an algorithm where the performance of the algorithm can be observed in real time. The Petri-Net is one of the power mechanism which provides the representation and analysis of concurrent systems [7]. The representation specifies the nature and workflow of the system, whereas analysis specifies the performance of the system in real time. Resource Allocation System (RAS) is mostly based on the analysis of the system, control techniques that are used to evaluate the performance. Most of these techniques are literature based on which a structural element approach is used to characterize the deadlock [8].

The Petri-Net models can be used to analyze interdependencies among the tasks, criticality which is occurred during system implementation. Moreover, it can also handle substitution, conflicting resource priorities which is common a scenario. In real time, the available resources are in different variants, so it is challenging one to allocate resources to users. The above challenges can be handled by Petri-Net [7]. A mechanism for conflict challenges in resource allocation is proposed using AHP [9] when there is more than one similar tasks or having similar priorities.

In this work, we used the Petri-Net to model the task scheduling algorithm. The aim of the work is to find out the performance of the backfilling algorithm which is used for resource allocation in cloud computing using Petri-Net. The rest part of the paper is organized as follows: the related work is discussed in Sect. 2; the proposed model for the backfilling algorithm using Petri-Net is available in Sect. 3; Sect. 4 presents the result and discussion of the proposed model; and the conclusion and future scope are presented in Sect. 5.

2 Related Work

Resource allocation is a challenging issue for researchers in cloud computing due to elasticity property of it [10]. In a physical machine, different virtual machines (VMs) are created according to the user request. A single user may demand number of virtual machines [11]. In our previous work [5], we discussed the existing backfilling algorithm. In this paper, we found some disadvantages and proposed a mechanism for truthful resource allocation in cloud computing. Calheiros and Buyya [12] also proposed a mechanism for task scheduling and task replication for deadline-based task.

In last few years, a number of task scheduling algorithms and mechanisms are proposed by researchers by considering different task parameters. A resource cost optimization technique is proposed by Chaisiri et al. [13] in cloud computing environment. In recent year, Duan [14] proposed an energy aware scheduling algorithm. In this work, the authors proposed a mechanism to reduce the energy in cloud computing by scheduling the tasks in an efficient manner. Kalra and Singh [15] discussed different metaheuristic approach which is used for task scheduling in cloud computing. The work provides the basic review of metaheuristic approaches in cloud computing.

Moreover, task replication is proposed by Calheiros and Buyya [12]. The tasks are replicated if sufficient resources are not available within the deadline. Here, the tasks are associated with the deadline. In Open Nebula, backfilling algorithm is used to schedule the deadline-based tasks. Different mechanisms are also proposed for scheduling these types of tasks. Ergu et al. [16] proposed a scheduling algorithm using the analytical hierarchy process in cloud computing. The work performs better resource utilization as compared to other scheduling algorithm for deadline-based task.

Petri-Net is a basic model of parallel and distributed systems, in which resources are allocated parallel or in distributed fashion for computing the task. The basic Petri-Net model was designed by Carl Adam Petri in year 1962 [7]. The author used it for resource allocation in the project. Moreover, Petri-Net is also used for server consolidation for heterogeneous computing in cloud computing environment [17]. The Petri-Nets are also used to solve multi-criteria fuzzy rules by evaluating priorities for the conflicting process. In the real-time information-based system, the conflict problem is solved by using Petri-Nets [18].

Though Petri-Net is used to find, analysis and resolve the conflicts in a system, in this work, we try to find the performance of the backfilling algorithm by modeling the system. We also discussed how the different tasks are fired within the deadline with their required resources. We also observed some of the conflicts occurred during the system modeling and modified the firing principles. The detail of the proposed work is followed in the next section.

3 Proposed Model

In this section, we have discussed the basic principle of the Petri-Net and the modeling of backfilling algorithm using it.

3.1 Petri-Net

As we discussed in Sect. 2, it is a graphical and mathematical model for system to evaluate and analyze the system. Formally, a Petri-Net is represented in five-tuple as [7]:

$$PN = (P, T, A, W, M)$$

where,

$P = \{p_1, p_2, \ldots, p_M\}$ is a finite set of places.
$T = \{t_1, t_2, \ldots t_N \pi r^2\}$ is a finite set of transitions.
$A \subseteq (P \times T) \cup (T \times P)$ is a set of arcs.
$W: A \rightarrow \{1, 2, 3, \ldots\}$ is a weight function.
$M: P \rightarrow \{0, 1, 2, \ldots\}$ is the set of initial markings where $P \cap T = \emptyset$ and $P \cup T$ Ø.

Basically, the transition (firing) rules are applicable in the case of untimed Petri-Nets when we are simulating the dynamic behavior of a system. The Fig. 1 shows the different states used in Petri-Net. There is an initial position (outer circle) which contains a token (filled circle). An initial position may contain more than one

Fig. 1 States of a simple Petri-net

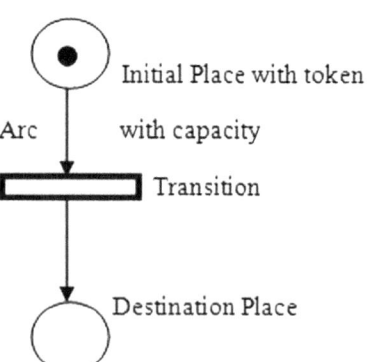

Initial Place with token
with capacity

Arc

Transition

Destination Place

token as per the system design requirement. There is an arc from the initial position to transition. It holds the weight or capacity. Based upon the firing rules, the transition is occurred from the initial state to the destination state.

3.2 Model of Backfilling Algorithm

Backfilling algorithm is the optimized first come first serve (FCFS) algorithm. It is basically used for parallel scheduling. In OpenNebula, a user request is represented using following parameters as follows: Lease number (Task serial number), Node (Number of VMs), Arrival Time, Start time, Duration, and Deadline. So a task T can be represented as:

$$T = \{n, A, N, S, E, D\}$$

In OpenNebula, user needs the number of VMs. So the challenge is that how all the leases or tasks could be scheduled within their deadline. As we discussed, the backfilling algorithm provides better scheduling of tasks along with better resource utilization. Formally, in backfilling algorithm, all the tasks are sorted according to their start time. The 1st task is scheduled first. Then, the free nodes are evaluated, and a task is selected from the queue whose start time is same with the 1st task where the required node is less than or equal to the free nodes. The existing backfilling is as below.

Backfilling Algorithm

Step 1: Initialize queue (Q) with n number of leases or tasks with their start time
Step 2: Schedule 1st task from Q
Step 3: Select a task T_i which can be scheduled parallel with 1st task as:

If $T_i(S) == T_1(S)$ and
$T_i(N) \leq N_{free}$ and meet its deadline at time slot t_i

Step 4: Schedule the next task at time slot t_i where N is maximum and S is equal.

And repeat step 3 to 4 until the Q is empty.
else
Schedule next task with different S and repeat step 3 to 4 until Q is empty

Step 5: end

The above backfilling algorithm is formulated in the Petri-Net with five-tuple form as:

$$PN = \{P, T, I, O, M\}$$

where

P   = task
T   = number of VMs required by P
I   = input function with start time S
O   = output function
M   = initial marking at time t

The transition firing is based on transition enable that initiated by input function and fire rules by which tasks are scheduled within the deadline. To model the backfilling algorithm in Petri-Net, we proposed following steps for scheduling the tasks. Moreover, we also proposed the firing rules by which the tasks changed the position that is scheduled, and VMs on the physical machine goes from busy state to free state and vice versa. The proposed algorithms are based on the earliest deadline of the task, number of VMs required by the task, and the duration of the execution.

3.2.1 Proposed Steps

Step 1: Select the task or lease according to its earliest deadline.
Step 2: Find number of VMs required by that task.
Step 3: If resource available, fire transition.
Step 4: Move allocated VMs to busy state from free state.
Step 5: Execute lease for predefined time.
Step 6: Fire another transition. VMs go to free state.

3.2.2 Proposed Rules for Firing

Rule 1. Select transition delay (T_{dl}) as minimum execution time (E_i) of the leases in queue.

$$T_{dl} = \min (E_i)$$

Rule 2. If $E_i > T_{dl}$, return the lease to original place else move it out.

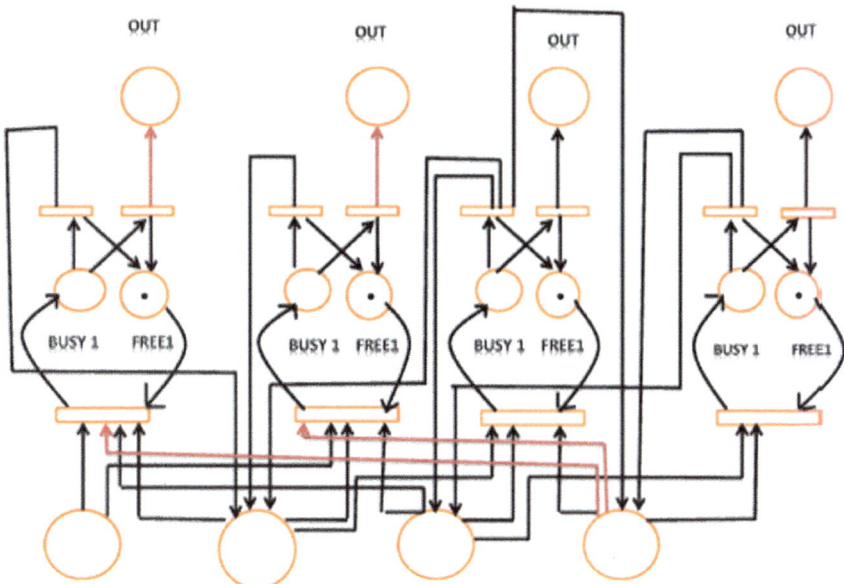

Fig. 2 Proposed Petri-net model

The Fig. 2 shows the proposed model for backfilling algorithm in Petri-Net. In each stage of the proposed model, VM has two states: busy state and free state. When the VM is allocated to any task, it goes to busy state for the time duration of the execution duration of the task. When the task is completed, the VM goes to free state. For better understanding, we considered the below illustration.

3.3 Illustration

Table 1 shows the leases or tasks information. There are four tasks with maximum node requirement is four by the task four. The tasks are having their start time, duration, and deadline. Though max (N) is 4, so in the physical machine, four VMs will be created to schedule these tasks.

Table 1 Information of leases or tasks

Lease no.	Nodes	Submit time (AM)	Start time (PM)	Duration	Deadline (PM)
1	2	11.10	12.00	20	12.30
2	3	11.20	12.00	40	01.00
3	2	11.30	12.00	30	01.50
4	4	11.40	01.00	20	01.50

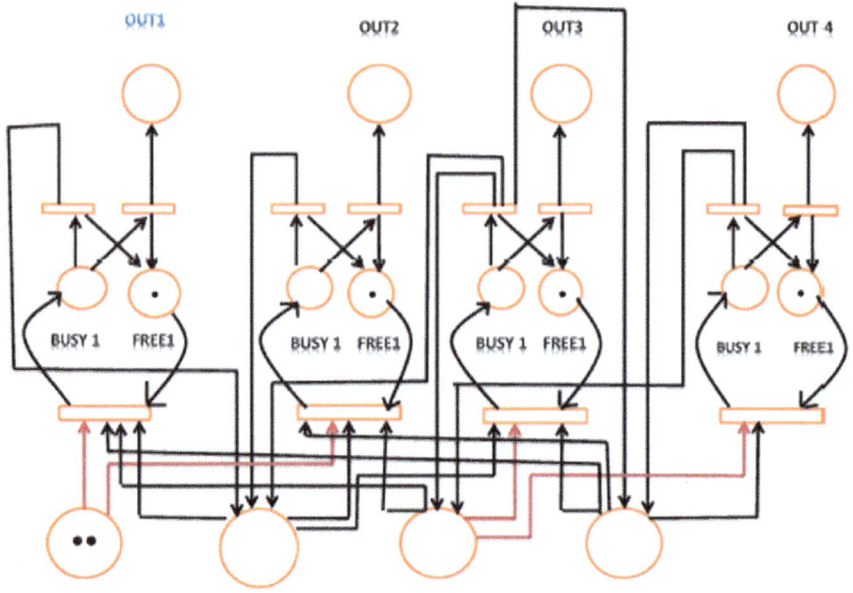

Fig. 3 Initial State of the model before task 1 fired

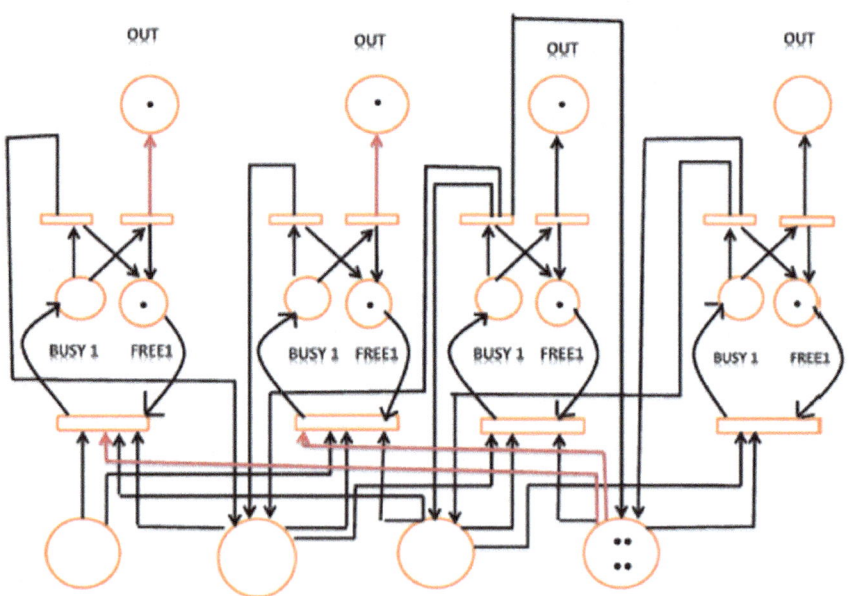

Fig. 4 Final State of the model before task 4 fired

According to the proposed mechanism task, 1 has the earliest deadline 12.30 PM. It needs two VMs, so Task 1 will be fired and two VMs go to busy state for the duration of twenty minutes. When Task 1 is fired two more, VMs are in free state. So, Task 3 is selected and fired for twenty minutes out of its execution duration thirty minutes. Similarly, all the tasks are fired for scheduling within their deadline. The Fig. 3 shows the initial stage of the model, whereas the Fig. 4 shows the final stage of the illustration.

In Fig. 3, all the VMs are in the free state. The Task 1 is going to be fired, and the out1 is highlighted. After firing, if task 1, 2, and 3 all four VMs are in free state, so task four can be fired, though it needs four VMs, as shown in Fig. 4.

4 Simulation and Result Analysis

In this work, we modeled the system to schedule the deadline-based task using Petri-Net. We designed the model using PIPE (Platform Independent Petri-Net Editor) which is an open source Petri-Net tool. In this work, we proposed the model which is shown in Fig. 2. Similarly, we performed a number of experiments in the designed model to schedule the deadline-based tasks as discussed in Sect. 3.3. The Figs. 3 and 4 are the different states of the model when it schedules the tasks.

The performance of the proposed model is evaluated by number of task input, number of tasks scheduled, and number of task rejected. The proposed work aims to model backfilling algorithm using Petri-Net. In the Table 2, the comparison results are shown. The number of tasks scheduled and rejected is same in both backfilling algorithm and its model in Petri-Net. The Petri-Net model is designed according to the basic scheduling policy of backfilling algorithm. The Fig. 5 shows the comparisons of number of tasks scheduled in backfilling algorithm and its proposed Petri-Net model. The Fig. 6 also shows the comparison of number of tasks rejected in backfilling and its proposed Petri-Net model of Table 2.

The Figs. 5 and 6 show the comparison results of the existing backfilling algorithm and its proposed Petri-Net model. Though we have modeled the backfilling algorithm in Petri-Net results of number of scheduled tasks and rejected tasks are same. The proposed Petri-Net model performance is same as backfilling performs. So the proposed Petri-Net model is the model of backfilling algorithm.

Table 2 Comparison result task scheduled and rejected

Experiment sl no	Number tasks input	Backfilling		Proposed Petri-net	
		Allocated	Rejected	Allocated	Rejected
1	4	4	0	4	0
2	13	11	2	11	2
3	20	16	4	16	4
4	30	23	7	23	7
5	40	28	12	28	12

Fig. 5 Comparison of tasks
scheduled

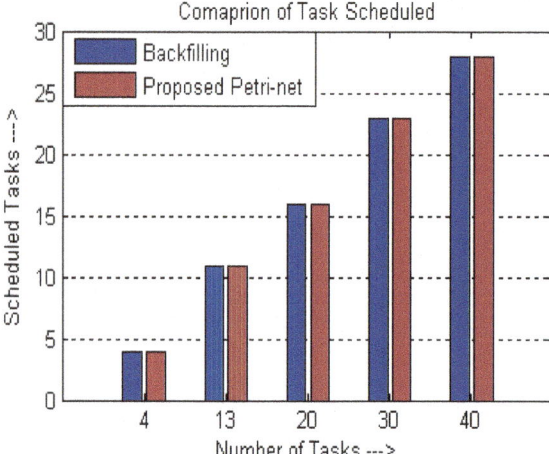

Fig. 6 Comparison of tasks
rejected

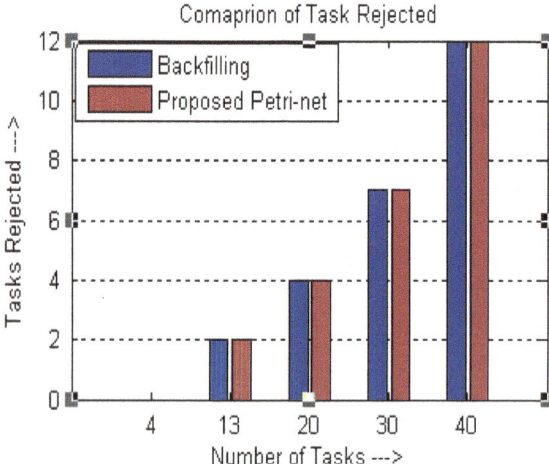

5 Conclusion and Future Work

In this proposed work, we designed the Petri-Net model for the existing backfilling
algorithm. The existing backfilling algorithm is used for scheduling deadline-based
tasks in OpenNebula. The performance of the backfilling algorithm and its proposed Petri-Net model is same. The number of tasks both scheduled and rejected is
same in both the mechanism. Here we have conducted five experiments in the
proposed model, more experiments can be also conducted, and the performance can
be observed in future.

References

1. Alkhanak, E.N.: Cost optimization approaches for scientific workflow scheduling in cloud and grid computing: a review, classifications, and open issues. J. Sys. Softw. **113**, 1–26 (2016). http://www.sciencedirect.com/science/article/pii/S0164121215002484
2. Li, X., Cai, Z.: Elastic resource provisioning for cloud workflow applications. IEEE Trans. Automa. Sci. Eng. (Jan), 1–16 (2015). http://ieeexplore.ieee.org/lpdocs/epic03/wrapper.htm?arnumber=7352380
3. Sotomayor, B.: Resource leasing and the art of suspending virtual machines. In: 2009 11th IEEE International Conference on High Performance Computing and Communications, pp. 59–68 (2009). http://www.computer.org/portal/web/csdl/doi/10.1109/HPCC.2009.17
4. Nathani, A., Chaudhary, S., Somani, G.: Policy based resource allocation in IaaS cloud. Futur. Gener. Comput. Syst. **28**(1), 94–103 (2012). https://dx.doi.org/10.1016/j.future.2011.05.016
5. Parida, S., Nayak, S., Tripathy, C.: Truth Full Resource Allocation Detection in Cloud Computing. WCI, ACM, New York, NY, USA, pp. 487–49 (2015)
6. Parida, S., Nayak, S.C.: Study of deadline sensitive resource allocation scheduling policy in cloud computing **3**(12), 521–528 (2014)
7. Kumar, A.V., Ganesh, L.S.: Use of Petri Nets for resource allocation in projects **45**(1), 49–56 (1998)
8. Colom, J.: The resource allocation problem in software applicationss: a Petri Net perspective, 219–233 (2012)
9. Nayak, S.C., Tripathy, C.: Deadline sensitive lease scheduling in cloud computing environment using AHP. J. King Saud. Univ.—Comput. Inf. Sci. (2016). https://dx.doi.org/10.1016/j.jksuci.2016.05.003
10. Di, S., Kondo, D., Wang, C.L.: Optimization of composite cloud service processing with virtual machines. IEEE Trans. Comput. **64**(6), 1755–1768 (2015)
11. Sotomayor, B.: Capacity leasing in cloud systems using the opennebula engine, 1–5 (2008). http://scholar.google.com/scholar?hl=en&btnG=Search&q=intitle:Capacity+Leasing+in+Cloud+Systems+using+the+OpenNebula+Engine#0
12. Calheiros, R.N., Buyya, R.: Meeting deadlines of scientific workflows in public clouds with tasks replication. IEEE Trans. Parallel Distrib. Syst. **25**(7), 1787–1796 (2014)
13. Chaisiri, S., Member, S.: Optimization of resource provisioning cost in cloud. Computing 5 (2), 164–177 (2012)
14. Duan, H.: Energy-aware scheduling of virtual machines in heterogeneous cloud computing systems. Futur. Gener. Comput. Syst. (2016). http://linkinghub.elsevier.com/retrieve/pii/S0167739X16300292
15. Kalra, M., Singh, S.: A review of metaheuristic scheduling techniques in cloud computing. Egypt. Inform. J. **16**(3), 275–295 (2015). http://linkinghub.elsevier.com/retrieve/pii/S1110866515000353
16. Ergu, D.: The analytic hierarchy process: task scheduling and resource allocation in cloud computing environment. J. Supercomput., 835–848 (2011)
17. Al-azzoni, I.: Server consolidation for heterogeneous computer clusters using Colored Petri Nets and CPN Tools. J. King Saud. Univ.—Comput. Inf. Sci. **27**(4), 376–385 (2015). http://dx.doi.org/10.1016/j.jksuci.2015.02.001
18. Kato, E.R.R.: A Conflict solution manufacturing system modeling using fuzzy coloured petri net, 3983–3988 (2010)

Cuckoo Search on Parallel Batch Processing Machines

Arindam Majumder and Dipak Laha

Abstract This paper presents a new version of discrete cuckoo search algorithm to minimize makespan on parallel batch processing machines. In the proposed algorithm, we consider a modified Lévy flight based on job interchange and job insertion techniques to generate new cuckoos by random walk. The initial population of the algorithm is generated using best-fit heuristic approach. Results of computational experimentation with a large set of random instances of non-sparse parallel batch processing with unequal job ready times show that the proposed algorithm significantly performs better than some state-of-the-art algorithms.

Keywords Scheduling · Batch processing machines · Makespan
Cuckoo search algorithm

1 Introduction

Each batch processing machine executes a set of jobs in a batch simultaneously. This research is motivated by the testing burn-in operation of assembled printed circuit boards (PCBs) in an electronic manufacturing unit. In batch processing, since a set of jobs are processed in a batch in the machine at a time, the processing time of each batch is taken as the longest processing time among all jobs assigned to the batch. Therefore, the batch scheduling is a special case of classical scheduling

A. Majumder (✉)
Mechanical Engineering Department, National Institute of Technology,
Barjala, Jirania, Agartala 799046, India
e-mail: arindam2012@gmail.com

D. Laha
Mechanical Engineering Department, Jadavpur University, Kolkata 700032, India
e-mail: dipaklaha_jume@yahoo.com

© Springer Nature Singapore Pte Ltd. 2018
K. Saeed et al. (eds.), *Progress in Advanced Computing and Intelligent Engineering*,
Advances in Intelligent Systems and Computing 563,
https://doi.org/10.1007/978-981-10-6872-0_62

where each machine processes one job at at a time. In batch processing, the efficiency of the processing depends on both sequencing of batches in the machines and allocation of jobs in the batches and it is important to find the optimal sequences of batches in the machines along with the allocation of jobs in the corresponding batches.

There has been growing number of studies on scheduling problems related to parallel batch processing problems. Researchers proposed various heuristics to minimize the makespan of flowshop with parallel batch processing machines [1–9] apart from implementing exact optimization approaches, namely, mixed integer programming and branch and bound algorithms for solving these problems [10–12]. Also, researchers applied different metaheuristics for solving these problems [13–16] and have shown that these algorithms are effective for solving batch scheduling problems. Recently, cuckoo search (CS) algorithm has drawn a lot of attention of researchers in different areas of engineering and management. A number of studies [17–24] have shown its promising efficiency in solving different optimization problems including the discrete optimization problems. Zhou et al. [25] proposed a new discrete cuckoo search algorithm to solve traveling salesman problems. The unique features of this algorithm consist of learning operation, 'A' operator, and 3-opt operator to the bulletin board. Based on a number of benchmark examples, they have shown that the proposed algorithm performs better than the existing algorithms. Dasgupta et al. [26] presented a discrete inter-species CS algorithm to solve two significantly important flowshop scheduling problems. In order to convert the continuous inter-species CS algorithm into discrete form, a heuristic rule known as smallest position value was implemented in their study. The computational results indicate the superiority of discrete CS algorithm over the existing metaheuristic algorithm. Ouaarab et al. [27] introduced an improved discrete CS algorithm and illustrated its use by solving NP-hard combinatorial optimization problem [28]. In their algorithm, the improvement was done by restructuring its population and by implementing a new class of cuckoo. The computational results show the superiority of the proposed algorithm with respect to some other metaheuristics. Therefore, based on the literature review, it is found that application of cuckoo search algorithm to parallel batch processing machines was not reported.

The objective of the present study is to develop a discrete cuckoo search algorithm to solve parallel batch processing scheduling problems. Two mutation techniques, namely, job interchange and job insertion are used for Lévy flights random walk with a probability of P_t. In addition to that, the initial population for the proposed algorithm is generated using best-fit heuristic algorithm. A set of test problems related to parallel batch processing machines with unequal job ready times are considered for evaluating the performance of the proposed algorithm. In this study, the test problems are generated randomly. The solutions generated by the proposed discrete cuckoo search algorithm are compared with the solutions of two existing heuristics, namely modified delay (MD) [1] and greedy randomized adaptive search procedure (GTASP) [8] and simulated annealing (SA) [15].

2 Problem Description

In this study, a scheduling problem related to parallel batch processing machines with unequal job ready times is taken under consideration. Firstly, we describe some notations used in the problem under study. We consider a set (J) of n jobs to be processed on m identical machines having batch processing capability. Each set of job j ∈ J is associated with three parameters namely, processing time, ready time, and size. For each batch, the processing time is considered as the longest processing time taken by the job among all jobs assigned to the batch. Similarly, the ready time for each batch is the maximum job ready time among the jobs in the batch. Two independent decisions, namely (i) grouping of jobs into batches and (ii) sequencing of batches, are taken into consideration. We assume that splitting of job and preemptions of machines are not permitted.

3 Discrete Cuckoo Search Algorithm

Cuckoo Search algorithm proposed by Yang and Deb [29, 30] is one of the newly introduced metaheuristic algorithms, inspired by brooding behavior of some cuckoo species. In these species, cuckoos do not build their own nest and lay their own eggs in the others' nests, providing the host the responsibility of and nurturing the young cuckoos. However, if the host birds discover these alien eggs then, either they threw them from their own nest or build a new nest somewhere by simply getting rid of the nest. Thus, the cuckoo search algorithm is constructed based on the three following basic rules:

- Each of the cuckoos lays one egg in a nest randomly.
- The nest with higher quality egg is chosen as the best nest for the next generation.
- The total number of host nests is fixed and a host bird can identify a foreign egg with a probability of Pa ∈ (0, 1).

3.1 Initial Population Generation

In the proposed DCS algorithm, a one-dimensional search space has been considered, where the position of each cuckoo is updated by mutation operation. The encoding of each cuckoo is carried out by considering the allocation of job in batches, i.e., $b \in [b_1\ b_2\ b_B\ b_3...b_2]^J$, while the ordered list of batch allocation in machines, i.e., $m \in [m_1\ m_2\ m_1\ ...m_1]^B$ is taken as constant. For example, if $J = 10$ and $B = 4$, then the string can be represented by [1-2-1-3-3-4-1-2-2-4]. To obtain the population at the initial stage, a well-known heuristic known as best-fit heuristic of classical bin packing [31] is used for this algorithm.

3.2 Discrete Random Walk Using Lévy Flight

In the cuckoo search algorithm, Yang et al. [29, 30] utilized Lévy flight random walk for generating new cuckoo. In this step, each of the nests, excluding the best one, is substituted by the new nest with higher quality eggs. The equation used to generate new cuckoo by implementing Lévy flight is as follows:

$$\text{Nest}_i^{t+1} = \text{Nest}_i^t + \alpha \cdot S \cdot \left(\text{Nest}_i^t - \text{Nest}_{\text{Best}}^t\right) r \tag{1}$$

where Nest_i^t is the current position of ith nest at tth iteration. $\text{Nest}_{\text{Best}}^t$ is the position of best nest at tth iteration. α indicates the step size parameter, r is the random number generated from standard normal distribution, and s is the step length based on Lévy distribution. However, for the proposed algorithm, the step length S has been calculated using Mantegna's algorithm [30]. The equation used for determining the step length is as follows:

$$S = \frac{u}{|v|^{\frac{1}{\beta}}} \tag{2}$$

where the parameter β is in between [1, 2]. u and v are calculated from normal distribution as:

$$u \sim N\left(0, \sigma_u^2\right), v \sim N\left(0, \sigma_v^2\right) \tag{3}$$

where

$$\sigma_u = \left[\frac{\Gamma(1+\beta)\sin\left(\frac{\pi\beta}{2}\right)}{\Gamma\left\{\frac{(1+\beta)}{2}\right\}\beta^{2\left(\frac{1-\beta}{2}\right)}}\right]^{\frac{1}{\beta}}, \sigma_v = 1 \tag{4}$$

A new approach has been proposed to produce new cuckoo efficiently by implementing Lévy flight. In this approach, the Lévy flight is performed considering two mutation operations, job insertion and job interchange with a probability P_L, while the round value of the term $[\alpha \cdot S \cdot (\text{Nest}_i^t\text{-Nest}_{\text{Best}}^t)r]$ is treated as the number of random job insertion and job interchange operation. The differences between the positions of two cuckoos are obtained by calculating the total number of elemental difference between the two sequences. Suppose [1-2-1-3-3-4-1-2-2-4] is the location of C_1 cuckoo and accordingly, [1-1-2-3-4-3-1-2-2-4] is the location

```
Begin
(σ¹_B, σ²_B, σ³_B, ... ... σᴾ_B) ← Initial Best − fit heuristic solution
    while F^Current best < F^Previous best
        F^Current best = F^previous best
            while iter < iter_Max | | F^Current best > LB | | F^iter_Current best > F^(iter+Max Count)_Current best
                for j=1 to P
                    σ^new_B ← Levy flight(σʲ_B, σ^Current best_B)
                    F^new=Fitness Function(σ^new_B)
                    if F^new<Fʲ
                        σʲ_B←σ^new_B
                    end if
                end for
                Rank the fitness of the solutions
                Abandon a fraction P_a of worse nests
                Keep nests with top quality solutions
                Find the current best (σ^Current best_B)solution
            end while
            σ_B ← Insert new batch
    end while
    Post process results
Stop
```

Fig. 1 Proposed discrete cuckoo search algorithm

of C_2 cuckoo, then the calculated discrete difference is 4. Figure 1 shows the pseudo-code of proposed discrete cuckoo search algorithm (DCS).

4 Computational Results

In this section, in order to determine the efficiency of the proposed algorithm, 200 instances of randomly generated non-sparse parallel batch processing problems are considered. The parameter values used to generate these problem instances taken from Damodaran et al. [8] are n = [10, 20, 50, 100, 500], P_{Max} = [10, 30], S_{Max} = [5, 20], and ρ = [0.05, 0.1], while the following input parameters used by DCS are: P_L = 0.3, β = 1.5, α = 0.01, and P_a = 0.25. The proposed algorithm is coded in MATLAB 2009a programming environment and executed in a PC equipped with Intel i5-2450 M CPU with 4 GB in RAM running at 2.50 GHz. The performance parameter used for this study is the percentage of gap between the optimal solution generated by the algorithm (C^A_{Max}) and the lower bound (C^{LB}_{Max}) and it is defined as follows:

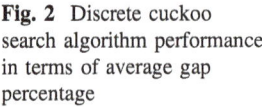

Fig. 2 Discrete cuckoo search algorithm performance in terms of average gap percentage

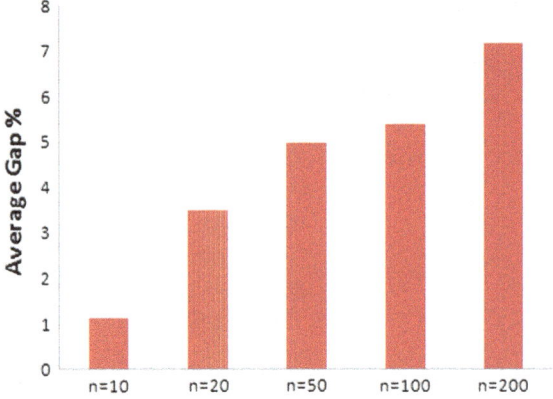

$$\text{Gap}\% = \left[\frac{C_{\text{Max}}^{A} - C_{\text{Max}}^{LB}}{C_{\text{Max}}^{LB}} \times 100\right]\% \qquad (5)$$

Figure 2 presents the computed average gap % between the optimal solution generated by the DCS and the lower bound for 10, 20, 50, 100, and 200 job instances. From the figure, it is seen that the change in gap percentage for all the instances varies from 1.11 to 7.14%, which indicates stability in the gap % even for the increased part size.

For more illustration, the solution achieved by the proposed DCS is compared with the solutions achieved by existing MD [1], GRASP [8], and SA [15] in terms of average gap% between the achieved optimal solution and the lower bound. Figure 3 shows the average gap percentage obtained by the DCS and existing MD, GRASP and SA. From the figure, it has been observed that the proposed DCS produces significantly improved solution as compared to MD, GRASP, and SA. On an average, 25.87% of the improvement has been seen by using DCS as compared to existing SA, while comparing with the solutions of MD and GRASP, an average improvement of 22.25 and 60.79% has been observed by using DCS.

5   Conclusion

In the present research work, we have presented a new discrete cuckoo search algorithm to minimize makespan for the parallel batch processing machines scheduling problem. The motivation to develop this algorithm came from the limitation of commercial solvers to solve parallel batch scheduling problems. In this algorithm, two mutation techniques such as job interchange and job insertion are used to generate new cuckoos based on Lévy flight-based discrete random walk, while the best-fit heuristic is implemented to obtain the initial population of the proposed DCS algorithm. A number of instances related to non-sparse batch

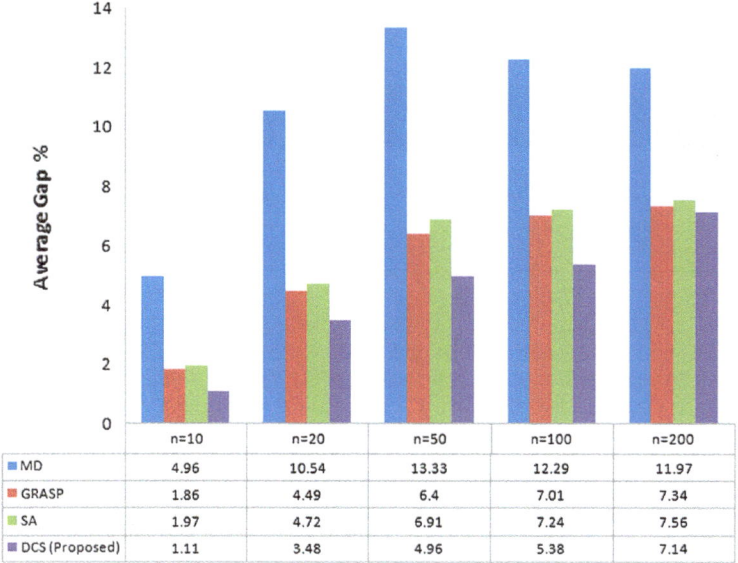

	n=10	n=20	n=50	n=100	n=200
MD	4.96	10.54	13.33	12.29	11.97
GRASP	1.86	4.49	6.4	7.01	7.34
SA	1.97	4.72	6.91	7.24	7.56
DCS (Proposed)	1.11	3.48	4.96	5.38	7.14

Fig. 3 Performance of MD, GRASP, SA, and proposed DCS in terms of average gap percentage

scheduling problems with unequal job ready times are considered to assess the efficiency of this algorithm. The performance measuring parameter considered to evaluate the efficiency of the algorithm is the gap percentage between the obtained solution and lower bound. The results produced by proposed method are compared with those given by modified delay, greedy randomized adaptive search procedure as well as simulated annealing and it is found that the proposed algorithm out-performs the existing algorithms.

References

1. Chung, S.H., Tai, Y.T., Pearn, W.L.: Minimising makespan on parallel batch processing machines with non-identical ready time and arbitrary job sizes. Int. J. Prod. Res. **47**(18), 5109–5128 (2009)
2. Gallego, V., Cesar, M.: Algorithms for scheduling parallel batch processing machines with non-identical job ready times (2009)
3. Li, X., et al.: Heuristics to schedule uniform parallel batch processing machines with dynamic job arrivals. Int. J. Comput. Integr. Manuf. **26**(5), 474–486 (2013)
4. Damodaran, P., Velez-Gallego, M.C.: Heuristics for makespan minimization on parallel batch processing machines with unequal job ready times. Int. J. Adv. Manuf. Technol. **49**(9-12), 1119–1128 (2010)
5. Dong, J., Jueliang, H., Lin, G.: A note on the algorithm LPT-FF for a flowshop scheduling with two batch-processing machines. Optim. Lett. **10**(1), 109–118 (2016)
6. Sáenz-Alanís, C.A., et al.: A parallel machine batch scheduling problem in a brewing company. Int. J. Adv. Manuf. Technol 1–11

7. Lei, D., Guo, X.P.: Variable neighbourhood search for minimising tardiness objectives on flow shop with batch processing machines. Int. J. Prod. Res. **49**(2), 519–529 (2011)
8. Damodaran, P., Vélez-Gallego, M.C., Maya, J.: A GRASP approach for makespan minimization on parallel batch processing machines. J. Intell. Manuf. **22**(5), 767–777 (2011)
9. Jia, Z.-h., Li, K., Leung Joseph, Y.-T.: Effective heuristic for makespan minimization in parallel batch machines with non-identical capacities. Int. J. Prod. Econ. **169**, 1–10 (2015)
10. Ozturk, O., Begen, M.A., Zaric, G.S.: A branch and bound based heuristic for makespan minimization of washing operations in hospital sterilization services. Eur. J. Oper. Res. **239** (1), 214–226 (2014)
11. Parsa, N.R., Karimi, B., Husseinzadeh Kashan, A.: A branch and price algorithm to minimize makespan on a single batch processing machine with non-identical job sizes. Comput. Oper. Res. **37**(10), 1720–1730 (2010)
12. Kosch, S., Christopher Beck, J.: A New MIP Model for Parallel-Batch Scheduling with Non-identical Job Sizes. Integration of AI and OR Techniques in Constraint Programming, pp. 55–70. Springer International Publishing (2014)
13. Wang, D.-Y., Grunder, O., Zhu, K.-J.: A hybrid coding SA method for multi-item capacity-constrained production and delivery scheduling problem with arbitrary job volumes and customer inventory considerations. Int. J. Ind. Syst. Eng. **22**(1), 17–35 (2016)
14. Al-Salamah, M.: Constrained binary artificial bee colony to minimize the makespan for single machine batch processing with non-identical job sizes. Appl. Soft Comput. **29**, 379–385 (2015)
15. Purushothaman, D., Vélez-Gallego, M.C.: A simulated annealing algorithm to minimize makespan of parallel batch processing machines with unequal job ready times. Expert Syst. Appl. **39**(1), 1451–1458 (2012)
16. Bayi, C., et al.: An improved ant colony optimization for scheduling identical parallel batching machines with arbitrary job sizes. Appl. Soft Comput. **13**(2), 765–772 (2013)
17. Chifu, V.R., Pop, C.B., Salomie, I., Suia, D.S., Niculici, A.N.: Optimizing the semantic web service composition process using cuckoo search. In: Intelligent Distributed Computing, pp. 93–102. Springer, Berlin, Heidelberg (2012)
18. Choudhary, K., Purohit, G.N.: A new testing approach using cuckoo search to achieve multi-objective genetic algorithm. J. Comput. **3**(4), 117–119 (2011)
19. Dhivya, M., Sundarambal, M., Anand, L.N.: Energy efficient computation of data fusion in wireless sensor networks using cuckoo based particle approach (CBPA). Intl. J. Commun. Netw. Syst. Sci. **4**(04), 249 (2011)
20. Dhivya, M., Sundarambal, M.: Cuckoo search for data gathering in wireless sensor networks. Int. J. Mobile Commun. **9**(6), 642–656 (2011)
21. Durgun, İ., Yildiz, A.R.: Structural design optimization of vehicle components using cuckoo search algorithm. Mater. Test. **54**(3), 185–188 (2012)
22. Gandomi, A.H., Yang, X.S., Alavi, A.H.: Cuckoo search algorithm: a metaheuristic approach to solve structural optimization problems. Eng. Comput. **29**(1), 17–35 (2013)
23. Kaveh, A., Bakhshpoori, T.: Optimum design of steel frames using cuckoo search algorithm with Lévy flights. Struct. Des. Tall Spec. Build. **22**(13), 1023–1036 (2013)
24. Yildiz, A.R.: Cuckoo search algorithm for the selection of optimal machining parameters in milling operations. Int. J. Adv. Manuf. Technol. **64**(1–4), 55–61 (2013)
25. Zhou, Y., Ouyang, X., Xie, J.: A discrete cuckoo search algorithm for travelling salesman problem. Int. J. Collab. Intell. **1**(1), 68–84 (2014)
26. Dasgupta, P., Das, S.: A discrete inter-species cuckoo search for flowshop scheduling problems. Comput. Oper. Res. (2015)

27. Ouaarab, A., Ahiod, B., Yang, X.S.: Discrete cuckoo search algorithm for the travelling salesman problem. Neural Comput. Appl. **24**(7–8), 1659–1669 (2014)
28. Graham Ronald, L., et al.: Optimization and approximation in deterministic sequencing and scheduling: a survey. Ann. Discret. Math. **5**, 287–326 (1979)
29. Yang, X.S., Deb, S.: Cuckoo search via Lévy flights. In: World Congress on Nature & Biologically Inspired Computing, 2009. NaBIC 2009, pp. 210–214. IEEE (2009
30. Yang, X.S.: Nature-Inspired Metaheuristic Algorithms. Luniver press, Frome (2010)
31. Rieck, B.: Basic Analysis of Bin-Packing Heuristics. Publicado por Interdisciplinary Center for Scientific Computing. Heidelberg University(2010)

Energy Saving Task Consolidation Technique in Cloud Centers with Resource Utilization Threshold

Mahendra Kumar Gourisaria, S. S. Patra and P. M. Khilar

Abstract The data centers are the world's biggest consumers of electricity. The consumption of energy in the cloud is proportional to the CPU utilization of the virtual machines (VMs). As the size of the cloud infrastructure increases the complexity of the resource allocation problem increases and becomes very difficult to solve it efficiently. This is an NP-Hard problem. There are several heuristics that may be used to solve the problem. Through task consolidation, we can get many benefits such as maximizing cloud computing resource, utilization of resources in a better way, efficient use of power, customization of IT services, Quality of Service, and other reliable services, etc. We find from the literature review that there is a high level of coupling between energy consumption and resource utilization. This paper presents the resource allocation problem in cloud computing with the objective to minimize energy consumed in computation. The simulation results show that a 70% principle of CPU utilization is the most energy efficient threshold for task consolidation in a virtual cluster. It has been verified with MaxUtil and ECTC (Energy Conscious Task Consolidation) algorithms.

Keywords Cloud computing · Virtual cluster · MaxUtil · ECTC
Energy efficient

M. K. Gourisaria (✉)
School of Computer Engineering, KIIT University, Bhubaneswar 751024, Odisha, India
e-mail: mkgourisaria2010@gmail.com

S. S. Patra
School of Computer Application, KIIT University, Bhubaneswar 751024, Odisha, India
e-mail: sudhanshupatra@gmial.com

P. M. Khilar
Department of Computer Science and Engineering, National Institute of Technology
Rourkela, Rourkela 769008, Odisha, India
e-mail: pmkhilar@nitrkl.ac.in

© Springer Nature Singapore Pte Ltd. 2018
K. Saeed et al. (eds.), *Progress in Advanced Computing and Intelligent Engineering*,
Advances in Intelligent Systems and Computing 563,
https://doi.org/10.1007/978-981-10-6872-0_63

1 Introduction

The development of the technologies, network devices, software applications, and hardware capacities made cloud computing a popular computing paradigm. In all these computing systems, the resources can be dispersed widely and the taking part of resources can be ranged from numerous physical servers to an intact data center. Proficient methods of management are required to join together and make the resource use optimally at different scales to make the environment of the cloud computing more proficient. Therefore, in current years, the center of attention of much research has been shifted to exploit resources efficiently and to minimize the power consumption done by data centers.

Virtualization is one of the key technologies in cloud computing can create virtual machines (VMs) dynamically as per the dynamic require on the physical machines. Therefore, numerous techniques have been proposed and developed which has been improved the resource utilization such as task allocation, memory firmness, defining threshold for use of resources, and request inequity among VMs. There has been several research which aims to improve the resource utilization as well as to reduce the energy consumption in different conditions. But there needs a substitution between the entities. The overall goals of both the situation are to reduce the cost for data centers. There exists a linear relationship between the parameters of power consumption and energy consumption. Utilization of CPU is related to the energy consumption, as the utilization of CPU is directly proportional to the energy consumption. However, if there is a higher CPU utilization of the system, that does not equate to the energy efficiency with in that system. Thus, this process has been motivated the idea to maintain a threshold level for the CPUs with high levels of utilization for saving the energy. The cloud environment has been becoming popular by reducing energy consumption and increasing profit without any eroding of service-level agreements mentioned by the users.

Power management has been broadly classified into two types, static management and dynamic management where static power management has been dealing with the fixed power, and dynamic power management has the dynamic behavior for additional degree of capability in virtualized data centers [1]. Software-as-a-Service (SaaS), Platform-as-a-Service (PaaS), Infrastructure-as-a-Service (IaaS), and Database-as-a-Service (DaaS) are the four level of access in which clouds has been deployed to the clients. The task has been originated by the different type of customer according to their requirements. There are several heuristic algorithms that have been proposed by a local cloud for the centralized controller which has been power aware. Based on the structure and the characteristic of the cloud infrastructures, a function between the resources of cloud and the combinatorial allocation task has been proposed, as an economic-based optimization model.

The concept of virtualization which encapsulates the variety of services that has been met the user need in the scenario of cloud computing [2]. VMs have been designed to run on various servers which provide the multiple operating system environments for different application. Particularly, executing an application which requires resource has been made available for resource provisioning and VM provisioning. Resource provisioning is scheduling the request for the physical resources where as VM provisioning creates the instance of VM as required by the different applications [3].

The task consolidation allows the servers on a single physical server for minimization of energy consumed by a cloud data center. In the present paper, task consolidation problem has been addressed to allot n task to a set of resources, and the utilization of nodes and distributed VMs are maintained by the energy efficiency and load management. The availability of computer nodes during the power consumed by the cloud is the prime concern of the algorithm which has been derived [4, 5].

In this paper, it has been evaluated and implemented the greedy heuristic algorithm for 3 basic task consolidations which assigns tasks to the servers in order to reduce the total energy. It has also been shown the performance improvement based on the different tasks. In Sect. 2, it has been discussed the various related work concern with the task consolidation and in Sect. 3, it defines general model of cloud computing environment, energy consumption, and task model of the system. On the basis of the system model, it has defined the problem to minimize the energy. Sections 4 and 5 have been dealt with the heuristic algorithms which are used and illustrated the algorithm by taking an example. Section 6 has been illustrated the setup for the simulation and analyzes the results generated by the simulation. At last, the conclusions have been discussed in Sect. 7.

2 Related Work

In the energy-aware task consolidation technique, the developers particularly in cloud environment, it requires to know the information on energy consumption of various components in the incorporated systems to manage efficient task allocation techniques. CPU is the most important component which affects the most in the energy consumption [6, 7].

Recent studies have been viewed that the power usage of thousands of servers and discover an identifiable gap in power in well-running applications. It is compared with activity range, and peak performance should be assumed for power efficiency [8]. It has been analyzed with the five varieties of high-level full system power models over a laptop to a server and analyzed that models based on utilization of OS and performance of CPU [9]. It has also been proposed PowerNap, which has two-state energy conservation approach, to simplify the complex power-performance states of systems [10]. Though the advancements in hardware technologies, such as energy efficient computer monitors, low power CPUs, and solid state drives have helped ease in the energy issues of energy to a certain degree, but research applying software practices such as resource allocation, task consolidation, and scheduling also plays an important role.

Energy efficient utilization of resources method provisionally reduces the level of voltage supply at the cost of lowering dispensation speed. Sagging repossession is completed likely for the most part by the parallel nature of the deploy tasks and DVFS-enabled processors. While the majority DVFS-based energy-aware scheduling and source allotment method are fixed (offline) algorithms with a hypothesis of stiff pairing between everyday jobs and resources (i.e., local tasks and dedicated resources), their function to the cloud situation is not noticeable, if not likely. Disk usages with CPU are the two major characteristics in traditional bin-packing for task consolidation [11]. The performance and the task consolidation balancing in energy consumption have been performed with the optimal points. So the proposed algorithm has incorporated with two steps in which initially determination of optimal points from energy-aware resource allocation and profiling data using the Euclidean distance between the current allocation and the optimal point at each server [5].

Again, a utility model has been proposed to consider the task being requested for service in web services of e-commerce [12]. The major aim of performance is the maximization of the utilization of resources which reduces the consumption of energy with the equivalent quality of service guarantee in the use of dedicated servers. Task consolidation mechanisms have been developed with energy fall using different methods. One approach is memory compression and other is request discrimination has been proposed for task consolidation [13].

The Virtual Power approach has been proposed which incorporates consolidation of task into its power management with hard and soft scaling methods. Both these techniques are stands on power management facilities equipped with VMs and physical processors. It has also been approached the game theoretic approaches for energy-aware scheduling for in grid environment [9]. The major intention is to reduce the energy utilization though maintaining a particular time and equality check value. The job offered is like on the way to that they together pact with autonomous jobs through the fixed and semi-fixed development mode [14].

3   Cloud System Model

The current section depicts the cloud, its function with the energy models. It also defines the job consolidation problem. The high-level architecture of the cloud system is shown in Fig. 1 [16]. Virtualization has been allowed the cloud providers to create multiple VMs on a single physical machine that improving the return on investment (ROI). The energy consumption may be reduced by switching off the idle nodes, which eliminates the idle power consumption of the given system [11].

In the present work, the target system has been used which consists of a set R of r resources which can be interconnected in the sense that a common route exists between whichever two individual resources that have been shown in Fig. 2. It assumes that the resources are identical in terms of their potential of computing. It can also be validated by using virtualization technologies. In the present study,

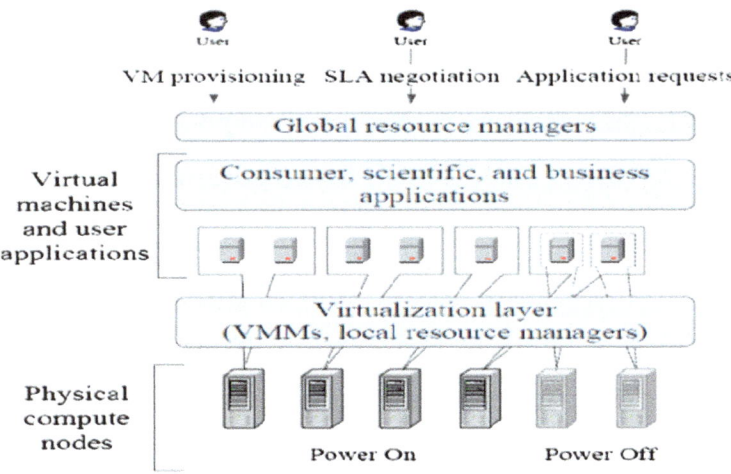

Fig. 1 High-level system view [15]

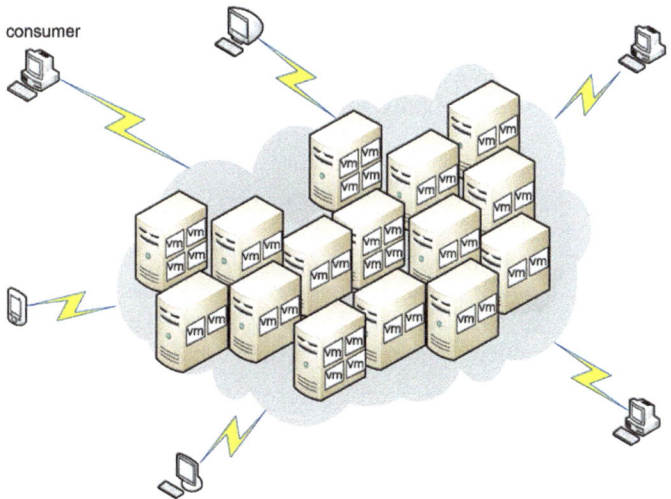

Fig. 2 Cloud model [3]

it has not considered the federated cloud environment in which the data centers can be placed at different physical locations and the client requests may be processed at various geographical locations.

Cloud providers provide the services in the form of DaaS, SaaS, PaaS, and IaaS. When instances of these desired services are running, they may be termed as computational tasks. Application consists of several tasks which are allocated to the different resources. IaaS responsible for the request with predefined point in time frame i.e., pay-per-hour, whereas DaaS, SaaS, and PaaS be frequently not stalwartly joined through unchanging quantity of instance i.e., pay-per-utilize. On the other hand, it can live to have an estimate for check desires for DaaS, SaaS, and PaaS based on past data. It assumes that the utilization of CPU of every service request can be classifiable. It is also presumed that the disk and memory use correlates with r utilization of processor in which task, service, and application are used interchangeably [6].

The energy consumption by data centers worldwide has risen by 56% from 2005 to 2010 which has been shown in Fig. 3. Furthermore, CO_2 emissions of the ICT industries are currently estimated to be 2% of the global emissions. It has been observed that global emissions are equivalent to the emissions of the aviation industries. The energy model is conceptualized on the basis that energy consumption has a linear relationship with processor utilization [7, 17].

This means, for a particular task, the processing time of a task and the processor utilization are the required parameters to determine the energy consumption for that task. For a resource ri at any given time, we can define the utilization Ui as

$$U_i = \sum_{j=1}^{n} u_{i,j} \tag{1}$$

where a is the number of tasks running at the current time, and $u_{i,j}$ is the usage of resource of a task tj. The consumption of energy Ei of a resource ri at a given time is defined as

$$E_i = (P_{max} - P_{min}) \times U_i + P_{min}$$

where pmax is the power consumption at 100% utilization or the peak load, and pmin is the minimum power consumption as low as 1% consumption or in dynamic mode. In this paper, we assume that the resources in the objective arrangement are comprised through an efficient power saving method in favor of inactive time slot. Particularly, the energy use of an inactive source at any specified time is set to 10%

Fig. 3 Worldwide data center energy consumption 2000–2010 [16]

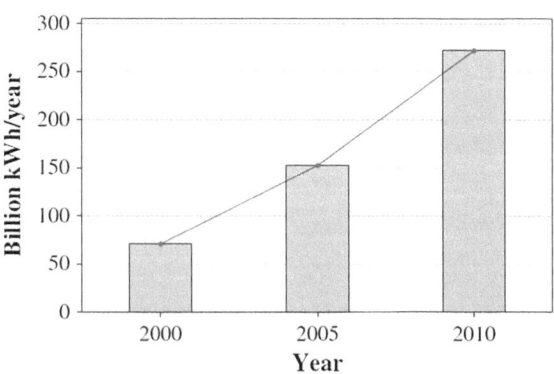

Fig. 4 Five levels of CPU
utilization [4]

$$E_t(V_i) = \begin{cases} \alpha \text{ watts/s, if idle} \\ \beta+\alpha \text{ watts/s, if } 0\% < \text{CPU utilization} \le 50\% \\ 2\beta+\alpha \text{ watts/s, if } 50\% < \text{CPU utilization} \le 70\% \\ 3\beta+\alpha \text{ watts/s, if } 70\% < \text{CPU utilization} \le 80\% \\ 4\beta+\alpha \text{ watts/s, if } 80\% < \text{CPU utilization} \le 90\% \\ 5\beta+\alpha \text{ watts/s, if } 90\% < \text{CPU utilization} \le 100\% \end{cases}$$

of pmin. VM can be partitioned into 6 different levels with the energy consumption, an idle state and five levels of CPU utilization at running state as shown in Fig. 4.

In the present study, the task consolidation problem is the process of assigning a set N of n tasks to a set R of r cloud resources without the violating time constraints which aiming to maximize resource utilization.

4 Task Consolidation Heuristic Algorithms

Task allocation is an NP-Hard problem in the cloud. Heuristic and meta-heuristic algorithms are the two effective and efficient technologies for scheduling in cloud due to the ability to deliver and distribute the optimized solutions. In this section, it presents the proposed task consolidation algorithm and has been compared the performance of the proposed algorithm with the existing ECTC and MaxMaxUtil energy conscious task consolidation algorithms.

4.1 MaxMaxUtil Algorithm

This algorithm selects all the tasks at an instance of time and stores them in a task queue. The task having the maximum need of CPU utilization will be scheduled to the VM which has currently maximum CPU utilization and by the inclusion of the task the CPU utilization of the VM must be below 100% or if there is any defined threshold value. If no such VM is available then the task will wait in the task queue until any suitable VM will get available.

4.2 ECTC Algorithm

ECTC has calculated the energy consumption of a given task and subtracts the idle consumption if other tasks are allocated on this physical machine at the same time. It uses the cost function as follows:

$$f_{i,j} = \{(P_{\max} - P_{\min}) * u_i + P_{\min}\} * t_1$$
$$- [(P_{\max} - P_{\min}) * u_i + P_{\min}\} * t_2 + (P_{\max} - P_{\min}) * u_i * t_3] \tag{2}$$

where t1 is the execution time of new task, t2 is the time new task executing alone and t3 is the time new task running in parallel with other tasks.

4.3 MaxMaxUtil Algorithm

```
Input : Task Matrix (mat)
Output : Allocation Table (Alloc)
1. [minArrivalTime maxArrivalTime]=
        FindMinimumArrivalTimeMaximumArrivalTime(mat)
2.time= minArrivalTime
3. while (time <= maxArrivalTime) do
4.       Tasklist = GetTasksatArrivalTime(mat , time)
5.       sort the tasklist in descending order of required CPU utilization
6.       for each task in tasklist do
7.            find the VM which has currently highest CPU Utilization
8.            Assign the task to the VM and update the Alloc table
9.    time= time +1
10. End Algorithm
```

4.4 ECTC Algorithm

```
Input : Task Matrix (mat)
Output : Allocation Table (Alloc)
1.   for each task in tasklist do
2.   for each vm in vmlist do
3.            max=-1
4.            E= EnergyConsumedIncludingTheTask(task,vm)
     //Allocate the task into the VM to which it will be maximum energy
     efficient
5.            if E > max
6.                max=E
7.                allocatedVm =Vm
8.            end if
9.    end for
10.   if allocatedVm !=NULL
11.           allocate task to allocatedVM
12.           Update the Alloc table
13.   end if
14. end for
15. End Algorithm
```

5 An Illustration

Consider a set of 10 VMs V = {V1, V2,…V10} and a set of 20 independent tasks T = {T1, T2, T3,…T20} in which each task Ti has 4 tuples {TaskId, Arrival Time, Processing Time, and CPU Utilization}. We have considered the threshold value of the CPU utilization as 100%. The task table has been shown in Table 1. Table 2 shows the energy used for MaxMaxUtil and ECTC algorithms for different CPU utilizations.

6 Simulation Results

It has been simulated the behavior of 2 task consolidation heuristic with 1000 tasks. The tasks bring out for different group of VMs with the use of incompatible ETC algorithm [6]. MATLAB 2012 has been used for simulation for 1000 task. The tasks are arriving to the central server queue with a rate of λ having unlimited queue

Table 1 Example of task table

Task Id	Arrival time	Processing time	Utilization
1	1	12	24
2	1	5	32
3	1	7	31
4	1	12	41
5	1	9	37
6	2	8	19
7	2	11	37
8	2	8	31
9	2	10	44
10	2	10	26
11	2	17	11
12	3	17	45
13	3	13	43
14	3	9	19
15	3	7	13
16	3	12	40
17	4	12	23
18	4	11	22
19	4	6	18
20	4	14	33

Tabel 2 Energy used in MaxUtil, ECTC algorithms

Algorithm	CPU utilization	Energy used
MaxMaxUtil	100	1.7033e+04
MaxMaxUtil	90	1.6328e+04
MaxMaxUtil	80	1.5873e+04
MaxMaxUtil	70	1.4873e+04
MaxMaxUtil	60	1.7366e+04
ECTC	100	1.7722e+04
ECTC	90	1.6243e+04
ECTC	80	1.5873e+04
ECTC	70	1.4873e+04
ECTC	60	1.6243e+04

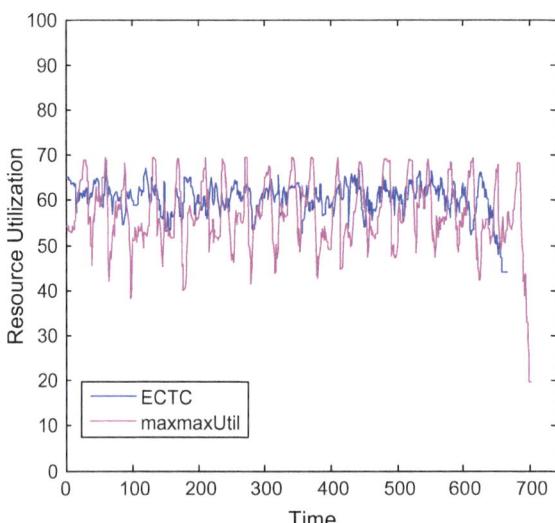

Fig. 5 Resource utilization comparison for 1000 tasks on 10 VMs

length. We have considered the task arrival interval as 1 and arrival rate to be 30 for our studies. The behavior of task consolidation algorithms is presented for 10 VMs in Fig. 5. Figure 6 shows the consumption of energy on 15 VMs by varying task size from 500 to 1500.

Fig. 6 Energy consumption
for number of tasks on 15

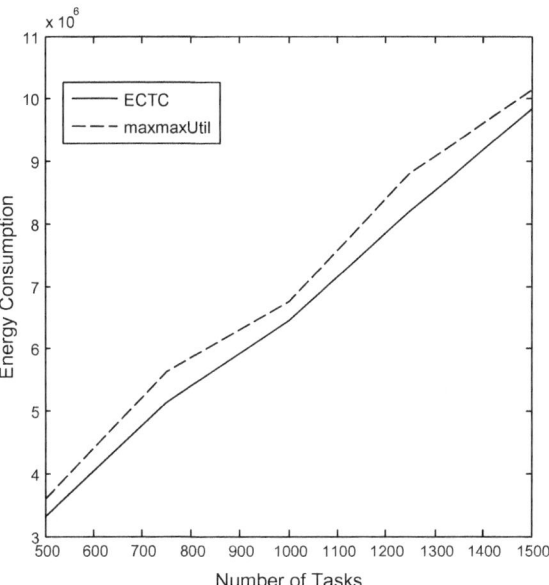

7 Conclusion

The simulation experiments have been successfully carried out which examines the performance of heuristic-based task consolidation algorithms. It has also been optimized the energy consumption in cloud environment. The performance analysis has been presented for the different task consolidation algorithm with ETC matrix. The present simulation results have shown that the CPU utilization threshold is 70% for energy efficiency.

References

1. Wen, G., Hong, J., Xu, C., Balaji, P., Feng, S., Jiang, P.: Energy-aware hierarchical scheduling of applications in large scale data centers. In: International Conference on Cloud and Service Computing (2009)
2. Kim, K.H., Buyya, R., Kim, J.: Power aware scheduling of bag-of-tasks applications with deadline constraints on DVS-enabled clusters. In: Proceedings of the Seventh IEEE International Symposium on Cluster Computing and the Grid (CCGrid'07), pp. 541–548 (2007)
3. Lee, Y.C., Zomaya, A.Y.: Energy efficient utilization of resources in cloud computing systems. J. Supercomput. **60**, 268–280 (2012)
4. Hsu, C.H., Chen, S.C., Lee, C.C., Chang, H.Y., Lai, K.C., Li, K.C., Rong, C.: Energy-aware task consolidation technique for cloud computing. In: 2011 IEEE Third International Conference on In Cloud Computing Technology and Science, pp. 115–121 (2011)

5. Srikantaiah, S., Kansal, A., Zhao F.: Energy aware consolidation for cloud computing. In: International Conference on PowerAware Computing and Systems (2008)
6. Ali, S., Siegel, H.J., Maheswaran, M., Hensgen, D.: Task execution time modeling for heterogeneous computing systems. In: Heterogeneous Computing Workshop, 2000 (HCW 2000) Proceedings, pp. 185–199 (2000)
7. Hsu, C., Chen, S., Lee, C., Chang, H., Lai, K., Li, K., Rong, C.: Energy-aware task consolidation technique for cloud computing. In: Third IEEE International Conference on Cloud Computing Technology and Science (2011)
8. Fan, X., Weber, X.-D., Barroso, L.A.: Power provisioning for a warehouse-sized computer. In: Proceedings of the 34th Annual International Symposium on Computer Architecture (ISCA'07), pp. 13–23 (2007)
9. Lee, Y.C., Zomaya, A.Y.: Energy efficient utilization of resources in cloud computing systems. J. Supercomput. (Springer) **60**, 268–280 (2012)
10. Meisner, D., Gold, B.T., Wenisch, T.F.: PowerNap: eliminating server idle power. In: Proceedings of the 14th International Conference on Architectural Support for Programming Languages and Operating Systems (ASPLOS '09), pp. 205–216 (2009)
11. Hsu, C.H., Slagter, K.D., Chen, S.C., Chung, Y.C.: Optimizing energy consumption with task consolidation in clouds. Inf. Sci. **258**, 452–462 (2014)
12. Lee, Y.C., Zomaya, A.Y.: Minimizing energy consumption for precedence-constrained applications using dynamic voltage scaling. In: Proceedings of the International Symposium on Cluster Computing and the Grid (CCGRID '09), pp. 92–99 (2009)
13. Tian, W., Xiong, Q., Cao, J.: An online parallel scheduling method with application to energy-efficiency in cloud computing. J. Supercomput. (Springer) **66**, 1773–1790 (2013)
14. Kim, N., Cho, J., Seo, E.: Energy-credit scheduler: an energy-aware virtual machine scheduler for cloud systems. Future Gener. Comput. Syst. (Elsevier) **32**, 126–137 (2014)
15. Beloglazov, A.: Energy-efficient management of virtual machines in data centers for cloud computing. Ph.D. thesis, Department of Computing and Information Systems, The University of Melbourne (2013)
16. Koomey, J.: Growth in data center electricity use 2005–2010. A report by Analytical Press, completed at the request of The New York Times 9 (2011)
17. Bojanova, I., Samba, A.: Analysis of cloud computing delivery architecture models. In: Proceedings of International Conference on Advanced Information Networking and Applications, pp. 45–458 (2011)

Cloud Service Ranking Using Checkpoint-Based Load Balancing in Real-Time Scheduling of Cloud Computing

Mohammad Riyaz Belgaum, Safeeullah Soomro, Zainab Alansari and Muhammad Alam

Abstract Cloud computing has been gaining popularity in the recent years. Several studies are being proceeded to build cloud applications with exquisite quality based on user's demands. In achieving the same, one of the applied criteria is checkpoint-based load balancing in real-time scheduling through which suitable cloud service is chosen from a group of cloud services candidates. Valuable information can be collected to rank the services within this checkpoint-based load balancing. In order to attain ranking, different services are needed to be invoked in the cloud, which is time-consuming and wastage of services invocation. To avoid the same, this chapter proposes an algorithm for predicting the ranks of different cloud services by using the values from previously offered services.

Keywords Load balancing · Checkpoint · Cloud services

M. R. Belgaum (✉) · S. Soomro · Z. Alansari
College of Computer Science, AMA International University, Salmabad
Kingdom of Bahrain
e-mail: bmdriyaz@amaiu.edu.bh

S. Soomro
e-mail: s.soomro@amaiu.edu.bh

Z. Alansari
e-mail: zeinab@amaiu.edu.bh; z.alansari@siswa.um.edu.my

M. Alam
Institute of Business and Management (IoBM), Korangi Creek, Karachi, Pakistan
e-mail: malam@iobm.edu.pk

M. R. Belgaum · M. Alam
Universiti Kuala Lumpur (IPS UniKL), Kuala Lumpur, Malaysia

Z. Alansari
University of Malaya, Kuala Lumpur, Malaysia

© Springer Nature Singapore Pte Ltd. 2018
K. Saeed et al. (eds.), *Progress in Advanced Computing and Intelligent Engineering*,
Advances in Intelligent Systems and Computing 563,
https://doi.org/10.1007/978-981-10-6872-0_64

1 Introduction

The term cloud originated from the network diagram that shows the Internet as a schematic cloud [1]. Cloud computing being a network-based computing connects heterogeneous systems in different types of the network like private, public, and hybrid infrastructure. The applications and services can be accessed over the network and Internet. The cloud computing resources like storage, processing, memory network bandwidth, and virtual machines are efficiently managed by the cloud service provider (CSP) [1–4] by making use of the available computing tools. Various cloud deployment models have been discussed as follows [1, 5].

Private cloud is restricted to one particular organization or institution. Moreover, the same organization or the third party is responsible for organizing and managing the cloud [1, 5]. In public cloud, the term public makes it accessible to all through the cloud service provider. The service providers follow "pay-as-you-use" as the service of the resources is rendered by them. Regularly, the services are offered to the users [1, 5] without the knowledge of the resource location. Community clouds have a comparable group of organizations functioning with the same interests to access the resources. The services are restricted to particularly those organizations in that group. One of the organizations in such a group will be responsible for organizing and managing the services [1, 5]. In hybrid cloud, the organizations use both private and public cloud based on the requirements. The services of such clouds are accessed using standardized interfaces [1, 5] (Fig. 1).

There are various common characteristics of cloud including massive scale, homogeneity, virtualization, resilient computing, low-cost software, geographic distribution, service orientation, and advanced security technologies [6]. Cloud services placed a number of challenges [7] to IT. Prominent of those are service level agreement, load balancing, security, the integrity of cloud services, costs, etc. (Fig. 2).

The checkpoint is a locally stable state of a process [8] which determines the consistency. Checkpoints are paramount and provide a point in time (PIT) copies of machines (VMs). Point in time (PIT) allows VMs to be rolled back to any point of the user's request. Checkpoints can be automatic and manual. Automatic checkpoints are taken regularly at every few seconds or as early as possible. These checkpoints are crash consistent and are useful during recovery. Manual checkpoints can be created by the users manually by allowing to control date and time. So it gives the user an option to do PIT recovery using manual checkpoints regularly. Executing the checkpoints helps to make the system be in a consistent state. And failure caused in any instance does not lead restarting of a process or to a long rollback. Checkpoints can be integrated with load balancing algorithms in order to provide high availability [9] to the requests of the clients in a no preemptive real-time scheduling environment.

The load balancer implements the virtualization technique [10] where the resources of the physical server at the data centers [11] are allocated to a wide range of virtual machines each of which performs the function of a physical server.

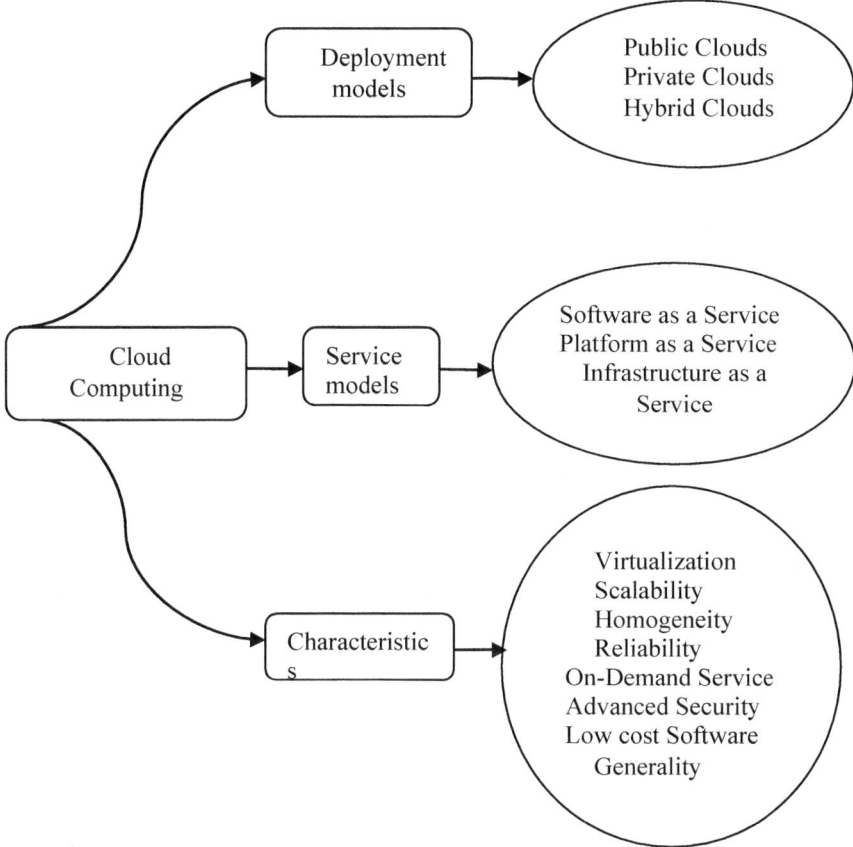

Fig. 1 Cloud computing overview

Whenever the checkpointing occurs, the load of the physical server is analyzed, and without preempting the resources from the available resources, they are allocated to the virtual machine to accomplish the task. At any instance of time, the resources required or allocated must not be more than available.

2 Related Work

In [2], Pandey has followed multi-agent system (MAS) in order to do the load balancing in cloud computing. MAS is a loosely coupled network of the software agents which work to solve issues which could not be solved using individual capacity. Authors used different parameters for load balancing which are reliability, configurability, and ability to modify.

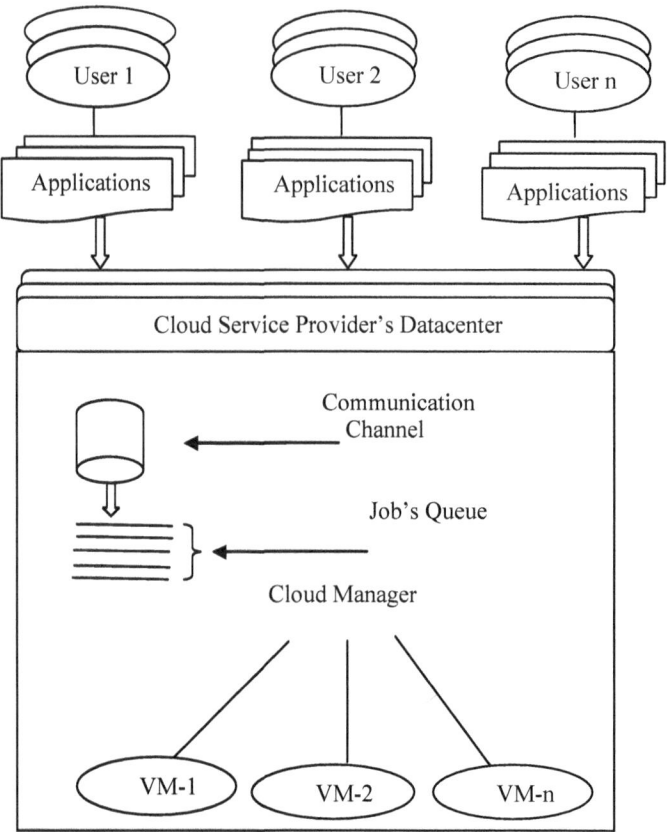

Fig. 2 Cloud architecture for load balancing

A checkpointing infrastructure for visualized service providers [1] was proposed by the authors. The authors used Another Union File System (AUFS) to bifurcate the read only from read and write parts in the virtual machine image. This idea almost works similar to the dirty bit (modified bit) architecture of the page replacement algorithm of operating system. They have used Hadoop Distributed File System for the checkpointing architecture.

Cooperative checkpointing discussed in [12] was compared with periodic checkpointing with an experimental analysis which shows the cooperative check-pointing helps the application or a process to proceed further under several group of failure distributions. Also, the cooperative checkpointing is used to improve the reliability techniques like QoS, fault tolerance, thereby making the system robust and increase the performance.

The authors proposed an algorithm [13] for reducing the execution time of tasks which are migrated from one VM to other VM when the deadline is not met by the task. Considering the metrics like throughput, profit, and loss, a simulation is carried out to show the non-preemptive real-time scheduling checkpointing decreases the execution time.

The virtual disk-based check-restart [14] reduces the overhead of checkpointing by implementing a technique known as "Selective copy-on-write" taking the snapshots of the disks. Checkpoint-restart proves to be more advantageous for the HPC applications as the execution resumes from the intermediate states and shows gain in the performance also reducing the bandwidth and storage.

For preventing the security threats in cloud computing, a design of internal traffic checkpoint model [3] was proposed. This model has three components which are used to identify and prevent threats to make the cloud resources secure.

The load balancing techniques [15] along with the metrics are discussed to meet the challenges in the cloud computing. Different types of load balancing algorithms are categorized based on system load and system topology. Considering the first category, the load can be centralized, distributed, and also mixed. Where as in the second category as the topology can be static, dynamic and adaptive so are the load balancing algorithms.

The identity-based authentication for cloud computing [16] focuses on the security and trust issue of the cloud computing. A new mathematical scheme based on encryption and signatures was proposed for the hierarchical model for cloud computing. A comparative analysis was made between SSL authentication protocol and the proposed identity-based hierarchical model for cloud computing (IBHMCC) authentication protocol model to prove that the IBHMCC was more efficient and was certificate free.

The authors in [17] have proposed a new real-time scheduling algorithm for cloud computing whose aim is to have a maximum utility of the resources using the time utility function. Two-time utility functions, namely profit and penalty, have been used in their work. The penalty was used to punish the tasks that have missed deadlines, and the profit was used to reward the tasks which have met the deadlines.

The CPU scheduling algorithms in cloud using CloudSim were analyzed by the authors in [18]. Using three of the CPU scheduling algorithms, the results were simulated to show which will better suit the requirements of the user in the virtual environment of cloud. The virtual machine manager decides to use which allocation policy and on to which virtual machine is it to be assigned. These experimental results help the cloud providers to charge the users based on their usage.

A preemptive cloud scheduling algorithm was used in [19] with a fixed priority assigned to every task in order to improve the QoS. Two variants of preemptive scheduling algorithms were mathematically proved to be fit for service-oriented tasks. In these algorithms, a dispatcher plays the role of preempting the low-priority task when a high-priority task arrives with less overhead and maintaining optimality to achieve QoS.

The authors in [20] made a survey of various load balancing algorithms in cloud computing presenting the challenges to be considered, while the tasks were to be assigned to the virtual machine in the cloud. The heuristic algorithms were proposed considering the load of the server as the load was monitored to disseminate the overload having a replica of the task at the machine where it was originated. The proposed approaches focused to improve the efficiency and satisfy the users.

In order to efficiently use the resources and to balance the load, a genetic algorithm [21] was proposed by the authors. They classified the algorithms as static, dynamic, and mixed scheduling algorithms. The methodology of considering the historical data and current state used calculates the load before it is deployed to a particular virtual machine. Using set theory, the relationship between physical machines and virtual machines was shown to know the load on each physical machine.

The authors in [22] considered load balancing as the main challenge along with achieving green computing with the various surveys. Because of the exponential increase of the cloud computing, the need for the data centers increased which in turn is resulting in excess of carbon emissions contaminating the environment. Various metrics to evaluate the load balancing algorithms along with carbon emission and energy consumption metrics were used to show which algorithm is efficient.

The authors in [23] proposed an algorithm named "Load Balance Max-Min and max algorithm" for efficient load balancing in less completion time. A threshold is calculated to show the average completion time in all the machines. A case study is taken to make a comparison between the proposed algorithm and other algorithms to prove that the proposed algorithm completes the task in less time compared to others.

Load balancing in conjunction with availability is discussed in [24] by the authors along with a hospital data management system. In that case study, the data of a patient need to be accessed by different doctors and nurses globally from different systems when the information of the patient is available. A resource manager is responsible for the complete operations like monitoring, availability, and performance.

3   Proposed Algorithm

The following section explains computation of the cloud service ranking using checkpoint-based load balancing. As checkpoints can be either automatic or manual, here in our context we have considered automatic checkpoints; as in manual-based system, it can be fixed by the user. But in the automatic checkpoints, they take place regularly to have a proper load balancing in a real-time environment. It is divided into two parts as selection of services based on user's requirements with checkpoint-based load balancing and the other ranking of corresponding services.

3.1   Service Selection Based on User's Requirements with Checkpoint-Based Load Balancing

The user will access the system based on the ranks obtained according to the services rendered before or have been accessed earlier. So initially, when all the jobs are submitted by the clients to the cloud service provider (CSP), the CSP evaluates the load on each of the sub-cloud and the number of times the jobs have been migrated from one sub-cloud to another. And whenever the jobs are migrated from one sub-cloud to other, it is sent along with the previously saved checkpoint as the job has to resume from that point. In order to calculate correspondent services and rank at each sub-cloud, a simple technique has been used. Based on the degree of correspondence between two services, the ranking can be calculated considering checkpoint-based load balancing according to user's requirements. For doing this, we will take the response time of a set of services with corresponding service nodes, and then, we will calculate the numbers of times these response times are inverted to convert one rank to other. The correspondence value of node is evaluated by

$$CV(\chi, y) = \frac{a - b}{n(n-1)/2} \tag{1}$$

where n is total services, a is total number of consistent pairs, and b is total number of variant pairs among two lists, $n(n-1)/2$ are the total number of pairs in the cloud with n services.

After evaluating the correspondence values among the node's service and the previously accessed similar services from the node, the corresponding node can be identified. For improving the accuracy of finding the corresponding node, we ignore the services with negative correspondence values and only include positive corresponding values.

Preferred nodes among the correspondent nodes are selected by subtracting the ranks of services,

$$P(\chi, y) = S_\chi - S_y. \tag{2}$$

where

$P(\chi, y)$ prefer value among node x and y,
S_χ rank of node x's service,
S_y rank of node y's service

The greater prefer value indicates that the service is more reliable than the other service.

3.2 Ranking Services

The system then arranges the ranks of all the services in an order. Then, the prefer values of each service are added with all the preferred values of the other services. The added value is priority value which means that the service at a particular node with greater priority value will be prioritized higher in the list.

$$PV = \sum_{y \in S} P(\chi, y). \tag{3}$$

where

PV   priority value of service x

The system then arranges the list having services with greater priority values higher in the list.

To improve the accuracy of rank prediction of services, the system prefers the greater priority values of implicit services which the user has already accessed.

Algorithm 1. Proposed Algorithm

```
Input: A set of service S, x is a cloud service and π
stacks in the ranking.
 Output: ranked service list x
Step1: for each service from 1 to n
Step 2:           calculate correspondence value of each
     service based on user's requirements using eq. (1)
Step 3: end for
Step 4: for each service from 1 to n
Step 5: calculate prefer value of each service using eq.
     (2)
Step 6: end for
Step 7: R=S;
Step 8: for each service from 1 to n
Step 9: rank each service on the basis of checkpoints and
     the load balancing, present on the cloud,
                x = max rank in S,
                π(x) = S-R+1;
                R=R-x;
Step 10: end for
Step 11: for each service from 1 to n
Step 12:    calculate the priority value of each service
     using eq. (3)
Step 13:    rank  the  services  on  the  basis  of  their
     priority values,
                R=μ(i)
                     a= max rank in priority value set,
           μ(i),
           π(x) = μ(i)-R+1,
           R=R-x,
Step 14: prioritize the implicit services with greater
                rank.
Step 15:    update  the  service  set  S  with  the  ranked
           services and save it in ranked service list
           x
Step 16: end for
```

4 Conclusion and Future Work

In this paper, cloud service ranking has been done in checkpoint-based load balancing in the real-time scheduling. In this framework, the user can access the services with higher rank by communicating with the users which have accessed the services in the past. By applying this technique, ranking of services can be improved and there will be less overhead in the cloud.

Simple ranking technique has been used in this framework. To improve the service ranking, other approaches can also be included that are based on rating the services. Other techniques can also be included to negate the checkpoint-based load balancing in real-time environment which reduce the prediction of services.

References

1. Rejinpaul, N.R., Visuwasam, M.L.: Check point based intelligent fault tolerance for cloud service providers. Int. J. Comput. Distrib. Syst. **2**(1), 59–64 (2012)
2. Pandey, R., Ranjan, R.M.S.: Distributed load balancing in cloud computing. In: International Conference on Computer Science and Information Technology, pp. 32–36 (2013)
3. Eom, J.H., Park, M.W.: Design of internal traffic checkpoint of security checkpoint model in cloud computing. Int. J. Secur. Appl. **7**(1), 119–128 (2013)
4. Adari, V.R., Diwakar, Ch., Varma, P.S.: Cloud computing with service oriented architecture in business applications. Int. J. Comput. Sci. Technol. **3**(1), 452–455 (2012)
5. Ahmed, M., Abu Sina, Md., Chowdhury, R., Ahmed, M., Rafee, M.H.: An advanced survey on cloud computing and state-of-the-art research issues. Int. J. Comput. Sci. **9**(1), 201–207 (2012)
6. Okuhara, M., Shiozaki, T., Suzuki, T.: Security architecture for cloud computing. FUJITSU Sci. Technol. J. **46**(4), 397–402 (2010)
7. Buyya, R., Garg, S.K., Calheiros, R.N.: SLA-oriented resource provisioning for cloud computing: challenges, architecture and solutions. In: IEEE International Conference on Cloud and Service Computing (2011)
8. Singh, D., Singh, J., Chhabra, A.: Evaluating overheads of integrated multilevel checkpointing algorithms in cloud computing environment. Int. J. Comput. Netw. Inf. Secur. 29–38 (2012)
9. Cao, N., Yu, S., Yang, Z., Lou, W., Hou, T.Y.: LT codes-based secure and reliable cloud storage service. In: Proceedings IEEE INFOCOM (2012)
10. Rashmi, K.S., Suma, V., Vaidehi, M.: Enhanced load balancing approach to avoid deadlocks in clouds. Spec. Iss. Int. J. Comput. Appl. Adv. Comput. Commun. Technol. HPC Appl. 31–35 (2012)
11. Tsai, W.T., Sun, X., Balasooriya, J.: Service oriented cloud computing architecture. In: IEEE Seventh International Conference on Information Technology, pp. 684–689 (2010)
12. Oliner, A.J., Rudolph, L., Sahoo, R.K.: Cooperative checkpointing: a robust approach to large-scale systems reliability. In: ICS 06: Proceedings of the 20th Annual International Conference on Supercomputing, pp. 14–23 (2006)
13. Santosh, R., Ravichandran, T.: Non preemptive realtime scheduling using checkpointing algorithm for cloud computing. Int. J. Comput. Appl. **80**(9) (2013)
14. Nicolae, B., Cappello, F.: BlobCR: virtual disk based checkpoint restart for HPC applications on Iaas clouds. J. Parallel Distrib. Comput. **73**, 5 (2013)

15. Sidhu, A.K., Kinger, S.: Analysis of load balancing techniques in cloud computing. Int. J. Comput. Technol. **4**(2), 737–741 (2013)
16. Hongwei, L., Dai, Y., Tian, L., Yang, H.: Identity-based authentication for cloud computing. In: CloudCom. LNCS, vol. 5931, pp. 157–166 (2009)
17. Liu, S., Quan, G., Ren, S.: Online scheduling of real-time services for cloud computing. In: 6th World Congress on Services, pp. 459–464 (2010)
18. Gahlawat, M., Sharma, P.: Analysis and performance assessment of CPU scheduling algorithms in cloud using Cloud Sim. Int. J. Appl. Inf. Syst. **5**(9), 5–8 (FCS, USA) (2013)
19. Dubey, S., Agrawal, S.: QoS driven task scheduling in cloud computing. Int. J. Comput. Appl. Technol. Res. **2**(5), 595–600 (2013)
20. Al Nuaimi, K., Mohamed, N., Al Nuami, M., Al-Jaroodi: A survey of load balancing in cloud computing: challenges and algorithms. In: Second Symposium on Network Cloud Computing and Applications, pp. 137–142 (2012)
21. Rawat, S., Bindal, U.: Effective load balancing in cloud computing using genetic algorithm. Int. J. Comput. Sci. Eng. Inf. Technol. Res. **3**(4), 91–98 (2013)
22. Kansal, N.J., Chan, I.: Cloud load balancing techniques: a step towards green computing. Int. J. Comput. Sci. Iss. **9**(1), 238–246 (2012)
23. Hung, C.L., Wang, H., Hu, Y.C.: Efficient Load Balancing Algorithm for Cloud Computing Network. Supported by NSC, pp. 251–253
24. Chaczko, Z., Mahadevan, V., Aslanzadeh, S., Mcdermid, C.: Availability and load balancing in cloud computing. Int. Conf. Comput. Softw. Model. **14**, 134–140 (2011)

E2G: A Game Theory-Based Energy Efficient Transmission Policy for Mobile Cloud Computing

Joy Lal Sarkar, Chhabi Rani Panigrahi, Bibudhendu Pati, Rajani Trivedi and Shibendu Debbarma

Abstract The mobile users in mobile cloud computing (MCC) environment suffer due to less energy and resources of mobile devices. Although there are several works have been proposed, those are not sufficient to overcome these problems. To accommodate this claim, in this work, a game theory-based energy efficient transmission policy for MCC has been proposed named as *E2G*. *E2G* works based on the subgame perfect Nash equilibrium (SPNE) where each mobile device works as a player and selects a best mobile device by choosing best strategy among all players. The mobile device which is chosen based on SPNE decision will be responsible for communication with the cloud. The simulation results show that *E2G* helps to minimize the energy consumption of mobile devices as compared to the existing approaches.

Keywords Mobile cloud computing · Offloading · Energy-efficiency · SPNE

J. L. Sarkar (✉) · C. R. Panigrahi
Central University of Rajasthan, Rajasthan, India
e-mail: joylalsarkar@gmail.com

C. R. Panigrahi
e-mail: panigrahichhabi@gmail.com

B. Pati
C.V. Raman College of Engineering, Bhubaneswar, India
e-mail: patibibudhendu@gmail.com

R. Trivedi
Biju Patnaik University of Technology, Rourkela, India
e-mail: rtrivedi76@rediffmail.com

S. Debbarma
Tripura University, Agartala, India
e-mail: shibendu@gmail.com

© Springer Nature Singapore Pte Ltd. 2018
K. Saeed et al. (eds.), *Progress in Advanced Computing and Intelligent Engineering*,
Advances in Intelligent Systems and Computing 563,
https://doi.org/10.1007/978-981-10-6872-0_65

1   Introduction

The rate of growth of mobile users has been increasing day by day. Different mobile users have different demands. Among them, some of the users may have interest in online gaming, sports, etc. In such scenarios, MCC takes a vital role for solving the users demands. Typically, MCC is used for execution of rich mobile applications. For example, if a mobile user wants to simulate his code in NS-2, then he can simulate it without installing the required software that means he can write his code only and all the processing can be done on the cloud [3]. But, there are certain difficulties with the mobile devices such as having limited battery power and less resources [4–6]. In this work, a game theory-based energy efficient transmission policy for MCC named as *E2G* has been proposed.

The rest of the paper is organized as follows: Sect. 2 presents the state of the art. Section 3 describes the energy consumption model for *E2G*. Section 4 presents the proposed energy efficient transmission policy of *E2G*. Section 5 presents the results obtained along with the analysis of results. Finally, Sect. 6 concludes the paper.

2   Related Work

The development of wireless technology creates huge interest to the users [7]. There are a few works have been proposed for minimizing the energy consumption of mobile devices [1–3, 9]. One approach is by offloading the rich application from the mobile devices to the cloud [1–3]. In [2, 3], authors used remote sensing to reduce the energy consumption. *Thinkair* [8] is used for method-level computation offloading to the cloud and also supports cloud scalability. Context-aware decision engine as proposed in [9] supports offloading scheme from mobile devices to cloud, but it does not support multiple cloud resources, the availability of the wireless channel, and information of the geographical location is incorrect when a device is indoor.

3   Energy Consumption Model

E2G only considers that situation when the mobile devices send necessary information to the cloud. However, the offloading strategy for each mobile device also needs to be considered. The mobile devices take the decision whether the application is to be offloaded to the cloud or not based on the following decision rule:

$$Exec(ME, CE) = \begin{cases} ME, \text{if } E_c(M) \le E_c(C) \\ CE, \text{if } E_c(M) > E_c(C) \end{cases}$$

where, $E_c(M)$ and $E_c(C)$ denote the energy consumption by CPU in mobile devices and cloud, respectively. There are also different factors where a mobile device consumes its energy and are discussed as follows:

Energy consumption by CPU: The energy consumed by CPU is denoted by κ_{CPU} and is computed by using Eq. (1).

$$\kappa_{CPU} = A_p * T_e \qquad (1)$$

where A_p denotes the consumption of average power by CPU for an application and T_e denotes the execution time for CPU. The A_p can be further calculated as follows:

$$A_p = (P_{RN} + P_{BPP} + P_{IWP} + P_{LSP}$$
$$+P_{RG} + P_{REB} + P_{CLK} + P_{ALU})/CycleTime \qquad (2)$$

In Eq. (2), P_{RN}, P_{BPP}, P_{IWP}, P_{LSP}, P_{RG}, P_{REB}, P_{CLK}, and P_{ALU} denote Rename Unit Power, Branch Predictor Power, Instruction Window Power, Load Store queue Power, Register File Power, Result Bus Power, Clock Power, and ALU power, respectively.

Energy consumption by I/O operation: During I/O operation, a mobile device can consume its energy and is computed by using Eq. (3).

$$\kappa_{I/O} = (t_{elapsed} - t_{CPU}) * P_{I/O} \qquad (3)$$

where $t_{elapsed}$ and t_{CPU} denote the total time and CPU time required for I/O operation, respectively, and $P_{I/O}$ denotes the power consumption by I/O operation.

Energy consumption by memory: For read and write memory access, the mobile device consumes its energy and is computed by using Eq. (4).

$$\kappa_{memory} = (\kappa_{read} * read) + (\kappa_{write} * write) \qquad (4)$$

In Eq. (4), κ_{read} and κ_{write} denote the consumption of energy to perform read and write operation, respectively. The *read* and *write* denote the number of reads and writes respectively.

Energy consumption by cache: A mobile device consumes its energy for cache read and write is computed by using Eq. (5).

$$\kappa_{cache} = \kappa_{IL1cache} + \kappa_{DL1cache} + \kappa_{DL2cache} \qquad (5)$$

where $\kappa_{IL1cache}$, $\kappa_{DL1cache}$, and $\kappa_{DL2cache}$ denote the consumption of energy due to the Instruction cache, level 1 cache, and level 2 cache, respectively.

Now, the total energy consumption denoted by E_{total} is computed by using Eq. (6).

$$E_{total} = \kappa_{CPU} + \kappa_{I/O} + \kappa_{memory} + \kappa_{cache} \qquad (6)$$

The residual energy of each mobile device can be calculated by using Eq. (7).

$$E_{res} = E_{total} - E_{init} \tag{7}$$

where E_{res} and E_{init} denote the residual energy and initial energy of a mobile device, respectively. The energy consumption of mobile devices depends on various factors and is described as in [1].

4 Energy Efficient Transmission Policy

In this section, the proposed energy efficient transmission policy between cloud and mobile devices is described briefly. There may be several devices available in a network, but it will be less efficient if all mobile devices will be connected with the cloud all the time. So, for energy efficient transmission $E2G$ takes an intelligent decision by selecting a master mobile device which can initiate to communicate with the cloud.

Let us consider a game consists of four tuples (P, S, E, A) [10]. In this game, each mobile device is considered as a player denoted as P and for each player let S be the number of strategies that a player can use, E be the initial energy of the mobile device, and A denotes a finite set of actions. The mobile devices play game according to the Subgame Perfect Nash Equilibrium (SPNE) decision of game theory. The backward induction is used for determining SPNE. According to Fig. 1, there are two players Player-1 and Player-2. The corresponding strategies of Player-2 are denoted by a, b, c, respectively. Suppose Player-1 follows the Player-2 strategy and based on

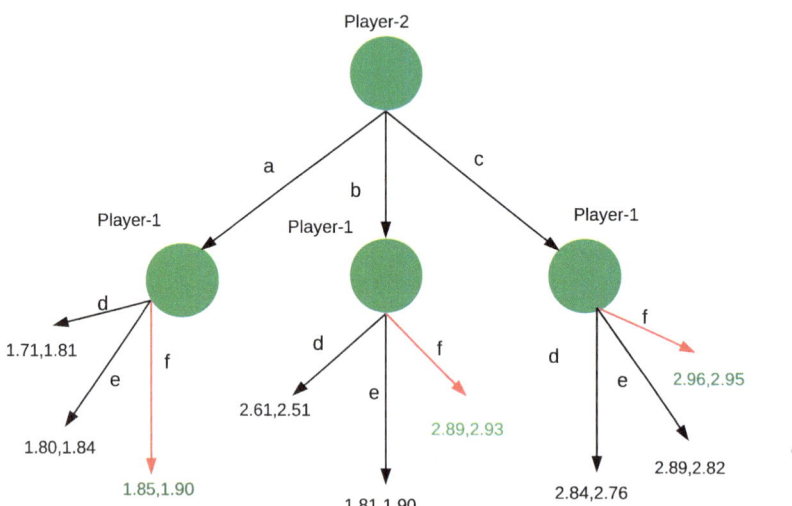

Fig. 1 An example of extension form of game tree

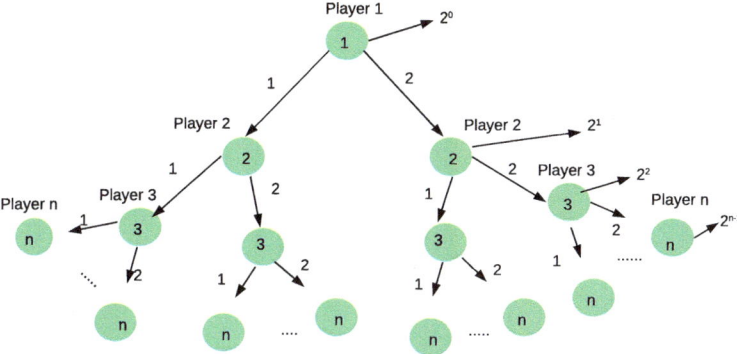

Fig. 2 n-player game

this action can be taken by Player-1. In Fig. 1, initially Player-2 chooses strategy c first. Player-1 will follow the Player-2 strategy where Player-1 also has three strategies d, e, f but it selects strategy f as f is having the highest payoff. The payoff is determined by computing the residual energy of the mobile devices. Similarly, for strategy b and a Player-1 will choose f for both. The n-player game is shown in Fig. 2. The mobile device that is selected based on SPNE decision then communicates with the cloud.

If a player does not deviate from his strategy, then the following calculation is required. Let us consider that for the lower bound o_i, the player does not deviate from his strategy. If a player will deviate from his strategy in stage x and the player moves to other strategy in $y > x$, then

$$(1 - o_i) \sum_{y=0}^{\infty} o_i^y . \alpha_i^* \geq (1 - o_i)(\sum_{y=1}^{x-1} o_i^y . \alpha_i^* + o_i^y . \beta_i + \sum_{y=x+1}^{\infty} o_i^y . \alpha_i$$

$$\iff \sum_{y=x}^{\infty} o_i^y . \alpha_i^* \geq o_i^y . \beta_i + \sum_{y=x+1}^{\infty} o_i^y . \alpha_i$$

$$\iff \sum_{y=0}^{\infty} o_i^y . \alpha_i^* \geq \beta_i + \sum_{y=1}^{\infty} o_i^y . \alpha_i \text{ (Cancel first } x \text{ terms and cancel } o_i^x)$$

$$\iff \frac{\alpha_i^*}{1-o_i} \geq \beta_i + \frac{o_i}{1-o_i} \alpha_i$$

$$\iff \alpha_i^* \geq \beta_i(1 - o_i) + o_i . \alpha_i$$

$$\iff \alpha_i^* - \beta_i \geq o_i(\alpha_i - \beta_i)$$

$$\iff o_i \geq \frac{\beta_i - \alpha_i^*}{\beta_i - \alpha_i}$$

That means during a game, a player will never deviate from his strategy at lower bound o_i.

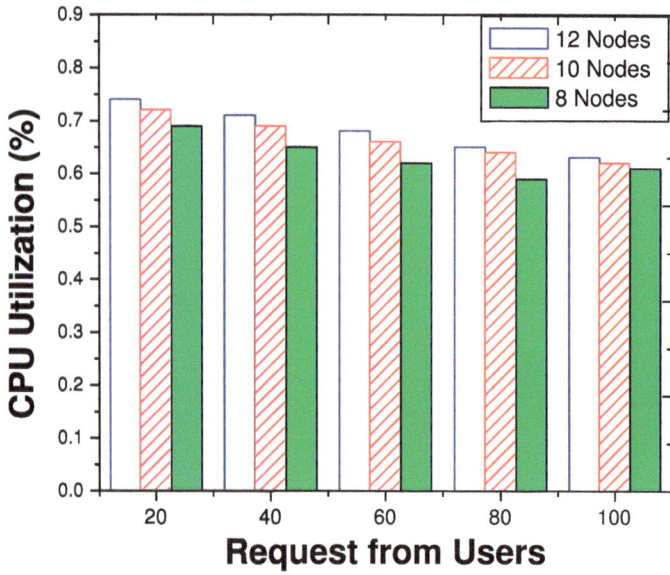

Fig. 3 Average CPU utilization percent for mobile nodes

5 Results and Analysis

The proposed approach *E2G* was evaluated based on Android operating system. Android x86 was installed on Intel I3 laptop. Samsung I997 was used for deploying the applications. Android X86 clone was setup in the Amazon EC2 t2.micro instance and energy consumption was monitored by PowerTutor. *E2G* was run in a standalone environment, where unwanted applications were completely closed and background jobs were also shutdown. For calculating residual energy, we used mathematical model as described in Sect. 3.

In Figs. 3 and 4, X-axis represents the requests from users and Y-axis represents the CPU and memory utilization percentage, respectively. Figures 3 and 4 show the average CPU and memory utilization of mobile nodes with respect to requests from users. From Figs. 3 and 4, it can be found that both CPU and memory utilization percentage increase when the number of mobile nodes increase. Figure 5 shows the network lifetime of *E2G* with two existing baselines approaches that are *eTime* [9] and *Thinkair* [8]. The results from Fig. 5 indicate that the network lifetime is large in *E2G* with respect to two baselines.

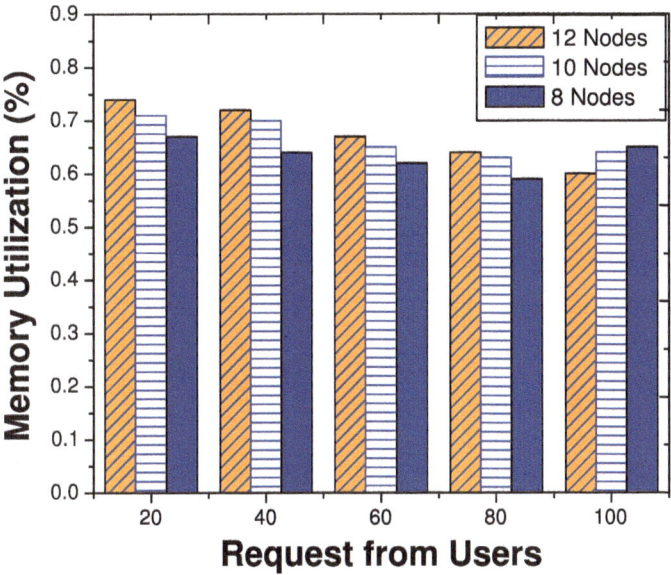

Fig. 4   Average memory utilization percent for mobile nodes

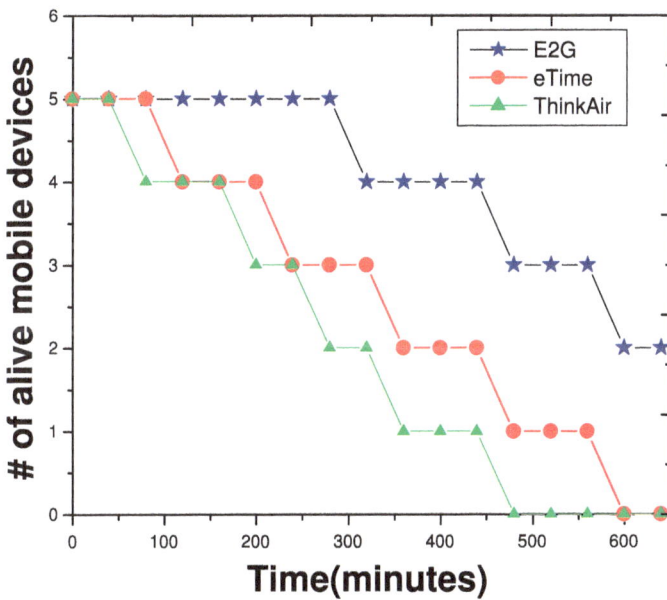

Fig. 5   Network lifetime of *E2G* with respect to *eTime* and *Thinkair*

6 Conclusion

In this work, a game theory-based energy efficient transmission policy named as *E2G* is proposed. *E2G* chooses best mobile device from all available mobile devices using SPNE decision of game theory. The selected mobile device then sends necessary information to the cloud.

References

1. Panigrahi, C.R., Pati, B., Tiwary, M., Sarkar, J.L.: EEOA: improving energy efficiency of mobile cloudlets using efficient offloading approach. In: Proceedings of IEEE International Conference on Advanced Networks and Telecommunications Systems, pp. 1–6 (2015)
2. Rudenko, A., Reiher, G.P.P., Kuenning, G.: Saving portable computer battery power through remote process execution. Mob. Comput. Commun. Rev. **2**, 19–26 (1998)
3. Rudenko, A., Reiher, G.P.P., Kuenning, G.: The remote processing framework for portable computer power saving. In: Proceedings of 1999 ACM Symposium on Applied Computing, pp. 365–372 (1999)
4. Mitra, K., Saguna, S., Ahlund, C.: A mobile cloud computing system for emergency management. IEEE Cloud Comput. **43**(4), 30–38 (2014)
5. Bowen, Z. , Dastjerdi, A.V., Calheiros, R.N., Srirama, S.N., Buyya, R.: A context sensitive offloading scheme for mobile cloud computing service. In: Proceedings of the IEEE 8th International Conference on Cloud Computing, pp. 869–876 (2015)
6. Rong, P., Pedram, M.: Extending the lifetime of a network of battery powered mobile devices by remote processing: a Markovian decision based approach. In: Proceedings of 2003 Annual Design Automation Conference, pp. 906–911 (2013)
7. Pati, B., Sarkar, J.L., Panigrahi, C.R.: ECS: an energy-efficient approach to select cluster-head in wireless sensor networks. Arab. J. Sci. Eng. 1–8 (2016). https://doi.org/10.1007/s13369-016-2304-2
8. Kosta, S., Aucinas, A., Hui, P., Mortier, R., Zhang, X.: Thinkair: Dynamic resource allocation and parallel execution in the cloud for mobile code offloading. In: Proceedings of 31st IEEE International Conference on Computer Communications, pp. 945–953 (2012)
9. Shu, P., Liu, F., Jin, H., Chen, M., Wen, F., Qu, Y., Li, B.: eTime: energy-efficient transmission between cloud and mobile devices. IEEE Infocom, pp. 195–199 (2013)
10. Pati, B., Sarkar, J.L., Panigrahi, C.R., Tiwary, M.: ECHSA: an energy-efficient cluster-head selection algorithm in wireless sensor networks. In: Proceedings of 3rd International Conference on Mining Intelligence and Knowledge Exploration, pp. 184–193 (2015)

Development of Educational Geospatial Database for Cloud SDI Using Open Source GIS

Rabindra K. Barik, R. K. Lenka, S. R. Sahoo, B. B. Das and J. Pattnaik

Abstract Open source software (OSS) can be used for the development of geospatial database in cloud computing environment. Open source GIS (OSGIS) yields strong spatial solutions in a cost-effective manner and used for work connected to geospatial database creation, geospatial Web services, geospatial database storage, etc. The present work critically analyzes two popular OSGIS software i.e., Quantum GIS and Map Window GIS for geospatial database creation to give better comprehension of procedure and execution in education sector with cloud SDI. Presently, geospatial database provides detailed information about the entire technical institute of Odisha in test case. The objective is to deliver spatial statistics at reasonable costs and anticipated to be advantageous for stake holders such as students, parents, faculty members, government organizations. The same may be scaled up in future for technical institutes of other states in India.

Keywords GIS · Open source · Geospatial database · Cloud SDI

R. K. Barik (✉)
School of Computer Application, KIIT University, Bhubaneswar, India
e-mail: rabindra.mnnit@gmail.com

R. K. Lenka
Department of Computer Science & Engineering, IIIT Bhubaneswar,
Bhubaneswar, India
e-mail: rakeshkumar@iiit-bh.ac.in

S. R. Sahoo · B. B. Das
Department of Computer Science & Engineering,
C.V. Raman College of Engineering, Bhubaneswar, India
e-mail: mlswagat@gmail.com

B. B. Das
e-mail: banee.bandana@gmail.com

J. Pattnaik
CSM Technologies Pvt Ltd, Bhubaneswar, India
e-mail: jashikapattnaik24@gmail.com

© Springer Nature Singapore Pte Ltd. 2018   685
K. Saeed et al. (eds.), *Progress in Advanced Computing and Intelligent Engineering*,
Advances in Intelligent Systems and Computing 563,
https://doi.org/10.1007/978-981-10-6872-0_66

1 Introduction

The energetic progression degree of the economic system and technological engineering of any nation certifies a quick improvement in the education sector, by which maximum of the manpower helps to serve the country. This qualified manpower contributes extensively to the economy of nation in terms of financial and technological development. In order to yield excellent manpower, educational segments which is vital and faced by an elevated race on the level of technology employed and on account of globalization; hence demanding fresh innovations to mitigate the forthcoming challenges. A coordinated effort can result in people encouragement and extract improved information on the eminence of educational details in a much advanced manner by the integration of current tools. The technological platforms like GIS, remote sensing, and GPS used to evacuate the hindrance which hampers efficiency and proficiency in the area [1].

Fortunately, in the field of GIS applications, the proprietary software competed by many open source software [2, 3]. Developers have created many open source libraries and GIS sets to handle huge data of GIS and its layouts. Open Geospatial Consortium (OGC) aims to strengthen utility of development of community-led projects and open source GIS standards [4, 5]. The geospatial Web services, database creation, spatial modeling services can use the open source GIS software [6]. This software is also utilized for development of geospatial database which contains Map Window GIS and Quantum GIS. The geospatial database is one of the most vital components for spatial data infrastructure (SDI) model. SDI is implemented in cloud computing environment. A specific Cloud SDI Model has been described in the next section.

2 Cloud SDI Model

Cloud SDI Model delivers a platform in which organizations interrelate with technologies, tools, and expertise to nurture deeds for producing, handling, and using geographical statistics and data. SDI also defines the cumulative of technology, standards, strategies, policies, and manpower required to attain, allot, sustain, process, use, and reserve spatial data. The basic constituents of SDI have been observed as data, networking, public, policy, and standards [7–10]. Further, SDI Model can be implemented through service-oriented architecture (SOA) or cloud computing technologies for better and efficient use. The SOA tries to construct dynamic, distributed, and flexible facility system over the Web in order to see data and required services for development of SDI. Components in the service-oriented architecture-based spatial data infrastructure are geospatial Web services i.e., structured collections of activities which are stateless, self-confined, and independent upon the state of other services [11–17].

Likewise, Cloud SDI Model deploys a unique-instance, multitenant design, and permitting more than one client to contribute assets without disrupting each other. This integrated hosted service method helps installing patches and application advancements for user's transparency. Geospatial Cloud another characteristic is embrace of web services and SOA, a wholly established architectural methodology in the engineering [18]. Many cloud platforms uncover the applications statistics and functionalities via Web service. This permits a client to query/update different types of cloud services and applications data programmatically, along with the

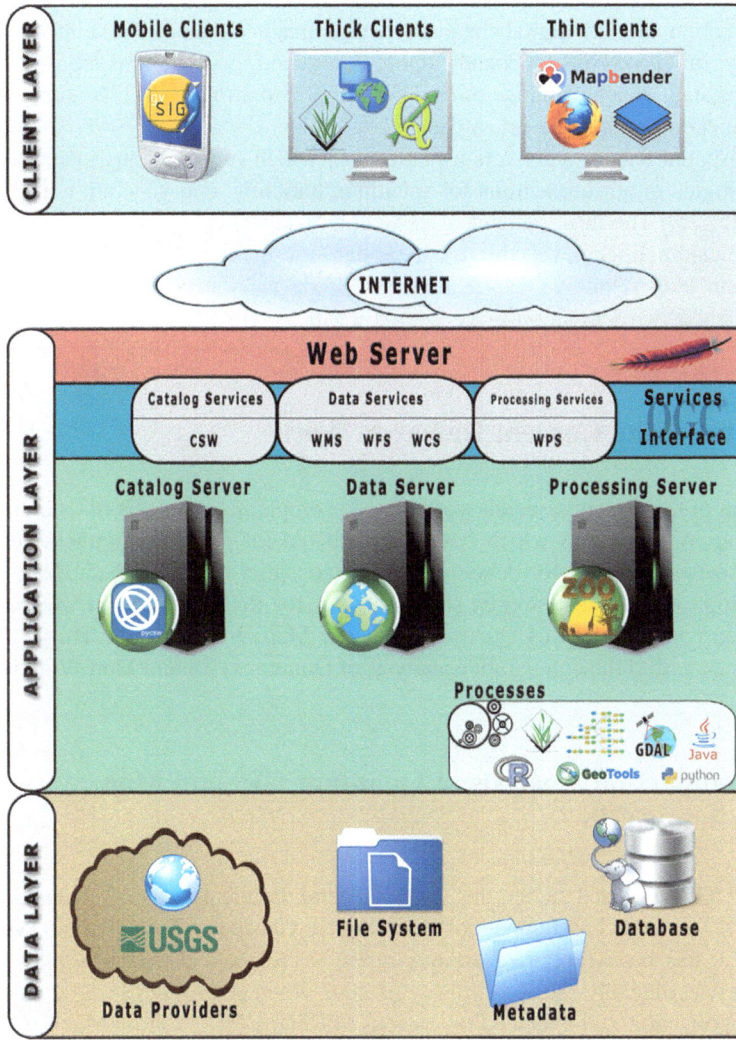

Fig. 1 System architecture for Cloud SDI Model

provision of a standard mechanism to assimilate different cloud applications in the software cloud with enterprise SOA infrastructure [19–23]. Figure 1 illustrates the system architecture for Cloud SDI Model [24].

It has been shown from the system architecture of Cloud SDI Model where geospatial database is a vital module in data layer in Cloud SDI Model. Thus, next section has been emphasized on the geospatial database creation.

3 Geospatial Database Creation

The creation of GIS digital database is significant & tedious assignment where efficacy in GIS project depends upon. Integrated geospatial database creation includes stages such as inputs of data on spatial and attribute and its authentication by connecting with same set of data.

Geospatial database delivers a platform in which organizations interrelate with technologies to nurture actions for spending, handling, and generating geographic data [25, 26]. The development of geospatial database supports in various administrative and political levels through these decision-making functions. Quantum GIS 1.6.0 and Map Window GIS 4.8 are two OS GIS software selected to examine the competences w.r.t. creation of geospatial database.

4 Aim of the Current Research Work

The aim of the current research work is to do comparative analysis of Quantum GIS and Map Window GIS which have been utilized for geospatial database creation and also broadly used for development of geospatial educational database. It has also proposed a robust step-by-step approach for the development of geospatial database with the help of Quantum GIS and Map Window GIS. Thus, the next section describes the comparative analysis of Quantum GIS and Map Window GIS.

5 Comparison Analysis of Quantum GIS and Map Window GIS

An OS GIS licensed under the GNU General Public License is Quantum GIS (QGIS). The formal assignment of Open Source Geospatial Foundation (OSGeo) is QGIS. It has been supported various raster, vector, and database formats. QGIS project was officially released in May of 2002 when coding began. It is a multiple stage application and executed on various OS like UNIX, Linux, Mac OS X, and Microsoft Windows. It can be utilized as GUI to GRASS and having trivial size of

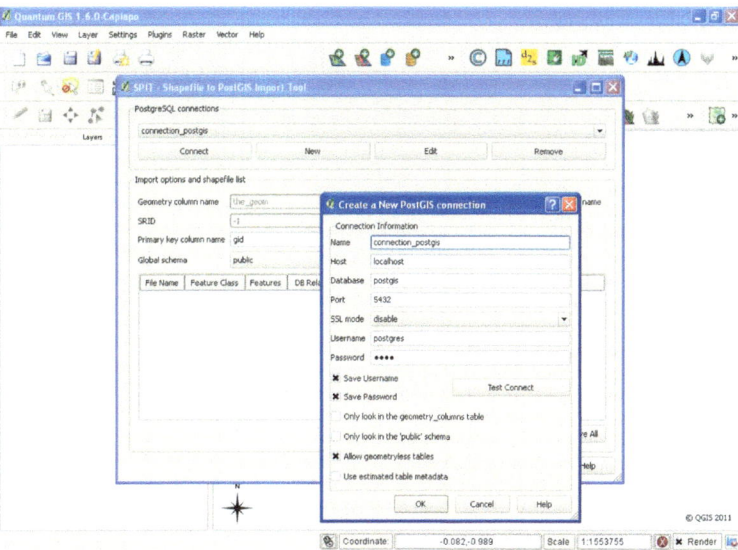

Fig. 2 SPIT plug-in tool

file in comparison with commercial GIS software. SPIT (Shapefile to PostGIS Import Tool) plug-in tool has been used in QGIS and can be used to load multiple shapefiles at one time and includes support for schemas. To use SPIT in QGIS, it needs to open the plug-in manager from the plug-in menu and check the box to the SPIT plug-in. Figure 2 shows the snapshot of SPIT plug-in tool with filed input box [27].

MapWindow GIS is an OSGIS software which has collections of programmable planned modules. It is adopted as the primary GIS platform for its BASINS (Better Assessment Science Integrating Point and Nonpoint Sources) watershed analysis and modeling software by the United States Environmental Protection Agency. It is purely developed under Microsoft.NET Technology platform. It includes plug-ins for several geoprocessing jobs like accessing online data sources, buffer and merge, watershed delineation, and an experimental geodatabase plug-in. Map Window GIS is provisioned with wide variety of data format and read/write ESRI outline records. The structures and aspects of spatial have been revised in accordance with the requirements for managing the database. Map Window GIS has also unique plug-ins 'Spatial Converter' to create geospatial database from excel datasheet. Spatial converter also has the features to import and export ESRI shapefiles from different file formats. Figure 3 shows the snapshot of spatial converter tool in which Excel file imported to create ESRI shapefile in point data [28, 29].

The following Table 1 has been summarized the analysis of two OSGIS in terms of various parameters associated with the user points of level.

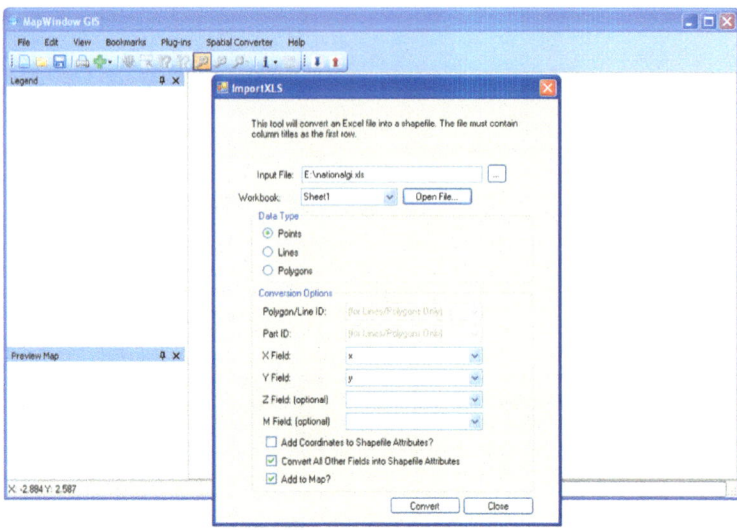

Fig. 3 Spatial converter tool

Table 1 Comparison analysis of quantum GIS and Map window GIS

Geospatial database creation software	Open source	Free software	Web	Linux/ BSD/ Unix	Mac	Windows	Other
Quantum GIS	Yes	Yes	Yes	Yes	Yes	Yes	Google Earth Plugin, KML, WMS
Map window GIS	Yes	Yes	No	No	No	Yes	No

From, the above comparison analysis, it has been considered to be used these two OSGIS software according to the user requirement and the platform which has to be demand for the development of geospatial database creation. Thus, the next section describes about the methodology adopted for the geospatial database creation.

6 Methodology Adopted for Geospatial Database Creation

For creation of geospatial database, the prime emphasis has been on the real-world approach to discover and spread the thought of geospatial database creation in academics sector. The established geospatial database has to provide a proficient means of allocation of geospatial and non-spatial data in Cloud SDI Model. The prototype is based on Object-Oriented Software Engineering (OOSE) proposed by

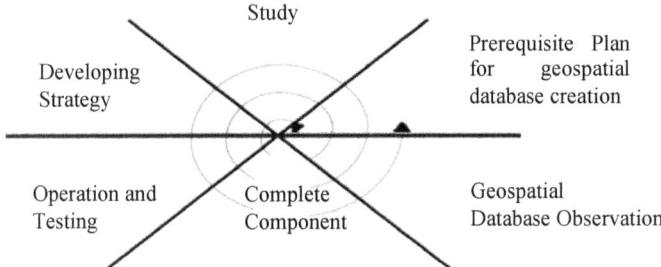

Fig. 4 Win-Win spiral model for geospatial database creation

Jacobson's method to combine the time critical nature and strong user focus [30, 31]. Figure 4 represents the fully win-win procedure model for creation of geospatial database creation.

The procedure model of geospatial database creation is recurring or frequent in nature and each operation improves the study and strategy steps through assessment and testing of a completed component. In complete component, Quantum GIS Open Source GIS software has set up an educational geospatial database by the help of political map of India. QGIS is also used for integrated geospatial database creation. The educational geospatial database has been nominated to illustrate the capabilities of developed framework. Geospatial database for educational sector has been prepared by Quantum GIS Ver. 1.6.0 and Map Window GIS Ver. 4.8.1.

Initially, the base image of India has been downloaded from the Google Earth. The downloaded image is geo-referenced with the help of Geo-referencer tool in Quantum GIS Ver. 1.6.0. For geo-referencing, the base map of India, 10 numbers of GCPs have been taken. The GCPs have been selected at the intersection of latitude and longitude lines. For universal coordinate system, WGS-84 with EPSG:4326 coordinate reference system has been chosen. Now, the image is ready for geo-referencing. After geo-referencing, the generated image is used to extract the thematic maps. Figure 5 shows the snapshot of geo-referencing of India map in Geo-referencer Tool from Quantum GIS Ver. 1.6.0.

In the present application case study, the entire technical institutes of Odisha have been taken. These have been categorized into the different layers with schema definition. Figure 6 shows the layer name with respect to schema definition.

After schema definition, six thematic layers have been created. First layer has been created which indicates the whole India state boundary. For this layer, WGS-84 with EPSG:4326 coordinates reference system has been chosen. India state boundary has been created by on-screen digitization process in Quantum GIS in ESRI shapefile format. The next two thematic layers have been created by Map Window GIS Ver. 4.8.1 with spatial converter tool.

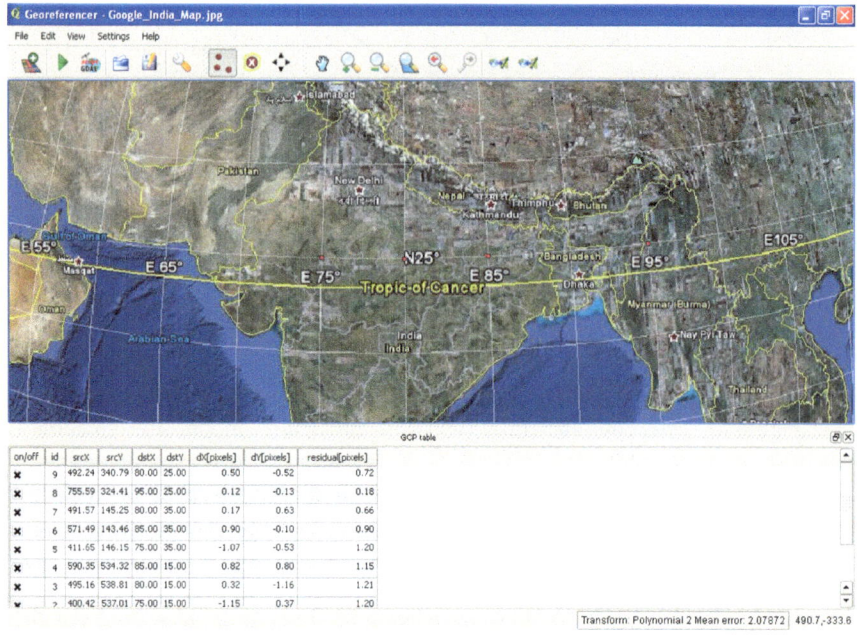

Fig. 5 Geo-referencing: Map of India

Technical Institute of Odisha

UID	Name	D_Name	D_Contact No	Website	State	Longitude (X)	Latitude (Y)

India State Boundary

UID	State Name	State Capital Name

Fig. 6 Schema definitions for educational geospatial database layers

From the spatial converter tool, two thematic layers have been generated, namely:

- Technical Institute of Odisha
- India State Boundary

Finally, these two layers have been overviewed with India State Boundary. Figure 7 shows the snapshot of two layers in Quantum GIS 1.6.0.

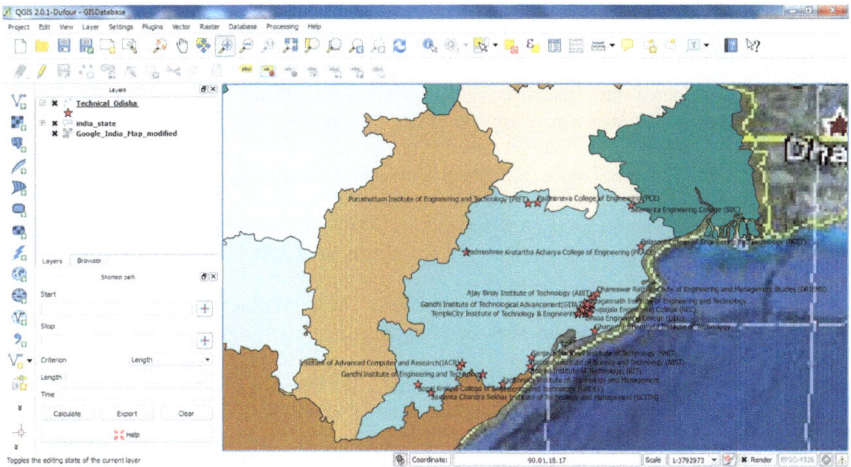

Fig. 7 Integration of educational geospatial database

7 Concluding Remarks

The current research work endeavors to link the information between the real merits and demerits of OS GIS software via comprehensive exploration & evaluation of particular aspects correlated with functionality and complete execution. It is recognized as competent for delivering vigorous proficiencies to form the geospatial database.

In regards to the creation of geospatial database, both Quantum GIS and Map Window GIS software invented as appropriate. However, the core emphasis of the current assignment is to cultivate the educational geospatial database particularly at state level, and further, it will implement for Cloud SDI Model. Therefore, the database which has been established is analytical and does not include complete structures. This database may be made more ample in forthcoming studies. Currently, it has been planned to extend for other state level and the equivalent may be deployed on the cloud environment in imminent studies.

References

1. Barik, R.K., Samaddar, A.B.: Service oriented architecture based SDI model for education sector in India. In: International Conference on Frontiers of Intelligent Computing: Theory and Applications, pp. 555–562 (2014)
2. Barik, R.K., Samaddar, A.B., Gupta, R.D.: Investigations into the efficacy of open source GIS software. In: International Conference, Map World Forum, on Geospatial Technology for Sustainable Planet Earth, 10–13 Feb 2009

3. Raghunathan, S., Prasad, A., Mishra, B.K., Chang, H.: Open source versus closed source: software quality in monopoly and competitive markets. Syst. Man Cybern. Part A: Syst. Hum. **35**(6), 903–918 (2005)
4. Harper, E.: Open source technologies in web-based GIS and mapping, Master's Thesis, Northwest Missouri State University, Maryville Missouri (2006)
5. Nasr, M.R.: Open Source Software: the use of Open Source GIS Software and its Impact on Organization, Master's Thesis, Middlesex University, UK (2007)
6. Kim, D.-H., Kim, M.-S.: Web GIS service component based on open environment. In: IEEE Geoscience and Remote Sensing Symposium, IGARSS'02, vol. 6, pp. 3346–3348 (2002)
7. Rajabifard, A., Williamson, M.-E.F.: Future directions for SDI development. Int. J. Appl. Earth Obs. Geoinf. **4**(1), 11–22 (2002)
8. Mansourian, A., Rajabifard, A., Valadan Zoej, M.J., Williamson, I.: Using SDI and web-based system to facilitate disaster management. Int. J. Comput. Geosci. **32**, 303–315 (2005)
9. Puri, S.K., Sahay, S., Georgiadou, Y.: A Metaphor-Based Sociotechnical Perspective on Spatial Data Infrastructure Implementations: Some Lessons from India, Research and Theory in Advancing Spatial Data Infrastructure Concepts, pp. 161–173. ESRI Press (2007)
10. Ramachandra, T.V., Kumar, U.: Geographic resources decision support system for land use, land, cover dynamics analysis. In: Proceedings of the FOSS/GRASS User Conference, Bangkok Thailand (2004)
11. Rawat, S.: Interoperable Geo-Spatial Data Model in the Context of the Indian NSDI, Thesis (Master), ITC, The Netherlands (2003)
12. Barik, R.K., Samaddar, A.B., Shefalika, G.S.: Service oriented architecture based SDI model for geographical indication web services. Int. J. Comput. Appl. **25**(4), 42–49 (2011)
13. Li, H., Lu, J., Cai, B., Yao, S.: Study on SOA-Orient WebGIS framework. In: 14th IEEE International Conference on Automation and Computing (2008)
14. Vaccari, L., Shvaiko, P., Marchese, M.: A geo-service semantic integration in Spatial Data Infrastructure. Int. J. Spat. Data Infrastruct. Res. **4**, 24–51 (2009)
15. Lu. X.: An investigation on service oriented architecture for constructing distributed WebGIS application. In: IEEE International Conference on Services Computing (SCC'05), vol. 1, pp. 191–197 (2005)
16. Lenka, R.K., Barik, R.K., Gupta, N., Ali, S.M., Rath, A., Dubey, H.: Comparative analysis of SpatialHadoop and GeoSpark for geospatial big data analytics. In: 2nd International Conference on Contemporary Computing and Informatics (IC3I), pp. 484–488 (2016)
17. Barik, R.K., Dubey, H., Samaddar, A.B., Gupta, R.D., Ray, P.K.: FogGIS: Fog Computing for geospatial big data analytics. In: IEEE Uttar Pradesh Section International Conference on Electrical, Computer and Electronics Engineering (UPCON), pp. 613–618 (2016)
18. Morris, S.P.: Geospatial Web services and geoarchiving: new opportunities and challenges in geographic information service. Libr. Trends **55**(2), 285–303 (2006)
19. Leidig, Mathias, Teeuw, Richard: Free software: a review, in the context of disaster management. Int. J. Appl. Earth Obs. Geoinf. **42**, 49–56 (2015)
20. Yang, C., Raskin, R., Goodchild, M., Gahegan, M.: Geospatial cyberinfrastructure: past, present and future. Comput. Environ. Urban Syst. **34**(4), 264–277 (2010)
21. Wu, B., Wu, X., Huang, J.: Geospatial data services within Cloud computing environment. In: 2010 IEEE International Conference on Audio Language and Image Processing (ICALIP), pp. 1577–1584 (2010)
22. Schäffer, B., Baranski, B., Foerster, T.: Towards spatial data infrastructures in the clouds. Geospatial Thinking Lecture Notes in Geoinformation and Cartography, vol. 0, pp. 399–418 (2010)
23. Barik, R.K., Das, P.K., Lenka, R.K.: Development and implementation of SOA based SDI model for tourism information infrastructure management web services. In: 6th International Conference on Cloud System and Big Data Engineering, pp. 748–753 (2016)

24. Evangelidis, K., Ntouros, K., Makridis, S., Papatheodorou, C.: Geospatial services in the Cloud. Comput. Geosci. **63**, 116–122 (2014)
25. Pandey, S.: Cloud computing technology & GIS applications. In: The 8th Asian Symposium on Geographic Information Systems from Computer & Engineering View (ASGIS 2010), ChongQing, China, pp. 1–2 (2010)
26. Yang, C., Goodchild, M., Huang, Q., Nebert, D., Raskin, R., Xu, Y., Bambacus, M., Fay, D.: Spatial cloud computing: how can the geospatial sciences use and help shape cloud computing? Int. J. Digit. Earth **4**(4), 305–329 (2011)
27. Quantum GIS User Guide, Version 1.6.0 (2010)
28. Croft, T.: Quick Guide to MapWindow (2007)
29. MAPWINGIS Reference Manual (2007)
30. Jessica, S., William, M., Allison, K., Ian, W.: Spatial data infrastructure requirements for mobile location based journey planning. Trans. GIS **8**(1), 23–44 (2004)
31. Mall, R.: Fundamentals of Software Engineering, Rev, 2nd edn. Prentice-Hall of India Private Limited, India (2004)

M2C: An Energy-Efficient Mechanism for Computation in Mobile Cloud Computing

Rajesh Kumar Verma, Bibudhendu Pati, Chhabi Rani Panigrahi,
Joy Lal Sarkar and Subhashish Das Mohapatra

Abstract Recently, the development of mobile devices creates lots of interest to the users, where a mobile user can run various kinds of applications. In mobile cloud computing (MCC) environment, mobile devices suffer from less battery power when performing operation for longer duration. To overcome this problem, in this work, an energy-efficient mechanism for computation in mobile devices named as Mobile to Cloud (*M2C*) is proposed. In this approach, offloading the resource-intensive computation is done by the cloud. *M2C* can reduce the overall latency for processing various kinds of data from the mobile devices by considering a cloudlet layer. The simulation results indicate that the execution time of applications is improved by *M2C* as compared to without using cloudlet layer.

Keywords MCC · Offloading · Energy-efficient

R. K. Verma (✉)
Infosys Limited CIS Lab, Hyderabad, India
e-mail: rajeshverma_chicago2004@yahoo.com

B. Pati
C.V. Raman College of Engineering, Bhubaneswar, India
e-mail: patibibudhendu@gmail.com

C. R. Panigrahi · J. L. Sarkar
Central University of Rajasthan, Ajmer, Rajasthan, India
e-mail: panigrahichhabi@gmail.com

J. L. Sarkar
e-mail: joylalsarkar@gmail.com

S. D. Mohapatra
Mahashakti Techno Lab, Bhubaneswar, India
e-mail: subhashish@mstlindia.org

© Springer Nature Singapore Pte Ltd. 2018   697
K. Saeed et al. (eds.), *Progress in Advanced Computing and Intelligent Engineering*,
Advances in Intelligent Systems and Computing 563,
https://doi.org/10.1007/978-981-10-6872-0_67

1 Introduction

Mobile devices are being manufactured at a tremendous pace in the world today and it is a known fact that there are greater number of mobile devices (around 7.22 billion and growing every day) than the entire population (7.19 billion). The power of these mobile devices (smart phones, tablets, robots, sensors, etc.) can be leveraged to accomplish many aspects such as computation, safety, health monitoring, etc. to have a better lifestyle by being seamlessly integrated with devices. MCC comprises of mainly three components like mobile devices, cloud, and network [1]. The main player in MCC is the mobile device which is used to initiate the activity of computation. The various mobile devices are connected together in a network and have good power to execute a task which may be logically split across these devices. Cloud helps toward accomplishing the activity of computation and storage as mobile devices may offload a relatively large task or storage in the cloud [2]. Networks are important communication media as they help to establish the link between the various devices and also the cloud.

Wireless networks are implemented at the physical layer of the OSI model. Cloud computing (CC) can be used by mobiles for offloading as it is elastic and dynamically scalable. It is benefited by use of virtualization technology, network bandwidth and CC infrastructure. The reason for offloading is that sometimes the mobile device may not be able to run computational-intensive jobs or store huge olumes of data (e.g., in case of programs such as big data related computation and storage, it is easily performed via computation offloading process). The decision to do offloading depends upon the type of problem to be solved which involves the computation power required and it also helps to solve problems faster thereby saving energy of the mobile devices which have limited battery life [3]. In case of complex computations required which may also be near real time and can be done by offloading jobs to the cloud (e.g., in case of robots working in electronic chip manufacturing plant of an automobile factory where there is a need to quickly respond to tasks for completing the entire manufacturing process. Internet of Things (IoT) can also involve complex tasks and offloading can be used here for complex computation and near real-time tasks.

In this work, an architecture for MCC is proposed where the mobile devices communicate first to the cloudlet for any computationally intensive job and subsequently leverages the cloud in case the power of the cloudlets still does not serve the purpose. This architecture is different from the architecture proposed in [2] where there is direct communication to the cloud from the mobile. In the proposed architecture, there is less latency and more efficiency as many of the tasks can still be accomplished by the cloudlets. Hence, in case of MCC, offloading is done to enhance the speed of computation and enable storage of huge volumes of data on the cloud, thereby enabling mobile devices to work efficiently and deliver results at a cheaper cost.

The rest of the paper is organized as follows: Sect. 2 describes the related work and Sect. 3 presents the proposed architecture for *M2C*. Section 4 describes the experimental setup used for *M2C* and Sect. 5 presents the results obtained along with the analysis of results. Finally, Sect. 6 concludes the paper.

2 Related Work

The desire to perform computation anywhere in the world can be made possible by offloading the work to a relatively free machine from another machine having scarce resources [4]. In [5], authors proposed a tool for smart mobile devices (SMDs) that will help to compare whether the local execution at the device end or at the cloud is going to be profitable. In [6], authors suggested an algorithm which divides the total computation into different parts such that the average cost is kept at the minimum. In [7], offloading is decided based on the previous nature of the running and consumption of resources by the particular type of application. In [8], authors came up with an engine which can decide the portions of the program that need to be offloaded for faster execution. In [1], the mixed nature of the applications is considered and the designed system does the task of appropriately assigning the right target platform for execution of the different code pieces in the big application or software.

In [9], authors discussed the ever-changing requirements for running various applications and have come up with an approach for identifying the right system in a dynamic fashion. In [10], an approach to break a big application into smaller chunks such that these tiny pieces can be executed effectively has been discussed. In [11], the authors discussed the model which tells how the systems which offload their work are performing for the SMDs. In [12], the authors suggested a framework for making decisions about offloading such that cost is optimized. In [3], offloading of computation required in case of robots to improve their performance has been discussed.

3 Proposed Architecture

In this section, the proposed architecture *M2C* is described. The proposed architecture for *M2C* is shown in Fig. 1. *M2C* mainly incorporates four layers which are described as follows:

The Service Requester/Consumer Layer: This layer consists of the various consumers who tend to use the different applications provided by the mobile. Also, the services can be consumed via Web services (WS) by the different consumers. The *mobile devices and App layer* as shown in Fig. 1 contain mobile devices and applications (Apps). The service requester/consumer layer communicates to this layer through user interface (UI) layer. The mobile devices include smart phones, tablets,

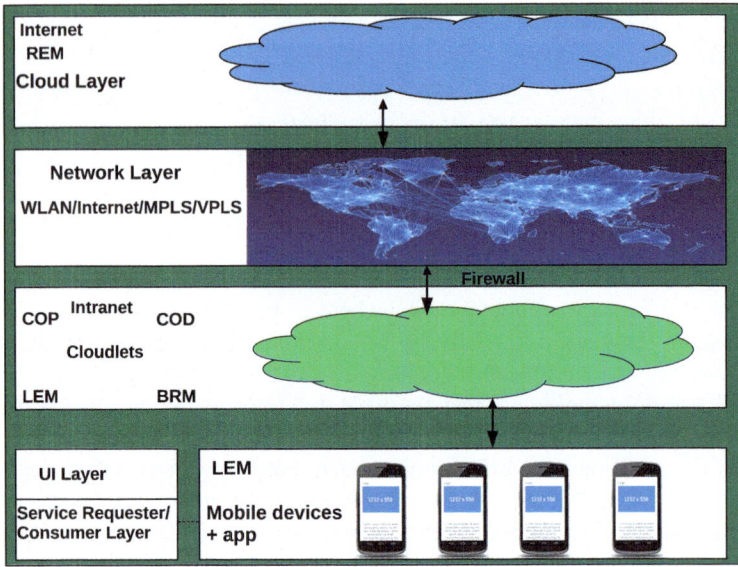

Fig. 1 Proposed *M2C* architecture

laptops, etc. and various mobile applications can be downloaded into these mobile devices which help to serve different functionality for the users. The component *Local Execution Manager (LEM)* in mobile devices+app layer does the task of execution of the applications.

Cloudlets Layer: This layer consists of components such as computation offloading proxy (COP) which does the task of offloading the particular task to the cloud. The computation offloading decision (COD) component decides whether to offload the task or not. The LEM does the management of the execution of the task either in the cloudlet or in the cloud. Bandwidth and resource manager (BRM) component monitors the current network bandwidth and the resource usage (i.e., CPU) of the mobile devices.

Network Layer: This layer is used to communicate with the cloud layer via Internet, multiprotocol label switching (MPLS), or virtual private LAN services (VPLS). MPLS or VPLS is more secure and hence preferred; however, it will be a slightly costly option. The network layer is essential for wireless sensor networks applications [13, 14]. For example, the data communication between sensor nodes and base station and is out scope of this work.

Cloud Layer: The offloading of the application(s) is done depending on the computation type and subsequently the result of the computation from the cloud is sent across to the mobile device. Private or public cloud can be used for this purpose and the percentage of split of application (s) to be sent to the cloud can be decided. This is useful as it also helps us to exploit the cloud as per our needs. This layer consists of remote execution manager (REM) component which gets the task from the cloudlet and helps for running the application (or piece of code) on the cloud. The result derived from computation that has occurred on the cloud will then be sent across to the cloudlet and subsequently to the mobile user.

4 Experimental Setup

The proposed approach *M2C* was evaluated based on android operating system. Android x86 was installed on Intel I3 laptop. Samsung I997 was used for deploying the applications. *M2C* was run in a standalone environment where unwanted applications were completely closed and background jobs were also shutdown. An Android X86 clone was setup in the Amazon EC2 t2.micro instance, and energy consumption was monitored by PowerTutor.

5 Results and Analysis

This section presents the results obtained from the simulation study. In Fig. 2a, X-axis represents the percentage of CPU workloads and Y-axis represents the energy consumption. Figure 2a shows that *M2C* performs better as compared to without considering cloudlet layer. Figure 2a is validated only during offloading the tasks. Figure 2b shows the energy consumption with respect to the input size of the applications. In Fig. 2b, X-axis represents the input size of applications in KB and Y-axis represents the energy consumption. Figure 2b indicates that the energy consumption is less in case of *M2C* as compared to without considering cloudlet layer. Figure 3

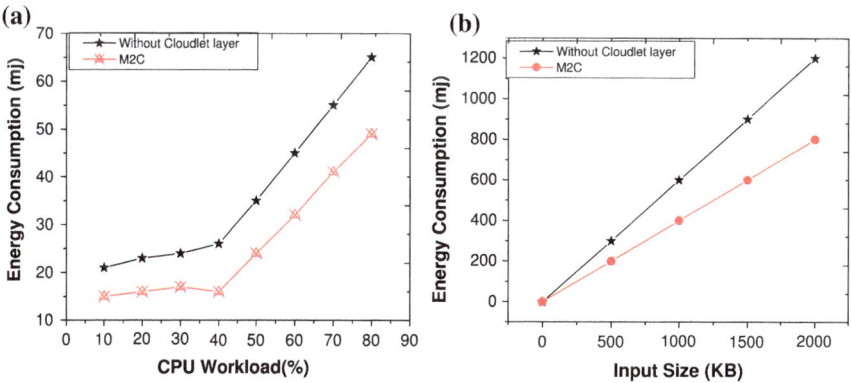

Fig. 2 a Energy consumption of mobile devices with respect to the percentage of CPU workloads. **b** Energy consumption of mobile devices with respect to the input size of the applications

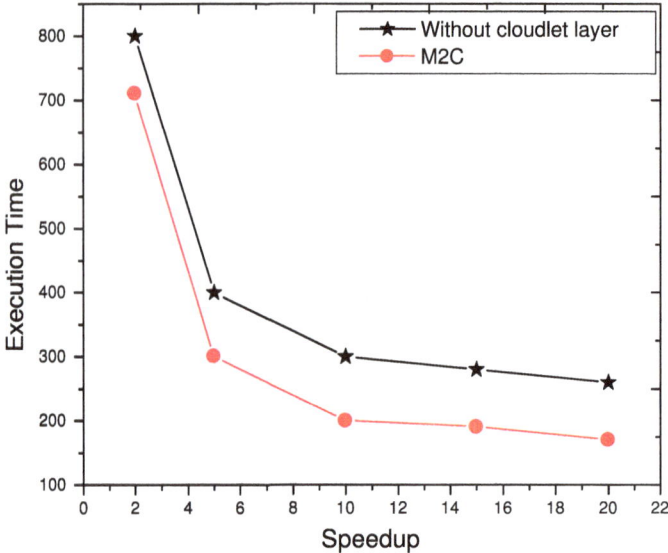

Fig. 3 Execution time when speedup varied

shows the comparison of execution time of applications with speedup. From Fig. 3, it is observed that the execution time is improved by using cloudlet layer.

6   Conclusion

In this work, an architecture named as *M2C* is proposed for MCC which ensures that the computation of applications is faster by using the cloudlet layer via the mobile devices. Offloading is an important mechanism that is mostly leveraged by mobile devices to exploit the huge computational potential of the cloud. The experimental results indicate that the proposed *M2C* approach outperforms in terms of energy consumption for both CPU workload and input size as compared to without using cloudlet layer.

References

1. Chu, H., Song, H., Wong, C., Kurakake, S., Katagiri, M.: Roam, a seamless application framework. J. Syst. Softw. **69**(3), 209–22 (2004)
2. Kumar, K., Liu, J., Lu, Y.-H., Bhargava, B.: A survey of computation offloading for mobile systems. Mob. Netw. Appl. **18**(1), 129–140 (2012)

3. Panigrahi, C.R., Pati, B., Tiwary, M., Sarkar, J.L.: EEOA: Improving energy efficiency of mobile cloudlets using efficient offloading approach. IEEE International Conference on Advanced Networks and Telecommunications Systems, pp. 1–6, (2015)
4. Messer, A., Greenberg, I., Bernadat, P., Milojicic, D., Chen, D., Giuli, T., GM2Cu, X.: Towards a distributed platform for resource-constrained devices. In Proceedings of International Conference on Distributed Computing Systems, pp. 43–51 (2002)
5. Gurun, S., Krintz, C., Wolski, R.: NWSLite: a lightweight prediction utility for mobile devices. In: Proceedings of International Conference on Mobile Systems, Applications, and Services, pp. 2–11 (2004)
6. Ou, S., Yang, K., Liotta, A.: An adaptive multi-constraint partitioning algorithm for offloading in pervasive systems. In: Proceedings of IEEE International Conference on Pervasive Computing and Communications, pp. 116–125 (2006)
7. Huerta-Canepa, G., Lee, D.: An adaptable application offloading scheme based on application behavior. In Proceedings of International Conference on Advanced Information Networking and Applications Workshops, pp. 387–392 (2008)
8. Gu, X., Nahrstedt, K., Messer, A., Greenberg, I., and Milojicic, D.: Adaptive offloading inference for delivering applications in pervasive computing environments. In proceedings of IEEE International Conference on Pervasive Computing and Communications, pp. 107–114 (2003)
9. Sivavakeesar, S., Gonzalez, O., Pavlou, G.: Service discovery strategies in ubiquitous communication environments. IEEE Commun. Mag. **44**(9), 106–113 (2006)
10. Balan, R.: Simplifying cyber foraging. Ph.D. thesis, School of Computer Science, Carnegie Mellon University (2006)
11. Ou, S., Wu, Y., Yang, K., Zhou, B.: Performance analysis of fault-tolerant offloading systems for pervasive services in mobile wireless environments. In: Proceedings of IEEE International Conference on Communications, pp. 1856–1860 (2008)
12. Nimmagadda, Y., Kumar, K., Lu, Y-H., Lee, C.: Real-time moving object recognition and tracking using computation offloading. In: Proceedings of IEEE International Conference on Intelligent Robots and Systems, pp. 2449–2455 (2010)
13. Pati, B., Sarkar, J.L., Panigrahi, C.R., Tiwary, M.: ECHSA: An energy-efficient cluster-head selection algorithm in wireless sensor networks. In: Proceedings of 3rd International Conference on Mining Intelligence and Knowledge Exploration, pp. 184–193 (2015)
14. Pati, B., Sarkar, J.L., Panigrahi, C.R.: ECS: An energy-efficient approach to select cluster-head in wireless sensor networks. Arab. J. Sci. Eng. pp. 1–8 (2016). https://doi.org/10.1007/s13369-016-2304-2
15. Abolfazli, S., Sanaei, Z., Gani, A., Shiraz, M.: MOMCC: market-oriented architecture for mobile cloud computing based on service oriented architecture. In: proceedings of IEEE MobiCC'12, pp. 8–13 (2012)
16. Chun, B., Ihm, S., Maniatis, P., Naik, M., Patt, A.: Clonecloud: elastic execution between mobile device and cloud. In: Proceedings of the Sixth Conference on Computer Systems, pp. 365–372 (1999)
17. Wolski, R., Gurun, S., Krintz, C., Nurmi, D.: Using bandwidth data to make computation offloading decisions. In: Proceedings of IEEE International Symposium on Parallel and Distributed Processing, pp. 1–8 (2008)

Author Index

© Springer Nature Singapore Pte Ltd. 2018
K. Saeed et al. (eds.), *Progress in Advanced Computing and Intelligent Engineering*,
Advances in Intelligent Systems and Computing 563,
https://doi.org/10.1007/978-981-10-6872-0

Printed by Printforce, the Netherlands